STRUCTURAL CHEMISTRY OF LAYER-TYPE PHASES

PHYSICS AND CHEMISTRY OF MATERIALS WITH LAYERED STRUCTURES

Managing Editor

E. MOOSER, *Laboratoire de Physique Appliquée, CH-1003, Lausanne, Switzerland*

Advisory Board

E. J. ARLMAN, *Bussum, The Netherlands*

F. BASSANI, *Physics Institute of the University of Rome, Italy*

J. L. BREBNER, *Department of Physics, University of Montreal, Montreal, Canada*

F. JELLINEK, *Chemische Laboratoria der Rijksuniversiteit, Groningen, The Netherlands*

R. NITSCHE, *Kristallographisches Institut der Universität Freiburg, West Germany*

A. D. YOFFE, *Department of Physics, University of Cambridge, Cambridge, U.K.*

VOLUME 5

F. HULLIGER
Laboratorium für Festkörperphysik ETH, CH-8093 Zürich, Switzerland

STRUCTURAL CHEMISTRY OF LAYER-TYPE PHASES

Edited by

F. LÉVY
Laboratoire de Physique Appliquée, EPF, Lausanne, Switzerland

D. REIDEL PUBLISHING COMPANY

DORDRECHT-HOLLAND/BOSTON-U.S.A.

Library of Congress Cataloging in Publication Data

Hulliger, F. 1929-
    Structural chemistry of layer-type phases.

    (Physics and chemistry of materials with layered structures; v. 5)

    Includes bibliographical references and indexes.
    1. Layer structure (Solids). I. Title. II. Series.
QD478.P47 vol. 5 [QD471]   530.4′1s   [548′.81]   76-26635
ISBN 90-277-0714-6

Published by D. Reidel Publishing Company
P.O. Box 17, Dordrecht, Holland

Sold and distributed in the U.S.A., Canada, and Mexico
by D. Reidel Publishing Company, Inc.
Lincoln Building, 160 Old Derby Street, Hingham,
Mass. 02043, U.S.A.

All Rights Reserved
Copyright © 1976 by D. Reidel Publishing Company, Dordrecht, Holland
No part of the material protected by this copyright notice may be reproduced or
utilized in any form or by any means, electronic or mechanical,
including photocopying, recording or by any informational storage and
retrieval system, without written permission from the copyright owner

Printed in The Netherlands

TABLE OF CONTENTS

PREFACE	IX
I. INTRODUCTION AND DEFINITION	1
II. GEOMETRICAL DERIVATION OF SIMPLE LAYER-TYPE STRUCTURES	3
1. Coordination Numbers 9 and 8	3
2. Coordination Number 6: Trigonal Prism	9
3. Coordination Number 6: Octahedron	11
4. Coordination Number 4: Tetrahedron	19
5. Coordination Number 3: Trigonal Pyramid and Triangle	26
III. REPARTITION OF THE LAYER STRUCTURES	28
IV. DISCUSSION OF SPECIAL COMPOUNDS	40
1. Compounds with Coordination Number 2	40
a. AuI	40
b. $HgCl_2$	41
c. $HgBr_2$	42
2. Compounds with Coordination Number 3	43
a. Sassolite $B(OH)_3$	43
b. Metaboric Acid HBO_2	43
c. The B_2S_3 Structure	44
d. The BCl_3 Structure	47
e. Graphite	48
f. Boron Nitride	50
g. Non-symmetrical Graphite Analogs, B_2O	51
h. Graphite Intercalation Compounds	52
α. Metal—Graphite Compounds	52
β. Graphite Salts	57
3. Compounds with Lone-Pair Cations (CN 3 · · · 6)	62
A. Phosphorus-Group Chalcogenides	62
a. M_2X_3 Compounds with CN3	62
b. Claudetite $As_2O_3(h)$	67
c. Orpiment As_2S_3	67
d. As_2Te_3	69
e. Kermesite Sb_2S_2O	71
f. Getchellite $AsSbS_3$	72

g.	Wolfsbergite-Type Compounds	72
h.	Livingstonite $HgSb_4S_8$	75
i.	$CuTe_2Cl$ (see Appendix)	77
B.	Group V Elements and Isoelectronic Compounds	77
a.	The Arsenic Structure	77
b.	Superstructures Derived from the As Structure: GeTe, SnP_3	82
c.	The Black Phosphorus and SnS Structure	87
d.	Hittorf's Phosphorus	94
e.	Yellow PbO (Massicot)	95
f.	Red PbO (Litharge)	97
g.	Red SnO	99
C.	Various Compounds with s^2 Cations	100
a.	Na_2PbO_2, $LiBiO_2$, $NaAsO_2$,...	100
b.	Lanarkite $Pb_2O(SO_4)$	105
c.	$Pb_3(PO_4)_2(r)$	107
d.	$SbPO_4$ and Related Phases	109
e.	Tellurite β-TeO_2	114
f.	$2\,TeO_2\cdot HNO_3$	115
g.	Mixed Te^{IV} Oxides: Te_2O_5, $H_2Te_2O_6$ and Te_4O_9	116
h.	Li_2TeO_3	117
i.	A Layered Sulfate, $(IO)_2SO_4$	119
j.	$SnCl_2$	120
k.	Tl Chalcogenides Tl_2X and Tl_4X_3	122
4. MX and MX_2 Phases with Low Coordination Number		125
a.	In and Tl Monohalides and Alkali Hydroxides	125
b.	LiOH and Related Structures	131
c.	CuTe	133
d.	β-NiTe	135
e.	Tetrahedral MX_2 Structures (HgI_2, $AlPS_4$, $GaPS_4$, GeS_2)	136
5. Polycationic Tetrahedral Layer Structures		144
a.	$GeAs_2$-Type Compounds	144
b.	Layer-Type Group III Monochalcogenides and Group IV Monopnictides	145
6. Compounds with Polycationic Layers		153
7. Compounds with d^8 Cations in Square-Planar Coordination		158
a.	PdS_2 and PdPS	158
b.	$Pd_3(PS_4)_2$	162
c.	$AuTe_2Cl$	163
d.	$AuCl_3$ and $AuBr_3$	164
e.	AuSe (see Appendix)	
8. Oxides with Cation-Coordination Numbers $4\cdots 6$		166
a.	P_2O_5	166
b.	The d^0 Transition-Element Oxides and Related Hydroxides	167

c.	Tc_2O_7	168
d.	Re_2O_7	169
e.	MoO_3 and Related Phases	169
f.	$MoO_3 \cdot 2H_2O$ and $MoO_3 \cdot H_2O$	173
g.	$Mo_{18}O_{52}$	176
h.	Layered Vanadium Oxyhydroxides	177
i.	V_2O_5	181

9. Close-Packed Structures with Group II and III Cations in Tetrahedral and Octahedral Coordination ... 182
10. Layer Compounds with B Metal Cations in Octahedral Coordination ... 194
    a. Non-Transition-Element CdI_2-Type Chalcogenides ... 195
    b. Tetradymite (Bi_2Te_2S)-Type Compounds ... 195
    c. Layer Structures Based on NaCl—CdI_2-Type Mixtures ... 201
    d. Tetradymite-Bismuth Composites ... 207
11. Transition-Element Compounds with Octahedral Cation Coordination ... 210
    a. Smythite Fe_3S_4 ... 210
    b. $(Fe,Ni)_3Te_2(h)$ ... 211
    c. Pt_2Sn_3 ... 213
    d. The Structures of PtTe (ZrCl), Pt_3Te_4 and Pt_2Te_3 ... 213
    e. The $FePS_3$ and the $FePSe_3$ Structures ... 217
    f. CdI_2-Type Transition-Element Chalcogenides ... 219
    g. The $NbTe_2$ Structure ... 226
    h. Calaverite, Krennerite and Sylvanite ... 228
    i. $PtBi_2(h_2)$ ... 229
    j. Pt_4PbBi_7, $MoTe_2(h)$ and WTe_2 ... 230
    k. $ReSe_2$ and TcS_2 ... 233
12. Group V and VI Dichalcogenides with Trigonal-Prismatic Cation Coordination ... 234
13. Compounds with Cation Coordination Number $\geqslant 8$ ... 242
    a. Layer-Type Rare-Earth Polytellurides ... 242
    b. $PdBi_2(h)$ ... 246
    c. The $ZrSe_3$ Structure ... 247
    d. The $HfTe_5$ Structure ... 250
    e. $TaSe_3$ ... 251
    f. $NbSe_3$ ... 253
14. Mixed-Anion Compounds MXY ... 254
    a. The PbFCl (Matlockite) Structure ... 258
    b. The SmSI Structure ... 263
    c. The β-ZrNCl Random Structure ... 265
    d. The FeOCl and the γ-FeO(OH) Structure ... 265
    e. The AlOCl Structure ... 268

	f. The InTeCl Structure	269
15.	Layer-Type Dihalides	270
	a. Halides with Trigonal-Prismatic Cation Coordination	272
	b. $CdCl_2$- and CdI_2-Type Halides	273
	c. Cr and Cu Dihalides	277
	d. The Brucite-Type Hydroxides	279
	e. Mixed Hydroxyhalides	285
	f. Composite Layer Structures Derived from Brucite	291
	g. White Lead $Pb(OH)_2 \cdot 2\,PbCO_3$	294
	h. Composite Layer Structures with Incommensurate Sublattices	295
	i. The Composite Layer Structure of Lithiophorite	297
16.	MX_3 Halides	298
	a. The $PuBr_3$ Structure	301
	b. Rare-Earth Hydroxychlorides	302
	c. $CdCl_2$-/CdI_2-Type Derivatives ($TiCl_3$, BiI_3, $AlCl_3$, $CrCl_3$)	307
	d. $MoCl_3$	315
	e. The NbS_2Cl_2 Structure	316
	f. Nb_3Cl_8, Nb_3Br_8, $NbCl_4$	318
	g. $Al(OH)_3$	320
	h. Rhenium Trihalides	322
	i. The $AsBr_3$ Structure	323
	j. Group III Halides ($AlBr_3$, $GaCl_3$, InI_3)	325
17.	Layer-Type Halides of Composition MX_4, MX_5 and MX_6	328
	a. The UBr_4 Structure	329
	b. The ThI_4 Structure	330
	c. $SrCl_2 \cdot 2\,H_2O$ and $BaCl_2 \cdot 2\,H_2O$	332
	d. The $ZrCl_4$ Structure	333
	e. The SnF_4 Structure	334
	f. $UO_2(OH)_2$ and UO_2F_2	335
	g. WO_2Cl_2 and WO_2I_2	338
	h. The NbI_5 Structure	338
	i. Triclinic UCl_5	339
	j. The α-WCl_6 Structure	340
	k. $Cr(OH)_3 \cdot 3\,H_2O$	341
18.	Layered Silicates Derived from Brucite and Bayerite	342

APPENDIX 349

REFERENCES 352

MINERAL NAME INDEX 375

FORMULA INDEX 377

PREFACE

This monograph is intended to give the reader an appreciation of the wealth of phases, elements and inorganic compounds, which crystallize in layer-type or two-dimensional structures. Originally this work was planned as a short review article but the large number of phases made it grow out to the size of a book. As is evident from the arrangement of the chapters our point of view was gradually transmuting from geometric to chemical. Moreover, the decision about the compounds that should be discussed was taken only during the course of the work, as is partly evident from the sequence of the references. For chemical or geometrical reason we have included also certain layered chain and molecular structures as well as some layered structures whose layers are linked by hydrogen bonds, thus are in fact three-dimensional.

Instead of writing only a review with pseudo-scientific interpretations that later turn out to be wrong anyway we thought it more profitable to include the crystallographic data which are scattered in various original articles and hand books but never in one single volume. We have transcribed many of the data in order to make them correspond with the standard settings of the International Tables for X-Ray Crystallography. The figures are consistent with the data given in the tables. We apologize for errors and hope that their number is at a reasonably low level in spite of the time pressure. In a vain attempt to keep the list of references at a manageable size we have omitted certain citations of original work and instead refer to standard books [1–23].

Scientific books can hardly be written without the generous support of an institution. Thus I am greatly indebted to the Laboratory of Solid State Physics of ETH, Zürich, and to its director, Professor G. Busch, for the permission to spend so much time for this book. Moreover, I have profitted of private communications of many experts in this field. Particularly, I am grateful for discussions and unpublished data to Professor H. Bärnighausen (Karlsruhe), Dr T. A. Bither (DuPont, Wilmington), Dr M. J. Duggin (CSIRO, North Ryde, Austr.), Professors J. Flahaut (Paris), H. Hahn (Stuttgart), B. Krebs (Bielefeld), V. Kupčík (Göttingen), L. Pauling (Menlo Park, Calif.), H. Schäfer (Darmstadt) and H. G. v. Schnering (Stuttgart).

CHAPTER I

INTRODUCTION AND DEFINITION

Layer-type compounds have received considerable attention in recent years, not only as lubricants but above all because of their anisotropic physical properties. It is an attractive feature of many of these compounds that their quasi-two-dimensional character can be gradually weakened or enforced by intercalation with metal atoms or with organic radicals, respectively.

In this review we have made an attempt to compile and discuss the various layer-type compounds in relation to the Periodic System. As is demonstrated by the existence of two- and three-dimensional modifications for the same substance, e.g. $Al(OH)_3$ and $TiBr_3$, it may be tedious to find necessary and sufficient conditions which separate the fields of existence of the various types. In any case this investigation does not provide safe criteria to predict the occurrence of two-dimensional compounds.

The crystal structure of a solid is determined by forces acting between its various components. These forces may be of the ionic, covalent, metallic or Van der Waals type. The relative strength of each contribution depends upon the position of the elements within the Periodic Table, i.e. upon their valence electron configuration and size. The higher the principal quantum number of the valence electrons, the less important is the specific character of the valence electrons and the structures then are mainly determined by the size and the radius ratio of the elements.

According to the spatial extension of the bonds, we distinguish between molecular structures, chain structures, layer structures and three-dimensional or framework structures. In magnetism it is common to talk of one-dimensional and two-dimensional systems. However, those dimensional specifications refer to the magnetic sublattice only and the corresponding chemical crystal lattice may well form a three-dimensional network, though layered in a certain sense (as in K_2NiF_4, $Ba_5Ta_4O_{15}$, $La(OH)CO_3$, $Cs(FeF_4)$, $Cs_4Ni_3F_{10}$, Mn_2Hg_5 or $NaCrS_2$). A true layer-type structure according to our definition, however, is one that consists of neutral sandwiches held together by Van der Waals forces only. This almost necessarily requires that the top and bottom layers of the sandwiches consist either of anions only, or, in some rare cases, of cations only. Our definition severely reduces the number of representatives. Compounds such as $K_2^+[Pt_4S_6]^{2-}$ or $Na^+[CrS_2]^-$, though crystallizing in the form of thin plates or flakes, are not true laminar compounds. The same is in fact true for the metallic phases of the γ-CuTi type and those which crystallize in an antitype of a layer structure, such as

Ti$_2$O of the anti-CdI$_2$ structure. Although identical sandwiches are also stacked upon each other in these phases, and even though the crystals will cleave preferentially between these sandwiches, there are nevertheless chemical bonds between the metal atoms of adjacent sandwiches. Since these phases form layered structures in a geometrical sense, we will list them for comparison and also discuss a series of improper layer-type compounds.

Many of the true layer compounds can be modified into three-dimensional ones by introduction of additional atoms into the interlayer interstices. This may occur in true solid solution, as in the system MgCl$_2$—LiCl: Li$_x$[Mg$_{1-x}$Li$_x$Cl$_2$], or by a bronze-type intercalation of additional metal atoms as in K$_x$[MoS$_2$], or by isoelectronic substitution as is the replacement of one quarter of the Si atoms by KAl that transforms pyrophyllite Al$_2$(OH)$_2$Si$_4$O$_{10}$ into three-dimensional muscovite K$^+$[Al$_2$(OH)$_2$AlSi$_3$O$_{10}$]$^-$. In this way we continuously move into the field of perfectly three-dimensional compounds in which the layer character becomes gradually less pronounced. On cleaving such pseudo-layer compounds bonds are cut and the resulting unpaired electrons of the surface layer will react with the atmosphere. In true (non-metallic or two-dimensional metal) layer-type compounds, cleaving does not create dangling bonds and the freshly cleaved surface may stay clean for a long time, as these substances are chemically relatively inert.

CHAPTER II

GEOMETRICAL DERIVATION OF SIMPLE LAYER-TYPE STRUCTURES

From a geometrical point of view, a crystal structure may be considered as composed of coordination polyhedra. In inorganic compounds one usually discusses the cation coordination, i.e. the polyhedra formed around the cations by the bonded neighbors which are anions in normal valence compounds, but also include cations in polycationic compounds. The number of vertices of these polyhedra, the coordination number, depends largely on the cation-to-anion radius ratio. It is quite obvious that a minimum coordination number is necessary for the formation of a layer structure. With a coordination number 2, only chains or finite molecules are possible. A high coordination number, on the other hand, will lead to a three-dimensional or framework structure.

1. Coordination Numbers 9 and 8

For inorganic compounds, 9 is a high coordination number that is realized with neighbors forming a tricapped trigonal prism. If the three neighbors in the equatorial plane are different from the central atom, then it is hard to construct separate layers with these polyhedra. Infinite columns, formed by connecting the trigonal prisms via shared triangular faces, can be joined into layers, if they are shifted by half a prism height relative to each other. The composition of these layers will be MX_3X'. Stacking of the layers is somewhat problematic; one suggestion is given in Figure 1. We ignore whether or not such a structure occurs in nature.

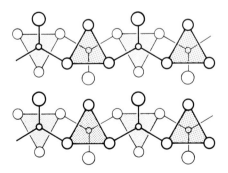

Fig. 1. Possible MX_3X' layer structure projected along the columns. Dark and light elements are shifted by half a prism height relative to each other.

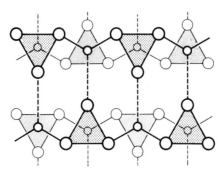

Fig. 2. Improper MX$_3$ layer structure (idealized PuBr$_3$ type). Weakened bonds are indicated by dashed lines.

If we use the X' sites to join the different slices, we obtain a three-dimensional structure of formula MX$_3$ which is realized in Re$_3$B and CuMgAl$_2$. An improper or pseudo-layer structure can be derived from this arrangement by weakening this latter bond as occurs in the PuBr$_3$-type family (Figure 2). In this way the coordination number has been reduced to ~8, but the layer character is accidental or incomplete and based on the weakness or absence of the M—X bonds between the puckered MX$_3$ layers. Other closely related structures are those of the ZrSe$_3$ and TaSe$_3$ type.

If we admit identical neighbors in the equatorial plane, we can easily construct a layered assembly with trigonal prisms that share all square faces. If we characterize the three possible positions of the X atoms by A, B, C (each letter thus defines a planar layer of spheres), and those of the central M atoms by a, b, c, then the sandwich unit is described by $A(bc)A$. A geometrically perfect layer structure of composition MX results from stacking these sandwiches in the following manner: $\cdots A(bc)A\ A(bc)A\ A(bc)A \cdots$. We arrive at exactly this structure if we eliminate every second boron layer in the AlB$_2$ structure. This structure type has indeed been observed in δ'-MoN [24] which, however, is not a genuine layer-type compound since it is metallic and the rim layers are formed by the metal atoms (Figure 3). Instead of placing the prisms exactly on top of each

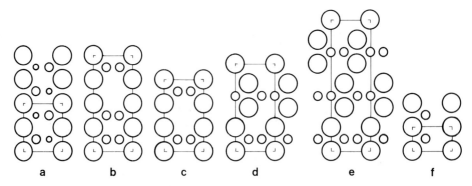

Fig. 3. Trigonal-prism stacking variants (a) ZrBeSi = ordered AlB$_2$ type; (b) hypothetical M$_3$X$_4$ structure; (c) δ'-MoN; (d) δ^{IV}-WN; (e) rhombohedral δ-WN; and (f) WC structure.

other we may stack the layers in a close-packing manner: $A(bc)A\ B(ca)B\cdots$ or $A(bc)A\ B(ca)B\ C(ab)C\cdots$. Such a structure has been reported for δ^{IV}-WN [25]. But in this case the layer character depends on the axial ratios of the prisms and the cell. If the prisms and the empty layers are squeezed, bonding across the layers may evolve, leading to a framework structure.

The relation between the AlB_2 and the δ'-MoN structure demonstrates how layered framework structures offer a simple way to create true layer-type structures. By removing periodically a complete layer of atoms M, various structure types can be derived, the composition being given by the periodicity. Nature eventually will prove whether such a speculation is reasonable or not. For example, instead of removing every second boron layer in AlB_2, we might remove only every third or fourth, so that a layer compound M_3X_4 or M_2X_3, resp., would result (Figure 3). On the other hand, we may as well remove certain of the rippled vertical boron layers of the AlB_2 structure. If every second layer is kept empty, we arrive at the MX structure shown in Figure 4a. Layering is now perpendicular to the plane of the close-packed X atoms. This structure model represents an idealized version of the NaHg structure. In NaHg the empty prisms are compressed and the additional metallic bonds destroy the layer character.

If we shift neighboring prism layers of Figure 4a by half a prism height, we arrive at another hypothetical layer structure (Figure 4b), which may be regarded as the idealized CrB structure, adopted by many borides, silicides, gallides and even true intermetallic compounds such as LnNi. In this structure the central M atom (B, Si, Al, ... in the CrB type) can acquire an additional X neighbor in the equatorial plane, as shifting of neighboring prism slabs allows a closer approach. In certain representatives, such as ScSi, this seventh M—X bond distance is actually the shortest of all. Thus, within the same family of compounds, there are typical layer-type representatives, e.g. BaSi, as well as true three-dimensional phases. Another deformed version, also with metallic bonds from slab to slab and hence no longer of a layer-type, is realized in the PdBi structure. Here, the distortions interrupt certain of the zig-zag M—M bonds.

In the TlI family, which represents some kind of CrB antitype, the central atom is displaced towards the outer square face which leads to a reduction of the coordination number to 5 (or 1+4). These small shifts of the M atoms change the character of the structure quite essentially. It is still a laminar structure, but the crystals will cleave within the original M zig-zag chains.

In the MoB structure, every second prism slab is rotated by 90°. The coordination of the B atoms is the same as in the CrB structure, so that again a three-dimensional network results.

Quite a series of structure types are somehow related to the CrB structure and none of them represents a true layer type. Some may be interpreted as intercalation types. Thus in $ZrNiH_{2.7}$ hydrogen atoms are inserted between the layers as well as into the triangular faces of the CrB-type framework of ZrNi. Layers of Si zig-zag chains are inserted between the CrB-type prism slabs in the structures of both $ZrSi_2$ and $Th_{0.9}Ge_2$.

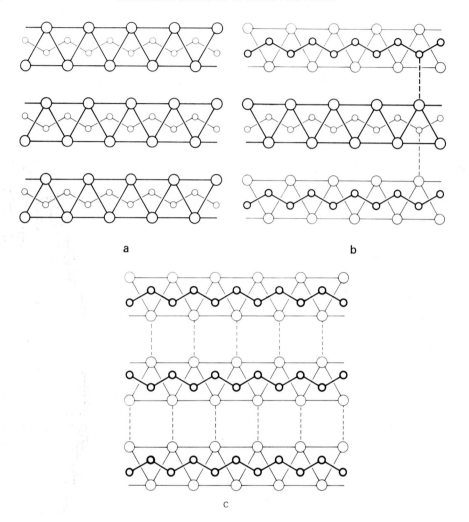

Fig. 4. Three possibilities for the stacking of trigonal-prism slabs.

In the Mn$_2$B$_2$Al structure the stacking of the CrB-type prism slabs is again the same as in Th$_{0.9}$Ge$_2$, but here the square-prism holes between the slabs are filled with Al atoms.

Finally, in Mo$_2$BC the prism slabs of composition MoB are intergrown by an MoC layer.

In all these phases the bonding is three-dimensional.

A coordination number of 9 can be attained with a CrB-type slab if we allow for additional single atoms attached to the equatorial zig-zag chains. A layer-type structure then may result if the slabs are stacked as in CrB (Figure 5) provided the single atoms are halogens that form only Van der Waals bonds towards the adjacent slabs. A hypothetical representative of such a structure type might be KSiCl with Si in the zig-zag chains. Deformed metallic versions (with loss of the

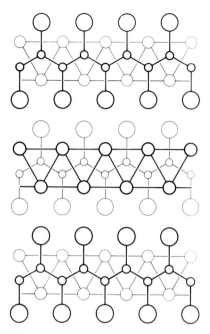

Fig. 5. MXX' layer structure derived from trigonal-prism slabs.

layer character as a consequence) are found in the UBC-type and the closely related MoAlB-type structures.

If in the AlB$_2$ structure every third prism slab is freed from B atoms, we arrive at a layer structure of stoichiometry M$_3$X$_4$ with nonequivalent M and X sites M$_2$M'X$_2$X$'_2$. Various orientations of adjacent twin slabs are again possible as with the simple CrB slabs (Figure 6a, b). A stacking as in CrB is realized in the Ta$_3$B$_4$ structure and its ordered super-structure Cr$_2$NiB$_4$. But again, the pronounced interslab bonding is responsible for the fact that a three-dimensional network results. The layer character could again be preserved by attaching additional ions in place of these interslab bonds (Figure 6c). K$_2$CaSi$_4$Cl$_2$ might be a representative of this hypothetical structure, with Ca atoms at the sites within the mirror plane of each layer.

Instead of the $(2 \times 3) + 2$ neighbors we may choose (2×4) or $(4 + 4)$ neighbors with fourfold symmetry in order to obtain a coordination number of 8. Thus, planar assemblies of cubes or square prisms may be stacked on each other in such a way that the atoms of the neighboring sandwich lie on top of the centers of the square prisms of the lower layer. The empty space between these sandwiches then consists of tetrahedra and square pyramids ($\frac{1}{2}$ octahedra). Obviously, the character of this structure type – the MoSi$_2$ or CaC$_2$ structure – is strongly dependent upon the axial ratio of the square-prism unit as well as upon the distance between the sandwiches (i.e. c/a of the unit cell). For low c/a, the 'anion' above and below the central 'cation' both belong to the coordination of the cation as in MoSi$_2$

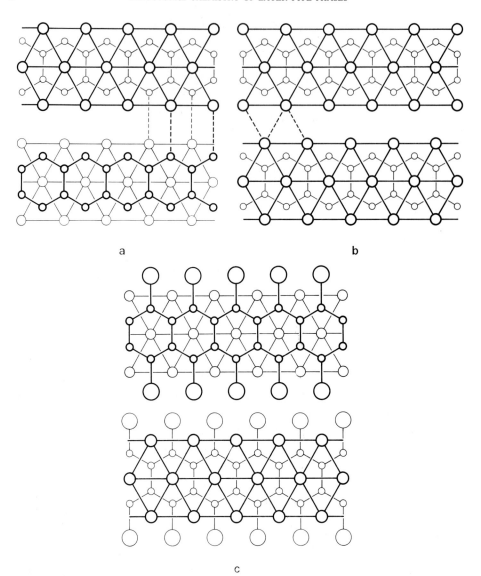

Fig. 6. Structures constructed with trigonal-prism twin slabs. (a) and (b) stacking variants of composition M_4X_3; (c) section perpendicular to the layers of a hypothetical structure of composition $M_4X_3X'_2$.

($c/a = 2.46$). By elongating the MX_8 square prism and increasing c/a of the unit cell one gradually arrives at the layer structure of $PdBi_2(h)$.

For steric reasons it is impossible to link square antiprisms MX_4Y_4 in the same way as in the case of cubes since the periodicity length is different for the upper (X) and the lower (Y) sublattice. We can, however, restore the periodicity by reducing the area of one square (Y_4) by a factor of 2 so that its diagonal equals the edge of the opposite (X_4) square. This yields sandwiches of composition MXY_2 which can be stacked in various ways. We cannot give examples of such structures, but the PbFCl structure represents a modified version of one of these hypothetical layer types. If in the arrangement of the layers according to the sequence —XMY_2—Y_2MX—XMY_2—XMY_2—Y_2MX— we let the Y atoms coincide in the mirror plane, we obtain a double layer —XMY_2MX— as is formed in the PbFCl structure. The usual stacking of these twin sandwiches again opens the possibility of bonding across the layers and any degree of transition between true layer-type and three-dimensional framework structure can occur.

As mentioned above, an easy method to create layer-type structures is by ordered elimination of M layers in layered compounds —MXMX— and such a possibility is offered by the $CuAl_2$ structure in which connection of the square antiprisms leads to X_2 pairs. Omission of every second M layer generates, at least geometrically, a true layer structure, which is realized in the $PtPb_4$ type and the $PtSn_4$ type. Every third M layer is missing in the $PdSn_3$ structure. However, the representatives of all these types are metallic, so that there exists some metallic bonding between the sandwiches.

2. Coordination Number 6 – Trigonal Prism

Trigonal-prismatic sandwiches XMX can be formed by joining the trigonal prisms [MX_6] along their vertical edges such that each edge is shared by 3 prisms. Both X layers then form close-packed trigonal nets, and the sandwich $(XMX)_\infty$ may be described by the stacking symbol AbA or AcA (layer of opposite orientation). Stacking of these sandwiches appears to occur only in a close-packing manner. Different kinds of superposition give rise to the formation of various structures and polytypes of the MoS_2 family. Possible polytypes have been derived by various authors [26–29]. For the description and derivation of polytypes it is convenient to use shorter symbols for the trigonal-prism sandwiches:

$AbA \to A$ and $AcA \to A'$ (opposite orientation.)
$BcB \to B$ $BaB \to B'$
$CaC \to C$ $CbC \to C'$

The checking of the individuality of the different structure symbols is greatly facilitated if we use the empty polyhedra between the sandwiches for further characterization. As the contacting anion layers of neighboring sandwiches are in close-packed arrangement the 'empty' space is in fact composed of a layer of octahedra (Ω) and two layers of tetrahedra (τ), one pointing upwards (τ_1, lying below the anions of the upper layer) and one pointing downwards (τ_2, lying above the anions of the lower layer). In describing a certain polytype it is useful to indicate the type of empty polyhedra that are attached above and below the occupied trigonal prisms. With close packing as the only stacking condition for the sandwiches, we have the choice of an infinity of polytypes. Zvyagin and Soboleva

[28] introduced the condition of periodicity and uniformity: If two successive layers have the same orientation, then all layers must have this orientation. If two orientations are present in a structure, these orientations must alternate with one another along the sequence. If the layers have like orientations, then the structure $ABAB\ldots$ (or $ACAC$, which is equivalent) with a repeat unit of two layers and the three-layer structures $ABC\,ABC\ldots$ and $ACB\,ACB\ldots$ are the only possibilities. With opposite orientations, the possible two-layer sequences are $AB'AB'\ldots$ and $AC'AC'\ldots$. The uniformity condition prevents formation of a three-layer structure and leads automatically to a six-layer sequence $AB'CA'BC'\ldots$ or $AC'BA'CB'\ldots$ which represents a rhombohedral stacking of the MoS_2(C7)-type cell. These layer types are listed in Table 1, where the

TABLE 1
Periodic and uniform stacking sequences of trigonal-prism sandwiches

Case number	Symbol of prism and empty layers	Repeat unit	Space group	Example
1	$_{\tau_2}A_{\tau_1}\Omega B_{\Omega\tau_2}A_{\tau_1}\Omega B_{\tau_1}\ldots$	2	P$\bar{6}$m2	$2H$—$Nb_{1+x}Se_2$
2	$_\Omega A_{\tau_1}\Omega B_{\tau_1}\Omega C_{\tau_1}\ldots$	3	R3m	rhomboh. MoS_2
3	$_\Omega A_{\tau_1}\Omega C_{\tau_1}\Omega B_{\tau_1}\ldots$	3	R3m	
4	$_{\tau_2}A_{\tau_1\tau_2}B'_{\tau_1\tau_2}A_{\tau_1\tau_2}B'_{\tau_1}\ldots$	2	P6$_3$/mmc	MoS_2(C7)
5	$_\Omega A_\Omega C'_\Omega A_\Omega C'_\Omega\ldots$	2	P6$_3$/mmc	NbS_2
6	$_\Omega A_{\tau_1\tau_2}B'C_{\tau_1\tau_2}A'_\Omega B_{\tau_1\tau_2}C'_\Omega\ldots$	6	R$\bar{3}$m	$6R$—$Ta_{1+x}S_2$
7	$_\Omega A_{\tau_1\tau_2}C'B_{\tau_1\tau_2}A'_\Omega C_{\tau_1\tau_2}B'_\Omega\ldots$	6	R$\bar{3}$m	

contacting empty layers are also added below each symbol. We note that the cations in the prism centers as well as in the centers of the empty octahedra lie above each other if the octahedral layers contact both neighboring sandwiches ($A_{\Omega\Omega}C' \equiv A_\Omega C'$). This detail is of importance in intercalation compounds.

The sequences 2 and 3, as well as 6 and 7 of Table 1 can be transformed into each other by reflection at a plane parallel to the layers; in other words, they are enantiomorphs. Types 1, 2, 4, 5 and 6 are known as $2H$—$Nb_{1+x}Se_2$, rhombohedral MoS_2, C7-type MoS_2, NbS_2 and $6R$—$Ta_{1-x}S_2$ [30] resp.

If we drop the uniformity condition, many more polytypes can be created. Brown and Beerntsen [27] list the sequences with three and four layers (Table 2). To our knowledge, only two of these possible polytypes have been detected, no. 10 in $NbSe_2$ and $TaSe_2$, and no. 18 in $TaSe_2$. Wickman and Smith [29] also list the 16 five-layer and the 81 six-layer polytypes.

Compounds crystallizing as trigonal-prismatic sandwiches seem to be especially well suited for intercalation of additional atoms or radicals within the empty octahedral layers. A large number of examples are known at present.

As already mentioned, the MX_2 double layers in all real structures are always found to be close packed. A structure derived from the WC type by emptying every second (or third or fourth) prism layer seems to be unknown. In fact, we

TABLE 2
Non-uniform stacking sequences of trigonal-prism sandwiches.

Case number	Symbol of prism and empty layers	Repeat unit	Space group	Example
8	$_\Omega A_{\tau_1 \Omega} B_{\tau_1 \tau_2} C'_\Omega \ldots$	3	P3m1	
9	$_\Omega A_{\tau_1 \Omega} B_{\Omega \tau_2} A_{\tau_1 \Omega} C_{\tau_1} \ldots$	4	P$\bar{6}$m2	
10	$_\Omega A_{\tau_1 \Omega} B_{\Omega \tau_2} A_\Omega C'_\Omega \ldots$	4	P$\bar{6}$m2	4H$_a$—NbSe$_2$
11	$_{\tau_2} A_{\tau_1 \tau_2} B'_{\tau_1 \tau_2} A_{\tau_1 \Omega} B_\Omega \ldots$	4	P$\bar{6}$m2	
12	$_\Omega A_{\tau_1 \tau_2} B'_{\tau_1 \tau_2} A_{\tau_1 \Omega} C_{\tau_1} \ldots$	4	P$\bar{6}$m2	
13	$_\Omega A_{\tau_1 \tau_2} B'_{\tau_1 \tau_2} A_\Omega C'_\Omega \ldots$	4	P$\bar{6}$m2	
14	$_\Omega A'_\Omega B_{\tau_1 \tau_2} A'_{\Omega \tau_2} B'_{\tau_1} \ldots$	4	P$\bar{6}$m2	
15	$\Omega A_{\tau_1 \tau_2} B'_{\tau_1 \Omega} A'_\Omega B_\Omega \ldots$	4	P$\bar{3}$m1	
16	$_\Omega A_{\tau_1 \Omega} B_{\tau_1 \tau_2} A'_{\tau_1 \tau_2} C_{\tau_1} \ldots$	4	P3m1	
17	$_\Omega A_{\tau_1 \Omega} B_{\Omega \tau_2} A_\Omega C'_\Omega \ldots$	4	P3m1	
18	$_\Omega A_{\tau_1 \tau_2} B'_{\tau_1 \Omega} A'_{\tau_1 \tau_2} C_{\tau_1}$	4	P3m1	4H$_c$—TaSe$_2$

need not choose our generating sandwich as simply MX$_2$. It might be feasible to play the whole exercise with a double layer XMXMX or XMYMX (X and Y designating different kinds of anions), as in hypothetical Zr$_2$NCl$_2$, Ta$_2$SSe$_2$ or Mo$_2$SN$_2$. Such WC-type double layers do exist indeed in δ^1-W$_{1+x}$N. They are stacked in the same way as the sandwiches in C7-type MoS$_2$, namely *AbAbA BaBaB*... [31]. The outer layers are said to be $\frac{2}{3}$ occupied only, corresponding to a stoichiometry W$_{14}$N$_{12}$.

Double layers derived from the ZrBeSi structure by omitting, say, the atoms at the Be sites are cubically close packed in δ^{IV}-W$_{1+x}$N: *AcAbA BaBcB CbCaC*... [32]. Other polytypes might exist as well.

3. Coordination Number 6 – Octahedron

Most of the structures containing cations in octahedral coordination can be derived from a close-packed anion array. In the generating octahedron layers of a close-packed array, the octahedra rest on a triangular face and share 6 of the 12 edges. In cubic close-packed arrays the layers are stacked in such a way that the basis triangle of the upper layer lies above a tetrahedral hole of the lower octahedron layer. If all possible octahedra are occupied, then these octahedra have all edges in common. In a hexagonal close-packed array the octahedron layers are connected through common faces. The number of common elements can be reduced by picking some of the octahedra out. Thus, any structure built of octahedra is somehow related to a close-packed arrangement and we may either construct layer structures by connecting coordination octahedra [MX$_6$] into laminar arrays or by eliminating the central atoms (and simultaneously the corresponding coordination polyhedra) from close-packed structures.

In hcp arrays, the plane parallel with the close-packed layers (i.e. parallel with the basal faces of the octahedra) is the preferred possibility for ordering of the

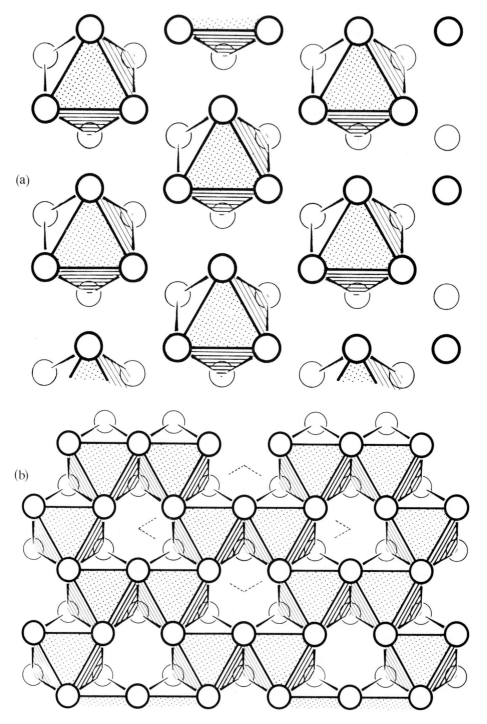

Fig. 7. Octahedron layers. (a) Octahedra $\frac{1}{3}$ occupied: layered molecular structure; (b) Octahedra $\frac{2}{3}$ occupied, all octahedra sharing three edges.

occupied (octahedral) interstices. Ordering of the cations in layers parallel with the stacking direction leads to puckered anion rim layers. In cubic close-packed arrays, however, the planes parallel with any of the triangular faces and, furthermore, parallel with any of the three square cross sections of the octahedron may be chosen for the generation of sandwiches.

We may try to find in a systematic way the possible layer structures by increasing the contacts between occupied octahedra. A layer of discrete octahedra leads to a layer compound MX_6 (Figure 7a). Different stacking of the sandwiches will give rise to polytypes. A laminar arrangement parallel with the square section of the octahedra in cubic packing is impossible as certain anions then would not be bound to any cation. If one element – vertex, edge, face – is shared by all octahedra with another one, then dimers are the only possibility. Pairs of octahedra can be arranged in just one way to form a sandwich of composition MX_5 (Figure 8). This

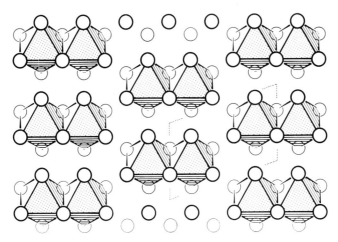

Fig. 8. Layer of double octahedra of a layered molecular compound MX_5. The unit cell for hexagonal anion stacking is indicated.

might be a possibility to form non-magnetic $MoCl_5$. With two shared vertices, edges or faces, chain structures MX_5 (e.g. trans-chain: α-UF_5, $WOCl_4$; cis-chain: VF_5, $MoOF_4$), MX_4 (e.g. NbI_4, $TcCl_4$ types) or MX_3 (e.g. ZrI_3 type), resp., will result. Two kinds of chains made up by edge-sharing octahedra are shown in Figures 9a, b. These sandwiches may be close-packed to form structures which are layered in two directions. In the structure which is complementary to MX_6 (i.e. in which $\frac{2}{3}$ of the octahedra are occupied), the octahedra share three edges (BiI_3, $CrCl_3$, $AlCl_3$, $Al(OH)_3$ types). Four common equatorial vertices lead to a chess-board-like layer of composition MX_4 (Figure 10) as met in the structures of SnF_4, $Sn(CH_3)_2F_2$, $UO_2(OH)_2$, WO_2Cl_2 and WO_2I_2. In these cases, however, the layer character and hence the cleavability of the crystals strongly depends upon the manner in which these layers are stacked. In SnF_4, where the F ions form a

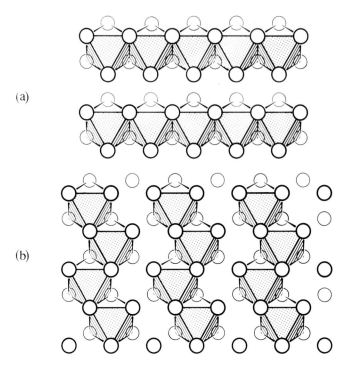

Fig. 9. Two kinds of octahedron layers of composition MX_4. Each $[MX_2X_{4/2}]$ octahedron is sharing two edges.

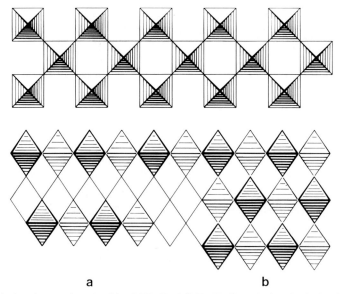

a b

Fig. 10. Octahedron layers of composition MX_4. Each $[MX_2X_{4/2}]$ octahedron is sharing four corners.
(a) true layer structure with the two free vertices on each octahedron opposite (trans);
(b) improper layer structure with the free vertices trans (SnF_4 type);
above: part of one layer
below: section perpendicular to the layers

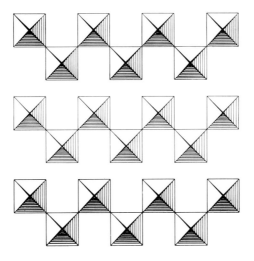

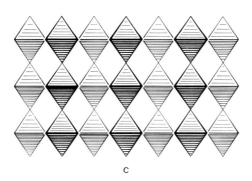

Fig. 10c. Layer structure with the free vertices adjacent (cis) as realized in the $[NbO_2O_{4/2}]_\infty$ sheets of $BiNbO_4(h)$.

               above: section perpendicular to the layers
               below: part of one corrugated layer

ccp array, only one intermediate layer of octahedral holes in empty, while two octahedron layers are unoccupied in WO_2Cl_2 (corresponding to Figure 10a). A puckered layer structure in which the octahedra $[MX_2X_{4/4}]$ are joined by four corners but with the two free vertices adjacent (cis) is shown in Figure 10c.

The transition from the SnF_4 structure to the ReO_3 structure is illustrated in Figure 11. If two SnF_4 layers are stuck together and these units are stacked as in SnF_4 we end up with a layered structure of composition M_2X_7, reminiscent of the $SmZrF_7$ structure [33]. Each octahedron $[MXX_{5/2}]$ now shares five corners. If we build up units with more than two SnF_4 sheets then the additional octahedra share

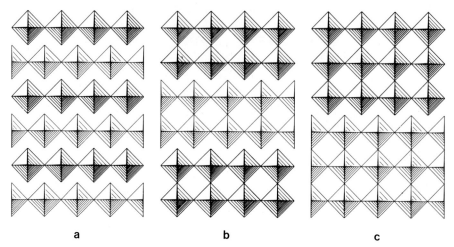

Fig. 11. Transition from the SnF$_4$ structure to the three-dimensional ReO$_3$ structure. Sections perpendicular to the layers. (a) SnF$_4$ type, same as in Figure 10b; (b) double layers of octahedra [MXX$_{5/2}$]; and (c) sheets composed of octahedra [MX$_{6/2}$] and [MXX$_{5/2}$], chemical formula M$_n$X$_{3n+1}$ = M$_3$X$_{10}$.

all six corners as in ReO$_3$. Units with n layers lead to layered structures of composition M$_n$X$_{3n+1}$ and finally to the three-dimensional ReO$_3$ structure.

Octahedra with four common edges give rise to a layer-like double-chain structure MX$_3$ (Figure 12) analogous to MX$_4$ or a layer structure M$_3$X$_8$ (Figure 13). With the M$_3$X$_8$ units a hexagonal layer structure is imperative, whereas the double chains may as well be equally distributed between all anion layers.

Sharing of all 6 vertices necessarily leads to a framework structure such as is met in the ReO$_3$, Sc(OH)$_3$ and FeF$_3$ types. Sharing of two vertices and two edges occurs in structures built up from chains connected through the free corners (Figure 14). The double-layer version is realized in MoO$_3$. Examples of stepped layers derived from these structures are shown in Figure 15 and Figure 16.

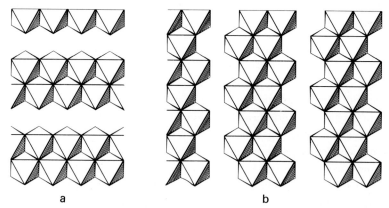

Fig. 12. Octahedral layers of composition MX$_3$ (a) double chains with each octahedron [MXX$_{2/2}$X$_{3/3}$] sharing four edges; (b) zigzag double chains with one octahedron [MX$_{2/2}$X$_{4/3}$] sharing four edges, the neighboring one [MX$_2$X$_{2/2}$X$_{2/3}$] sharing only three edges.

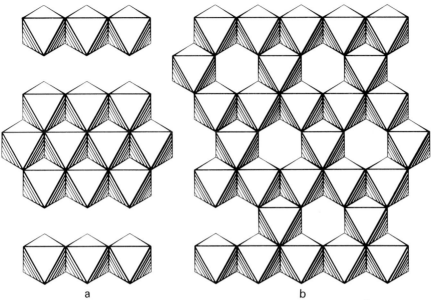

Fig. 13. Octahedral layers of composition M_3X_8. (a) artificial fibre-like model with $\frac{1}{3}$ of the octahedra sharing six edges and the remaining octahedra sharing four edges; (b) symmetrical arrangement of octahedra with four common edges (two pairs of opposite edges).

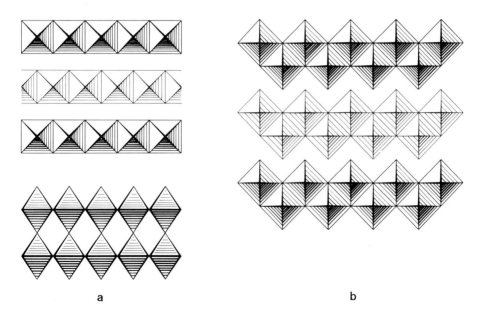

Fig. 14. Layer structures composed of octahedra which share two edges and two corners, (a) planar single layer, opposite edges in common, VO(OH)$_2$ structure. Above: section perpendicular to the layers; below: part of one layer; (b) adjacent edges shared, idealized MoO$_3$ structure. In both structures neighboring layers are shifted by half an octahedron height relative to each other.

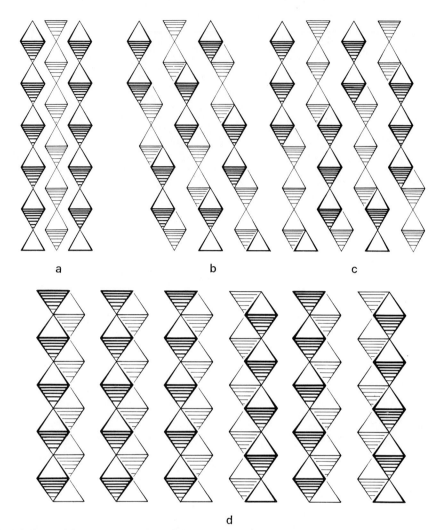

Fig. 15. Stepped layer structures derived from chains of (trans) edge-sharing octahedra by additional edge sharing. (a) The layer structure MXX'_2 of Figure 14a viewed perpendicular to the layers. (b) Stepped layers M_2X_5 built from double octahedron chains. (c) Combination M_nX_{3n-1} of (a) and (b) with $n = 3$. The octahedra of this figure are somewhat stretched. (d) Stepped double-layer structures MX_2,

left: idealized FeOCl structure
right: idealized boehmite γ-AlO(OH) structure.

Six edges are shared if all octahedral interstices between two close-packed X layers are occupied by M atoms. Hexagonal stacking of these sandwiches produces the CdI_2 type; cubic stacking leads to the $CdCl_2$-type structure. In contrast to the MoS_2-type family, the empty interlayer space here consists of the same kind of polyhedra as the sandwiches. All the octahedra are finally occupied in the NiAs and in the NaCl structure. But, whereas mixed NaCl–$CdCl_2$ type stacking can lead to new layer structures, any $M_{1+y}X_2$ ($1 < y < 2$) NiAs–CdI_2 type mixtures

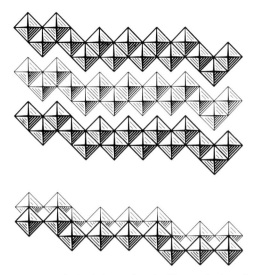

Fig. 16. Stepped layer structures derived from the double octahedron layers MX_3 (above) and $M_{6n}X_{18n-1}$ (below).

are invariably of the intercalation type and hence no longer true laminar compounds.

4. Coordination Number 4 – Tetrahedron

There are many fewer possibilities to create structures based on tetrahedra only. Not only are there geometrically fewer connecting elements, but face sharing is excluded as it would lead to unreasonably short M—M distances.

A laminar arrangement of discrete $[MX_4]$ tetrahedra in a close-packed X array is shown in Figures 17a, b. All known molecular MX_4 structures, such as those of OsO_4, $SnBr_4$ and SnI_4, are three-dimensional structures in which the occupied tetrahedra are equally distributed between all X layers.

Tetrahedra sharing three vertices, each with one other neighbor, may form fibrous layers (Figures 18a, b) or hexagonal nets (forming part of a pair of close-packed X layers) as occur as part of the kaoline layers. Sharing of three vertices with two other neighboring tetrahedra leads to sandwiches MX_2 that can be derived from a pair of close-packed layers (Figure 19a) or from three hcp X layers (Figures 19b, c). Only the third variant (Figure 19c) is realized in the AlOCl-type structure. The first one represents an analogue of the CdI_2 or $CdCl_2$ types. A wavy layer structure MX_2 can be constructed from a hcp arrangement by filling half the positions of one set of tetrahedral holes (i.e. all pointing in the same direction) in the form of double layers parallel with the stacking direction.

If we join two sheets of tetrahedra (Figure 18a) through their fourth vertex we obtain a double layer $A \beta B \beta A$ (using Greek letters to denote tetrahedral sites)

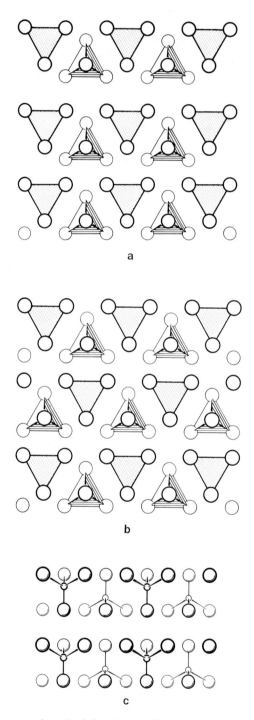

Fig. 17. Layered arrangement of tetrahedral molecular MX_4 compounds (a) and (b) projection of one layer; (c) section perpendicular to hcp layers.

GEOMETRICAL DERIVATION OF SIMPLE LAYER-TYPE STRUCTURES 21

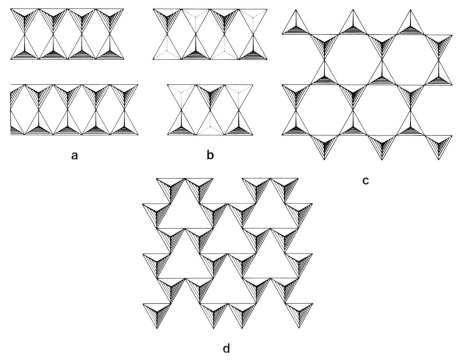

Fig. 18. Layers formed from tetrahedra [MXX$_{3/2}$] sharing three corners with a neighboring tetrahedron (a) and (b) fibrous structures; (c) and (d) true two-dimensional layers.

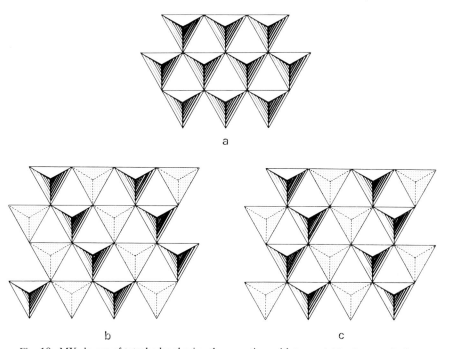

Fig. 19. MX$_2$ layers of tetrahedra sharing three vertices with two neighboring tetrahedra.

that could be used as a generating element for various layer structures M_2X_3:

$A\beta B\beta A \quad A\beta B\beta A\ldots; \quad A\beta B\beta A \quad B\gamma C\gamma B\ldots;$
$A\beta B\beta A \quad B\alpha A\alpha B\ldots,$ etc.

No example is known to us. If we form the double layers with the layers of Figure 18c or d, we obtain units MX_2 in which each tetrahedron is now sharing its four vertices with one other tetrahedron. Stacking these perforated sandwiches in AAA sequence would create a funny channel layer structure, e.g. as a hypothetical modification of SiO_2. There exist silicates such as $CaAl_2Si_2O_8$ [34] that do indeed contain these perforated double layers but the large alkaline-earth cations are located between adjacent layers, stuffing the channels.

A tetragonal layer structure MX_2 based on a ccp X array results if the tetrahedra share all four vertices (Figure 20), as in the HfI_2 structure. If the empty tetrahedra of the perforated MX_2 sandwiches are also occupied, we arrive at the CuTe and Li(OH) structures, in which the tetrahedra now have four out of the six edges in common. In the Li_2O or anti-CaF_2 structure, finally, the tetrahedra between the Li(OH) sandwiches are occupied as well and the tetrahedra then share all six edges. In a hypothetical layer structure M_4X_3, based

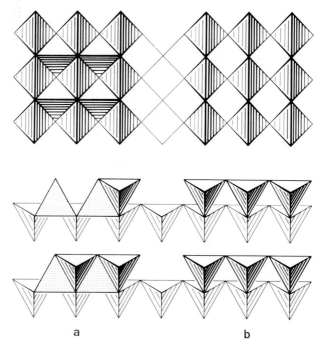

Fig. 20. Tetragonal layer structures based on ccp anion arrays (a) tetrahedra sharing four edges; (b) tetrahedra sharing four vertices.
        above: projection of one layer in tetragonal representation
        below: the layers in a hexagonal representation of the anion sublattice

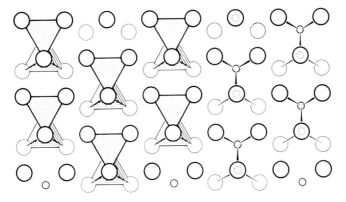

Fig. 21. Molecular layer structure MX_3 based on tetrahedra $[MX_2X_{2/2}]$ with one common edge.

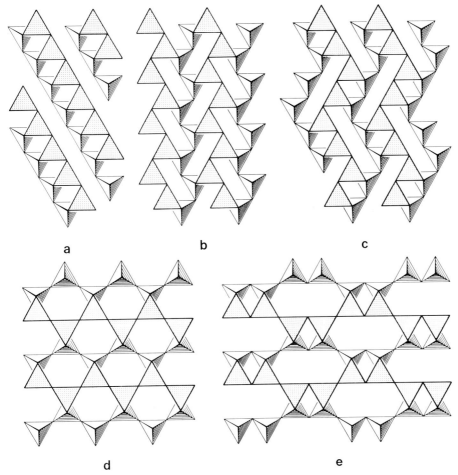

Fig. 22. Layers of tetrahedra sharing one edge and two corners (a) fibrous layer; (c) idealized $GaPS_4$ layer. a, b and c are based on close-packed X arrays, whereas in the layers illustrated in d and e the X sublattices are not close-packed.

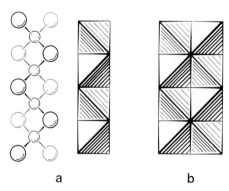

Fig. 23. (a) Chain of tetrahedra [$MX_{4/2}$] sharing two opposite edges; (b) double chain of composition M_2X_3, tetrahedra [$MX_{2/2}X_{2/4}$] sharing three edges.

on a mixture of the LiOH and Li_2O structures, where the empty layer follows only after 2 Li_2O-type occupied layers, each tetrahedron would share five edges.

A molecular layer structure MX_3 can be constructed from pairs of tetrahedra with a common edge. This structure is realized in the Al_2Br_6 type (Figure 21). Examples of tetrahedral layers in which each tetrahedron is sharing one edge and the other two corners are given in Figure 22. Only the first three types are based on a close-packed X array. Chains of composition MX_2 are formed if the tetrahedra have two opposite edges in common (e.g. SiS_2 or [FeS_2]$^-$ in $KFeS_2$) as is illustrated in Figure 23. Double chains M_2X_3 (as part of a ccp X array) result if

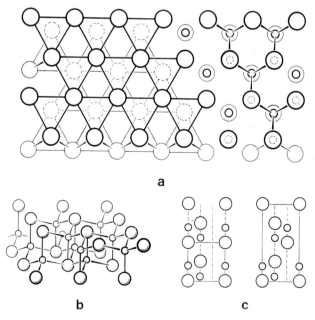

Fig. 24. Tetrahedron layers with each tetrahedron [$MX_{4/4}$] sharing three edges with a common corner. (a) projection of one layer; (b) perspective view of part of one layer; (c) (110) section of a hexagonal representation with two possible stacking arrangements.

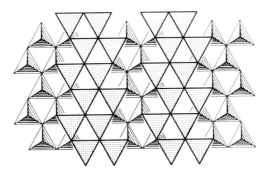

Fig. 24d. Puckered layer of a M_2X_2X' structure derived from the double layer of Fig. 24. Each cation has the coordination polyhedron $[MX_{3/3}X_{1/2}]$.

two such chains are connected through a third edge (Figure 23), as is met in the complex anion of $Cs^+[Cu_2Cl_3]^-$. If the fourth equivalent edge is used to connect the double layers we end up with the Li(OH) structure described above.

Three edges with a common corner (Figure 24) are shared if in a pair of close-packed X layers all the tetrahedra interstices are occupied (the empty space within the sandwich then corresponds to the octahedron layer). This arrangement with cubic close-packing of the anions X is claimed to occur in the rhombohedral structure of NiTe.

In close-packed arrays of anions X, both kinds of interstices, octahedral and tetrahedral ones, may simultaneously be occupied by cations M. In order to obtain a layer structure occupation of the tetrahedra and octahedra has to be such that periodically an empty layer alternates with a three-dimensional block. An example with only one kind of mixed layer is shown in Figure 25.

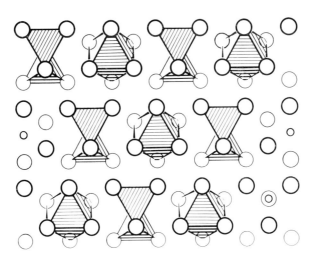

Fig. 25. Close-packed anion double layer with molecular layer structure MM'_2X_{12} based on a close-packed arrangement of octahedra $[MX_6]$ and double tetrahedra $[M'_2X_6]$.

5. Coordination Number 3 – Trigonal Pyramid and Triangle

By omitting the upper rim layers in Figure 18c and d and Figure 19a, we reduce the coordination polyhedron of the cations M to a trigonal pyramid and end up with layers shown in Figure 26. A triangular coordination results if all atoms are placed at the same height (Figure 24a). These model layers are found in graphite and BN (a), approximately in As and GeTe (b), and in As_2O_3 and As_2S_3 (c). On

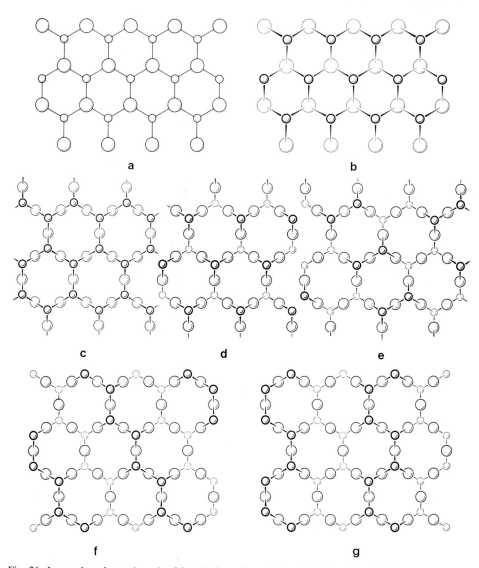

Fig. 26. Layers based on triangular (a) and trigonal-pyramid coordination (b–g). (a) BN-type layer; (b) GeTe-type layer; (c) As_2O_3 layers in KAs_4O_6I and $NH_4As_2O_3Cl.\frac{1}{2}H_2O$; (d) idealized layers of claudetite II, $As_2O_3(h_2)$; (e) idealized layer of orpiment, As_2S_3; (f) strongly idealized version of the layers found in claudetite I, As_2O_3.

connecting these 'polar' layers, however, the layer character may more or less get lost. Stacking of the layers in the same sense can provide additional interlayer bonds and thus increase the coordination number of the M atoms. A repulsion instead of the undesired attraction could result if the layers were stacked in the opposite sense. For this kind of stacking the only (though inadequate) example we know refers to the layer type of Figure 26b and the GaS family. Here, however, the two contacting cation layers are (and have to be) bonded, thus increasing the cation coordination number definitely to four.

CHAPTER III

REPARTITION OF THE LAYER STRUCTURES

Van der Waals bonding of layers, formed by identical or chemically similar ions, has been proposed as a criterion for layer-type structures. Therefore, ionic compounds are not expected to be included in this structure family. As can be checked partly in Tables 3 and 4, where we have compiled certain structural data, no alkali compound except Cs_2O is found among the true binary layer compounds. To our knowledge, the only binary fluorides that crystallize in a layer structure are TlF and Ag_2F. The metallic subfluoride Ag_2F is probably an interstitial compound like Ti_2O, both crystallizing in the CdI_2 structure. In both cases interlayer bonding contradicts our definition of true two-dimensional phases. In the tetragonal structure of SnF_4, PbF_4 and NbF_4, the individual layers are not well separated. Among the ternary phases, the rhombohedral LaOF-type fluorides possess a layered structure. In PbFCl, layering is not due to the F ions, but to the Cl ions. It is surprising that quite a number of simple oxides adopt layer structures:

Rb_3O, Cs_3O, Ti_2O (interstitial, metallic)
Cs_2O, $Ag_2O(p)$, Tl_2O; SnO, PbO (all with contacting cation layers)
V_2O_5, MoO_3, Re_2O_7, α-PtO_2, P_2O_5, $As_2O_3(h)$, β-TeO_2.

Of course, the important family of the layered silicates also belongs to this group.

Among the nitrides we are aware only of ThNF with the LaOF structure. The hexagonal structure of β-Be_3N_2, though layered, is not a true layer structure. $CdCl_2$-type Ca_2N and Sr_2N are interstitial metallic phases (the high resistivity reported for Ca_2N and Ba_2N [35] is odd. Was it Ca_2NH?) like many carbides such as Y_2C, Ln_2C.

As for the carbides, we hesitate somewhat to call Al_4C_3 a true two-dimensional compound, though it is non-metallic and crystallizes in hexagonal thin plates. The number of layer-type compounds increases drastically on going to chlorides, bromides and iodides, as well as sulfides, selenides and tellurides. On the other hand, only a few pnictides (SiP, GeAs, black phosphorus...) and tetrelides (BaSi,..., ~$CaSi_2$) are found with a layer structure. Moreover, the few examples met among the tetrelides are in fact 'polyanionic' compounds; the normal valence compounds all crystallize in three-dimensional structures (only a very restricted number of non-metallic representatives form at all: SiC, Be_2C, Al_4C_3, M_2X with M = Mg, Ca, Sr, Ba, (Ra?), Eu, Yb; and X = Si, Ge, Sn, Pb).

TABLE 3

Structure types occurring with binary tetrelides, pnictides, chalcogenides and halides. Layer structures are underlined.

Composition	Cation coordination number	Normal valence compounds and metallic phases without M—M or X—X bonds	Polycompounds
MX	2	HF, CO, ICl, HgO, HgS, Au, AuCl	HgCl, NS
	3	BN, SnS	AsS
	3+4		CuS
	4	ZnO, ZnS, NiTe, CuO, CuTe, PdS, PtS, TlF, α-PbO, β-PbO	CF, SiS
	5	TlI, SnP,	
	6	NaCl, NiAs (MnP, CrS, FeS), TiAs, NiP, NbP, WC, TiSi, PtB (anti-NiAs)	LiAs, NaSi, NaGe, KGe, NaPb, PtTe, LiO, NaO, KO, RbO
	4+8	TlSe (=Tl$^+$Tl^{III}Se$_2$)	
	~7	FeSi	
	8	CsCl	
	9	CrB, MoB, FeB, PdBi	
	>9	LiSn, CoGe, CoSn	
M$_3$X$_4$	4	Si$_3$N$_4$(r), Si$_3$N$_4$(h)	
	4+6	Fe$_3$O$_4$, Co$_3$S$_4$	
	6	Fe$_3$S$_4$, Fe$_3$Se$_4$	
	8	Th$_3$P$_4$	Nb$_3$Se$_4$
M$_2$X$_3$	3	B$_2$S$_3$, As$_2$O$_3$(r), As$_2$O$_3$(t), Sb$_2$O$_3$(r), As$_2$S$_3$	
	3–4	Sb$_2$S$_3$	
	4	B$_2$O$_3$, Ga$_2$S$_3$ (ZnS-deriv.), α-Ga$_2$S$_3$ (ZnO-deriv.)	
	3+6	As$_2$Te$_3$	
	4+6	mon. Ga$_2$O$_3$, β-In$_2$S$_3$ (ordered spinel), In$_2$Se$_3$(r)	
	6	Al$_2$O$_3$, Sc$_2$S$_3$, Sc$_2$Te$_3$, Cr$_2$S$_3$, Mo$_2$S$_3$, Mo$_2$As$_3$, Mn$_2$O$_3$, Rh$_2$S$_3$, Pt$_2$Ge$_3$, Pt$_2$Sn$_3$, Ba$_3$P$_2$, In$_2$S$_3$(h$_2$), Bi$_2$Te$_3$	Rb$_2$O$_3$, Pt$_2$Te$_3$
	8	La$_2$O$_3$	
MX$_2$	2	SO$_2$, CO$_2$, HgO$_2$, HgCl$_2$, SCl$_2$	HgO$_2$
	3	SeO$_2$, GeF$_2$, (SnCl$_2$), SnF$_2$	N$_2$O$_4$, P$_2$I$_4$
	4	Be(OH)$_2$, SiO$_2$, SiS$_2$, SiS$_2$(p), GeS$_2$, GeS$_2$(h) HgI$_2$, PdCl$_2$, PdBr$_2$, CuCl$_2$, CuBr$_2$, α-TeO$_2$, β-TeO$_2$	ZnP$_2$, ZnAs$_2$, CdP$_2$, GeAs$_2$, PdSe$_2$, PdP$_2$, CuP$_2$, NaS$_2$
	4–6		FeS$_2$, FeAs$_2$, CoSb$_2$, IrSe$_2$, Th$_{0.9}$Ge$_2$, PdSn$_2$,
	6	TiO$_2$ (rutile, anatase, brookite), CaCl$_2$, VO$_2$, CuF$_2$, PbO$_2$, HgBr$_2$, CdI$_2$, CrBr$_2$, CdCl$_2$, ReSe$_2$, NbTe$_2$, MoTe$_2$(h), WTe$_2$, AuTe$_2$ (calaverite), AuTe$_2$ (krennerite), MoS$_2$, NbS$_2$	
	6–8	SnI$_2$	
	7	Sr(OH)$_2$, SrBr$_2$, SrI$_2$, EuI$_2$, ZrO$_2$(mon.)	MoP$_2$, CaSi$_2$
	8	CaF$_2$, SnBr$_2$	PdBi$_2$(h), NdTe$_2$, SrS$_2$, SmSb$_2$, CeSe$_2$
	9	(PbCl$_2$), ~PdBi$_2$(r)	α-ThSi$_2$, α-GdSi$_2$, NbAs$_2$
	>9		CaC$_2$, ThC$_2$, AlB$_2$, ZrSi$_2$

(*Continued overleaf*)

Table 3 (Continued)

Composition	Cation coordination number	Normal valence compounds and metallic phases without M—M or X—X bonds	Polycompounds
M_2X_5	2+3	N_2O_5	
	4	P_2O_5 (rhomb.), P_2O_5(orth.), $\underline{P_2O_5\text{(orth.)}}$	
	4–6	V_2O_5	
	6–8	$\underline{Nb_2O_5}$, $\underline{Ta_2O_5}$. Pa_2O_5, Cr_2F_5, Mn_2F_5	$\underline{Nd_2Te_5}$
MX_3	3	NH_3, NCl_3, $\underline{BCl_3}$, SbF_3, $SbCl_3$, $\alpha\text{-}\underline{SbBr_3}$, ClF_3, BrF_3,	
	4	$\beta\text{-}SO_3$, $\gamma\text{-}SO_3$, $\underline{AlBr_3}$, $\underline{InI_3}$, CrO_3, AuF_3, $AuCl_3$, ICl_3	
	6	ReO_3, $Sc(OH)_3$, MoO_3, WO_3(h), AlF_3, $\underline{AlCl_3}$ VF_3, $\underline{CrCl_3}$, $\underline{BiI_3}$, $\underline{TiI_3}$, $RuBr_3$	SrS_3, RhS_3, $\underline{SnP_3}$, $CoAs_3$
	7		SrP_3, BaP_3
	8	$\underline{PuBr_3}$, $\alpha\text{-}UO_3$	KN_3, CaP_3, BaS_3, $\underline{ZrSe_3}$, $\underline{TaSe_3}$, $\underline{PdSn_3}$
	9	LaF_3, UCl_3	$\underline{NdTe_3}$
M_2X_7	4	Tc_2O_7, Mn_2O_7	
	4+6	Re_2O_7	
MX_4	4	CH_4, SiF_4, $SnBr_4$, SnI_4, OsO_4, $Ni(CO)_4$	CdP_4
	6	SnF_4, $ZrCl_4$, $NbCl_4$(h), $PtBr_4$, $\gamma\text{-}PtI_4$, TeI_4	$Os(CO)_4$
	~7	UBr_4	$NbCl_4$(r)
	8	ZrF_4, UCl_4	CsI_4, VS_4, $NbTe_4$, $RhBi_4$, $\underline{PtSn_4}$, $\underline{PtPb_4}$

Obviously the structural character of a compound is determined in a complex way by the electron configuration of cations and anions (including the principal quantum number of the valence electrons), their electronegativity difference and the cation/anion radius ratio. The valences of cation and anion define the stoichiometry and this in turn determines the coordination number. If in a compound MX_n the electron configuration of the cation and its size favor a coordination number $CN = n$, then the compound necessarily will adopt a molecular structure. With $n = CN - \frac{1}{2}$ dimers M_2X_{2n} and with $n = CN - 1$ chains or rings $[MX_{n-1}X_{2/2}]$ will form. These discrete structure elements, however, still may be arranged in such a way that a laminar structure results, as in $\alpha\text{-}WCl_6$. For a certain compound, the steric conditions usually permit different realizations and therefore many compounds possess different, though closely related modifications. P_2O_5 may serve as an example. In a covalent formula we describe the unit as $P^+O^-O_{3/2}$ which immediately shows the coordination. P has to be surrounded by a tetrahedron of O atoms, three of the corners of which are shared with a neighboring P atom. Thus, the metastable rhombohedral modification contains discrete P_4O_{10} molecules of the same type as in the vapor (the 4P themselves form a tetrahedron). The stable orthorhombic form consists of rings of $[PO_4]$ tetrahedra linked up to form a three-dimensional network. In the third modification the $[PO_4]$ tetrahedra are connected in a similar fashion, but here rings of six tetrahedra form layers of the type shown in Figure 16c.

TABLE 4

Repartition of layer-type structures among the chalcogenides, pnictides and tetrelides of composition MX_2. The layer structures are underlined. Chemical formulae are given where the structure type is not yet known. r, t, h, p: room-temperature, low-temperature, high-temperature and high-pressure, resp.

M	MO_2	MS_2	MSe_2	MTe_2	MP_2	MAs_2	MSb_2	MBi_2	MC_2	MSi_2	MGe_2	MSn_2	MPb_2
Li												$LiSn_2$	
Na	$r:NaCl$ $t_1:FeS_2$ $t_2:FeAs_2$	Na_2S_4	Na_2Se_4	Na_2Te_4						$NaSi_2$		$NaSn_2$	
K	$h:NaCl$ $r:CaC_2$ $t_1:$ $t_2:$monocl.	K_2S_4	K_2Se_4			KAs_2	KSb_2	Cu_2Mg				KSn_2	$MgZn_2$
Rb	$h:NaCl$ $r:CaC_2$ $t_1:$	Rb_2S_4	Rb_2Se_4				$RbSb_2$	Cu_2Mg					
Cs	$h:NaCl$ $r:CaC_2$ $t:$	Cs_2S_4						Cu_2Mg					
Mg	FeS_2												
Ca	CaC_2	FeS_2	FeS_2	FeS_2			$CaSb_2$		CaC_2 triclinic	$CaSi_2$ stack. var. p:tetrag.	$CaSi_2$		
Sr	CaC_2	$CuAl_2$			SrP_2		$CaSb_2$		CaC_2	$SrSi_2$ $BaSi_2$ $ZrSi_2$	$BaSi_2$		
Ba	CaC_2	ThC_2			BaP_2				CaC_2		$BaSi_2$		
Sc													
Y		LaS_2					$HoSb_2$			α-$GdSi_2$	α-$ThSi_2$	$ZrSi_2$	
La		LaS_2	$CeSe_2$	$\sim Cu_2Sb$	LaP_2	$r:NdAs_2,$ $h:LaP_2$	$SmSb_2$	$LaBi_2$	CaC_2 $r:CaC_2$	α-$ThSi_2$	α-$ThSi_2$		
Ce	CaF_2	LaS_2	$CeSe_2$	$\sim Cu_2Sb$	$r:NdAs_2$ $h:LaP_2$	$NdAs_2$	$SmSb_2$	$LaBi_2$	CaC_2	α-$ThSi_2$	α-$ThSi_2$		
Pr	CaF_2	LaS_2	$CeSe_2$	$\sim Cu_2Sb$	$NdAs_2$	$NdAs_2$	$SmSb_2$	$LaBi_2$	CaC_2	α-$GdSi_2$	α-$ThSi_2$		
Nd		LaS_2	$CeSe_2$	$\sim Cu_2Sb$	$NdAs_2$	$NdAs_2$	$SmSb_2$	$LaBi_2$	CaC_2	α-$ThSi_2$	α-$ThSi_2$		
Sm		LaS_2	$ErSe_2$	$\sim Cu_2Sb$			$SmSb_2$		CaC_2	α-$GdSi_2$	α-$ThSi_2$		
Eu					EuP_2	$EuAs_2$	$CaSb_2$		CaC_2	α-$ThSi_2$	$EuGe_2$		
Gd		LaS_2	$ErSe_2$	$\sim Cu_2Sb$			$SmSb_2$ $p:HoSb_2$		CaC_2	α-$GdSi_2$	$GdGe_2$	$ZrSi_2$	

(*Continued overleaf*)

Table 4 (Continued)

M	MO_2	MS_2	MSe_2	MTe_2	MP_2	MAs_2	MSb_2	MBi_2	MC_2	MSi_2	MGe_2	MSn_2	MPb_2
Tb	CaF_2	LaS_2	$ErSe_2$	$\sim Cu_2Sb$			$SmSb_2$ $p:HoSb_2$		CaC_2	α-$GdSi_2$	$GdGe_2$	$ZrSi_2$	
Dy		LaS_2	$ErSe_2$	$\sim Cu_2Sb$			$HoSb_2$		CaC_2	α-$GdSi_2$	$GdGe_2$	$ZrSi_2$	
Ho		LaS_2	$ErSe_2$	$\sim Cu_2Sb$			$HoSb_2$		CaC_2	α-$GdSi_2$	$GdGe_2$	$ZrSi_2$	
Er		LaS_2	$ErSe_2$ $h:UTe_2$	$\sim Cu_2Sb$			$HoSb_2$		CaC_2	AlB_2		$ZrSi_2$	
Tm		LaS_2	$\sim Cu_2Sb$	$\sim Cu_2Sb$			$HoSb_2$		CaC_2	AlB_2		$ZrSi_2$	
Yb		LaS_2	$\sim Cu_2Sb$	$\sim Cu_2Sb$			$ZrSi_2$		CaC_2	AlB_2		$ZrSi_2$	
Lu		LaS_2	$\sim Cu_2Sb$	$\sim Cu_2Sb$			$HoSb_2$		CaC_2	AlB_2		$ZrSi_2$	
Th	CaF_2	$PbCl_2$	$PbCl_2$	$ThTe_2$		$r:PbCl_2$ $h:Cu_2Sb$	Cu_2Sb	Cu_2Sb	ThC_2	α-$ThSi_2$	α-$ThSi_2$ $ZrSi_2$ $Th_{0.9}Ge_2$		
U	CaF_2	$r:US_2$ $h:PbCl_2$	$r:US_2$ $h:PbCl_2$	UTe_2	$r:UP_2$ $h:Cu_2Sb$	Cu_2Sb	Cu_2Sb		$r:CaC_2$	α-$ThSi_2$ AlB_2	$ZrSi_2$		
Np	CaF_2			$\sim Cu_2Sb$		Cu_2Sb				α-$ThSi_2$			
Pu	CaF_2	$CeSe_2$	$CeSe_2$	$\sim Cu_2Sb$					CaC_2	α-$ThSi_2$ AlB_2	α-$ThSi_2$		
Am	CaF_2	$\sim Cu_2Sb$	$\sim Cu_2Sb$	$\sim Cu_2Sb$									
Ti	rutile anatase brookite $p:\alpha$-PbO_2	CdI_2	CdI_2	CdI_2	$PbCl_2$	$TiAs_2$	$CuAl_2$			$TiSi_2$ $ZrSi_2$	$TiSi_2$		
Zr	monocl. tetrag. ZrO_2	CdI_2	CdI_2	CdI_2	$PbCl_2$	$PbCl_2$	$TiAs_2$	$TiAs_2$		$ZrSi_2$	$ZrSi_2$	$TiSi_2$	
Hf		CdI_2	CdI_2	CdI_2	$PbCl_2$	$PbCl_2$	$r:TiAs_2$ $h:Cu_2Sb$	$TiAs_2$		$ZrSi_2$	$ZrSi_2$	$CrSi_2$	
V	$r:VO_2$ $h:TiO_2$	$p:CdI_2$	CdI_2	$p:CdI_2(?)$	$NbAs_2$	$NbAs_2$	$CuAl_2$			$MoSi_2$			
Nb	$r:NbO_2$ $h:TiO_2$	NbS_2 polytypes	$NbSe_2$ polytypes $h:CdI_2$	$NbTe_2$	$NbAs_2$	$NbAs_2$	$NbAs_2$			$CrSi_2$ $MoSi_2$	$CrSi_2$	$CuMg_2$	
Ta	$r:NbO_2$ $h:TiO_2$	NbS_2 polytypes	NbS_2 polytypes $h:CdI_2$	$NbTe_2$	$NbAs_2$	$NbAs_2$	$NbAs_2$			$CrSi_2$	$CrSi_2$		

REPARTITION OF THE LAYER STRUCTURES

Cr	TiO$_2$	p:CdI$_2$	p:CdI$_2$(?)	p:CdI$_2$	NbAs$_2$	NbAs$_2$	FeAs$_2$		MoSi$_2$	
Mo	VO$_2$	MoS$_2$	MoS$_2$	r:MoS$_2$	MoP$_2$	NbAs$_2$			MoSi$_2$	
	h:TiO$_2$	polytypes		h:MoTe$_2$					PbCl$_2$	
W	VO$_2$	MoS$_2$	MoS$_2$	WTe$_2$	NbAs$_2$	NbAs$_2$			MoSi$_2$	
	h:TiO$_2$				h:MoP$_2$					
Mn	TiO$_2$	FeS$_2$	FeS$_2$	FeS$_2$						CuAl$_2$
	diaspore orthorh.									
Tc	VO$_2$	TcS$_2$	TcS$_2$	TcTe$_2$					TcSi$_2$	
Re	VO$_2$	ReSe$_2$	ReSe$_2$	ReTe$_2$					MoSi$_2$	
	h:α-PbO$_2$									
Fe	FeO$_2$	r:FeAs$_2$	FeAs$_2$	FeAs$_2$	FeAs$_2$	FeAs$_2$	FeAs$_2$		r:OsSi$_2$	CuAl$_2$
		h:FeS$_2$	p:FeS$_2$	p:FeS$_2$					h:α-FeSi$_2$	
Ru	TiO$_2$	FeS$_2$	FeS$_2$	FeS$_2$	FeAs$_2$	FeAs$_2$			NbAs$_2$	CuAl$_2$
Os	TiO$_2$	FeS$_2$	FeS$_2$	FeS$_2$	FeAs$_2$	FeAs$_2$			OsSi$_2$	CuAl$_2$
Co		FeS$_2$	FeAs$_2$	FeAs$_2$	CoAs$_2$	CoAs$_2$			CoGe$_2$	CuAl$_2$
			h:FeS$_2$	(CdI$_2$)					CaF$_2$	
				p:FeS$_2$						
Rh	TiO$_2$		r:IrSe$_2$	FeS$_2$	CoAs$_2$	CoAs$_2$	CoAs$_2$			r:CoGe$_2$	CuAl$_2$
			h:FeS$_2$	CdI$_2$			h:RhBi$_2$			h:CuAl$_2$	
Ir	TiO$_2$	IrSe$_2$	IrSe$_2$	CdI$_2$	CoAs$_2$	CoAs$_2$	CoAs$_2$				
		p:FeS$_2$									
Ni		FeS$_2$	FeS$_2$	CdI$_2$	PdP$_2$	r:NiAs$_2$	FeAs$_2$		CaF$_2$		
				p:FeS$_2$	p:FeS$_2$	h:FeAs$_2$					
						p:FeS$_2$					
Pd		PdSe$_2$	PdSe$_2$	CdI$_2$	PdP$_2$	FeS$_2$	FeS$_2$	r:PdBi$_2$		CoGe$_2$	CuAl$_2$
								h:MoSi$_2$			
Pt	CaCl$_2$	CdI$_2$	CdI$_2$	CdI$_2$	FeS$_2$	FeS$_2$	FeS$_2$	r:dist. FeS$_2$	~FeAs$_2$		
	CdI$_2$							h_1:FeS$_2$			
	polytypes							h_2:PtBi$_2$			
Cu		FeAs$_2$	FeAs$_2$	FeS$_2$	CuP$_2$						
		h:FeS$_2$	h:FeS$_2$	p:CdI$_2$							
				calaverite	CuP$_2$		FeS$_2$				
				krennerite							
Ag											
Au									brookite	AuPb$_2$	

(*Continued overleaf*)

Table 4 (Continued)

M	MO$_2$	MS$_2$	MSe$_2$	MTe$_2$	MP$_2$	MAs$_2$	MSb$_2$	MBi$_2$	MC$_2$	MSi$_2$	MGe$_2$	MSn$_2$	MPb$_2$
Zn	FeS$_2$	FeS$_2$	FeS$_2$	p:CdI$_2$	β-CdP$_2$ ZnAs$_2$	ZnAs$_2$ p:							
Cd	FeS$_2$	FeS$_2$	FeS$_2$		r:α-CdP$_2$ h:β-CdP$_2$	CdAs$_2$ p:							
Hg	α-HgO$_2$ β-HgO$_2$												
Si	quartz(r, h) cristobalite(r, h) tridymite(r, h) rutile keatite SiS$_2$ p:coesite	SiS$_2$ p:α-ZnCl$_2$	SiS$_2$	SiS$_2$(?)	GeAs$_2$ p:FeS$_2$	GeAs$_2$ p:FeS$_2$							
Ge	TiO$_2$ cristobalite	GeS$_2$ p:α-ZnCl$_2$ h: monocl. CdI$_2$ polytypes	GeS$_2$ mon. GeS$_2$ CdI$_2$			GeAs$_2$ p:FeS$_2$							
Sn	TiO$_2$	p:KHF$_2$ p:CdI$_2$											
Pb	TiO$_2$ α-PbO$_2$ p:CaF$_2$		p:α-KHF$_2$										

TABLE 5
Influence of the principal quantum numbers on coordination

O_3	Molecular structure monomer CN 2	SO_2	Molecular structure monomer CN 2 and 1	SeO_2	Chain structure $[SeOO_{2/2}]_\infty$ CN 3 and 1 or 2	TeO_2	(α, stable) three-dimensional CN 4 and 2 (β) layer structure CN 4 and 2
As_2O_3	(t) claudetite 'polar' layer structure CN 3 and 2 (h) senarmontite type dimers As_4O_6 CN 3 and 2	As_2S_3	orpiment layer structure similar to claudetite CN 3 and 2	As_2Se_3	orpiment type CN 3 and 2	As_2Te_3	layer structure CN 6 or 3 and 3
Sb_2O_3	(r) valentinite double chains $[Sb_2O_{6/2}]_\infty$ CN 3 and 2 (h) senarmontite dimers Sb_4O_6 CN 3 and 2	Sb_2S_3	stibnite ribbons $[Sb_4S_6]_\infty$ CN>3 and >2	Sb_2Se_3	stibnite type	Sb_2Te_3	Bi_2Te_2S type layer structure CN 6 and 6 or 3
PI_3 'pure' single p bonds CN 3 and 1		AsI_3 increasing admixture of d character, resonant bonds CN increases gradually		SbI_3		BiI_3 CN 6 and 3	

The influence of the principal quantum number of the ions may be demonstrated with the chalcogenides of As and Sb and some oxides and iodides in Table 5.

An increase of the principal quantum number of one or even both ions leads to an increase of the coordination number. The single p bonds of the anions (and the cations, eventually) gradually change into mixed (p, d) bonds when the p-d energy differences become smaller.

Layer structures that are based on p single bonds are quite different from the usual layer structures. They consist of two atomic sheets only, each forming a boundary of the layer. The layer units are in a certain sense polar and represent infinite molecules. These polymeric structures develop with cations that are able to form three single p bonds (hence a coordination number 3) namely P, As, (Sb), and anions that prefer CN≥3 (P, O, S, Se, ev. Cl, Br).

The common feature of both layer types is the character of the boundary sheets. The boundary sheets of a sandwich compound as well as both sheets of a two-dimensional polymer consist of ions that are able to form covalent bonds directed into half-space. This condition is satisfied for the three p bonds. Therefore, the boundaries of true layer-type compounds contain anions from Group V to Group VII and possibly cations such as Rb^+, Cs^+, Tl^+, Sn^{2+}, Pb^{2+}. (Pt_2Sn_3, $PdSn_3$, $PtSn_4$ and $PtPb_4$ seem to be exceptions. Geometrically, they do have layer structures, but in these metallic compounds bonding between adjacent layers due to additionally available bond orbitals spoils the layer character physically.) Moreover, the cations below the boundary layers are preferentially located in octahedral coordination: only a few layer compounds are known with the cations in tetrahedral coordination. With cations in octahedral interstices of a close-packed anion array the anion coordination number CN is ≥ 3 for a composition MX_n (with $n \geq 1$) of the outermost (cation + anion) layer. Indeed, a great number of halides and chalcogenides crystallize in layer structures, the exceptions being mainly fluorides and oxides. For sandwich compounds of composition MX_3 the anion coordination number is 2, so that geometrically the p^2 bonds may link cations all lying on one side of the anion layer or cations lying on either side. For close-packed MX_6 compounds with anion CN 1, this reasoning holds analogously for the p bond of neighboring anions. In fluorides the three-dimensional distribution is preferred. In certain compounds, such as TiI_3 and WCl_6, both two- and three-dimensional structures occur.

The coordination is not only (though mainly) determined by the size of the constituent atoms but the coordination polyhedron frequently depends on the electron configuration of the atoms. Thus, the structure-decisive role of the lone electron pair of Pb^{2+}, Sb^{3+} and Bi^{3+} and of the d-hole of Cr^{2+} and Cu^{2+} is known well enough.

Octahedral coordination is to be expected with the following cations:

d^0	(Sc^{3+}, Y^{3+}, Lu^{3+}, Ti^{4+}, Zr^{4+}, Hf^{4+}, ...)
d^3	(V^{2+}, Cr^{3+}, Mn^{4+}, ...)
d^5	(Mn^{2+}, Fe^{3+}, ...) (if not tetrahedral)
d^6	(Fe^{2+}, Co^{3+}, Rh^{3+}, Ir^{3+}, ...)
d^8	(Ni^{2+}, ...)
d^{10}	(Zn^{2+}, Cd^{2+}, In^{3+}, Sn^{4+}, Pb^{4+}, ...) (if not tetrahedral);

distorted octahedral (if non-metallic) with:

d^1	(Ti^{3+}, V^{4+}, ...)
d^2	(Ti^{2+}, V^{3+}, ...)
d^4	(Cr^{2+}, Mn^{3+}, ...)
d^7	(Co^{2+}, ...)
d^9	(Cu^{2+}, Ag^{2+}).

The distortions may be so pronounced that a square planar coordination results in d^8 cations (Ni^{2+}, Pd^{2+}, Pt^{2+}, Au^{3+} ...).

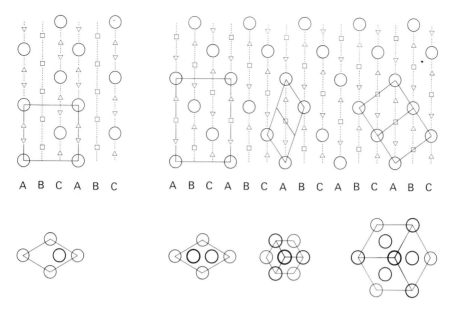

Fig. 27. (110) section through a hexagonal (left) and a cubic (right) close-packed anion array with all octahedral and tetrahedral sites given. The two sets of tetrahedral holes are distinguished by triangles pointing either upwards (τ_1) or downwards (τ_2). For the ccp array the most common cells (hexagonal, primitive rhombohedral and fcc) are indicated.

A trigonal-prismatic coordination is energetically favored in certain compounds with d^0, d^1 and d^2 cations:

d^0 (Zr^{4+}, Hf^{4+}, Th^{4+}, ...)
d^1 (Nb^{4+}, Ta^{4+}, ...)
d^2 (Mo^{4+}, W^{4+}, Re^{5+}, ...).

A trigonal pyramid (half an octahedral coordination) is adequate around lone-pair s^2 cations (Sn^{2+}, Pb^{2+}, P^{3+}, As^{3+}, Sb^{3+}, Bi^{3+}, Te^{4+}). Coordination numbers >6 are met with the large cations Sr^{2+}, Ba^{2+}, Ra^{2+}, Ln^{2+}, Ln^{3+}, Ac^{3+}, Th^{4+}, U^{3+}, U^{4+}, ..., Tl^+, Pb^{2+}, Bi^{3+} ....

Each dense packing of spheres is built up of hexagonal layers and is therefore somewhat layered. Moreover, each close-packed array of spheres contains 1 octahedral and 2 tetrahedral holes per sphere (in other words, space can be filled completely with n octahedra and $2n$ tetrahedra). Therefore, we are not surprised that the structures of a great part of the layer-type compounds with cation coordination number 4 and 6, are based on or can be derived from close-packed anion arrays. Many of the close-packed structures are fully characterized by giving the (110) section in hexagonal representation as is illustrated in Figure 27 for the simple hexagonal and cubic close-packing. The letter under each row describes the anion stacking sequence. In order to define the occupancy of the

TABLE 6
Occurrence of layer types among structures based on anion close packing

Occupancy of τ_1	τ_2	Ω	ccp (c)	hcp(h)	Mixed ccp/hcp
1	1	0	Mg$_2$Sn (AgMgAs, Li$_3$AlN$_2$, Li$_5$GeP$_3$, ...)		
1	$\frac{3}{4}$	0	Na$_6$PbO$_4$		
$\frac{3}{4}$	$\frac{3}{4}$	0	Mg$_3$As$_2$ (anti-Mn$_2$O$_3$), Zn$_3$P$_2$, Li$_5$GaO$_4$, Cu$_5$FeS$_4$,		Li$_8$CoO$_6$
$\frac{3}{4}$	$\frac{1}{4}$	0	Cu$_3$VS$_4$		
1	0	0	ZnS (LiSH, CuFeS$_2$, Cu$_2$GeS$_3$, Cu$_2$FeSnS$_4$, Cu$_3$SbS$_4$)	ZnO (BeSiN$_2$, Cu$_2$SiS$_3$(h), Cu$_2$ZnGeS$_4$, Cu$_3$AsS$_4$)	
$\frac{1}{2}$	$\frac{1}{2}$	0	FeS (tetrag.), CuTe, β-NiTe, LiNH$_2$	CuFe$_2$S$_3$	
$\frac{3}{4}$	0	0	β-Cu$_2$HgI$_4$, CdGa$_2$S$_4$, HgGa$_2$Te$_4$, Zn$_3$PI$_3$, CuIn$_2$Se$_3$Br, CdIn$_2$Se$_4$, AgIn$_5$Te$_8$	ZnAl$_2$S$_4$	
$\frac{1}{3}$	$\frac{1}{3}$	0	In$_2$Te$_3$ (disord.)	β-Ga$_2$S$_3$	
$\frac{1}{2}$	0	0	α-ZnCl$_2$, **C19τ**, InPS$_4$		
$\frac{1}{4}$	$\frac{1}{4}$	0	**HgI$_2$**, SiS$_2$ (BPS$_4$), AlPS$_4$, Cd(CN)$_2$, Mg(NH$_2$)$_2$	β-ZnCl$_2$, **GaPS$_4$**	
$\frac{1}{6}$	$\frac{1}{6}$	0		AlBr$_3$, InI$_3$	
$\frac{1}{8}$	$\frac{1}{8}$	0	SnI$_4$, OsO$_4$,	SnBr$_4$	
0	0	1	NaCl (NaCrS$_2$, Li$_2$TiO$_3$, Li$_2$ZrO$_3$, Li$_2$SnO$_3$, Li$_2$MnO$_3$, Li$_3$NbO$_4$, Na$_3$BiO$_4$,....)	NiAs (LiCrS$_2$) dist.: CrS, MnP,	TiAs, TiS,
0	0	$\frac{7}{8}$	Mg$_6$MnO$_8$	Fe$_7$S$_8$, Fe$_7$Se$_8$	
0	0	$\frac{5}{6}$		Cr$_5$S$_6$	Ge$_3$Bi$_2$Te$_6$
0	0	$\frac{4}{5}$	Ti$_4$O$_5$		Ti$_4$S$_5$, **Pb$_2$Bi$_2$Se$_5$**
0	0	$\frac{3}{4}$	Yb$_3$Se$_4$, Mg$_3$NF$_3$,	Cr$_3$S$_4$(CrTi$_2$S$_4$), CoMo$_2$S$_4$,	**Fe$_3$S$_4$ (smythite)**, GeBi$_2$Te$_4$
0	0	$\frac{3}{5}$	Ti$_3$O$_5$(r)		
0	0	$\frac{5}{7}$			GeSb$_4$Te$_7$
0	0	$\frac{2}{3}$	**In$_2$S$_3$(h$_2$)**, Sc$_2$S$_3$, ScCrS$_3$, K$_3$La(NH$_2$)$_6$,	α-Al$_2$O$_3$(FeTiO$_3$, LiSbO$_3$, Ni$_3$TeO$_6$) rhomboh. Cr$_2$S$_3$, trig. Cr$_2$S$_3$	(Pt$_2$Sn$_3$), Bi$_2$Te$_3$ (Bi$_2$Te$_2$S)
0	0	$\frac{1}{2}$	**CdCl$_2$, ReSe$_2$**, TiO$_2$(anatase) Cu$_2$(OH)$_3$Cl(paratacamite) Cu$_2$(OH)$_3$Cl (atacamite),	**CdI$_2$ (Li$_2$Pt(OH)$_6$)** AuTe$_2$ (calaverite), (CaCl$_2$) **Cd(OH)Cl**, str. dist.: TiO$_2$ (rutile) (InOOH), NiWO$_4$, Nb$_2$Mn$_4$O$_9$, Li$_2$ZrF$_6$ α-PbO$_2$ (LiNb$_3$O$_8$, Fe$_2$WO$_6$, α-NaTiF$_4$, LiFeW$_2$O$_8$, NaNbO$_2$F$_2$)	HgBr$_2$, WTe$_2$, MoTe$_2$(h) strongly distorted: SmSI
0	0	$\frac{3}{8}$		Nb$_3$Cl$_8$	β-Nb$_3$Br$_8$
0	0	$\frac{1}{3}+\delta$		Re$_{1+\delta}$O$_3$ (=partly-filled anti-Ni$_3$N)	
0	0	$\frac{1}{3}$	CrCl$_3$, AlCl$_3$(α-MoCl$_3$) β-IrCl$_3$	BiI$_3$, TiI$_3$, β-MoCl$_3$, FeF$_3$	
0	0	$\frac{1}{4}$	SnF$_4$, TeCl$_4$(str. def.), ZrCl$_4$, γ-PtI$_4$	NbCl$_4$, MoCl$_4$, PtBr$_4$	α-PtI$_4$

Table 6 (*Continued*)

τ_1	τ_2	Ω	ccp (c)	hcp (h)	Mixed ccp/hcp
0	0	$\frac{1}{5}$	NbF_5, α-UF_5, UCl_5(monocl.)	RuF_5, $TaCl_5$, $NbCl_5$, NbI_5, **UCl_5(triclinic)**	$ReCl_5$
0	0	$\frac{1}{6}$		**α-WCl_6**, UCl_6	orthorh. UF_6
1	1	1	Li_3Bi (Li_2MgSn)		
$\frac{1}{2}$	0	1	Cu_2Te (~Cu_2Sb)		
$\frac{1}{4}$	0	1	Pt_3Sb_2		
$\frac{1}{2}$	$\frac{1}{2}$	1			Mg_3Sb_2(anti-La_2O_3) (Li_2ZrN_2)
$\frac{1}{2}$	$\frac{1}{2}$	$\frac{1}{2}$	Cu_2MnTe_2, **$Fe_{1.5}Ni_{1.5}Te_2$**		
$\frac{1}{2}$	0	$\frac{1}{2}$	$CuCrS_2$(r)	$CuScS_2$	
$\frac{1}{4}$	$\frac{1}{4}$	$\frac{1}{2}$	$CuCrS_2$(h)		
$\frac{1}{4}$	0	$\frac{1}{2}$		**$FeAl_2S_4$**	
$\frac{3}{28}$	$\frac{3}{28}$	$\frac{1}{2}$		$Mg_7Si_3O_{12}(OH)_2$(humite)	
$\frac{1}{8}$	$\frac{1}{8}$	$\frac{1}{2}$	$MgAl_2O_4$(spinel) (Mn_3O_4, Li_5AlO_8, $CuCr_2S_3Cl$), β-Mg_2SiO_4, $Mn_5Sb(Mn_2As)O_{12}$ (manganostibite)	Mg_2SiO_4(olivine) ($LiMnPO_4$, $Mg_3(PO_4)_2$)	
$\frac{1}{8}$	$\frac{1}{16}$	$\frac{1}{2}$	$Al_{3/4}Mo_2S_4$ (defect spinel)		
$\frac{1}{8}$	$\frac{1}{24}$	$\frac{1}{2}$	$Ga_{2/3}Mo_2S_4$		
$\frac{1}{8}$	0	$\frac{1}{2}$	$Ga_{1/2}V_2S_4$		
$\frac{1}{10}$	$\frac{1}{10}$	$\frac{1}{2}$		$Mg_5Si_2O_8(OH)_2$(chondrotite)	
$\frac{1}{24}$	$\frac{1}{24}$	$\frac{1}{2}$		Fe_3BO_6(norbergite)	
$\frac{1}{16}$	0	$\frac{1}{2}$		$Zn_2Mo_3O_8$	
$\frac{1}{2}$	$\frac{1}{2}$	$\frac{1}{3}$	Cu_4Te_3		
$\frac{1}{3}$	0	$\frac{1}{3}$			α-In_2Se_3
$\frac{1}{12}$	0	$\frac{1}{3}$			$Al_2SiO_4F_2$(topaz)
$\frac{1}{10}$	$\frac{1}{10}$	$\frac{2}{5}$	Al_2SiO_5 (cyanite)		
$\frac{1}{5}$	$\frac{1}{5}$	$\frac{1}{5}$	Al_2SiO_5 (sillimanite)		
$\frac{1}{4}$	$\frac{1}{4}$	$\frac{1}{4}$	$MgGa_2S_4$	**$ZnIn_2S_4$(I)&(II)**	**$ZnIn_2S_4$(IIb)&(IIIa)**
$\frac{1}{8}$	$\frac{1}{8}$	$\frac{1}{4}$	$VCrO_4$		
$\frac{1}{5}$	$\frac{2}{5}$	$\frac{1}{5}$		**$Zn_2In_2S_5$(IIa)**	**$Zn_2In_2S_5$(IIIa)**
$\frac{1}{3}$	$\frac{1}{3}$	$\frac{1}{6}$		**$Zn_3In_2S_6$**	
$\frac{5}{8}$	0	$\frac{1}{24}$		**In_2Se_3(r)**	
$\frac{1}{8}$	$\frac{1}{8}$	$\frac{1}{8}$		$CoAl_2Cl_8$	

octahedral (Ω) and tetrahedral holes (τ) we use the corresponding lower-case and Greek letters, respectively.

In Table 6, we give a survey on the structures of halides, chalcogenides, pnictides and tetralides that are based on anion close-packing. Those which belong to a layer type are set in bold type.

We are now going to discuss certain layer structures in some detail beginning with the lowest coordination and simple compositions.

CHAPTER IV

DISCUSSION OF SPECIAL COMPOUNDS

1. Compounds with Coordination Number 2

a. AuI

With a coordination number 2, no true layer structure but at best a layered chain structure is possible. In other words, macroscopically, such a crystal can well possess one perfect cleavage plane, though the bonding is linear.

The tetragonal structure of AuI [36, 37] can be derived from an elongated ccp of iodine atoms. Instead of occupying the centres of every second layer of I tetrahedra, the Au atoms are located on the edges of these tetrahedra as is indicated in Figure 28 on the left. This leads to —Au—I—Au—I— zig-zag chains which are all parallel within one layer but which lie at right angles relative to those of the next layer. The bond angle at the anion is ~72.5° compared with 94° in AuCl. In addition to the two iodine neighbours within a chain, each gold atom has also 4 Au neighbours as close as 3.08 Å forming a square. These short Au—Au distances are a striking feature of the AuI structure which distinguishes it from that of AuCl [38, 39]. One might expect complete analogy between AuCl, AuBr and AuI [40] and in fact the structure of AuCl also contains the characteristic zig-zag chains (two sets of chains interpenetrating each other at right angles, hence no layer type) but in AuCl, the shortest Au—Au distances are 3.226 Å and 3.37 Å. We are thus tempted to assume 4 half bonds and Au^{III} with a diamagnetic d^8 configuration of its nonbonding valence electrons.

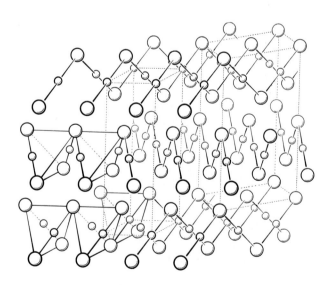

Fig. 28. The crystal structure of AuI. Left: unit cell and tetrahedra with the cations on the edges. Right: larger cell derived from the ccp iodine cell.

TABLE 7

AuI structure, tetragonal D_{4h}^{16}—P4$_2$/ncm (No. 138), $Z = 4$.
I in 4(e): $\pm(\frac{1}{4}, \frac{1}{4}, z; \frac{1}{4}, \frac{1}{4}, \frac{1}{2}+z)$
Au in 4(d): 0, 0, 0; $\frac{1}{2}, \frac{1}{2}, 0; \frac{1}{2}, 0, \frac{1}{2}; 0, \frac{1}{2}, \frac{1}{2}$.

AuI: $a = 4.36$ Å, $a = 13.71$ Å, $z = 0.153$ [36]
$a = 4.35$ Å, $c = 13.73$ Å, $z = 0.153$ [37]

TABLE 8

HgCl$_2$ structure, orthorhombic, V_h^{16}—Pnma (No. 62), $Z = 4$.
All atoms in 4(c): $\pm(x, \frac{1}{4}, z; \frac{1}{2}+x, \frac{1}{4}, \frac{1}{2}-z)$

HgCl$_2$: $a = 12.735$ Å, $b = 5.963$ Å, $c = 4.325$ Å [1]

	x	z
Hg	0.126	0.050
Cl$_I$	0.255	0.406
Cl$_{II}$	0.496	0.806

b. HgCl$_2$

The acicular crystals of HgCl$_2$ possess a layered structure formed of fishbone-like arrays of ClHgCl molecules as illustrated in Figure 29. Hg^{II} is isoelectronic with Au^I whose chloride exhibits the same low coordination numbers with linear sp-bonds. The covalency of the bonds is underlined by the low Cl—Cl distance. The nearest chlorine atoms are only 3.35 Å apart which is considerably less than the sum of the ionic radii.

HgCl$_2$ appears to be the only representative of this structure type. A strongly distorted version of the HgCl$_2$ structure is met in β-HgClBr. This mixed halide still contains the linear molecules but these are no longer layered.

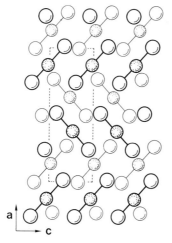

Fig. 29. The molecular layer structure of HgCl$_2$ projected on (010).
Dotted spheres: Hg

c. HgBr$_2$

The orthorhombic structure of HgBr$_2$ and yellow HgI$_2$ also is built up of linear molecules. Geometrically, however, this structure can be derived from a close-packed anion stacking of the hc type with the Hg atoms in every second octahedron layer as met also in WTe$_2$ and MoTe$_2$(h). The distances within the strongly distorted [HgBr$_6$] octahedron are

$$\text{Hg—Br} = 2.499, 2.504, 3.213(2) \text{ and } 3.216 \text{ Å}(2),$$

and correspondingly for the iodide

$$\text{Hg—I} = 2.615, 2.620, 3.507(2) \text{ and } 3.510 \text{ Å}(2).$$

Despite the molecular character of these phases we have added in Figure 29a the underlying close packing.

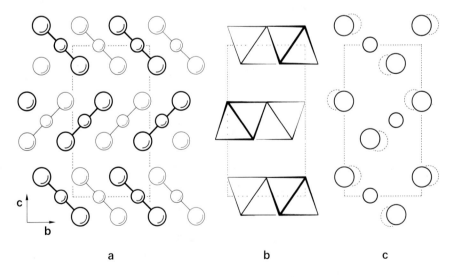

Fig. 29a. The HgBr$_2$-type structure of yellow HgI$_2$. (a) Projection on (100) of the real structure; (b) the anion sublattice idealized as close packing; (c) (110) section of the corresponding ideal close-packed structure. The actual positions are indicated by dotted circles.

TABLE 8a

$HgBr_2$ structure, orthorhombic, C_{2v}^{12}—$Cmc2_1$ (No. 36), $Z=4$.
$(0, 0, 0; \frac{1}{2}, \frac{1}{2}, 0)+$
All atoms in 4(a): $0, y, z; 0, \bar{y}, \frac{1}{2}+z$.

$HgBr_2$: $a = 4.624$ Å, $b = 6.798$ Å, $c = 12.445$ Å [4]

	y	z
Hg	0.334	0
Br_I	0.056	0.132
Br_{II}	0.389	0.368

$Hg(Br, I)_2$: Végard's rule obeyed [349]

HgI_2: $a = 4.674$ Å, $b = 7.320$ Å, $c = 13.76$ Å [1]
$a = 4.702$ Å, $b = 7.432$ Å, $c = 13.872$ Å [348]

	y	z	
Hg	0.3433 (1/3)	0 (0)	[348]
I_I	0.0916 (0)	0.1322 (1/8)	
I_{II}	0.4059 (1/3)	0.3678 (3/8) = $\frac{1}{2}-z(I_I)$	

(The figures added in brackets refer to an ideal close packing).

2. Compounds with Coordination Number 3

a. SASSOLITE $B(OH)_3$

Triclinic $B(OH)_3$ or orthoboric acid H_3BO_3, known as the mineral sassolite, crystallizes in pseudo-hexagonal needles. The structure contains planar BO_3 groups arranged in layers parallel to (001) with a layer separation of 3.18 Å (Figure 30). The BO_3 groups are held together by H bonds. The layers are not perfectly planar and it looks as if the small tilts of the BO_3 groups relative to the (001) plane were due to some weak interaction between B and O atoms of neighbouring layers. The shortest B—O distances between the layers, however, are 3.157 Å (B_{II}—O_{II}) and 3.184 Å (B_I—O_V) compared with the B—O bond distance of 1.36 Å within the layers. Perfect basal cleavage is observed. A trigonal planar coordination is expected for a low cation-to-anion radius ratio and/or covalent sp^2 bonds. Such BO_3 groups occur rather frequently, e.g. in calcite-type $LuBO_3$.

b. METABORIC ACID HBO_2

Metaboric acid is known to exist in three modifications. The stable room-temperature modification crystallizes in a cubic three-dimensional structure. Of the two modifications one has a monoclinic chain structure, while the other is orthorhombic of the layer type [43]. In this latter modification $B_3O_3(OH)_3$ molecules are fixed in a perfectly planar network by hydrogen bonds, O—H $\cdots$ O, of 2.68 to 2.83 Å length (Figure 31). Centre-related layers are separated by

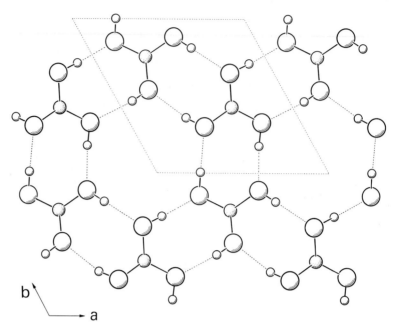

Fig. 30. One layer of the triclinic structure of sassolite B(OH)$_3$.

3.07 Å which accounts for the mica-like cleavage of the colorless platy crystals.
The difference of the B—O lengths within the ring compared with those outside the ring was shown by Coulson [44] to be due to the difference in π-bonding.

c. THE B$_2$S$_3$ STRUCTURE

The colorless crystals of B$_2$S$_3$ melt congruently at 563°C. Their monoclinic structure is built up of nearly planar layers (Figure 31a) running parallel to (102)

TABLE 9

B(OH)$_3$ structure, triclinic C_i^1—P$\bar{1}$ (No. 2), $Z = 4$.
All atoms in 2(i): $\pm(x, y, z)$

B(OH)$_3$: $a = 7.039$ Å, $b = 7.053$ Å, $c = 6.578$ Å,
$\alpha = 92.58°$, $\beta = 101.17°$, $\gamma = 119.83°$.

	x	y	z		x	y	z
B$_I$	0.646	0.427	0.258	H$_I$	0.347	0.361	(0.255)
B$_{II}$	0.307	0.760	0.242	H$_{II}$	0.671	0.172	(0.250) [41]
O$_I$	0.424	0.302	0.261	H$_{III}$	0.890	0.709	(0.252)
O$_{II}$	0.768	0.328	0.250	H$_{IV}$	0.600	0.817	(0.254)
O$_{III}$	0.744	0.650	0.261	H$_V$	0.083	0.469	(0.246)
O$_{IV}$	0.532	0.885	0.250	H$_{VI}$	0.297	0.006	(0.243)
O$_V$	0.214	0.540	0.244				
O$_{VI}$	0.180	0.856	0.233				

Compare [42] for B(OD)$_3$

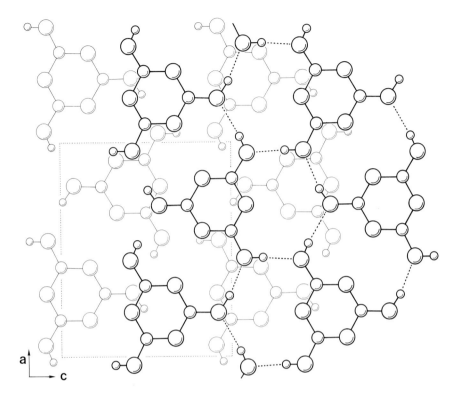

Fig. 31. Two layers of the orthorhombic modification of orthoboric acid $H_3B_3O_6$.

at a distance of 3.5 Å. Each boron atom is surrounded by 3S, all [BS$_3$] units being virtually planar. These [BS$_{3/2}$] units are joined to form [B$_2$S$_2$S$_{2/2}$] and [B$_3$S$_3$S$_{3/2}$] units and two of the former and six of the latter are connected to wavy rings. The B—S bond distances range from 1.78 to 1.84 Å. B$_2$S$_3$ crystals are usually twinned and therefore a twice as large orthorhombic cell had originally been

TABLE 10

Structure of α-HBO$_2$, orthorhombic, D_{2h}^{16}—Pnma (No. 62), Z = 12.
All atoms in 4(c): $\pm(x, \tfrac{1}{4}, z; \tfrac{1}{2}+x, \tfrac{1}{4}, \tfrac{1}{2}-z)$

α – HBO$_2$(25 °C): $a = 9.688$ Å, $b = 6.261$ Å, $c = 8.046$ Å [43]
(−130 °C): $a = 9.703$ Å, $b = 6.13$ Å, $c = 8.019$ Å

−130 °C:

	x	z		x	z
O$_I$	0.0531	−0.0909	B$_I$	0.1719	−0.0029
O$_{II}$	0.2943	−0.0885	B$_{II}$	0.4167	0.0017
O$_{III}$	0.5425	−0.0751	B$_{III}$	0.2926	0.2560
O$_{IV}$	0.1710	0.1688	H$_I$	−0.021	−0.029
O$_V$	0.4150	0.1704	H$_{II}$	0.522	−0.163
O$_{VI}$	0.2966	0.4239	H$_{III}$	0.231	0.471

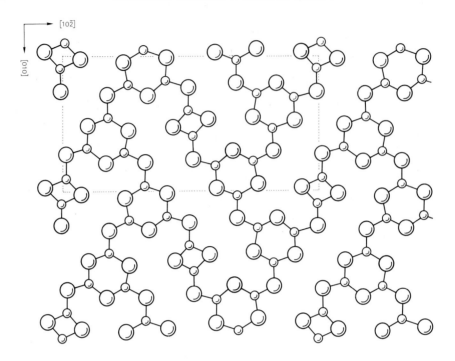

Fig. 31a. One layer of the monoclinic structure of B_2S_3.

TABLE 10a

B_2S_3 structure, monoclinic, C_{2h}^5—$P2_1/c$ (No. 14), $Z = 8$.
All atoms in 4(e): $\pm(x, y, z;\ x, \frac{1}{2}-y, \frac{1}{2}+z)$

B_2S_3: $a = 4.039$ Å, $b = 10.722$ Å, $c = 18.620$ Å, $\beta = 96.23°$ [44a]
$a = 4.048$ Å, $b = 10.73$ Å, $c = 18.655$ Å, $\beta = 96.23°$
calculated from pseudo-orthorhombic cell given in [44b]

	x	y	z	
B_I	0.586	0.4639	0.1798	
B_{II}	0.535	0.7169	0.2406	
B_{III}	0.741	0.6858	0.0960	
B_{IV}	0.045	0.3995	0.4902	
S_I	0.567	0.2929	0.1763	[44a]
S_{II}	0.491	0.5501	0.2565	
S_{III}	0.687	0.7950	0.1656	
S_{IV}	0.694	0.5168	0.0951	
S_V	0.108	0.2339	0.4887	
S_{VI}	0.083	0.5278	0.4291	

B_2Se_3: $a = 4.055$ Å, $b = 10.43$ Å, $c = 19.42$ Å, $\beta = 95.99°$
calculated from pseudo-orthorhombic cell given in [44c]

reported:

$$a_{orth} = a_{mon}$$
$$b_{orth} = 2c_{mon} - a_{mon}$$
$$c_{orth} = -b_{mon}$$

Obviously the selenide has the same structure.

A layer structure appears to occur also in the orthorhombic MnB_2S_4 which crystallizes in the form of yellow lamellae with a metallic lustre [1028].

d. THE BCl_3 STRUCTURE

This molecular layer structure can be derived from a hexagonal anion close packing. The cations are inserted in one third of the triangular sites within the anion planes. One $(BCl_3)_n$ layer thus represents a defective BN-type layer where only one third of the cation sites are occupied. This leads to a $\sqrt{3}\, a_0$ supercell. The stacking of the layers in BCl_3, however, is different from that of boron nitride. The axial ratios of the BCl_3-type compounds are distinctly higher than the ideal value though the anion close packing is only slightly disturbed by the formation of BX_3 molecular units. In BCl_3 the distances are:

B—3 Cl at 1.76 Å
Cl—2 Cl at 3.04 Å
    4 Cl at 3.75(2) and 3.78 Å(2) in neighboring molecules
    2 Cl at 3.76 Å in adjacent layers

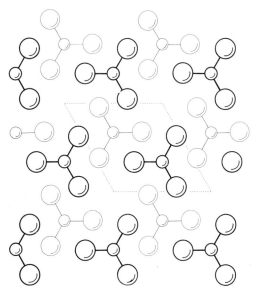

Fig. 31b. Projection along [001] of the hexagonal BCl_3 structure.

In BI_3 B—I = 2.12 Å [930] and I—2 I = 3.68 Å [930]. The covalent radius of boron is ~0.80 Å for trigonal coordination. Taking into account the electronegativity and π-bond contractions bond lengths of 1.71, 1.88 and 2.09 Å were calculated [928] for BCl_3, BBr_3 and BI_3, respectively, which are in fair agreement with the observed values.

For isostructural PI_3 and HCI_3 we would have expected a trigonal-pyramid rather than a planar coordination. In this connection we are curious about the structures of the adducts $PI_3 \cdot BBr_3$ and $PI_3 \cdot BI_3$ [935]. For PI_3 see [1049].

TABLE 10b
BCl_3 structure, hexagonal, C_{6h}^2—$P6_3/m$ (No. 176), $Z = 2$
Cl in $6(h)$: $\pm(x, y, \frac{1}{4}; \bar{y}, x-y, \frac{1}{4}; y-x, \bar{x}, \frac{1}{4})$
B in $2(c)$: $\pm(\frac{1}{3}, \frac{2}{3}, \frac{1}{4})$

Compound	a(Å)	c(Å)	$\sqrt{3}\,c/a$	x	y	Ref.
BCl_3						
108 K	6.08	6.55	1.87	0.052	0.372	[927]
~90 K	6.140	6.603	1.863	~0	~$\frac{1}{3}$	[929]
BBr_3						
~90 K	6.406	6.864	1.856	~0[a]	~$\frac{1}{3}$[a]	[929]
BI_3	7.00	7.46	1.85	0.03[b]	0.37[b]	[928]
	7.085 6	7.544 1	1.844			[1]

[a] Close to the values of BCl_3 [930].
[b] In order to obtain agreement with observed lattice-vibration frequencies it was necessary to rotate the two molecules in the unit cell by +2° and −2°, respectively, and increase the covalent B—I bond length to 2.12 Å from the reported value of 2.10 Å [930].

e. GRAPHITE

Graphite is the stable crystalline modification of carbon while cubic diamond, lonsdalite (wurtzite-type diamond), chaoite and two other hexagonal modifications [45, 46] are high-pressure or high-temperature phases. In graphite honeycomb-like layers are stacked in such a manner that half of the C atoms of the neighbouring layer are located above or below the empty centers of the hexagons. Thus, a hexagonal and a rhombohedral modification with stacking sequence $ABAB\cdots$ and $ABCABC\ldots$, resp., are possible, as shown in Figure 32. In pyrolytic graphite no periodic stacking order exists.

The carbon atoms in graphite form covalent sp^2 bonds to the three neighbours at a distance of 1.42 Å. The fourth valence electron (p_z) is involved in a π-bond to one of the three neighbours. One third of the bonds therefore are double bonds or, as the real state is a symmetric superposition of all possible mesomeric bond systems, these bonds correspond to $1\frac{1}{3}$ bonds (analogous to the $1\frac{1}{2}$ bonds in benzene C_6H_6). These resonant bonds are responsible for the metallic conductivity within the layer. Across the layers the C—C distances are 3.35 Å

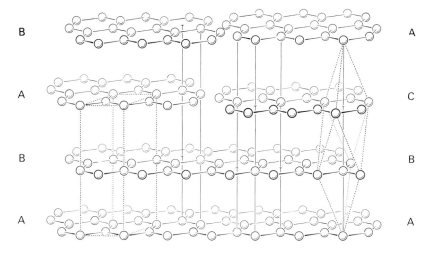

Fig. 32. Hexagonal and rhombohedral graphite.

hence there exists only extremely weak interlayer bonding of the Van der Waals type and, accordingly, the electrical resistivity [47] is high in this direction. Early band-structure calculations, therefore, used two-dimensional tight-binding approximations [48]. They led to two slightly overlapping energy bands for the π-electrons. If separated, the lower of these bands would be filled. Overlapping accounts for the high conductivity within the layer and the high diamagnetism reminiscent of bismuth. However, the three-dimensional Slonczewski-Weiss model [49] has been more successful in explaining the details of a large variety of experiments [50]. Reviews on graphite are found in refs. [51–54]. For a study of the X-ray photoemission spectrum of graphite discussed on the basis of recent LCAO band calculations [55] see refs. [56, 1001]. Thermal-expansion, specific-heat and thermal-conductivity data from 1000°C up to the sublimation temperature (3650°C) are reported in ref. [56a]. Hall and resistivity measurements in fields up to 500 kOe are discussed in terms of a 3-band model by Brandt et al. [1005]. A metal→semiconductor transition is expected for fields > 1 MOe [1006].

TABLE 11

Graphite, hexagonal, C_{6v}^4—P6$_3$mc (No. 186), $Z = 4$.

C_I in 2(a): 0, 0, z; 0, 0, $\frac{1}{2} + z$.
C_{II} in 2(b): $\frac{1}{3}$, $\frac{2}{3}$, z; $\frac{2}{3}$, $\frac{1}{3}$, $\frac{1}{2} + z$ (compare [984])

Graphite: $a = 2.456$ Å, $c = 6.696$ Å; $z_I = 0$, $z_{II} < 0.05$

Graphite, rhombohedral, D_{3d}^5—R$\bar{3}$m (No. 166), $Z = 2$.

C in 2(c): $\pm (x, x, x)$

Graphite: $a = 3.635$ Å, $\alpha = 39°30'$; $x \approx \frac{1}{6}$
$a_h = 2.456$ Å, $c_h = 10.044$ Å

f. BORON NITRIDE

BN is a polar isoelectronic version of carbon. Its stable modification has again a graphite-like structure and the high-pressure modifications are of the cubic sphalerite and hexagonal wurtzite type [57, 939]. In the hexagonal sheets of the normal phase of BN, each cation has only anion neighbors and vice-versa. The B—N separation in the layer is 1.45 Å and between layers 3.33 Å. With respect to graphite, the BN sheets are shifted laterally so that unlike atoms occur above one another in the consecutive layers (Figure 33).

TABLE 12

(a) *BN structure*, hexagonal, D_{6h}^4—P6$_3$/mmc (No. 194), $Z = 2$.
 B in 2(c): $\pm(\frac{1}{3}, \frac{2}{3}, \frac{1}{4})$
 N in 2(d): $\pm(\frac{1}{3}, \frac{2}{3}, \frac{3}{4})$
 BN (35 °C): $a = 2.5040$ Å, $c = 6.6612$ Å [4]
  $a = 2.504$ Å, $c = 6.674$ Å [57a]

(b) *Rhombohedral BN*, trigonal, C_{3v}^5—R3m (No. 160), $Z = 1(3)$.
 B and N in 1(a): x, x, x,
  with $x(B) = 0$, $x(N) \approx \frac{1}{3}$.
 hexagonal axes: $(0, 0, 0; \frac{1}{3}, \frac{2}{3}, \frac{2}{3}; \frac{2}{3}, \frac{1}{3}, \frac{1}{3})+$
 atoms in 3(a): 0, 0, z,
  with $z(B) = 0$, $z(N) \approx \frac{1}{3}$.
BN: $a = 2.504$ Å, $c = 10.01$ Å [57a]

The reaction of KCN with B$_2$O$_3$ at 1100°C leads to a rhombohedral BN modification which is an ordered β-graphite version [57a].

In BN all excess valence electrons not used for the sp^2 bonds are, in the ionic limit, well localized on a doubly occupied nitrogen p-orbital. Hence the one

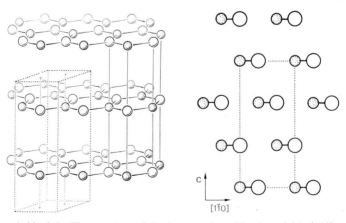

Fig. 33. Boron nitride. left: The structure of the hexagonal modification. right: (110) section of the rhombohedral modification.

TABLE 13

Symmetrical (A) and non-symmetrical (B) carbon analogs and corresponding (C, D) cross compounds with second- and third-row elements (SiC analogs, $\bar{n} = 2.5$).
Electronegativity differences in brackets.

(A)
- BN ($\Delta x = 1.0$) carbon-like
- BeO ($\Delta x = 2.0$) wurtzite structure
- LiF ($\Delta x = 3.0$) rocksalt structure

(C)
- SiC ($\Delta x = 0.7$) sphalerite and wurtzite structure
- BP ($\Delta x = 0.1$) sphalerite structure
- BeS ($\Delta x = 1.0$) sphalerite structure
- LiCl ($\Delta x = 2.0$) rocksalt structure

(B)
- B_2O ($\Delta x = 1.5$) graphite-like
- BeN_2 ($\Delta x = 1.5$) ?
- B_3F ($\Delta x = 2.0$) ?

(D)
- B_2S ($\Delta x = 0.5$) ?
- BeP_2 ($\Delta x = 0.6$) ?
- B_3Cl ($\Delta x = 1.0$) ?

empty p-orbital on each boron atom is well separated from the next by rare-gas-like nitrogen atoms so that no metallic resonance can develop. In other words, the two π-bands which overlap in graphite are now well separated. An energy gap of 3.85 eV has been deduced from diffuse-reflectance measurements [58].

g. NON-SYMMETRICAL GRAPHITE ANALOGS, B_2O

BN is not the only possible graphite analog, but with increasing electronegativity difference between anion and cation the polarity of the bonds and the coordination number increase. If we include elements from the third row of the Periodic Table, the ability to form π-bonds and trigonal planar coordination is lost, as is seen from Table 13.

Hall and Compton [59] were able to synthesize a graphite-like compound B_2O by high-pressure, high-temperature methods. A layer of the proposed structure is shown in Figure 34a. The unit cell contains three layers possibly shifted relative to each other along the a-axis by $a/3$. The structure contains two kinds of boron sites B_I and B_{II}. Each B_I is surrounded by $1\,B_I + 2\,O$ while the B_{II} atoms are arranged in zigzag chains, their coordination being $2\,B_{II} + 1\,O$. In order to form

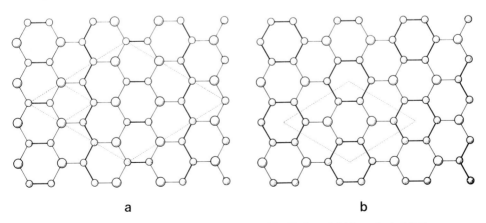

Fig. 34. Graphite analogs. (a) layer proposed for B_2O; and (b) hypothetical B_3F.

sp² bonds, oxygen has to transfer one of its valence electrons either to B_I or to B_{II} (or statistically to both). In all cases (empty or singly occupied p_z orbitals on B_{II}) the B_{II} chains act as metallic paths. B_2O was reported to be light reddish-brown in color which might contradict our expectation of metallic properties. Nonmetallic properties may be expected in hypothetical graphite-like B_3F (Figure 34b) in which the excess p_z-electrons in the benzene-like B_6 rings are well isolated by rare-gas-like fluorine ions.

h. Graphite intercalation compounds

Graphite reacts with a large number of substances to form lamellar compounds in which the reactant is present as ordered monolayers separated by single or multiple graphite layers. Since in graphite the p_z orbitals are only singly occupied and the p_z electrons are engaged only in non-localized metallic π bonds the carbon atoms in graphite can act as cations or anions. However, there exist no graphite-like compounds with saturated valence bonds. Whenever all four valence electrons of the carbon atoms are involved in single bonds, then sp³ bonds result leading to puckered C layers. These phases will be discussed together with the analogous Si compounds.

α. Metal-Graphite Compounds

With electropositive elements or radicals the carbon in graphite behaves as an anion. But although only certain definite equilibrium compositions MC_x are stable no compounds with saturated bonds (such as would be M^+C^- or $M^{2+}C_2^-$) are known. This is in contrast with the diborides of the AlB_2 type that one may consider as a sort of first-stage intercalation compound. As far as the electron concentration is concerned MgB_2 is the analog of graphite. Thus, π bands similar to those in graphite [60] though with somewhat stronger overlap and hence semimetallic properties can be expected in the golden yellow [61] MgB_2. AlB_2, ScB_2 and LnB_2 (Ln = Gd ··· Lu, Y) may correspond to hypothetical 'KC_2' and TiB_2 and ZrB_2 (ev. $BaSi_2$) could be analogous to hypothetical 'BeC_2'. The latter compositions should represent a limit since in these cases each carbon 'anion' would take up one electron and thus fill the p_z bond completely. The same should hold for the diborides of tetravalent cations and indeed Goodman [62] has speculated about semiconductor properties of TiB_2 and ZrB_2. Due to the higher electronegativity of C compared with B a beryllium or magnesium graphite 'BeC_2' or 'MgC_2' would more likely be semiconducting than the borides though we expect the graphite planes to become puckered as in CF. Up to now, however, nobody has succeeded in intercalating such a high Be concentration in graphite. Maximum concentrations are given in Table 14.

Metal-graphite intercalation compounds are best known with alkali elements. Since intercalation of the alkaline-earth metals is possible only in liquid ammonia those phases all contain NH_3. The same is true for Al and Ln (La, Ce, Sm, Eu).

TABLE 14

Metal-graphite intercalation compounds. n = stage, number of graphite layers per metal layer. I_c = chemical identity period along the hexagonal axis

	n	I_c (Å)		n	I_c (Å)
LiC_6	1	3.737 [66]	$Li(NH_3)_{1.6}C_{10.6}$	1	6.62 [63]
LiC_{12}	2	7.029 [66]	$Li(NH_3)_{2-4}C_{28}$	2	[63]
LiC_{18}	3	10.44 [66, 67]	$Li(CH_3NH_2)_2C_{12}$	1	6.9 [63]
LiC_{27}	4	[68]	$LiC_2H_4(NH_2)_2C_{28}$	2	11.85 [63]
LiC_{42}	7?	[67]			
NaC_{30}	?	[67]	$Na(NH_3)_2C_{13.6}$	1	6.63 [63]
NaC_{60-68}	8?	21.25 [63]	$Na(NH_3)_{2-4}C_{28}$	2	9.9 [63, 51]
NaC_{120}	?	[67]	$Na_{0.10}K_{0.90}C_{24.8}$	2	3.35 + 5.32 [69a]
			$Na_{0.13}K_{0.86}C_{36.1}$	3	6.70 + 5.32 [69a]
			$Na_{0.18}K_{0.82}C_{7.75}$	1	5.32 [69a]
			$Na_{0.12}Cs_{0.88}C_{15.9}$	1+2	5.89; 3.35 + 5.89 [69a]
			$Na_{0.19}Cs_{0.81}C_{28}$	2+3	(n−1) 3.35 + 5.89 [69a]
			$Na_{0.25}Cs_{0.75}C_{7.75}$	1	5.82 [69a]
			$Na_{0.37}Cs_{0.63}C_{52.2}$	4+5	(n−1) 3.35 + 5.26 [69a]
			$Na_{0.57}Cs_{0.43}C_{61}$	?	[69a]
KC_8 [70]	1	5.32 5.35 [71]	$K(NH_3)_{2.1}C_{12.5}$	1	6.56 [63]
		5.40 [51] 5.41 [63]	$K(NH_3)_{2-4}C_{28}$	2	9.9 [51]
KC_{8-10} (disordered)	1 (72, 73)		$KH_{2/3}C_8$	2	11.88 [69]
KC_{24}	2	$\frac{1}{2} \times 17.54$ [74] 8.75 [51]			
		$\frac{1}{3} \times 26.25$ [65]			
$KC_{22.2-27.5}$	2	[?]			
KC_{36}	3	12.12 [74] $\frac{1}{2} \times 24.20$ [65]			
KC_{48}	4	$\frac{1}{2} \times 30.98$ [74] $\frac{1}{3} \times 46.35$ [65]			
KC_{60}	5	18.84 [74]			
RbC_8	1	5.61 [63] 5.65 [51]	$Rb(NH_3)_2C_{12}$	1	6.58 [63]
RbC_{24}	2	9.02 [51] $\frac{1}{2} \times 18.08$ [74]	$Rb(NH_3)_{2-4}C_{28}$	2	[63]
		$\frac{1}{2} \times 17.95$ [1]			
RbC_{36}	3	12.34 [51] 12.37 [74]	$K_xRb_{1-x}C_8$	1	[75]
RbC_{48}	4	$\frac{1}{2} \times 31.50$ [51] $\frac{1}{2} \times 31.56$ [74]			
RbC_{60}	5	19.08 [51] 19.12 [74]			
CsC_8	1	$\frac{1}{4} \times 23.76$ [1]	$Cs(NH_3)_{2-4}C_{12.8}$	1	6.58 [63]
		$\frac{1}{4} \times 22.65$ [74]			
CsC_{24}	2	$\frac{1}{2} \times 18.51$ [1]	$Cs(NH_3)_{2-4}C_{28}$	2	[63]
CsC_{36}	3	[76]	Cs_8NaC_x [77]		
CsC_{48}	4	[76]	$K_xCs_{1-x}C_8$ [78]		
CsC_{60}	5	[76]			
$Be(NH_3)_2C_6$	3	12.24 [63]	SmC_6	1	4.58
$Ca(NH_3)_{2.2}C_{12.4}$	1	6.62 [63]	SmC_n	2	7.99
$Sr(NH_3)_{2.4}C_{11.3}$	1	6.4 [63]		3	11.32 [997]
$Ba(NH_3)_{2.5}C_{10.9}$	1	6.4 [63]		4	14.59
$Mg(NH_3)_{2-3}C_{32}$	4	15.95 [63]		5	17.53
$Eu(NH_3)_3C_{16-18}$	2	[63]		6	21.30
EuC_6	1	$\frac{1}{2} \times 9.745$	YbC_6	1	$\frac{1}{2} \times 9.147$
EuC_n	2	8.26	YbC_n	2	7.95
	3	11.60 [997]		3	11.21 [997]
	4	14.83		4	14.68
	5	18.19		5	17.90
	6	21.64		6	21.30
$Al(NH_3)_2C_{8-10}$	4	19.35 [63]	BaC_8	1	5.28 [1035]

In alkali-graphite compounds with the highest concentration, monolayers [63] of alkali atoms are inserted between every graphite layer. These phases are referred to as the first stage. Stages poorer in metal can e.g. be produced by thermal degradation of the first stage. There appears to be no continuous decrease of the alkali-metal concentration of all layers but in the ideal equilibrium form of the second stage every second metal layer is missing. According to Daumas and Hérold [64], this model, in which stage n compounds contain n perfect graphite layers between successive intercalate layers, is an idealization which cannot explain transformation of stages. They proposed a model depicted in Figure 35 which is able to explain the transformation of one stage into a higher stage. Nixon and Parry [65] observed that precursor stages invariably follow the sequence $n = 4 \rightarrow 3 \rightarrow 2 \rightarrow 1$.

The concentration of metal per alkali layer is highest in the first stage, and decreases to two-thirds in the higher stages of the K, Rb and Cs compounds. Li is exceptional as follows from Table 14. The color of these phases varies from brass-like for the first stage over dark copper-red (LiC_{12}) or steel-blue and blue-green (MC_{24}) to bluish-black and black for the higher stages, obviously depending upon electron concentration.

The number of metal atoms per inserted metal layer appears to be largely determined by the size of these atoms. Thus, the highest metal concentration is found in the Li compounds (hydrogen may form only the $(CH)_n$ phase). LiC_6 ($= Li\square_2 C_6$) crystallizes in an ordered defective AlB_2 superstructure [66]. The occupied hexagonal carbon prisms form columns along the c-axis. Thus, intercalation of metal layers does induce not only an expansion of the lattice in c-direction but also a lateral shift of the graphite nets. The structure of the second stage, LiC_{12}, seems to be identical except that every second metal layer is missing [66].

In the MC_8 compounds only $\frac{1}{4}$ of the centers of the hexagonal prisms are

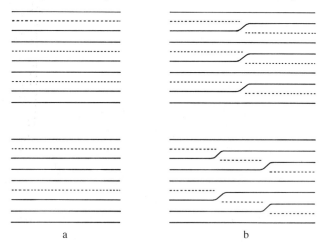

Fig. 35. Schematic representation of a third-stage alkali graphite. (a) classical model and (b) model proposed by Daumas and Hérold [64].

TABLE 15

LiC_6 structure, hexagonal D_{6h}^1—P6/mmm (No. 191), $Z=1$.

C in 6(j): $\pm(x, 0, 0; 0, x, 0; x, x, 0)$
Li in 1(b): $0, 0, \frac{1}{2}$

LiC_6: $a = 4.290$ Å, $c = 3.737$ Å, $x \approx \frac{1}{3}$ [66]
$a = 4.26$ Å, $c = 3.70$ Å, $x \approx \frac{1}{3}$ [1003]

---

occupied by an M atom. As indicated in Figure 36 this opens four possibilities (α, β, γ, δ) for locating a uniform M layer. Contrasting the case of LiC_6, all four sites are occupied once in KC_8 and RbC_8, and three sites in CsC_8, so that their stacking can be described by $A\alpha A\beta A\gamma A\delta$ and $A\alpha A\beta A\gamma$, resp. [65, 81]. These structures thus are again defect AlB_2-type superstructures $M\square_3C_8$. In the second and higher stages of K, Rb and Cs, where the metal layers are only $\frac{2}{3}$ occupied, the undisturbed carbon layers are stacked as in hexagonal or rhombohedral graphite. Thus, the following stacking has been derived: KC_{24} $A\alpha AB\beta BC\gamma CA\alpha' AB\beta' BC\gamma' C$, RbC_{24} and CsC_{24} $A\alpha AB\beta BC\gamma C^\cdot$ (triclinic cell) [81], in disagreement with earlier studies which had led to a stacking $A\alpha A\ B\beta B$ for all second stages MC_{24}. Whereas with K, Rb, Cs $\frac{1}{4}$ and with Li $\frac{1}{3}$ of the hexagonal prisms are occupied it was suggested that with the small Be^{2+} ion all hexagonal sites α, β, γ, δ are occupied at once. A first-stage Be graphite then simply would be BeC_2 with the AlB_2 structure. However, only a third-stage phase is known for which a hexagonal structure $AB(\alpha\beta\gamma\delta)BA$ was proposed, neglecting the incorporated NH_3 molecules.

Introduction of the metal layers leads, for all stages, to the same increase $\Delta I_c(M)$ in the distance between adjacent carbon planes:

	$\Delta I_c(M)$ in Å
Li	~0.38
Na	~1.15
K	1.97 ··· 2.06
Rb	2.26 ··· 2.32
Cs	2.58 ··· 2.60

Therefore, the identity period along the hexagonal axis can easily be calculated, or in turn the stage can be determined from the experimental c value:

1. stage $c = m\{3.355 \text{ Å} + \Delta I_c(M)\}$ $m = 1, \ldots, 4$
2. stage $c = m\{6.71 \text{ Å} + \Delta I_c(M)\}$ $m = 2, 3, \ldots$
3. stage $c = m\{10.26 \text{ Å} + \Delta I_c(M)\}$ $m = 1, 2, \ldots$
4. stage $c = m\{13.82 \text{ Å} + \Delta I_c(M)\}$ $m = 2, \ldots$
5. stage $c = m\{17.17 \text{ Å} + \Delta I_c(M)\}$ $m = 1, \ldots$

(m being determined by the stacking order of the metal and graphite sublattices.)

Experimental results are in agreement with the assumption that formation of these compounds involves an electron transfer from the alkali to the host graphite lattice. The anomalously high diamagnetism of pure graphite changes into Pauli

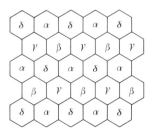

Fig. 36. Possible positions $\alpha,\beta,\gamma,\delta$ for metal layers intercalated in graphite.

paramagnetism. Optical studies [79] suggest that the intercalated metal atoms are fully ionized in the second and higher stages, but only $\frac{2}{3}$ ionized in the first-stage compounds. Thus the ionic formula can be written $M^+(C_{12n})^-$ for the higher stages and $M^+M^0_{1/2}(C_{12})^-$ or $M^{2/3+}_{1.5}(C_{12})^-$ (as the alkali atoms are equivalent) for the first stage. Each intercalate layer in any stage n thus transfers the same amount of charge to the graphite, the total charge being proportional to $1/n$. As the transferred electrons enter the upper π band (the conduction band) of graphite, i.e. go into antibonding states, the interaction between carbon atoms is weakened and an increase in the C—C bond length is expected. This has been verified by Nixon and Parry [71] who find an approximately linear increase in C—C bond length for KC_{12n}:

$$d_n(\text{Å}) = 1.4203 + 0.0113/n$$

It is noteworthy that superconductivity has been observed in the first-stage compounds of K, Rb and Cs [80] but not in the higher stages, where the metal layers are fully ionized, but separated by more than one graphite layer:

$KC_8: T_c = 0.39 \cdots 0.55$ K; $KC_{24}: T_n = 0.011$ K [80]
$RbC_8: T_c \sim 0.15$ K
$CsC_8: T_c \sim 0.135$ K

A quite new family of intercalation compounds has recently been discovered by Hérold and coworkers [69]. Treatment with hydrogen transforms the golden first-stage KC_8 into a blue second-stage phase $KH_{2/3}C_8$ which contains the metal atoms in the form of double layers. The empty interlayers can be filled with heavy alkali metals which leads to double-intercalation compounds $K_2H_{4/3}C_8 \cdot MC_8$ (M = K, Rb, Cs).

At liquid-nitrogen temperature the ternary metastable phases $Na_xK_{1-x}C_n$ and $Na_xCs_{1-x}C_n$ fix reversibly hydrogen [69a]. This is considered as a confirmation of the conclusion drawn from chemical and X-ray analysis that in the higher stages the metal layers are only $\frac{2}{3}$ occupied while in stage 8 NaC_{64} the metal layer is complete. It is interesting that MC_{24} compounds (M = K, Rb, Cs) sorb gases such as H_2, D_2, N_2, CH_4 and Ar to various extents whereas the first-stage compounds MC_8 are nonsorptive. Since in the MC_8 compounds the alkali atoms are 1.5 times more densely packed in an intercalated layer they obviously hinder the gas molecules sterically from entering the interlayer spacing.

Electron-donor acceptor complexes are potential catalysts. A review on catalytic reactions at alkali metal-graphite compounds has recently been given by Boersma [81a].

A recent publication [904] reports on the reaction of Na—Ba alloys with graphite, that leads to blue, airstable second-stage phases of compositions $Na_{1-x}Ba_xC_{\sim 7.5}$ with $x = 0.2 \cdots 0.4$. The separation of the carbon layers increases from 3.35 Å to 7.38 Å since the metal atoms are inserted in the form of triple layers Na—Ba—Na. The intercalated metal sublattice is hexagonal with $a = 6.36$ Å and parallel to the graphite lattice. Similar intercalated sandwiches have been observed with $FeCl_3$, $CrCl_3$ and alkali metals + hydrogen [904].

β. Graphite Salts

Towards strongly oxidizing chemicals, graphite will act as an electron donor forming compounds such as $C_{24}^+HSO_4^- \cdot 2H_2SO_4$ and $C_{24}^+NO_3^- \cdot 3HNO_3$ [82, 83]. The nature of the charge transfer has been confirmed by Hall-effect measurements [82, 84] revealing the expected p-type of these synthetic metals [1036/7].

Characteristically, the acids given as examples do not directly react with graphite but are electrolytically formed at a graphite anode. Further examples of acids, that form graphite salts are H_2SeO_4, $HClO_4$, H_3PO_4, H_3AsO_4, HF, CF_3COOH. Most of the graphite compounds, however, belong to the family of the spontaneous lamellar compounds [51] which are usually prepared by spontaneous reaction of graphite with the corresponding liquid or vapor. This family includes compounds such as CrO_2F_2, CrO_2Cl_2 (which is exceptionally stable), CrO_3, MoO_3 [51], N_2O_5 [85], AsF_5 [85a], SbF_5 [86], IF_5 [87], BrF_3 [87], $XeOF_4$ [88], XeF_6 [1007], MoF_6 [89], TiF_4, NbF_5 TaF_5, UF_6 [1048], Br_2 [90], ICl [90, 91], $FeBr_3$ [92], $AlBr_3$ [93], $GaBr_3$, $AuBr_3$ [94] and a large number of metal chlorides [51, 95]:

MCl_2 with $M^{2+} =$ Be, Co, Cu, Cd, Hg (not, however, Mg,
   Ca, Ba, Pb, Zn, Pd, Pt, Mn, Ni)
MCl_3 with $M^{3+} =$ Al, Ga, In, Tl, Cr, Fe, Ru, Rh, Au, Sm,
   $\cdots$ Dy, Y (not, however, La $\cdots$ Nd, P, As, Sb, Bi)
MCl_4 with $M^{4+} =$ Zr, Hf, Re, Pd, Pt (not, however, Si, Ge,
   Sn, Pb, Ti, V, Th)
MCl_5 with $M^{5+} =$ Nb, Ta, Mo, U, Sb (not P)
MCl_6 with $M^{6+} =$ W, U

Probably most work has been devoted to $FeCl_3$ [96–102a]. With transition elements usually only the halide of the highest oxidation state will react. One might expect that a necessary condition for intercalation would be the ability of the reactant to form a two-dimensional array. Obviously, there must exist an additional condition preventing $ZnCl_2$, $MnCl_2$, $NiCl_2$, $PdCl_2$ and $PtCl_2$ from being intercalated.

In most cases, the reaction of graphite with chlorides takes place only in the presence of free chlorine, which is intercalated simultaneously, acting as an electron acceptor [84]. The chlorine content varies with the compounds. Resistivity and Hall measurement revealed that in all cases the acceptor concentration is approximately equal to the amount of the additionally bounded chlorine. These ionic compounds therefore are often described by such formulae as $C_n^+ Cl^- \cdot 2AlCl_3$ or $C_n^+ (AlCl_4)^- \cdot AlCl_3$, though we feel that a formula $C_n^+ (2AlCl_3)^-$ might be more adequate as the anionic charge need not be localized.

In contrast to the above, Metz and Hohlwein [102a] found no indication of additionally intercalated Cl in their 'pure stages' prepared with the addition of 1 bar chlorine gas.

Not in every case is it possible to reach a first-stage compound. With WCl_6 only a fifth stage was attained. However, the fact that different graphite intercalation compounds have different thermal stabilities raises the possibility of formation of mixed layer complexes. Thus, heating the WCl_6 compound with $AlCl_3$ led to a first-stage $AlCl_3$–WCl_6-graphite. Similarly, a second-stage $CdCl_2$-graphite was converted into a first-stage $CdCl_2$–$InCl_3$ graphite [95]. On condensing N_2O_5 onto a fourth-stage $C_{31}FeCl_3$ at 0°C, a complex $C_{31}FeCl_3(N_2O_5)_{1.7}$ formed [103] which roughly corresponds to a second-stage N_2O_5 complex. The intercalated complex thus may show the sequence

$$-C-FeCl_3-C-N_2O_5-C-C-N_2O_5-C-FeCl_3-C-$$

with one layer empty.

Detailed structure determinations are rather rare ($FeCl_3$ [96, 100], Br_2 [91], $MoCl_5$ [104, 105]). In contrast to the alkali-metal compounds and the CF_3COOH compounds, the relative orientation $ABAB \cdots$ of the graphite layers is not altered by the intercalation of most of these metal chlorides, HF and bromine. For the graphite bisulfate compound apparently both types of stacking exist for the first-stage compound.

In the $FeCl_3$ and $MoCl_5$ compounds (and probably in most chlorides) the basic structure of the anhydrous chlorides is preserved. The Cl-ion layers are somewhat distorted [96, 102] due to the interaction with the graphite and since some of the Cl are ionized. Moreover, a certain fraction of the M sites (corresponding to the ionized Cl) in the central M layer must be empty compared with the free M chloride.

For intercalated layer compounds, particularly, it is easy to calculate the upper limit of composition as well as the periodicity length along the c-axis. As follows from Table 16, the apparent thickness of the reactant layers is always slightly

TABLE 16

Structural data of halide-graphite compounds.

n = stage number
I_c^{min} = chemical periodicity length along the c-axis
D = apparent intercalate thickness
D_0 = layer thickness of the free halide
m = calculated upper limit of composition $C_m MCl_x$

$C_m MCl_x$	n	I_c^{min} (Å)	D(Å)	D_0(Å)	m
$C_{11}MgCl_2$[1034a]	1	9.50	6.15	5.87	4.4
$C_{5.5}FeCl_{2+x}$ [107]	1	9.51	6.16	5.86	4.3
$C_{11}FeCl_{2+x}$ [107]	2	12.86	6.16	5.86	
$C_{5.6}CoCl_{2+x}$ [95]	1	9.50	6.15	5.80	4.2
$C_{4.9}CuCl_{2+x}$ [95]	1	9.40	6.05		
$C_{11.1}CdCl_{2.04}$ [95]	1	9.51	6.16	5.92	4.9
$C_{20}HgCl_{2.08}$ [95]	3	16.48	6.43		
$C_m CrCl_{3+x}$ [116]	2	12.80	6.10	5.78	6.0
$C_{22}CrCl_{3+x}$ [116]	3	16.15	6.10	5.78	
$C_{7.1}FeCl_{3.5}$ [95, 97]	1	9.37 [51]	6.02 ···	5.81	6.0
($C_{6.7}FeCl_{3.5}$ [117])		9.40 [95]	6.07		
		9.42 [99]			
$C_{18}FeCl_{3.5}$ [113]	2	{12.66 [99]	5.94 [92]	5.81	
($C_{13}FeCl_3$ [92])		12.73 [92]	6.03		
$C_{27}FeCl_{3.5}$	3	{16.21 [51]	6.15 [92]	5.81	
($C_{20}FeCl_3$ [92])		16.08 [92]	6.03		
$C_9AlCl_{3.5}$ [113]	1	9.54	6.19	5.84	
$C_{18}AlCl_{3.24}$ [118]	2	12.80 12.92 [118]	{6.09 6.21	5.84	
$C_9GaCl_{3.5}$ [113] ($C_{7.9}$ [117])	1	9.56 [95]	6.21		
$C_{18}InCl_{3.5}$ [113]	2	12.80 [51]	6.10	5.94	
$C_{27}InCl_{3.5}$ (?)	3	16.20 [51]	6.15	5.94	
$C_{12.6}AuCl_{3+x}$ [113]	1	6.80	3.45	3.22	
$C_{42}PtCl_{4.3}$ [114]	3	16.06	6.01		
$C_{40}NbCl_{5.1}$ [118]	3	16.21	6.15	5.89	10.5
$C_{32}TaCl_{5.1}$ [118]	3	16.19	6.13	5.89	10.5
$C_{10}MoCl_5$ [105]	1	9.31 (?)	5.96	5.74	9.8
$C_{20}MoCl_5$ [105]	2	12.54	5.84	5.74	
$C_{30}MoCl_5$ [105]	3	16.02	5.97	5.74	
$C_{40}MoCl_5$ [105]	4	19.37 [104]	5.97	5.74	
		19.42 [105]	6.02		
$C_{28}UCl_5$ [118]	1	9.62	6.27	5.98	11.5
$C_{37}UCl_5$ [118]	2	13.03	6.33		
$C_{41}MoOCl_{4.1}$ [118]	3	15.89	5.83		
$C_{70}WCl_6$ [95]	5	23.02	6.25	5.76	12.3
$C_{12}SbCl_5$ [1002]	1	9.42	6.06		
$C_{24}SbCl_5$ [1002]	2	12.72	6.01		
$C_{36}SbCl_5$ [1002]	3	16.08	6.01		
$C_{48}SbCl_5$ [1002]	4	19.45	6.03		
$C_m MBr_x$					
$C_{18}AlBr_3$ [94]	2	13.40	6.69		
$C_m AlBr_3$ [94]	4	20.10	6.68		
C_9AlBr_5 [93]	1	10.24	6.89		
$C_{24}AlBr_{3.3}$ [93]	2	13.35	6.64		
$C_m GaBr_3$ [94]	4	19.85	6.43		

(Continued overleaf)

Table 16 (Continued).

$C_m MBr_x$	n	I_c^{min}(Å)	D(Å)	D_0(Å)	m
$C_m GaBr_3$ [94]	5	23.20	6.43		
$C_m GaBr_{3+x}$ [94]	1	10.22	6.87		
$C_{13} GaBr_{5.5}$ [94]	2	13.55	6.84		
$C_{16.5} GaBr_{5.2}$ [94]	2	13.38	6.67		
$C_m AuBr_{3+x}$ [94]	1	6.90	3.55		
$C_m AuBr_{3+x}$ [94]	2	10.25			
$C_{21} TiF_4$ [1048]	3	15.10	5.05		
$C_{28} TiF_4$ [1048]	4	18.45			

larger than in the free halide. Estimated values for the carbon-to-halide ratio are also very reasonable as can be checked e.g. for $MoCl_5$ (from free halide: $m = 9.8$, from the structure of the intercalate: $m = 10.3$ [106]) for which the geometry of the intercalated $MoCl_5$ layer has been determined.

Pure higher stages form only under special conditions [92] and most samples reveal a disturbed periodicity of the layer sequence [100]. The structure of $C_{28}Br$ also bears a close resemblance to that of solid bromine, which demonstrates that Van der Waals interactions between Br_2 molecules are at least as important as charge transfer from the graphite layers in determining the configuration within a layer [91]. The regular sequence of carbon and intercalated layers must be due primarily to charge transfer but ionic interaction is not sufficiently strong to effect the ordering of successive intercalated layers within the (a, b) plane beyond ensuring that the molecules or ions lie in the same orientation in first-stage compounds, but not necessarily in higher stages.

Intercalation of magnetic compounds into graphite is of great interest with respect to the question of two-dimensional magnetic interactions. The difficulty in preparing well-defined phases is reflected in certain inconsistences of the experimental results.

$MoCl_5$ is a ferromagnetic insulator with a Curie temperature of 22 K, obeying a Curie-Weiss law at higher temperatures. Its first-stage graphite compound is still ferromagnetic with $T_C = 15$ K. In higher-stage compounds, i.e. as soon as more than one graphite layer separate the dimeric $MoCl_5$ array, the ferromagnetic phase transition disappears abruptly. For the second- to fourth-stage compounds, the magnetic interactions between Mo ions within the same layer are exactly the same as for the first-stage compound, as follows from the Curie-Weiss behavior of the susceptibility [105].

Anhydrous $FeCl_3$ is a helical antiferromagnet with a Néel temperature of 16 K, the magnetic moments lying in the (140) plane. Karimov et al. [99] found that in first-stage $FeCl_3$-graphite the interaction between the $FeCl_3$ layers is weakened by an order of magnitude but does not change sign ($T_N = 3.6$ K). For the second-stage compound the interaction between the $FeCl_3$ layers is negligibly small, but intralayer interactions lead to ferromagnetic ordering below 8.6 K. Ohhashi and Tsujikawa [102], however, describe both stages as anisotropic two-dimensional

TABLE 17

Structural data of graphite salts

n = stage number
I_c^{min} = chemical periodicity length along the hexagonal axis
D = apparent thickness of reactant layer

Intercalate or formula	n	I_c^{min} (Å)	D (Å)
$C_{24}NO_3 \cdot 3HNO_3$ [51]	1	7.84	4.49
HNO_3	2	11.14	4.44
$C_{6n}HNO_3$ [1036]	3	14.49	4.44
	4	17.84	4.44
	5	21.19	4.44
$C_{24}HSO_4 \cdot 2H_2SO_4$ [51]	1	7.98	4.63
	2	11.33	4.63
	3	14.72	4.60
	4	18.09	4.67
	5	21.46	4.69
$HClO_4$ [51]	1	7.94	4.59
	2	11.12	4.41
	3	14.30	4.24
	4	17.65	4.23
	5	21.00	4.23
H_3PO_4 [51]	2	11.3	4.6
$H_4P_2O_7$ [51]	2	11.5	4.8
CF_3COOH [51]	1	8.2	4.85
	2	11.52	4.81
	3	14.85	4.79
Br_2 [90]	2	10.40	3.70
C_5ICl [90]	1	7.14	3.79
IBr [90]	2	10.54	3.83
$C_{27.5}CrO_2Cl_2$ [112]	3	14.87	4.79
$C_9N_2O_5$ [85]	1	7.9	4.55
$C_{13}CrO_3$ [998]	3	14.64	7.94

antiferromagnets with spin flopping in the basal plane ($n = 1$: $T_N = 3.9$ K; $n = 2$: $T_N = 3.6$ K).

On the other hand, both the first- and second-stage $FeCl_2$-graphite compounds appear to be two-dimensional Ising ferromagnets [99, 107]. Anhydrous $FeCl_2$ is an antiferromagnet below $T_N = 23.5$ K. The strong anisotropy favors the c-axis as the easy axis. The antiferromagnetic interaction between the ferromagnetic layers is much weaker than in $FeCl_3$, being suppressed by a magnetic field of ~15 kOe. Separation of the $FeCl_2$ layers by graphite layers diminishes this interaction so strongly that the ferromagnetic interaction of the magnetic ions within the layer becomes dominant. Thus, the first- and second-stage compounds have practically the same Curie point (15.5 and 15.0 K, resp.) and close values of the paramagnetic Curie temperature (16.5 and 16.7 K, resp.) [99] (compare [108]).

The $FeCl_2$-graphite compounds are prepared by reduction of a corresponding $FeCl_3$-graphite phase. Strong reduction finally leads to layers of elemental iron, which may aggregate to form α-Fe [109]. According to Knappwost and Metz [110], careful reduction leads to monolayers of some 10^4 Fe atoms which are spontaneously magnetized. These ferromagnetic monolayers are antiferromagnetically coupled through the graphite layers (compare, however, [111]).

By reduction of $C_m MCl_n$ Vol'pin et al. [1009] prepared lamellar compounds of graphite with Cr, Mo, W, Mn, Fe, Co, Ni and Cu.

Interesting behavior can be expected from graphite compounds with $CrCl_3$ [23] and $CrBr_3$. For magnetically ordered $C_m NiCl_2$ see [1008].

For the latest review on graphite intercalation compounds see [1047].

There is some evidence that boron nitride is also able to form intercalation compounds. A characteristic buff color was observed on BN treated with alkali metals or $FeCl_3$ and $CuCl_2$ [119]. Though BN is structurally closely related to graphite it has to be borne in mind that BN is an insulator. Thus it is not surprising that BN reveals a somewhat different behavior. Not only is the amount of intercalated molecules much lower than in graphite, but moreover, with $FeCl_3$ there is a net transfer of charge to the BN layers from the $FeCl_3$ molecules, which is opposite to the effect observed in graphite-$FeCl_3$.

3. Compounds with Lone-Pair Cations (CN 3 · · · 6)

A. PHOSPHORUS-GROUP CHALCOGENIDES

a. M_2X_3 COMPOUNDS WITH CN 3

A coordination number 3 can be expected for the cations P^{3+}, As^{3+} and Sb^{3+}. In Table 18, we give a survey of the structures occurring in phosphorus-group chalcogenides M_2X_3. These compounds are all insulators or semiconductors in their normal-pressure modifications. Most of them crystallize in layer structures. As already mentioned the coordination numbers increase with increasing principal quantum numbers of the elements. Thus P_2O_3 will probably have a molecular structure similar to the one found in the cubic modifications of As_2O_3 and Sb_2O_3. The structure of the other known modification of Sb_2O_3, valentinite (Figure 37), is built up of double chains, analogous to those of Figure 18b (with the non-connected fourth corner of the tetrahedra missing). The bond angles at the antimony atom are 79.8°, 91.9° and 98.1°. The oxygen bond angle along the chain is 116.2°, while the one at the oxygen atom bridging the two chains is 130.8°. The double chains are stacked in such a way that, in addition to the p^3 bonds of Sb and the p^2—sp bonds of O (with Sb—O distances of 2.019 Å and 2.023 Å) within the chains, weak bonds between the double chains approximately in b-direction might develop. The distances, however, are definitely too long (2.52 Å) to correspond to bonds other than Van der Waals bonds. Valentinite crystals indeed grow as prismatic needles (needle axis parallel c) with perfect prismatic cleavage. If the double chains were not

TABLE 18

Structure types of M_2X_3 chalcogenides of As, Sb, and Bi. Layer types are printed in italics.

As₂O₃ (r) cubic arsenolite (CaF₂ deriv.) (h_1) monocl. *claudetite* (h_2) monocl. *claudetite II*	**Sb₂O₃** (r) cubic senarmontite (h, p) orth. valentinite [990]	**Bi₂O₃** (r) monocl. *layered* (α) (h) tetragonal (β) γ (?) δ (?) [122]
As₂S₃ (r) monocl. *orpiment* (h) *layer type* [120]	**Sb₂S₃** orthorh., stibnite, ribbons (ferroelectric $T_C = 20°C$)	**Bi₂S₃** Sb₂S₃ type
As₂Se₃ (r) *orpiment type* (>18kb/300°C) As₂Te₃ type (>35kb/700°C) new type	**Sb₂Se₃** Sb₂S₃ type	**Bi₂Se₃** (r) *tetradymite type* (p_1) [123a] (p_2) Sb₂S₃ type [123, 123b] (p_3) anti-Zn₃P₂ type (p_3') Mn₂O₃ type ?
As₂Te₃ (r) monocl. *layer type* (>10 kb) *layer type*	**Sb₂Te₃** *tetradymite type* [120a] piezoelectric, R3m or R32 [121]	**Bi₂Te₃** (r) *tetradymite type* (p_1) R3m, Sb₂Te₃ type [122a] (p_2) [124a] (p_3) (supercond.) [124]
AsSbS₃ monocl. *getchellite*	**Sb₂OS₂** triclinic *kermesite*	**Bi₂O₂Se** tetrag. *La₂O₂Te type, layered* **Bi₂Te₂S** rhomboh. *tetradymite*

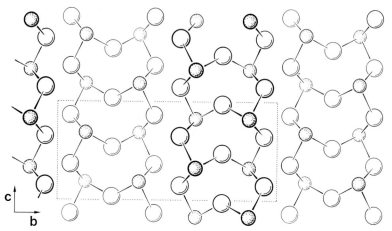

Fig. 37. The layered ribbons of the structure of valentinite Sb_2O_3 projected along $[\bar{1}00]$. Small stippled spheres: Sb

shifted relative to each other in the [100] direction then a layer structure with one-dimensional character would result, i.e. one with two perfect Van der Waals cleavage planes.

The structure of the sulfide Sb_2S_3 is composed of double ribbons of composition $(Sb_4S_6)_n$. The ribbons represent deformed portions of a rocksalt structure, somewhat reminiscent of the SnS structure. The coordination number of the Sb atoms

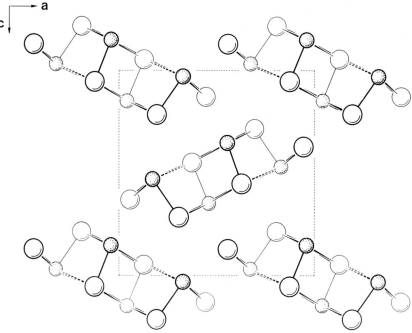

Fig. 38. Projection of the orthorhombic structure of Sb_2S_3 which is almost a layer structure. The (Sb_4S_6) ribbons extending parallel to the b-axis are clearly recognizable. Small dotted spheres: Sb.

TABLE 19

As$_2$O$_3$ claudetite I, monoclinic C$_{2h}^5$—P2$_1$/n (No. 14), Z = 4.
All atoms in 4(e): $\pm(x, y, z; \frac{1}{2}+x, \frac{1}{2}-y, \frac{1}{2}+z)$

As$_2$O$_3$ (h_1): a = 5.25 Å, b = 12.90 Å, c = 4.53 Å, β = 93.9°

	x	y	z	
As$_I$	0.335	0.351	0.026	
As$_{II}$	0.250	0.101	0.020	
O$_I$	0.432	0.219	0.949	[125]
O$_{II}$	0.452	0.344	0.411	
O$_{III}$	0.631	0.408	0.925	

(25°): a = 5.339 Å, b = 12.984 Å, c = 4.5405 Å, β = 94°16′ [1]

As$_2$O$_3$ (h_2) claudetite II, monoclinic, P2$_1$/n as above.
a = 7.99 Å, b = 4.57 Å, c = 9.11 Å, β = 78.3° [125]
a = 7.990 Å, b = 4.645 Å, c = 9.115 Å, β = 78.3° [125a]

	x	y	z	
As$_I$	0.6163	0.8311	0.3013	
As$_{II}$	0.1841	0.2920	0.3717	
O$_I$	0.677	0.459	0.291	[125a]
O$_{II}$	0.238	0.140	0.184	
O$_{III}$	0.966	0.349	0.367	

is hard to define, it is at least 3. The closest Sb—S contacts occur between the two ribbons that compose the double ribbon. The bonding may result from a superposition of a single-bond system and a NaCl-like mesomeric bond system. Since the Sb$_2$S$_3$ type is transitional between a molecular and a three-dimensional structure type, we thought it worthwhile to reproduce this structure in Figure 38 although it just fails to be two-dimensional. A very similar array of ribbons is found in the structure of Sn$_2$S$_3$(= Sn^{II}Sn^{IV}S$_3$) and PbSnS$_3$.

Table 18 demonstrates that a variation of the 'chemical parameters' (O → S → Se → Te and P → As → Sb → Bi) does not necessarily give rise to a smooth transition from molecular structures to chain structures and to layer structures and finally to three-dimensional structures, though the coordination number does increase. Thus, we find layer structures with As$_2$O$_3$ as well as with Bi$_2$Te$_3$, although of a quite different type. Taking into account the tendency of Bi to increase its coordination number due to the relative closeness of the 6p and 6d energies, we are not surprised that Bi$_2$O$_3$ does not adopt the structure of Sb$_2$O$_3$ or As$_2$O$_3$. Nevertheless, the low-temperature modification of Bi$_2$O$_3$ is rather incidently three-dimensional. The monoclinic structure of α-Bi$_2$O$_3$ [126] is made up of slightly puckered oxygen layers parallel to the bc plane at roughly $x = \frac{1}{4}$ and $x = \frac{3}{4}$. The Bi atoms are inserted between these O layers at roughly $x = 0$, and $x = \frac{1}{2}$. The resulting coordination number of the Bi atoms is hard to define but, at least in the central Bi layers, three short Bi—O distances (2.08, 2.17 and 2.21 Å) can clearly be distinguished from the remaining (2.54, 2.63 and 3.25 Å). However, instead of pointing towards the same O layer, one bond is directed towards the other O layer in a manner reminiscent of the Bi coordination in BiTaO$_4$.

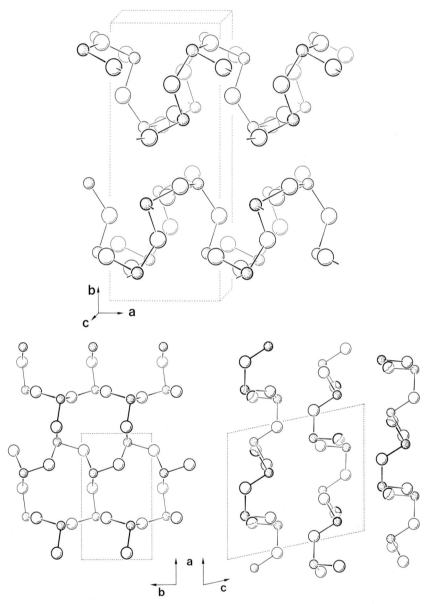

Fig. 39. The layer-type modifications of As_2O_3. Above: Claudetite I; As atoms dotted. Below: Claudetite II.
  right: Projection along [010] perpendicular to the layers.
  left: One layer projected onto the (a, b)-plane.

b. CLAUDETITE As$_2$O$_3$(h)

The claudetite structure shown in Figure 39 is a corrugated version of Figure 26f. Bond angles are such that together with the lone pairs an sp^3 bond configuration is approximated, closer than in the case of valentinite. The lone electron pairs frame (or connect) the layers (Table 19).

For both As$_2$O$_3$ and Sb$_2$O$_3$ the layer structure belongs to the high-temperature high-pressure modification [127] while the room-temperature modifications crystallize in a cubic molecular structure with As$_4$O$_6$ and Sb$_4$O$_6$ units, respectively.

For As$_2$O$_3$ another layer-type modification, monoclinic claudetite II, has been reported [125, 125a]. The layers in this structure are very similar to those in claudetite I but less compressed. The real structure of claudetite II is illustrated in Figure 39 but the difference between the two modifications is best seen in Figure 26. A related layer structure was found in orthorhombic As$_2$O$_3$·SO$_3$ [1021].

c. ORPIMENT As$_2$S$_3$

The monoclinic structure of orpiment As$_2$S$_3$ is shown in Figure 40 and in idealized form in Figure 26e. ··· As—S—As—S ··· spiral chains are linked by sulfur atoms

TABLE 20

Orpiment structure, monoclinic, C_{2h}^5—P2$_1$/n (No. 14), Z = 4.

All atoms in 4(e): $\pm(x, y, z; \frac{1}{2}-x, y+\frac{1}{2}, \frac{1}{2}-z)$

As$_2$S$_3$: $a = 11.475$ Å, $b = 9.577$ Å, $c = 4.256$ Å, $\beta = 90°41'$

	x	y	z	
As$_I$	0.264 7	0.191 7	0.862	
As$_{II}$	0.486 8	0.321 2	0.360 7	
S$_I$	0.401 5	0.121 3	0.508 1	[128]
S$_{II}$	0.347 4	0.397 2	0.010 1	
S$_{III}$	0.122 3	0.293 5	0.559 0	

As$_2$Se$_3$: $a = 12.053$ Å, $b = 9.890$ Å, $c = 4.277$ Å, $\beta = 90°28'$ [129]

Alternate description in P2$_1$/c (No. 14)
($\mathbf{a}' = -\mathbf{c}$, $\mathbf{b}' = \mathbf{b}$, $\mathbf{c}' = \mathbf{a}+\mathbf{c}$)
4(e): $\pm(x, y, z; \bar{x}, \frac{1}{2}+y, \frac{1}{2}-z)$

As$_2$S$_3$: $a' = 4.256$ Å, $b' = 9.577$ Å, $c' = 12.191$ Å, $\beta' = 109°45'$ [128]
As$_2$Se$_3$: $a' = 4.30$ Å, $b' = 9.94$ Å, $c' = 12.84$ Å, $\beta' = 109.1°$ [130]

	x	y	z	
As$_I$	0.148 3	0.197 7	0.263 7	
As$_{II}$	0.851 2	0.318 0	0.484 7	
Se$_I$	−0.069 9	0.114 3	0.398 7	[130]
Se$_{II}$	0.372 0	0.409 2	0.353 9	
Se$_{III}$	0.346 8	0.303 7	0.115 3	

As$_2$(S,Se)$_3$ continuous solid solution [130a]

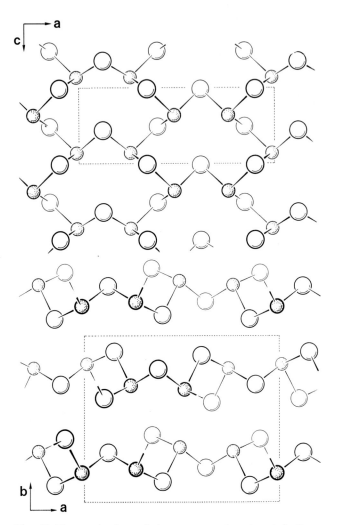

Fig. 40. Two projections of the structure of orpiment As_2S_3.
Arsenic atoms dotted.

to form corrugated layers perpendicular to the [010] direction. The As—S distances within the layers range from 2.243 Å to 2.308 Å. The shortest non-bonded contacts between the layers are S—S distances of 3.242 Å and As—S distances of 3.475 Å, which accounts for the perfect cleavage parallel to the (010) plane. The bond angle As—S—As of the sulfur atom bridging two parallel spiral chains is 87.9 Å, corresponding to covalent p^2 bonds, while the bond angles of the S atoms within the chains are 101.0° and 103.7°. The As bond angles range from 92.8 to 105.0°, reflecting the tendency to sp^3 bonding due to the s^2 lone pair.

The orpiment structure is found also in the selenide As_2Se_3.

Calculations of the density of valence states in As_2S_3, As_2Se_3 and As_2Te_3 have recently been reported [905].

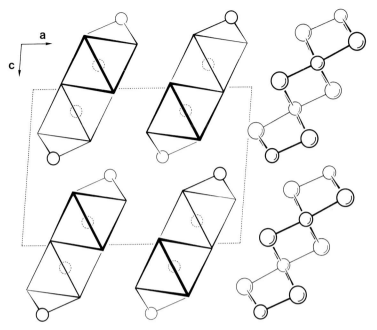

Fig. 40a. The monoclinic structure of As_2Te_3 projected along the double chains of the $[MXX_{2/2}X_{3/3}]$ octahedra.

small spheres: As.

d. As_2Te_3

The unique structure of semiconducting As_2Te_3 [128a] is the bridge between the orpiment structure which contains the cations in trigonal-pyramid coordination and the tetradymite structure with octahedrally coordinated cations. In As_2Te_3 half the cations are octahedrally coordinated. The coordination polyhedra form double chains which are capped by the remaining As atoms. The double-octahedron chains are very loosely connected via the As_I atoms ($As_I - 2Te'_{II}$, nearly parallel to the $(20\bar{1})$ plane). The actual distances are as follows:

As_I——1 Te_{II} at 2.68 Å Te_I——2 As_I at 2.77 Å
         2 Te_I at 2.77 Å   1 As_{II} at 2.93 Å
         2 Te_{II}, at 3.22 Å Te_{II}——1 As_I at 2.68 Å
                                          2 As_{II} at 2.90 Å
As_{II}——1 Te_{III} at 2.76 Å
          2 Te_{III}, at 2.85 Å Te_{III}——1 As_{II} at 2.76 Å
          2 Te_{II} at 2.90 Å   2 As_{II} at 2.85 Å
          1 Te_I at 2.93 Å

Whereas one would expect three single bonds for pure trigonal-pyramid coordination and six half bonds for octahedral coordination, the distances listed reveal that

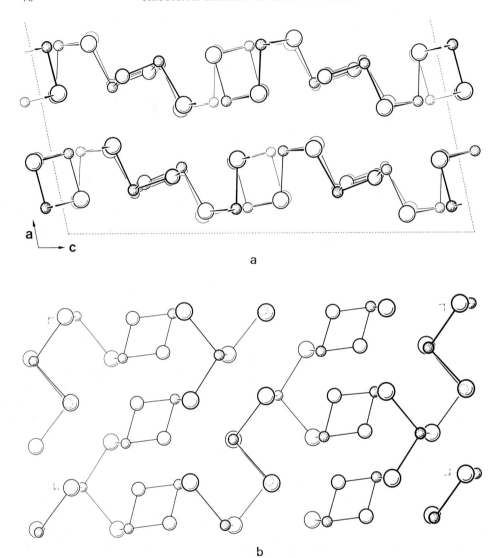

Fig. 41. The structure of kermesite Sb_2S_2O. (a) half the unit cell projected in b-direction (b) one layer projected in a-direction.

Largest spheres: S
medium spheres: O
smallest spheres (dotted): Sb

the situation is less clear-cut in the case of As_2Te_3. The octahedral coordination of As_{II} is in fact transitional, still revealing the underlying trigonal pyramid.

The only attractive forces between the staggered chains in the [100] direction are of the Van der Waals type in conformity with the observed (100) cleavage. The average Te–Te distance between the chains is 3.72 Å which is close to the 3.74 Å found in elemental tellurium.

TABLE 20a

As_2Te_3 structure, monoclinic, C_{2h}^3—C2/m (No. 12), $Z = 4$.
$(0, 0, 0; \frac{1}{2}, \frac{1}{2}, 0) +$
all atoms in 4(i): $\pm(x, 0, z)$

$As_2Se_3(p)$: $a = 13.37$ Å, $b = 3.73$ Å, $c = 9.31$ Å, $\beta = 95°$ [129]
As_2Te_3: $a = 14.339$ Å, $b = 4.006$ Å, $c = 9.873$ Å, $\beta = 95.0°$; [128a]

	x	z
As_I	0.615	0.445
As_{II}	0.205	0.145
Te_I	0.032	0.282
Te_{II}	0.780	0.337
Te_{III}	0.375	0.034

The As_2Te_3 structure type is met also in a high-pressure modification of As_2Se_3 [129]

e. KERMESITE Sb_2S_2O

Triclinic kermesite Sb_2S_2O has a rather complicated layer structure (Figure 41 and Table 21). Half the Sb atoms have a coordination number between 3 and 4, the other Sb atoms have a coordination similar to that in Sb_2S_3:

Sb_I——1 S (2.40 Å) + 3 O (1.99, 2.07 and 2.34 Å)
Sb_{II}——1 S (2.39 Å) + 3 O (1.99, 2.05 and 2.26 Å)
Sb_{III}——1 S (2.37 Å) + 2 S (2.61 and 2.64 Å) + 2 S (3.07 and 3.09 Å)
Sb_{IV}——1 S (2.35 Å) + 2 S (2.71 and 2.75 Å) + 2 S (2.91 and 3.02 Å).

TABLE 21

Kermesite structure, triclinic, $P\bar{1}$—C_i^1 (No. 2), $Z = 2$.
described in 8 times larger monoclinic cell:
$(0, 0, 0; \frac{1}{4}, \frac{1}{4}, 0; \frac{1}{2}, \frac{1}{2}, 0; \frac{3}{4}, \frac{3}{4}, 0; 0, \frac{1}{2}, \frac{1}{2}; \frac{1}{4}, \frac{3}{4}, \frac{1}{2}; \frac{3}{4}, \frac{1}{4}, \frac{1}{2}; 0, \frac{1}{2}, \frac{1}{2}; \frac{1}{4}, \frac{3}{4}, \frac{1}{2}) \pm x, y, z.$

Sb_2S_2O: $a = 20.972$ Å, $b = 8.16$ Å, $c = 20.378$ Å, $\beta = 101°50'$

	x	y	z	
Sb_I	0.101 7	0.750 3	0.184 4	
Sb_{II}	0.096 4	0.248 5	0.184 9	
Sb_{III}	0.194 5	0.000 4	0.072 7	
Sb_{IV}	0.190 9	0.502 2	0.041 9	
S_I	0.197 4	0.260 4	0.148 2	[131]
S_{II}	0.206 8	0.747 7	0.155 3	
S_{III}	0.078 8	0.003 1	0.044 9	
S_{IV}	0.079 9	0.488 7	0.046 1	
O_I	0.117 6	−0.017 2	0.212 5	
O_{II}	0.119 3	0.478 6	0.215 6	

The structure has a pronounced layer character. The shortest Sb—S distances between the layers are 3.40 Å.

f. GETCHELLITE AsSbS$_3$

The stucture of getchellite AsSbS$_3$ is made up of puckered (As, Sb)$_8$S$_8$ rings which are connected through 8 bridging S atoms to form infinite sheets parallel to (001). Within each such sheet, the structure is an open, glass-like network (Figure 42). The coordination of (As, Sb) and S is as expected for p^3 and p^2 single bonds with some s admixture. All metal atoms are three-coordinated by sulfur atoms, the MS$_3$ units forming trigonal pyramids with bond angles between 84.1° and 102.8°. All S atoms have a twofold coordination with bond angles between 99.4° and 106.8°. The shortest interlayer (As, Sb)—S separation is 3.326 Å compared with the largest bonding (As, Sb)—S distance of 2.430 Å.

The partial disorder observed on the synthetic sample used for the structure determination may be accidental. Crystals grown at lower temperatures and over longer periods of time may well be fully ordered. In that case, the crystallographic parameters will turn out to be somewhat different due to the different bond lengths of Sb—S and As—S.

g. WOLFSBERGITE – TYPE COMPOUNDS

Many ternary ABX_2 chalcogenides with B metals ($A^+B^{3+}X_2$ with A = Cu, Ag; B = Al, Ga, In, Tl; X = S, Se, Te) crystallize in the chalcopyrite (CuFeS$_2$)

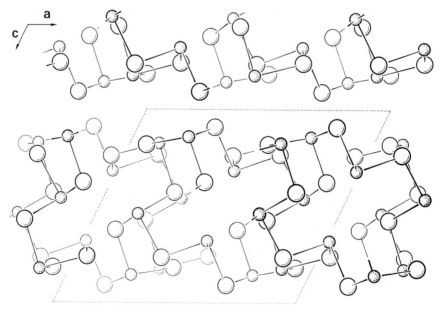

Fig. 42. The monoclinic structure of getchellite AsSbS$_3$.
Large spheres: S; small dotted spheres: As, Sb

TABLE 22

$AsSbS_3$ structure, monoclinic C_{2h}^5—P2$_1$/a (No. 14), $Z = 8$.
All atoms in 4(e): $\pm(x, y, z; \frac{1}{2}+x, \frac{1}{2}-y, z)$

$AsSbS_3$ (getchellite): $a = 11.85$ Å, $b = 8.99$ Å, $c = 10.16$ Å, $\beta = 116.45°$ [132]
(synthetic): $a = 11.857$ Å, $b = 9.015$ Å, $c = 10.194$ Å, $\beta = 116.37°$ [133]

	x	y	z
$As_{0.5}Sb_{0.5}$	0.122 4	0.358 1	0.174 8
$As_{0.75}Sb_{0.25}$	0.230 6	0.726 3	0.142 9
$As_{0.4}Sb_{0.6}$	0.484 9	0.438 9	0.317 3
$As_{0.35}Sb_{0.65}$	0.824 6	0.522 5	0.459 3
S_I	0.071 0	0.614 7	0.180 5
S_{II}	0.283 0	0.349 6	0.426 9 [133]
S_{III}	0.386 1	0.755 6	0.374 9
S_{IV}	0.317 9	0.519 1	0.091 3
S_V	0.450 6	0.185 5	0.238 6
S_{VI}	0.639 9	0.492 8	0.231 9

TABLE 23

Wolfsbergite-type structure, orthorhombic, D_{2h}^{16}—Pnma (No. 62), $Z = 4$.
All atoms in 4(c): $\pm(x, \frac{1}{4}, z; \frac{1}{2}+x, \frac{1}{4}, \frac{1}{2}-z)$.

$CuSbS_2$: $a = 6.020$ Å, $b = 3.792$ Å, $c = 14.485$ Å [1]
(wolfsbergite)

	x	z		x	z	
Cu	0.25	0.82		$\frac{1}{4}$	$\frac{33}{40}$	
Sb	0.23	0.06	[1] or	0.228	$\frac{1}{16}$	[261a]
S_I	0.62	0.10		$\frac{5}{8}$	0.097	
S_{II}	0.87	0.82		$\frac{7}{8}$	$\frac{33}{40}$	

$CuSbSe_2$: $a = 6.40$ Å, $b = 3.95$ Å, $c = 15.33$ Å

	x	z	
Cu	−0.738	−0.175	
Sb	0.225	0.062	
Se_I	0.652	0.097	[261a]
Se_{II}	−0.118	−0.176	

$CuBiS_2(r)$: $a = 6.137$ Å, $b = 3.898$ Å, $c = 14.541$ Å [1]
(emplectite) $a = 6.15$ Å, $b = 3.92$ Å, $c = 14.55$ Å [261b]
$a = 6.1426$ Å, $b = 3.9189$ Å, $c = 14.5282$ Å [1000]

	x	z		x	z	
Cu	0.250 5	0.830 8		0.2491	0.8281	
Bi	0.229 9	0.063 6		0.23156	0.06304	[1000]
S_I	0.636 7	0.097 0	[261b] or	0.6362	0.0980	
S_{II}	0.876 5	0.821 4		0.8742	0.8223	

$CuBiS_2$ (h): monoclinic, C_{2h}^3—C2/m (No. 12), $Z = 12$.
$a = 17.65$ Å (3×5.88 Å), $b = 3.93$ Å, $c = 15.24$ Å, $\beta = 100.5°$ [261c]

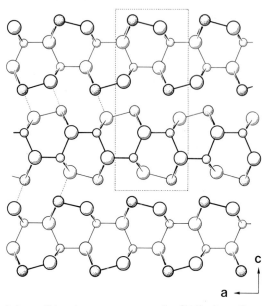

Fig. 43. Projection of the wolfsbergite structure onto the (010) plane. Some of the residual bonds between the layers are indicated by broken lines.

Cu: small spheres
Sb: dotted spheres
S: large spheres.

structure which is a zincblende superstructure. The strong preference of Cu for tetrahedral coordination and the nonbonding pair of valence electrons left on Sb^{3+} are responsible for the occurrence of the peculiar defect-tetrahedral structure of the wolfsbergite-type compounds (Table 23). The copper atoms and half of the sulfur atoms are in slightly distorted tetrahedral coordination while the antimony and the remaining sulfur atoms possess only three bonded neighbors, the space towards the fourth corner of the coordination tetrahedron being filled with the non-bonding lone electron pairs on Sb and S_I (Figure 43). A necessary condition for the formation of a layer surface is thus fulfilled. As follows from a comparison of the (probably not very accurate) Sb—S and Sb—Se distances:

$CuSbS_2$:
Sb—1 S_I at 2.42 Å [1] (2.44 Å [261a])
2 S_{II} at 2.64 Å (2.58 Å)
2 S_I at 3.12 Å (3.11 Å)

$CuBiS_2$:
Bi—1 S_I at 2.55 Å
2 S_{II} at 2.66 Å [261b]
2 S_I at 3.16 Å

$CuSbSe_2$:
Sb—1 Se_I at 2.78 Å [261a]
2 Se_{II} at 2.72 Å
2 Se_I at 3.23 Å

the separation of the layers is well above the bond distance and a perfect cleavage along the (001) plane confirms the layer character.

CuBiSe$_2$ has a rocksalt structure whereas for CuSbTe$_2$ and CuBiTe$_2$ a layer structure related to that of Bi$_2$Te$_3$ has been reported [1].

For the high-temperature modification of CuBiS$_2$ a monoclinic cell [261c] has been found with thrice the size of the orthorhombic cell of emplectite. Crystals of this variety were described as extremely fragile and very thin blades. These properties and the similarity of the cells make it probable that both modifications are closely related. The distortion and the enlargement of the cell possibly arise from a slight relative displacement of the individual layers.

h. LIVINGSTONITE HgSb$_4$S$_8$

The structure of livingstonite HgSb$_4$S$_8$ [261d, 811] is built up of alternating layers of compositions HgSb$_2$S$_4$ and Sb$_2$S$_4$. Both layers contain the same (Sb$_2$S$_4$)$_n$ double chains parallel with the b-axis. In one case these units are linked together by Hg atoms while in the other case the chains are directly connected via S—S bonds. Each Sb atom is thus trivalent and acquires 3 close S neighbors as well as 2 or 3 more distant S neighbors. In the HgSb$_2$S$_4$ sheet the average distances are:

$$\text{Sb}_\text{I}, \text{Sb}_\text{II} \text{------} 3\,\text{S} + 2\,\text{S at 2.48 Å}$$
$$2.55 \text{ Å and } 2.95 \text{ Å} \quad [261\text{d}]$$
$$2.66 \text{ Å and } 2.94 \text{ Å}$$

These S neighbors define a deformed square pyramid or half an octahedron around each Sb atom. In the Sb$_2$S$_4$ sheet the two crystallographically different Sb

TABLE 24

Livingstonite structure, monoclinic, C_{2h}^6—C2/c (No. 15), Z = 8.
$(0, 0, 0; 0, \tfrac{1}{2}, \tfrac{1}{2})+$
Sb and S in 8(f): $\pm(x, y, z; \tfrac{1}{2}+x, \bar{y}, z)$
Hg$_\text{I}$ in 4(b): $0, \tfrac{1}{2}, 0; \tfrac{1}{2}, \tfrac{1}{2}, 0$
Hg$_\text{II}$ in 4(e): $\pm(\tfrac{1}{4}, y, 0)$
HgSb$_4$S$_8$: $a = 21.48$ Å, $b = 4.00$ Å, $c = 30.25$ Å, $\beta = 104.2°$ [261d]
$a = 21.465$ Å, $b = 4.015$ Å, $c = 30.567$ Å, $\beta = 103.39$ [811]

	x	y	z		x	y	z	
Hg$_\text{II}$		0.001				0.0089		
Sb$_\text{I}$	0.075	0.063	0.120		0.0779	0.0563	0.1244	
Sb$_\text{II}$	0.425	0.064	0.131		0.4281	0.0586	0.1355	
S$_\text{I}$	0.092	0.493	0.062		0.0979	0.517	0.0642	
S$_\text{II}$	0.039	0.506	0.172		0.0352	0.520	0.1698	
S$_\text{III}$	0.407	0.494	0.189		0.4216	0.512	0.2060	
S$_\text{IV}$	0.460	0.507	0.078	[261d]	0.4625	0.516	0.0757	[811]
Sb$_\text{III}$	0.214	0.095	0.042		0.2115	0.0628	0.0395	
Sb$_\text{IV}$	0.287	0.078	0.208		0.2847	0.1091	0.2057	
S$_\text{V}$	0.180	0.028	0.229		0.1795	0.024	0.2177	
S$_\text{VI}$	0.222	0.521	0.149		0.2300	0.511	0.1527	
S$_\text{VII}$	0.318	0.021	0.022		0.3169	0.005	0.0148	
S$_\text{VIII}$	0.277	0.483	0.102		0.2881	0.501	0.1077	

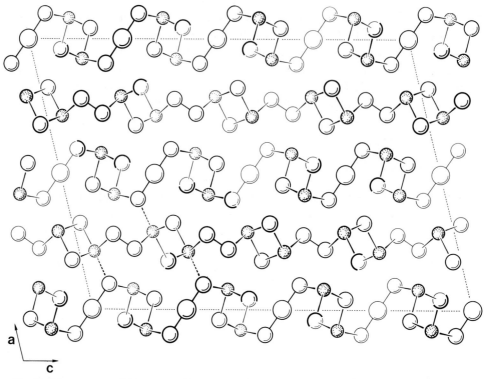

Fig. 44. The structure of livingstonite HgSb$_4$S$_8$ projected on (010). Some of the weak interlayer bonds are indicated by broken lines.

S: medium spheres
Sb: dotted spheres
Hg: large spheres

atoms show more variation in the distances:

Sb$_{III}$, Sb$_{IV}$ —— 3 S + 3 S at 2.47 Å   2.54 Å
2.44 Å and 3.11 Å   and   2.59 Å and 2.88 Å
2.52 Å and 3.15 Å   2.66 Å and 2.98 Å [261.d]
3.25 Å (interlayer)   3.24 Å (interlayer)

Each antimony atom of the Sb$_2$S$_4$ sheet thus is (though very weakly) bonded to a sulfur neighbor of the S—Hg—S unit of the other sheet (indicated by dashed lines in Figure 44). The remaining interlayer Sb—S distances are between 3.51 Å and 3.86 Å.

The Hg atoms have the same linear coordination as in cinnabar HgS. Four further remote S atoms complement the coordination to a strongly distorted octahedron:

Hg —— 2 S at 2.35 Å, 2S at 3.34 Å, and 2S at 3.37 Å.   [261d]

In spite of the higher accuracy in [811] the distances in [261d] appear to be more reasonable.

The relatively short interlayer Sb—S contacts spoil to some degree the layer character of livingstonite in spite of its perfect cleavage parallel with (001). $HgSb_4S_8$ does in fact crystallize in an elongated needle form.

Weak interlayer bonds are not uncommon to compounds with lone-pair cations. We would, however, expect the layer character to be more pronounced in hypothetical $HgAs_4S_8$. No examples of the livingstonite structure other than $HgSb_4S_8$ are known but the analogous semiconducting polyselenides $HgSb_4Se_8$ and $HgBi_4Se_8$ might also exist.

The mineral imhofite $Tl_{0.156}Cu_{0.025}As_{0.384}S$ might be another example of a layer-type polysulfide. It was reported to crystallize in very thin soft flakes [261e].

i. $CuTe_2Cl$ (SEE APPENDIX)

B. GROUP V ELEMENTS AND ISOELECTRONIC COMPOUNDS

a. THE ARSENIC STRUCTURE

Elements of the 5th Group of the Periodic Table carry five outer electrons, two s- and three p-electrons, and hence can act as trivalent anions as well as trivalent

TABLE 25
Modifications of Group V elements

Phosphorus	white	(t), <196 K: hexagonal? (molecular P_4?) [8]
		(r), >196 K: cubic, molecular P_4 [8]
	red:	cubic? [8]
		disordered As double layers? [134, 135]
		(h), >440°C and other (h) phases [8]
	Hittorf's:	monoclinic [137]
	black (p):	orthorhombic [138]
		p > 70 kbar: rhombohedral As type [139]
		>110 kbar: simple cubic [139], superconducting [140, 141]
		>170 kbar: (tetragonal?), superconducting [141, 142]
Arsenic	yellow:	cubic, molecular As_4, semiconducting
		$(t?)$: black-phosphorus type
		(r): rhombohedral A7 type
		100–140 kbar: monoclinic Bi II type? superconducting [143]
		>120–150 kbar: tetragonal, superconducting [142, 143]
Antimony		(r): rhombohedral As type
		>70 kbar: simple cubic, superconducting [144]
		>85 kbar: tetragonal, superconducting [142]
		(monoclinic, distorted SnS type, doubtful [145])
Bismuth		(r): rhombohedral As type
		>25 kbar: cubic (a = 6.354 Å, $x \neq \frac{1}{4}$) [146]
		simple cubic (a = 3.16 Å)
		>25.4 kbar: monoclinic, superconducting
		>27 kbar: tetragonal, superconducting [142]
		(monoclinic, distorted SnS type? [145])
		>~40 kbar: Bi IV [147]
		>~65 kbar: Bi V, bcc at 90 kb [148], superconducting [149]
		>92 kbar: Bi VI, superconducting [149]

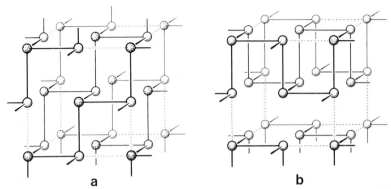

Fig. 45. Derivation from the rocksalt structure of (a) arsenic and GeTe structure; and (b) black phosphorus and SnS structure.

and pentavalent cations. In the elemental forms of P, As, Sb and Bi, covalent single bonds with pyramidal coordination might be expected. If we admit resonating half bonds then a simple cubic structure will fulfil the coordinative requirements. As can be seen in Table 25, simple cubic structures do indeed occur as high-pressure modifications of P, Sb and Bi. These phases are of course all metallic since the cell contains an odd number of valence electrons. The degeneracy is lifted in the binary analogs which crystallize in the rocksalt structure and therefore are low-gap semiconductors.

We can disturb or finally suppress the resonance of the bonds by distorting these cubic structures. Figure 45 visualizes how the structure of arsenic (or GeTe) and that of black phosphorus (or SnS) and yellow PbO evolve from the simple cubic (or NaCl) structure by loosening half the bonds. As the distortions lead to larger unit cells with an even number of atoms per cell, metallic properties are no longer compelling and the occurrence of semiconductivity just depends on the degree of the distortions. The transition from resonating half bonds to single bonds and hence non-metallic properties is complete in phases with the black phosphorus structure. In the arsenic structure (Figure 46), on the other hand, the distortions just fail to be sufficient so that the elements As, Sb and Bi are semimetals under normal conditions. Energy band overlap [150] in As-type phases, however, is very small and therefore sensitive to a variation of external parameters, such as pressure [151–156] or alloying – either addition of small amounts of neighboring elements [157] (Sb: Te or Sn) or mixing them with each other (As———Sb or Sb—Bi) [158–160]. Nevertheless, it is surprising that intermediate compositions ($Sb_{1-x}As_x$ with $x = 0.09 \cdots 0.40$ [161] and $Bi_{1-x}Sb_x$ with $x = 0.05 \cdots 0.23$ [162]) exhibit semiconducting properties just as does elemental bismuth under pressures of 6 to 24 kbar, and $Bi_{1-x}Sb_x$ ($x \leq 0.05$) alloys under correspondingly lower pressures [158], as if alloying compressed the cell in the same way as does hydrostatic pressure. Possibly a similar metal-to-semiconductor transition might be observed in the system $As_{1-x}P_x$, which is stable in the As structure up to $x = 0.13$ under ambient conditions.

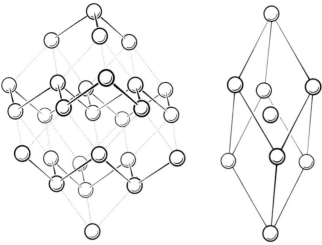

Fig. 46. The structure of rhombohedral arsenic. Left: the structure in relation to a rocksalt cell. Right: the primitive rhombohedral cell.

TABLE 26

Arsenic structure, rhombohedral D_{3d}^5—$R\bar{3}m$ (No. 166), $Z = 2$.

As in 2(c): $\pm(x, x, x)$
hexagonal axes: As in 6(c): $(0, 0, 0; \frac{1}{3}, \frac{2}{3}, \frac{2}{3}; \frac{2}{3}, \frac{1}{3}, \frac{1}{3}) \pm (0, 0, z)$
(face-centered rhombohedral cell derived from the rocksalt cell:
As at $(0, 0, 0; 0, \frac{1}{2}, \frac{1}{2}; \frac{1}{2}, 0, \frac{1}{2}; \frac{1}{2}, \frac{1}{2}, 0) \pm (x, x, x)$, x as above)

P (~100 kb/r.t.): $a = 3.524$ Å, $\alpha = 57.25°$ or $a = 3.377$ Å, $c = 8.806$ Å, $x = 0.21$–0.22 [139]

$As_{1-x}P_x$ (r.t.): $x \leq 0.13$ [170]

As (4.2 K): $a = 4.101\,8$ Å, $\alpha = 54.55°$ or $a = 3.759\,4$ Å, $c = 10.441$ Å; $x = 0.227\,6$ [171]
(78 K): $a = 4.106\,3$ Å, $\alpha = 54.48°$ or $a = 3.759\,1$ Å, $c = 10.458$ Å; $x = 0.227\,5$ [171]
(298 K): $a = 4.132\,1$ Å, $\alpha = 54.13°$ or $a = 3.760\,2$ Å, $c = 10.548$ Å; $x = 0.227\,1$ [171]
(677 K): $a = 4.186\,0$ Å, $\alpha = 53.37°$ or $a = 3.759\,5$ Å, $c = 10.738$ Å; $x = 0.226\,0$ [172]

As—Sb [166]

Sb (4.2 K): $a = 4.489\,8$ Å, $\alpha = 57.23°$ or $a = 4.300\,7$ Å, $c = 11.222$ Å; $x = 0.233\,6$ [173]
(78 K): $a = 4.492\,7$ Å, $\alpha = 57.20°$ or $a = 4.301\,2$ Å, $c = 11.232$ Å; $x = 0.233\,6$
(298 K): $a = 4.506\,7$ Å, $\alpha = 57.11°$ or $a = 4.308\,4$ Å, $c = 11.274$ Å; $x = 0.233\,5$

Sb ($p \leq 80$ kbar) [174]

Bi (4.2 K): $a = 4.723$ Å, $\alpha = 57.33°$ or $a = 4.541\,9$ Å, $c = 11.802\,5$ Å; $x = 0.234\,1$ [168]
(78 K): $a = 4.728\,8$ Å, $\alpha = 57.30°$ or $a = 4.534\,2$ Å, $c = 11.814\,2$ Å; $x = 0.234\,0$
(298 K): $a = 4.746\,0$ Å, $\alpha = 57.23°$ or $a = 4.546\,1$ Å, $c = 11.862\,3$ Å; $x = 0.233\,9$

Te (30 kbar/r.t.): $a = 4.690$ Å, $\alpha = 53.31°$ or $a = 4.208$ Å, $c = 12.036$; $x = 0.230$ [175]

Solid solutions:

$(AsAs)_{1-x}(GeSe)_x$: $x \leq 0.30$ at ambient conditions [176]
$(AsAs)_{1-x}(GeTe)_x$: $x \leq 0.37$
$(SbSb)_{1-x}(GeTe)_x$: $x \leq 0.37$
$(SbSb)_{1-x}(SnTe)_x$: $x \leq 0.23$

GeP_5 (disordered): $a = 3.885$ Å, $\alpha = 52.83°$ or $a = 3.457$ Å, $c = 10.001$ Å [177]

One might expect to detect a slight structural indication for the semimetal-to-semiconductor transition but the structural parameters vary rather continuously. In the Sb—As system, Skinner [166] observed a linear variation of the unit-cell volume with concentration. The abrupt deviations reported by Jain [162] near the critical concentrations in the Sb—Bi system were not confirmed [167]. The rhombohedral angle (and hence the c/a value) is a smooth curve with a minimum (maximum of c/a) near 70% Sb. Exact values for the positional parameter z have been determined in the $Bi_{1-x}Sb_x$ system for $x \leq 0.3$ by Cucka and Barrett [168]. They found z to remain approximately constant up to about 12% Sb and then to rise from 0.234 07 to 0.234 20 at 4.2 K, and from 0.234 00 to 0.234 13 at 78 K as the Sb content increases from 0 to 30% Sb. There is, however, no discontinuity in the interatomic distances through the transition (Figure 47): The difference between bonding and interlayer M–M distances remains practically constant (0.458 Å in Bi and 0.447 Å in Sb at 298 K). The parameter z decreases with increasing temperature, which means that the larger of the two M–M distances expands faster than the smaller. This has a large effect on the band parameters [169]. The temperature dependence is opposite to expectations, but apparently the effect of the thermally excited electrons is too small to provide a control of z. Alloying of acceptor (Sn, Pb) and donor (Te) elements also had no measurable effect on the value of z [168]. Obviously, this semimetal-to-semiconductor transition is a smooth higher-order transition.

It is worth noting that a high-pressure modification of tellurium was reported

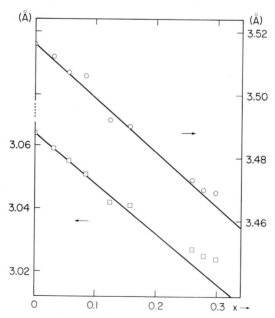

Fig. 47. Bond distances and interlayer M—M distances in $Bi_{1-x}Sb_x$ alloys at 78 K. The solid lines represent linear interpolations between the values for pure Bi and pure Sb.)

[175] to crystallize in the arsenic structure. Te has one excess electron and therefore must be metallic in this structure.

The distortion of the 'rocksalt' lattice increases the bond angle from 90° to 94.5° in Bi. Sb is very similar to Bi whereas, in As, the distortions are distinctly more pronounced as follows from Table 27. In the high-pressure modification of phosphorus, finally, the difference between the bond distance and the interlayer P–P distance is such that we conclude that this phase is a semiconductor provided that the reported z value $0.21 \cdots 0.22$ is reliable. The bond angle is rather close to the tetrahedral angle of 109.47° so that P II approximates an sp^3 bonding system with the non-bonding lone-pair s-electrons separating different layers. Therefore, we would expect A7-type phosphorus to represent a true layer-type phase, whereas the structures of As, Sb and Bi are transitional between framework and layer-type.

TABLE 27
Interatomic distances and distortions in arsenic-type elements

A7 type element	Bond distance (Å)	Interlayer M—M distance (Å)	Difference (Å)	Bond angle (°)
P (100 kb/r.t.)	2.13	3.27	1.14	105
As (298 K)	2.516	3.120	0.604	96.73
Sb (298 K)	2.908	3.355	0.447	95.60
Bi (298 K)	3.071	3.529	0.458	95.48
Te (30 kb/r.t.)	2.868	3.477	0.609	94.37
GeTe (300 K)	2.841	3.172	0.331	94.49

Though antimony is structurally very similar to bismuth, the band structures of the two elements differ in the details. Thus the pressure dependence of c/a in Sb [174] is opposite to the concentration dependence of c/a in $Bi_{1-x}Sb_x$ alloys [167]. Moreover, in both bismuth and arsenic the Fermi surface shrinks with increasing pressure [151] making them less metallic. In antimony, however, the number of charge carriers increases with pressure, making it more metallic. Pospelov [178] pointed out that although from the point of view of the crystal structure, the transition to the simple cubic high-pressure phase could be continuous, a basic change in the energy-band structure gives rise to a first-order transition. According to Jaggi [146] the transition in bismuth occurs in two steps. The semiconducting rhombohedral phase is first followed (discontinuously) by a cubic phase i.e. $\alpha_{rh} = 60°$, but $x \neq \frac{1}{4}$ remains. At this stage, bismuth is probably again semimetallic since the difference between the two sets of Bi—Bi distances decreases. The next transformation leads to a true metal with simple cubic structure ($x = \frac{1}{4}$).

According to Brugger et al. [179], however, Bi II has a monoclinic structure in which the Bi atoms form puckered layers parallel with the ab plane (Figure 48). Each Bi is bound to three others within a layer, one at 3.147 Å and two at

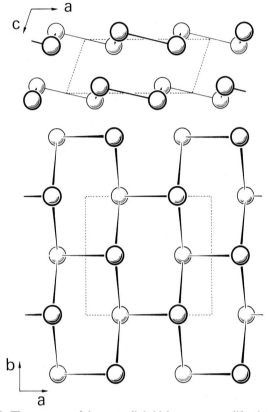

Fig. 48. The structure of the monoclinic high-pressure modification Bi II.

3.168 Å (bond angle 149.8° between the latter and 95.5° for the other angles) compared with 3.072 Å in Bi I. The layer character is less pronounced than in rhombohedral bismuth, as the interlayer distances are shorter (two Bi at 3.304 Å along c, one at 3.396 Å and one at 3.559 Å compared with 3 Bi at 3.529 Å in Bi I).

TABLE 28

Structure of Bi II, monoclinic, C_{2h}^3—C2/m (No. 12), $Z=4$.

Atoms in 4(i): $(0, 0, 0; \frac{1}{2}, \frac{1}{2}, 0) \pm (x, 0, z)$, with $x = \frac{1}{4}$, $z = \frac{1}{8}$.

$a = 6.674$ Å, $b = 6.117$ Å, $c = 3.304$ Å, $\beta = 110.33°$ [179]

b. SUPERSTRUCTURES DERIVED FROM THE As STRUCTURE: GeTe, SnP$_3$

As the primitive cell of the arsenic structure contains two atoms the simplest superstructure is realized in the GeTe type. Replacement of AsAs by GeTe removes the center of symmetry and therefore these phases can be piezoelectric and even ferroelectric. Removal of the inversion symmetry makes the three optical vibration modes both Raman and infrared active and it was indeed by Raman scattering that the ferroelectric nature of the rhombohedral modifications of GeTe and SnTe was verified [180, 181].

The GeTe type of ordering could also occur at equiatomic compositions in the alloy systems Bi—Sb and Sb—As (and As—P at high pressures). Such an ordering has been suggested in order to explain Raman spectra and X-ray diffraction data of Bi—Sb alloys [182].

TABLE 29

GeTe structure, rhombohedral, C_{3v}^5—R3m (No. 160), $Z = 1$.
  Both atoms in 1(a): x, x, x; $x_{Ge} = -x_{Te}$

GeTe (300 K): $a = 4.307$ Å, $\alpha = 57.94°$ or $a = 4.172$ Å, $c = 10.710$ Å
  NaCl-like cell: $a = 5.996$ Å, $\alpha = 88.18°$; $x = 0.237$ [203]
  α increases with T, reaching 90° at 400 °C [183]

$Ge_{1-\delta}Te$ (20 °C): $a = 5.9869$ Å, $\alpha = 88°16'$; $x = 0.233$ [184]
($T_t = 396$ °C): $a = 6.039$ Å, $\alpha = 89°44'$ [184]

$Ge_{1-\delta}Te$, Ge-saturated, r.t.
  annealed at 420°C, $\delta \approx 0.012$: $a = 5.988$ Å, $\alpha = 88°20'$ [208]
  annealed at 650°C: $a = 5.984$ Å, $\alpha = 88°10.2'$ [208]

  Te-saturated, r.t.
  annealed at 420°C, $\delta \approx 0.047$: $a = 5.956$ Å, $\alpha = 88°43.5'$ [208]
  annealed at 650°C: $a = 5.966$ Å, $\alpha = 88°10.0'$ [208]

SnTe (5 K): $a = 6.325$ Å, $\alpha = 89.895°$ [188]

$GeTe_{1-x}Se_x$ (r.t.): $x \leq 0.53$ [190]
  $x \leq 0.55$ [204]
  $x \leq 0.63$ [205]

$Ge_{1-x}Sn_xTe$ (r.t.): $x \leq 0.65$ [206]
  $x \leq 0.72$ [207]
α increases monotonically from 88.4° (GeTe) to 90° at $x = 0.72$ [207]

$Ge_{1-x}Sn_xTe_{1-x}Se_x$: $x < 0.4$ [207]
  $x \leq 0.2$ [205]

$Ge_{1-x}Pb_xTe$ (r.t.): $x < 0.82$ [208]

$Ge_{0.8}Pb_{0.2}Te$ (r.t.): $a = 6.095$ Å, $\alpha = 88°27'$
$Ge_{0.7}Pb_{0.3}Te$ (r.t.): $a = 6.144$ Å, $\alpha = 88°38'$
$Ge_{0.375}Pb_{0.625}Te$ (r.t.): $a = 6.293$ Å, $\alpha = 89°28'$
$Ge_{0.275}Pb_{0.725}Te$ (r.t.): $a = 6.338$ Å, $\alpha = 89°45'$
  (all samples annealed at 600°C)
The cubic → rhombohedral transition temperature is 130 K for $x = 0.953$, 60 K for $x = 0.02$ and drops to 0 K at $x \approx 0.01$ [208, 209]

$Ge_{1-x}Mn_xTe$: $x < 0.2$ [210, 211]
$Ge_{0.9}Mn_{0.1}Te$: $a = 5.962$ Å, $\alpha = 88.83°$ [210]
$Ge_{0.834}Mn_{0.166}Te$: $a = 5.944$ Å, $\alpha = 89.40°$ [210]
$Ge_{1-x}M_x^{II}Te$ (M = Zn, Cd, Hg): $x \leq 0.02 \cdots 0.03$ [212]
$Ge_{1-x}(TlSb)_{x/2}Te$, $Ge_{1-x}(TlBi)_{x/2}Te$: $x \leq 0.08$ [213]

Schiferl [228] who has discussed bonding and crystal structures of average-valence ⟨5⟩ compounds came to the conclusion that a critical ionicity appears to be the limit above which the rhombohedral GeTe and As structures cannot exist under any condition.

As follows from Table 27, the layer character in these phases is even less developed than in the elements bismuth, antimony and arsenic. Analogous to the case of antimony, the distortions from the rocksalt structure in GeTe smoothly decrease with increasing temperature [183, 184]. The nature of the phase transition [184], however, is extremely sensitive to stoichiometry and hence to charge-carrier concentration of the samples. Dilatometric studies [185] on $Ge_{50-x}Te_{50+x}$ with $x \geqslant 0.6$ revealed a marked volume contraction while a sample with $x < 0.6$ showed an abrupt expansion at the transition point (in contrast to alloys in which 6 and 40% of Ge was replaced by Sn [185a]). A first-order transition to the NaCl structure was observed in GeTe at room-temperature under a pressure of 35 kbar [186]. SnTe is even more delicate [181, 187]. Replacement of Ge by Sn is qualitatively equivalent to the application of high pressure [186]. Samples with carrier concentrations $p = 1.0 \times 20^{20}$ cm^{-3} were found to have a transition temperature $T_t = 70 \pm 5$K and recently a transition temperature as high as 98 K was reported for a sample with 0.88×10^{20} cm^{-3} [186a, 897]. The transition was, however, not observed at 5°K in samples with $p = 8.4 \times 10^{20}$ cm^{-3} and $p = 10.0 \times 10^{20}$ cm^{-3}. It was argued that a ferroelectric transition can take place in SnTe samples with carrier densities $p < 2.2 \times 10^{20}$ cm^{-3} only [181, 188]. Dilatation measurements of Novikova and Shelimova [187], on the other hand, clearly revealed a phase transition near 77 K in a sample of claimed composition $Sn_{49.6}Te_{50.4}$. [1038]: $p_{max} = 1.3 \times 10^{21}$ cm^{-3}.

A recent paper [188a] reports that the low-temperature modification of SnTe is not rhombohedral but orthorhombic (16K: $a = 6.274$ Å, $b = 6.288$ Å, $c = 6.309$ Å). An increase of the transition temperature with Te content was observed.

This behavior demonstrates that this group of compounds has rather peculiar properties. Deformation of a highly symmetric structure with resonating bonds usually induces or enhances a non-metallic character. It appears that the energy gaps between valence and conduction bands are also increased in this case, but this turns out to be extremely difficult to prove, as the samples all behave as degenerate semiconductors [189, 190]. In divalent Group IV compounds the cations are not completely ionized, two s-electrons being left on the cations. The band structure therefore, will be characterized by an additional valence band arising from these cation s-electrons. As these s-electrons become more strongly bound with increasing main quantum number, this band drops in energy on going from Si to Pb. Apparently, in GeTe, this band and the valence band originating from the (s, p) functions of the anions overlap, the difference between their upper levels being roughly $\frac{1}{4}$ eV [191]. In PbTe, the cation s^2 band merges into the anion valence band, whereas in the orthorhombic GeS the cation s^2

band will be separated. It is obviously this critical overlap that is responsible for the observed non-stoichiometry of these GeTe-type phases. Unlike the rocksalt-type lead chalcogenides, these phases crystallize with deviations from stoichiometry of the order of some %, compared with $\pm(0.1\cdots 0.3)\%$ of the former. Since the maximum in the Ge—Te phase diagram occurs at slightly higher Te concentrations than true stoichiometry, the GeTe samples usually show p-type conduction due to vacancies in the cation sublattice. Their composition should correctly be formulated as $Ge_{1-x}Te$ and the charge equilibrium as $Ge^{2+}_{1-2x}Ge^{4+}_{x}Te^{2-}$. Stoichiometric GeTe nevertheless should be an intrinsic semiconductor. Our 'chemical' interpretation is at variance with that of Lewis [191] (and Kolomoets et al. [189]) who assumes that each defect contributes one carrier and who defines from carrier-density measurements a 'stoichiometric' composition $GeTe_{1.014}$.

The most exciting property, that is coupled with the coexistence of a heavy-hole and a light-hole band, is the occurrence of superconductivity in GeTe [192] with hole concentration $p > 8 \times 10^{20}$ cm^{-3}, reaching a critical temperature T_c of 0.3 K at $p = 1.5 \times 10^{21}$ cm^{-3}. It is however not the lattice distortion that is responsible for superconductivity, as superconductivity was found also in the rocksalt-type SnTe.

It is tempting to ask for superconductivity in As-type phases. There, however, the cation and anion (s^2) valence bands coincide. Ordered BiSb or SbAs, on the other hand, should approximate GeTe and it might be worthwhile testing these phases (pure or doped) for superconductivity.

The GeTe structure is preserved in $GeTe_{1-x}Se_x$ alloys up to $x \approx 0.6$ (compare Table 29). In the region with $x = 0.65 \cdots 0.85$ a new yet unresolved structure type was observed [190]. These phases form peritectically, and single crystals were prepared by an iodine transport process. The crystals were described as very thin and extremely soft hexagonal plates having a metallic luster. Two polytypes were reported, one with $a = 3.82$ Å, $c = 15.62$ Å and the other with $a = 3.82$ Å, $c = 46.87$ Å at the composition $GeSe_{0.75}Te_{0.25}$. For the limiting compositions the lattice parameters were given as [190]

$GeSe_{0.65}Te_{0.35}$: $a = 3.852$ Å, $c = 47.14$ Å

$GeSe_{0.85}Te_{0.15}$: $a = 3.751$ Å, $c = 46.57$ Å.

The structure was said to be hexagonal, but not rhombohedral though the larger cell might be expected to represent a rhombohedral stacking of the simple unit. These are no doubt layer structures but they cannot be derived from a GeTe-type superstructure since the c-axis of these phases is $\frac{3}{2}c_0$ or $\frac{9}{2}c_0$, c_0 referring to the GeTe type.

In the Ge(Se, Te) alloy system, formation of a superstructure might well be possible at the composition Ge_2SeTe with Se and Te ordered in alternating layers. The same type of ordering would be feasible in the (Ge, Sn)Te system at the composition GeSnTe$_2$. Ordering does in fact occur in true ternary compounds. TlSbTe$_2$ [193], TlBiSe$_2$ [194] and TlBiTe$_2$ [193] (and probably TlSbS$_2$ [195, 196] and TlBiS$_2$ [196, 197, 198]) crystallize in the NaCrS$_2$ structure which evolves from the rocksalt structure by rhombohedral deformations (R$\bar{3}$m—D$_{3d}^5$; Tl in 1(a): 0, 0, 0; Sb, Bi in 1(b): $\frac{1}{2}$, $\frac{1}{2}$, $\frac{1}{2}$; Se, Te in 2(c): $\pm(x, x, x)$; $x = 0.243$ for TlSbTe$_2$ and $x = 0.250$ for TlBiTe$_2$ [193]). For a GeTe-type superstructure the anions should be at positions (x, x, x) and $(\frac{1}{2}+x', \frac{1}{2}+x', \frac{1}{2}+x')$ with x, $x' < 0.25$. The centrosymmetry of the NaCrS$_2$ structure leads to a coordination number 6, each cation having six equidistant anion neighbors. The NaCrS$_2$ structure therefore is a layered three-dimensional structure. The monoclinic structure of lorandite TlAsS$_2$ [199, 200] on the other hand, is no longer related to the GeTe structure. In lorandite, the AsS$_3$ pyramids are connected into (AsSS$_{2/2}$) spiral chains which are loosely held together by the Tl$^+$ ions. The Tl atoms are coordinated on the sides of the AsS$_3$ pyramid chains. Each Tl is more closely attached to one chain than to neighboring chains which explains the lamellar and fibrous habits of the dark red crystals and their excellent (100) and very good ($\bar{2}$01) cleavage.

The queer orthorhombic structure reported for TlAsSe$_2$ [201] and TlSbSe$_2$ [202] also bears no relation to that of rocksalt or GeTe. This structure is said to be characterized by planar centered layers of Sb+Tl between which lie planar zigzag chains of Se atoms, all parallel with one axis.

Although an A$_2$B ordering in the As structure would be feasible, no examples are known. Ordering was however observed at 3:1 compositions. In the monoclinic mineral paradocrasite Sb$_3$As [214] ordering is of the type Sb$_2$(SbAs), that means, its structure should be related to that of GeTe. Its cleavage plane is (010).

Another kind of superstructure type is realized in the compounds GeP$_3$ and SnP$_3$ (cf. Figure 49 and Table 30). In each As-type layer every fourth P atom is replaced by a cation in such a way that puckered P$_6$ rings result, which are held together by Ge or Sn (puckered B$_3$F type, compare Figure 34b). The foreign

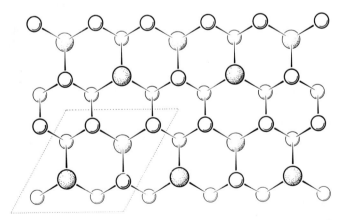

Fig. 49. The hexagonal structure of SnP$_3$ viewed along the c-axis.
Large dotted spheres: Sn.

atoms of course expand the P lattice, but the P rings themselves would define a hexagonal lattice parameter $a = 3.38$ Å, fairly close to the one observed in the As-type modification of phosphorus (3.378 Å). Interatomic distances are as follows:

GeP$_3$	SnP$_3$
Ge——3 P at 2.484 Å	Sn——3 P at 2.662 Å
3 P at 2.845 Å	3 P at 2.925 Å
P——2 P at 2.103 Å (98.25°)	P——2 P at 2.222 Å (99.08°)
1 Ge at 2.484 Å	1 Sn at 2.662 Å
1 Ge at 2.845 Å	1 Sn at 2.925 Å
2 P at 3.185 Å	2 P at 3.360 Å

TABLE 30

Arsenic-type superstructures
Sb$_3$As = Sb$_2$(SbAs), monoclinic, C$_2^3$—C2 (No. 5), Z = 1.
$a = 7.252$ Å, $b = 4.172$ Å, $c = 4.431$ Å, $\beta = 123.14°$ [214]

SnP$_3$ structure, rhombohedral, R$\bar{3}$m—D$_{3d}^5$ (No. 166), Z = 6.
hexagonal axes: $(0, 0, 0; \frac{1}{3}, \frac{2}{3}, \frac{2}{3}; \frac{2}{3}, \frac{1}{3}, \frac{1}{3})+$
P in 18(h): $\pm(x, \bar{x}, z; x, 2x, z; 2\bar{x}, \bar{x}, z)$
Sn in 6(c): $\pm(0, 0, z)$

GeP$_3$(p): $a = 6.989$ Å, $c = 9.986$ Å [177]
$z(\text{Ge}) = 0.269; x(\text{P}) = 0.515, z(\text{P}) = 0.282$ [215]

SnP$_3$: $a = 7.3785$ Å, $c = 10.5125$ Å;
$z(\text{Sn}) = 0.2576; x(\text{P}) = 0.5139, z(\text{P}) = 0.2828$ [215]

related (?): As$_4$S, orthorhombic
$a = 3.576$ Å, $b = 6.759$ Å ($\approx a\sqrt{3}$), $c = 10.074$ Å [216]

The difference between intralayer and interlayer distances is even smaller than in the As-type elements. In other words, the layer character is almost lost. This may be correlated with the metallic properties of these phases. Since an element with three bonding electrons is replaced by an element with (two or) four bonding valence electrons the metallic properties are by no means surprising. Less obvious, however, is the reason why these phases form at all. This structure appears to be much more suited for AsP$_3$ or SbP$_3$.

A mineral of queer composition, duranusite As$_4$S, is believed to have a layer-type crystal structure related to arsenic [216]. Its pseudohexagonal cell has in fact dimensions that are reasonably close to those of arsenic.

c. THE BLACK PHOSPHORUS AND SnS STRUCTURE

While the structures of the As and GeTe type arise from a rhombohedral distortion of the rocksalt structure, a quasi-tetragonal deformation leads to the structures of the black phosphorus and SnS types (Figs. 45, 50 and 51). The orthorhombic SnS cell (setting Pnma) evolves from the rocksalt cell according to the relations

$$a_{\text{orth.}} = 2 a_{\text{cubic}} - \delta_1$$
$$b_{\text{orth.}} = \frac{1}{\sqrt{2}} a_{\text{cubic}} - \delta_2$$
$$c_{\text{orth.}} = \frac{1}{\sqrt{2}} a_{\text{cubic}} + \delta_3$$

The distortion reduces the six half bonds acting in the rock-salt structure to three single bonds in the black phosphorus and SnS-type structure. This interpretation is substantiated by a comparison of the bond lengths. For NaCl-type SnSe films, a lattice constant $a = 5.99$ Å was reported [217] corresponding to an Sn—Se distance of 2.995 Å. In the ordinary orthorhombic modification of SnSe the distances are 2.77 Å ($1\times$) and 2.82 Å ($2\times$), the mean bond distance thus being 2.80 Å. The difference 0.19 Å agrees surprisingly well with Pauling's 0.18 Å difference between single-bond and half-bond distances.

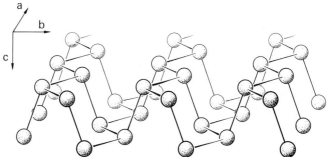

Fig. 50. One layer of the orthorhombic structure of black phosphorus.

As can be seen from Table 31, the SnS structural type occurs in high-pressure modifications of rocksalt-type chalcogenides with Group IV cations but not with alkaline earth cations. At first glance, it appears amazing that application of pressure results in a reduction of the coordination number, but there is indeed a negative volume change at the phase transitions [219, 224]. Furthermore, the phase transition from the rocksalt to the SnS structure is accompanied by an increase in the electrical resistivity of several orders of magnitude [225, 226]. This is again quite unusual for a high-pressure phase, but it is well accounted for by the transition from half bonds to single bonds. It was argued [224] that the pressure-induced structure sequence in Group IV chalcogenides would be:

arsenic type → NaCl type → SnS type,

as exemplified in the system GeTe—SnTe. This was also the reason why Kabalkina et al. [227] deduced a distorted SnS cell from their high-pressure X-ray studies on antimony and bismuth, though we do not accept their structure as well established. As further evidence for the analogy with PbS these authors men-

TABLE 31

Structures occurring among the Group IV monochalcogenides. Layer structures are underlined.

	S	Se	Te
SiO fibrous	SiS (a) fibrous (b) layer structure	SiSe	?
GeO amorphous	GeS (r) SnS type (>520°C hexagonal? [218]) (h) NaCl type?	GeSe (r) SnS type (>651°C : NaCl type [220])	GeTe (r) GeTe type (>440°C NaCl type (p_1) NaCl type (p_2) SnS type [221]
SnO (r) tetrag. PbO type (red) orthorhombic [223] (p) NaCl type?	SnS (r) SnS type, orthorh. (thin film) NaCl type [217] (>584°C) ? [222]	SnSe (r) SnS type (thin film) NaCl type [217] (514°C) ? [222]	SnTe (<70 K) GeTe type (r) NaCl type (p) SnS type
PbO (r) tetrag. PbO type (p, h) orthorhombic	PbS (r) NaCl type (p) SnS type	PbSe (r) NaCl type (p_1) SnS type (p_2) supercond. [1011]	PbTe (r) NaCl type (p_1) SnS type (p_2) supercond. [1011]

tioned a marked increase of the electrical resistivity. However, the high-pressure modifications Sb III and Bi III are in fact superconductors but not semiconductors as would be adequate for black phosphorus-type modifications. According to Duggin [142] the structure of both Sb III and Bi III are tetragonal. Without understanding the details, we feel that the pressure-induced structure sequences depend in a delicate way upon the fine properties of the involved valence electrons. In certain cases, the SnS structure appears to be less dense than the GeTe structure, as follows from a comparison of the atomic volumes of GeSe (22.94 Å^3) and orthorhombic As (22.00 Å^3) with that of A7-type As (21.52 Å^3). Therefore, in phosphorus and arsenic the structure sequence is quite different from that observed in (Ge, Sn)Te, namely

P: red → ~SnS type → As type → simple cubic → ···

As: ··· ~SnS type → As type (→ simple cubic ??) → ···

This may be connected with the degree of possible distortions of the lattice.

In Figure 51 the structures of SnS and of black phosphorus are opposed to each other for comparison. If in the unit cell of black phosphorus we interchange the

TABLE 32

Black phosphorus, orthorhombic, D_{2h}^{18}—Cmca (No. 64), $Z = 8$.
Atoms in 8(f): $(0, 0, 0; \frac{1}{2}, \frac{1}{2}, 0) \pm (0, y, z; \frac{1}{2}, y, \frac{1}{2} - z)$.

P: $a = 3.32$ Å, $b = 10.52$ Å, $c = 4.39$ Å; $y = 0.098$, $z = 0.090$ [4]
(22°C): $a = 3.3136$ Å, $b = 10.478$ Å, $c = 4.3763$ Å; $y = 0.10168$, $z = 0.08056$ [138]

$P_{0.75}As_{0.25}$: $a = 3.38$ Å, $b = 10.60$ Å, $c = 4.42$ Å [170]

$P_{0.55}As_{0.45}$: $a = 3.48$ Å, $b = 10.69$ Å, $c = 4.45$ Å [170]

$P_{0.31}As_{0.69}$: $a = 3.55$ Å, $b = 10.83$ Å, $c = 4.48$ Å [170]

$P_{0.3}As_{0.6}Sb_{0.1}$ [170]

As: $a = 3.62$ Å, $b = 10.85$ Å, $c = 4.48$ Å [170]
$a = 3.63$ Å, $b = 10.96$ Å, $c = 4.45$ Å [230]
$a = 3.65$ Å, $b = 11.00$ Å, $c = 4.47$ Å; $y = 0.110$, $z = 0.060$ [229]

Sb(p) SbIII type, claimed to be a distorted SnS type [145]
C_{2h}^2—$P2_1/m$ (No. 11), $Z = 4$.
Atoms in 2(e): $\pm(x, \frac{1}{4}, z)$.

Sb (140 kbar): $a = 5.56$ Å, $b = 4.04$ Å, $c = 4.22$ Å, $\beta = 86°$ [145]
$x_I = -0.25$, $z_I = 0.08$; $x_{II} = 0.33$, $z_{II} = 0.39$

Bi (35.5 kbar): $a = 6.05$ Å, $b = 4.20$ Å, $c = 4.65$ Å, $\beta = 85°20'$ [145]

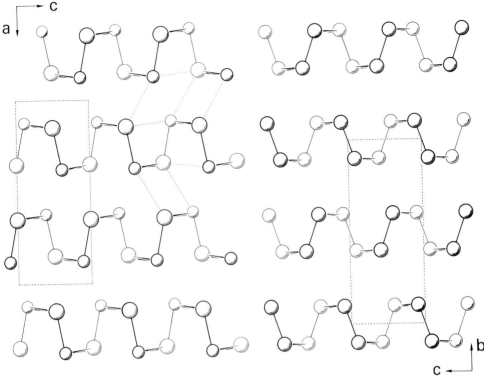

Fig. 51. The SnS-type structure of GeSe (left) and the structure of black phosphorus (right).

axes a and b then it can be described by a SnS cell with the parameters

$$P_I: x_{SnS} = \tfrac{1}{4} - y_P, \quad z_{SnS} = z_P$$
$$P_{II}: x_{SnS} = -\tfrac{1}{4} + y_P, \quad z_{SnS} = \tfrac{1}{2} + z_P.$$

In Table 33 the thus transformed phosphorus cell is added for comparison. Moreover, we have listed among the SnS-type representatives also the phase AsP with cell parameters interpolated from the data of Krebs et al. [170]. Although phosphorus has an only slightly higher electronegativity than arsenic, ordering might well occur and then our entry would be justified.

Each atom in black phosphorus has two neighbors at 2.224 Å and a third one at 2.244 Å. All other distances are at least as large as the a-axis. The bond angles are 96.34° (1×) and 103.09° (2×) (for As, the corresponding values are 2.49 and 2.48 Å, 94.1° and 98.5°). These values are comparable with those of the arsenic-type high-pressure modification. Again the distortions lead to a compromise between p^3 trigonal-pyramidal and sp^3 tetrahedral coordination with the lone pair s^2 pointing at an angle of approximately 45° towards the neighboring layer.

The distortions are more irregular in the SnS type, obviously as a consequence of the different size and charge of cations and anions. Although crystallographically, the SnS structure has a lower symmetry, it is in a certain sense less deformed, particularly around the cation, as follows from the bond angles given below. For the cation, the deformation mainly results in a reduction of the coordination number, and the three remaining anion neighbors still define a trigonal pyramid rather than three corners of a tetrahedron.

GeSe:
Ge——1 Se at 2.56 Å
      2 Se at 2.59 Å
      2 Se at 3.32 Å
      1 Se at 3.37 Å
Se—Ge—Se 91.3° (2×)
          96.1 (1×)
Ge—Se—Ge 96.1° (1×)
          103.4° (2×)

SnSe:
Sn——1 Se at 2.77 Å
      2 Se at 2.82 Å
      2 Se at 3.35 Å
      1 Se at 3.47 Å
Se—Sn—Se 88.7° (2×)
          95.9° (1×)
Sn—Se—Sn 95.9° (1×)
          101.4° (2×)

The existence of the lone electron pair localized on the cation is obviously essential for the occurrence of this structure type. Whereas S, Se and Te readily form three bonds this is no longer the case with Cl, Br and I. Thus, the analogs of SnTe and PbTe, InI and TlI crystallize in a geometrically related structure with a strong tendency towards isolation of linear molecules, whereas GaBr, the analog of GeSe, is unknown. Compounds of the type Ge_2AsBr with a possible SnS superstructure may not form either.

The rapid rise of the resistivity of $TlBiS_2$ observed at 65 kbar [197] is reminiscent of the resistance increase in PbS at the rocksalt type → SnS type transition. It is therefore tempting to speculate on a ternary analog of the SnS structure.

The SnS-type phases are semiconductors with energy gaps of 1–2 eV. For black phosphorus a gap of 0.33 eV was derived from resistivity and Hall-effect measurements [231]. This value is rather low compared with that of GeS. On the other hand, Keyes [231] reported black phosphorus to be transparent from 2 to 30 μ which implies an optical energy gap ≥ 0.63 eV. Another puzzle is the reported semiconductor to metal (or semimetal?) transition near 12 kbar [231]. However, we think that this interpretation of Bridgman's results is not well justified, since Bridgman [232] made his high-pressure resistivity measurements at 30 °C and 70 °C only, well below the range of intrinsic conduction (>150 °C) observed by Keyes [231]. Moreover, Bridgman observed a positive temperature coefficient of resistivity between 12 and 20 kbar (another sample revealed no sign reversal up to 12 kbar) and he extrapolated to a second sign reversal near 23 kbar. According to Keyes [231] the energy gap is squeezed to 0.21 eV at 8.1 kbar. If the energy gap really did vanish at pressures around 20 kbar, then it would be rather improbable that the As-type phase stable above 70 kbar would be non-metallic. It might be interesting to know the variation with pressure of the structural distortions.

TABLE 33
SnS structure, orthorhombic, D_{2h}^{16}—Pnma (No. 62), $Z = 4$.
All atoms in 4(c): $\pm(x, \frac{1}{4}, z; \frac{1}{2}+x, \frac{1}{4}, \frac{1}{2}-z)$

MX	a(Å)	b(Å)	c(Å)	x_M	z_M	x_X	z_X	Ref.
P	10.478	3.3136	4.3763	0.1483	0.0806	−0.1483	0.5806	
AsP	10.70	3.51	4.45					
GeS	10.44	3.65	4.30	0.121	0.106	−0.148	0.503	[227]
	10.45	3.649	4.301					[804]
GeSe	10.82	3.852	4.403	0.1213	0.1097	−0.1467	0.5020	[235]
	10.82	3.85	4.40	0.121	0.111	−0.146	0.500	[233]
	10.79	3.82	4.38	0.121	0.106	−0.148	0.503	[237]
GeTe$_{1+\delta}$	11.76	4.15	4.36	(sub-cell)				[234]
Sn$_{0.65}$Ge$_{0.35}$Te	11.53	4.29	4.36					[224]
(37 kbar)								
Sn$_{0.8}$Ge$_{0.2}$Te	11.65	4.30	4.42					[224]
(36 kbar)								
(110 K, $p = 0$)	11.95	4.30	4.51					[236]
SnS	11.20	3.99	4.34	0.118	0.115	−0.150	0.478	[4]
	11.202	3.988	4.349					[484]
herzenbergite	11.190	3.978	4.328					[1]
PbSnS$_2$	11.35	4.05	4.29					[5]
teallite	11.419	4.090	4.266					[1]
SnSe	11.57	4.19	4.46	0.118	0.103	−0.145	0.479	[237]
(Sn, Pb)Se								[238]
SnTe (20 kbar)	11.95	4.37	4.48					[217, 226]
	11.93	4.39	4.51					[224, 234]
(110 K, $p = 0$)	12.10	4.36	4.57					[236]
PbS (25 kbar)	11.28	3.98	4.21					[217]
	11.5	3.9	4.27					[219]
(80 kbar)	11.35	3.75	4.17					[219]
PbSe (43 kbar)	11.61	4.00	4.39					[217]
PbTe								
(42 kbar)	11.71	4.36	4.42					[217]
(75 kbar)	11.91	4.20	4.51					[224, 234]
(90 kbar)	11.83	4.05	4.67					[219]

Since the SnS type is a distorted version of a highly symmetric parent structure, one would expect the distortions to diminish with increasing temperature, as happens in the case of the GeTe structure. High-temperature X-ray diffraction studies on GeSe [220] indeed revealed such a behavior. The thermal expansion of the unit-cell axes was observed to be linear with a distinct change of the expansion coefficients above 400 °C. The relative changes of the axes indicate a rearrangement of the structure towards cubic symmetry. A first-order transition to the NaCl structure was observed at 651 °C with a volume change of 0.5%. GeSe then remains cubic up to its incongruent melting point of 670°C. No indication of a hexagonal phase was found. We therefore suspect that the hexagonal high-temperature phase reported for GeS might also turn out to be cubic.

d. Hittorf's phosphorus

Whereas single-crystalline flakes of black phosphorus can be grown from a bismuth solution [138], crystallization from liquid lead leads to red platy crystals of another modification, called Hittorf's phosphorus. Its monoclinic structure [137] (C_{2h}^4—$P2c$, $Z = 84$, all atoms in 4g) consists of cage-like P_8 and P_9 groups which are linked alternately by pairs of P atoms to form tubes of pentagonal cross section. Parallel tubes form double layers in which tubes of different layers are approximately perpendicular (at 89.6°) to each other. In one layer, the tubes are parallel with [110], in the crossed layer, parallel with [1$\bar{1}$0]. Each tube is connected through a P——P bridge with every second tube of the crossed layer. Thus, a double layer consists of two interpenetrating systems of tubes with no chemical bonds between them (Figure 52). These double layers are sort of close-packed in c-direction, the (001) plane thus being the cleavage plane. Bond angles lie between 84.5° and 115.9°, with a mean value of 101°. The bond lengths are between 2.196 Å and 2.299 Å with a mean value of 2.219 Å. The distance between P atoms of different tubes is 3.06 Å and larger.

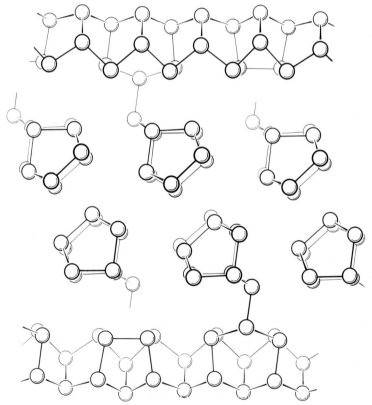

Fig. 52. Projection of the structure of monoclinic phosphorus along [110].

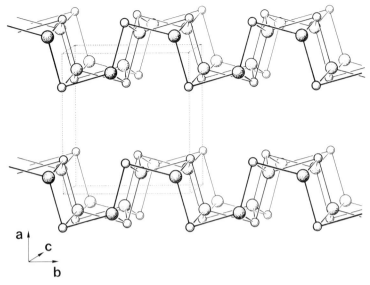

Fig. 53. The orthorhombic structure of yellow PbO (massicot).
Atoms at $z = -\frac{1}{4}, \frac{1}{4}, \frac{3}{4}$ and $\frac{5}{4}$ are shown.

A related fibrous P modification [137] forms on annealing white phosphorus at temperatures between 545° and 580° for several months. In this simpler structure each tube is linked to only one other tube which is no longer crossed but parallel.

e. YELLOW PbO (MASSICOT)

The structure of yellow lead oxide is another version of a strongly distorted NaCl structure but now with coordination number 4 (Figure 53). The structure is made up of layers containing crumpled —Pb—O— chains along the b axis of the orthorhombic cell. All atoms of a chain lie in a plane parallel with (a, b). These chains are wrinkled in a manner that each Pb atom can form two p bonds to oxygen neighbors at right angles (90.4°), whereas the corresponding angle at the O atoms is 120.9°. Neighboring chains of the same layer are linked in such a way that loose zig-zag chains can be seen to run in b-direction, the atoms forming angles of 147.5°. The coordination thus becomes strongly distorted tetrahedral around the oxygen atoms and nearly pyramidal around the lead atoms. The coordination polyhedra of the atoms are thus similar to those in red PbO though less symmetric. The Pb—O lengths in the chains (2.214 Å and 2.223 Å) are roughly the sum of the covalent radii and those between the chains (2.487 Å) approximately the sum of the ionic radii. A sophisticated bond scheme has been proposed by Dickens [239]. He came to the conclusion that interlayer bonding is van der Waals bonding between the filled 6s orbitals of adjacent lead atoms. A fascinating description of the structure is offered by Andersson and Åström [240], who demonstrated that in oxides and oxyfluorides, the inert electron pair on the

TABLE 34
Yellow PbO, orthorhombic, D_{2h}^{11}—Pbcm (No. 57), $Z = 4$
All atoms in 4(d): $\pm(x, y, \frac{1}{4}; \bar{x}, \frac{1}{2}+y, \frac{1}{4})$

β-PbO: $a = 5.876$ Å, $b = 5.476$ Å, $c = 4.743$ Å
   Pb: $x = 0.231$, $y = -0.014$; O: $x = -0.132$, $y = 0.082$ [241]
   $a = 5.891$ Å, $b = 5.489$ Å, $c = 4.775$ Å
   Pb: $x = 0.230\,9$, $y = -0.020\,8$; O: $x = -0.130\,9$, $y = 0.088\,6$ [242]
   27°: $a = 5.891$ Å, $b = 5.489$ Å, $c = 4.755$ Å [1]
for comparison:
α-PbO$_2$: $a = 5.951$ Å, $b = 5.497$ Å, $c = 4.947$ Å (Pcan)

---

cation requires space comparable with that of an anion and that it behaves somewhat like an anion. Thus, lone pairs are found in positions of the lattice which are normally occupied by anions. In the case of yellow PbO, the anions, together with the Pb lone pairs, form a very regular hexagonal close packing [240] as is shown in Figure 54. And it is quite amazing that on replacing the lone pairs by oxygen atoms only small shifts of the Pb atoms are required to obtain the structure of α-PbO$_2$ (the resulting Pb^{4+} is pushed into octahedral coordination by oxygen) [240]. In Table 34 we have added the lattice parameters of α-PbO$_2$ for comparison.

Yellow PbO seems to be the only known representative of this structure type. Isoelectronic TlF has a similar orthorhombic cell, also derived from a rocksalt cell. It is discussed with the Tl$^+$ halides.

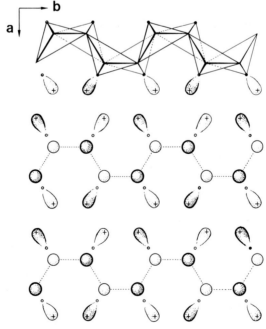

Fig. 54. The hexagonal close-packed arrangement of oxygen atoms plus lone electron pairs of the Pb atoms in yellow PbO.

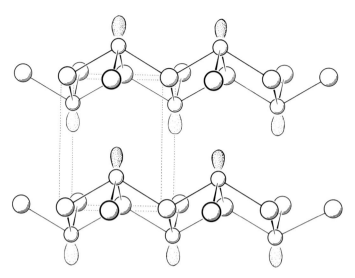

Fig. 55. The tetragonal structure of red PbO (litharge) with the lone electron pair on Pb^{II} indicated.

The isoelectronic ternary analog TlBiO$_2$ (and TlSbO$_2$) might crystallize in a PbO superstructure.

f. Red PbO (litharge)

The tetragonal structure of the mineral litharge or red PbO (Figure 55) belongs to the B10 type. The variable axial ratio and the free parameter of one kind of atom permit this structure to occur with chemically quite unrelated compounds. For $c/a = \sqrt{2}$ and $z = \frac{1}{4}$, the atoms in 2(c) are cubic close-packed with the other atoms in tetrahedral holes. This structure can also be described in terms of a CaF$_2$ lattice with alternate layers of anions missing. In PbO with $c/a = 1.27$, the close packing

TABLE 35
Red PbO, tetragonal, D_{4h}^7—P 4/nmm (No. 129), $Z = 2$.
Pb in 2(c): $\pm(\frac{1}{4}, \frac{1}{4}, z)$
O in 2(a): $\pm(\frac{1}{4}, \frac{3}{4}, 0)$

α-PbO: $a = 3.96$ Å, $c = 5.01$ Å; $z = 0.237$ [243]
$a = 3.964$ Å, $c = 5.008$ Å; $z = 0.238\ 5$ [244]
$a = 3.947$ Å, $c = 4.98\ 8$ Å; $z = 0.233$ [245]

SnO: $a = 3.796$ Å, $c = 4.816$ Å; $z = 0.235\ 6$ [245]
$a = 3.793$ Å, $c = 4.833$ Å [246]
(26 °C): $a = 3.802$ Å, $c = 4.836$ Å [1]

$p \leqslant 40$ kbar: $\dfrac{1}{a_0}\left(\dfrac{\Delta a}{\Delta p}\right) = -3.3 \times 10^{-10}$ kbar^{-1} [247]

$\dfrac{1}{c_0}\left(\dfrac{\Delta c}{\Delta p}\right) = -12.9 \times 10^{-10}$ kbar^{-1}

of the lead atoms is somewhat distorted, the layer character being enforced by the mutual repulsion of the lead atoms. Such an arrangement is quite unusual for a semiconducting compound and it is solely due to the lone electron pair of Pb^{2+}. In normal semiconductors it is the anions that are in a close-packed arrangement and therefore would form the outer layer. Since Pb^{2+} itself carries an inert electron pair it can act somewhat like an anion pushing the cations of the neighboring layers apart. In fact the oxygen atoms together with the lone pairs of Pb^{2+} also form a somewhat distorted hexagonal close-packed [240] arrangement.

In red PbO, the coordinations are much more regular, but otherwise are like those in the yellow modification. Here Pb occupies the apex of a regular tetragonal pyramid with Pb—O = 2.31 Å and O—Pb—O angles of 118.1° (2×) and 77.7° (4×), the former being the angle within the deformed [OPb_4] tetrahedron.

It is not possible to obtain a stable modification of PbO_2 by replacing the inert electron pair by an additional oxygen atom, but if we push the layers somewhat apart and place an additional anion above each lone electron pair, we end up with the layer structure of BiOF or PbFCl [240].

The only compound truly isomorphous with red PbO is the common modification of SnO, crystalline GeO being unknown. X-ray studies of the structure of SnO at room temperature and pressure up to 100 kbar were reported by Vereshchagin et al. [247]. A linear decrease of the lattice constants as well as of the axial ratio c/a was found up to 40 kbar where a first-order phase transition accompanied by a 7% volume decrease was observed. The high-pressure phase has not, as one might perhaps expect, the orthorhombic PbO structure (which appears to be about 2% denser than the tetragonal PbO structure) but is claimed to adopt the hexagonal wurtzite structure. A tetrahedral coordination for a lone pair cation, however, appears to be rather peculiar. The assumption of a second high-pressure phase with a rocksalt structure, on the other hand, seems to be quite plausible.

In the system $x PbO(1-x)Bi_2O_3$ a homogeneous ternary phase with a PbO-like structure exists from $x = 0.50$ to $x = 0.70$ [248]. The phase boundaries correspond to formulae $PbBi_2\square O_4$ and $Pb_7Bi_6\square_3O_{16}$. With increasing Bi_2O_3 content, the c-axis decreases while the a-axis increases linearly so that the cell volume remains constant. The powder diagram of $PbBi_2O_4$ might be indexed by doubling a and c. Simple superstructures can be constructed by eliminating from the Pb_{4n} $(O_2)_{4n}$ Pb_{4n} sandwiches a quarter of the cations from each layer or a half from every second layer and ordering Bi and Pb atoms.

As mentioned above, the PbFCl structure can be constructed by intercalating an anion double layer between each sheet of tetragonal PbO. The layer character is preserved as long as the intercalated sheet of anions is sufficiently corrugated so that each intercalated atom makes bonds only towards one neighboring layer. If only a single anion layer is located between the metal-oxygen sheets, we arrive at a layered three-dimensional structure as is found in $Pb_2O_2CO_3$ or $BaBiO_2Cl$. The same is true if we introduce a symmetrical block such as $M_{3/4}X_3$ in the Sillén phase $Ca_{1.25}Bi_{1.5}O_2Cl_3$.

g. RED SnO

Whereas the stable blue-black SnO is isotypic with the tetragonal red PbO, a red modification of SnO has been reported to be orthorhombic [233]. This second form of SnO, however, is not isostructural with orthorhombic PbO. Its structure derives from that of red PbO by simple distortions and its unit cell is related to the tetragonal cell

$$a_{orth.} = \sqrt{2}a_{tetr.} - \delta_1$$
$$b_{orth.} = \sqrt{2}a_{tetr.} + \delta_2 \quad (\delta_1 \sim \delta_2)$$
$$c_{orth.} = 2c_{tetr.} + \delta_3.$$

The positive value of δ_3 indicates that the empty Pb tetrahedra which separate the layers of occupied [OPb$_4$] tetrahedra are larger, the distance between adjacent Pb layers being 2.82 Å in orthorhombic SnO compared with 2.55 Å in tetragonal SnO. Accordingly, the red modification is roughly 15% less dense than tetragonal SnO. The small displacements of the atoms that transform the tetragonal SnO into the orthorhombic SnO structure are indicated in Figure 56. The atomic parameters are modified in the following way (compare Table 36):

$$Sn_{II}: y = \tfrac{1}{2} - \varepsilon, \quad z = \tfrac{1}{2} - z_{tetr.} - \varepsilon_2 \quad (y_{id.} = \tfrac{1}{2}, \; z_{id.} = \tfrac{1}{4})$$
$$O: x = \tfrac{1}{4} - \varepsilon_3, \quad y = \tfrac{1}{4}, \quad z = \tfrac{1}{4} + \tfrac{1}{2}z_{Sn} \quad (x_{id.} = \tfrac{1}{4}, \; y_{id.} = \tfrac{1}{4}, \; z_{id.} = \tfrac{3}{8})$$

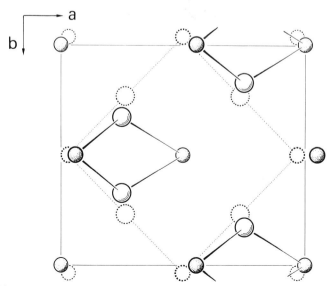

Fig. 56. The relation between the tetragonal (dotted) and the orthorhombic structure of SnO.

TABLE 36
Red SnO, orthorhombic, C_{2v}^{12}—Cmc2$_1$ (No. 36), $Z = 8$.
$(0, 0, 0; \frac{1}{2}, \frac{1}{2}, 0) +$
O in 8(b): $(x, y, z; \bar{x}, y, z; \bar{x}, \bar{y}, \frac{1}{2}+z; x, \bar{y}, \frac{1}{2}+z)$
Sn in 4(a): $(0, y, z; 0, y, \frac{1}{2}+z)$

---

red SnO: $a = 5.00$ Å, $b = 5.72$ Å, $c = 11.12$ Å [223]
Sn$_I$: $y = 0$, $z = 0$; Sn$_{II}$: $y = 0.441$, $z = 0.254$
O: $x = 0.18$ (?), $y = 0.25$, $z = 0.38$

The values added in brackets refer to an ideal cubic close-packing of Sn atoms. The reduction in symmetry is paralled by a reduction of the number of bonded Sn—O contacts and an enhancement of the non-metallic character. The two-dimensional structure thereby transforms into a molecular structure with ring-shaped Sn$_2$O$_2$ molecules. This transition is quite analogous to the sequences from three-dimensional to molecular structures observed in compounds with other lone-pair cations such as Bi → P and Te → S. The four equivalent Sn—O distances of 2.21 Å in tetragonal SnO go over into two sets of pairs in the orthorhombic modification, namely

$$\text{Sn}_I\text{—O} = 2.15 \text{ Å } (2\times) \text{ and } 2.53 \text{ Å } (2\times)$$
$$\text{Sn}_{II}\text{—O} = 1.99 \text{ Å } (2\times) \text{ and } 2.77 \text{ Å } (2\times)$$

At the same time, the angle between Sn and the bonded O is decreased from 74.7° to 49.2° (Sn$_I$) and 53.7°. A reduction of the Sn—O distance by 0.2 Å would well account for a transition from coordination number 4 to one of 2. However, the O—O distance is reduced from 2.68 Å in tetragonal SnO to the unreasonably low value of 1.80 Å in the distorted modification. We therefore think that the positional parameters given for the O atoms are fairly inaccurate and need refinement.

It is said [5] that an orthorhombically deformed structure variant of red PbO exists at room temperature. It is possible that this structure is identical with that of red SnO.

C. VARIOUS COMPOUNDS WITH s^2 CATIONS

a. Na$_2$PbO$_2$, LiBiO$_2$, NaAsO$_2$, . . .

The layer-forming tendency of the s^2 cations leads to two-dimensional structures also among ternary compounds. We have to restrict our selection to a few examples.

We may enhance the insulator character of PbO by combining it with Na$_2$O. One of the possible ternary phases is Na$_2$PbO$_2$. This colorless compound has a relatively simple structure [249] (Figure 57) which consists of layers perpendicular to the [100] direction:

$$\cdots (Na_{1/2}Pb_{1/2})-O-Na-O-(Na_{1/2}Pb_{1/2}) \cdots$$
$$(Na_{1/2}Pb_{1/2})-O-Na-O-(Na_{1/2}Pb_{1/2}) \cdots$$

The idealized structure can be interpreted as a close-packing of oxygen atoms with every third oxygen layer missing. The complete packing can be described by

$\boxed{C}bAcBa\boxed{C}aBcAb\boxed{C}$.

Capitals denote oxygen layers, those at C positions being empty. Cations at c positions are Na only, those at a and b are mixed Na+Pb. The stacking order is thus characterized by $cc\boxed{h}cc\boxed{h}$, the empty h layers being emphasized. Only the Na atoms in the middle of the sandwiches are octahedrally coordinated, the remaining cations can have only half an octahedral surrounding, i.e. a trigonal pyramidal coordination. Between the sandwiches each Pb^{2+} lies opposite a Na^+ ion, the distance being 3.38 Å which is larger than the shortest Na—Pb distance within the sandwich (3.26 Å).

The structure-determining role of the lone electron pair of Pb^{2+} is thus less obvious, but we would expect Na_2SrO_2 to have no layer structure. On the other hand, it would be tempting to replace the outer Na^+ by Tl^+ by forming a compound $NaTlPbO_2$.

A few ternary oxides with Sn^{2+} or Pb^{2+} and alkali and alkaline-earth metals have been reported [250] such as Li_2PbO_2, $Li_2Pb_2O_3$, $Na_2Pb_2O_3$, K_2PbO_2, $K_2Pb_2O_3$, $Rb_2Pb_2O_3$, $CaSnO_2$, $SrSnO_2$, $BaSnO_2$, $SrPbO_2$, $BaPbO_2$, but structural data are in most cases missing. $K_2Pb_2O_3$ is reported to be cubic, hence is not a

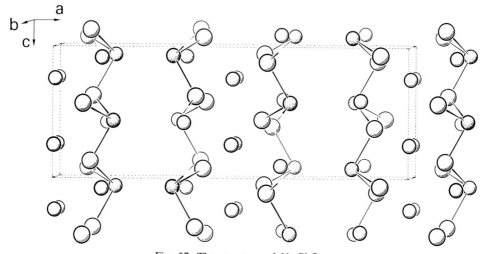

Fig. 57. The structure of Na_2PbO_2.
Large spheres: O
dotted spheres: Pb

TABLE 37

Na_2PbO_2 structure, orthorhombic, D_{2h}^{14}—Pbcn (No. 60), $Z=8$.
Na_{III}, Pb, O_I and O_{II} in 8(d): $\pm(x, y, z; \frac{1}{2}-x, \frac{1}{2}-y, \frac{1}{2}+z; \frac{1}{2}+x, \frac{1}{2}-y, \bar{z}; \bar{x}, y, \frac{1}{2}-z)$
Na_I and Na_{II} in 4(c): $\pm(0, y, \frac{1}{4}; \frac{1}{2}, \frac{1}{2}+y, \frac{1}{4})$

Na_2PbO_2: $a=16.83$ Å, $b=6.939$ Å, $c=5.882$ Å [249]

	x	y	z
Na_I		0.17	
Na_{II}		0.67	
Na_{III}	0.364	0.360	0.083
Pb	0.1634	0.3649	0.0671
O_I	0.100	0.384	0.380
O_{II}	0.418	0.341	0.433

layer compound. $Rb_2Pb_2O_3$ appears to be related though less symmetric. From its X-ray pattern Li_2PbO_2 is seen to have a structure which is different from that of Na_2PbO_2.

The structure of Ag_2PbO_2 [251] bears some relation to that of red PbO. The coordination figure of the Pb atom is of the same type as in tetragonal PbO, but somewhat distorted and instead of forming a chess-board like array, the [PbO_4] pyramids are connected in chains. These $(PbO_2)_\infty$ chains are linked by Ag atoms to a three-dimensional network.

The monoclinic structure of quenselite $PbMnO_2(OH)$ is characterized by the superposition of sheets in the sequence —Mn—O—Pb—OH—Pb—O—Mn—. Mn^{3+} is octahedrally surrounded by oxygen. Each Pb^{2+} ion is in contact with 1 O at 3.09 Å and 2 OH at 2.93 Å forming a very flat pyramid. The (MnO_2) layers are necessarily linked via the PbOH sheets. A layer structure would require sheets of composition —Mn—(O, OH)—Pb—O—Pb—(O, OH)—Mn—.

The so-called lead sesquioxide Pb_2O_3, which in fact is a double oxide $Pb^{II}Pb^{IV}O_3$, has a layered monoclinic structure [252]. Edge- and corner-sharing [$Pb^{IV}O_6$] octahedra form layers parallel with (001). The Pb^{II} atoms are arranged in double layers between these octahedron layers displaying their characteristic short and longer bonds: Pb^{II}—O = 2.31, 2.43, 2.44, 2.64, 2.91 and 3.00 Å. While the two shortest bonds and the medium bond are directed towards the adjacent $(PbO_6)_\infty$ layer, the third bond couples the opposite octahedron layer. Thus, a three-dimensional structure results whereas without this trans-layer bond the lone s^2 pair could help to separate the Pb^{II} double layer in order to create a cleavage plane between the two Pb^{II} layers. A layer structure possibly occurs in metastable Sn_2O_3 formed on disproportionating SnO at temperatures above 300 °C. To our knowledge, a structure determination is still lacking but this oxide was always obtained in the form of thin single crystal platelets. Lawson [253] reported for this phase (which he assumed to be Sn_3O_4) a triclinic cell ($a=4.86(16)$, $b=5.88(14)$, $c=8.20(17)$ Å,: $\alpha=93.0(3)°$, $\beta=93.35(35)°$, $\gamma=91.0(4)°$). A different triclinic cell was found by Murken and Trömel [254] ($a=5.457(4)$, $b=8.179(6)$,

$c = 3.714(3)$ Å, $\alpha = 93.8(2)°$, $\beta = 92.3(2)°$, $\gamma \approx 90.0(2)°$; $Z = 2$). Of course the crystal habit does not rule out the presence of a layered three-dimensional structure similar to Pb_2O_3 and Murken and Trömel [254] indeed point to a certain similarity of the triclinic cell of Sn_2O_3 with the monoclinic cell of Pb_2O_3.

Although the mineral minium, $Pb_3O_4 = Pb_2^{II}Pb^{IV}O_4$ has a higher concentration of s^2 cations, its tetragonal structure [255] is not even layered. It is made up of discrete $(PbO_2O_{2/2}O_{2/2})$ chains along c, held together by the divalent cations in trigonal-pyramid coordination. As this coordination is typical for s^2 cations, various other ternary compounds are found to crystallize with this structure, such as Pb_2SnO_4, As_2NiO_4 (magnetic susceptibility see [256]) and $Sb_2M^{II}O_4$, with M^{II} = Mn, Fe, Co, Ni, Cu, Zn, Mg, whereas the metastable Bi_2MnO_4 and Bi_2NiO_4 are cubic and $CuBi_2O_4$ is tetragonal [257].

Diluted versions of PbO with respect to the s^2 cation are the ternary oxides $ASbO_2$ and $ABiO_2$ with A representing an alkali metal. While the structures of these phases all show a low symmetry around the s^2 cation, only $LiSbO_2$ and $LiBiO_2$ [258] really have layer structures. The orthorhombic structure of $LiBiO_2$ (Figure 58 and Table 38) can be derived from a rocksalt structure 'BiO' by removing the cations from every third and fourth Ω-layer parallel to (100) and filling all the tetrahedral holes τ between these emptied double layers with Li atoms. The resulting distortion of the structure is such that the Bi—O distance towards the O atom of the τ sandwich becomes shorter while the opposite Bi—O distance is 1 Å larger and thus will no longer represent a bond (Bi—O = 2.20, 2.40, 2.49 (2×), 2.79 and 3.19 Å). The structure therefore is layered perpendicular to [100].

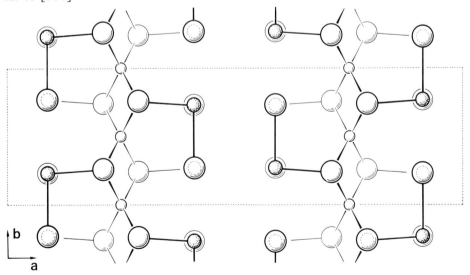

Fig. 58. The structure of $LiBiO_2$.
Large spheres: O
smallest spheres: Li
dotted spheres: Bi

TABLE 38
LiBiO$_2$ structure, orthorhombic, D_{2h}^{26}—Ibam (No. 72), $Z = 8$.

$(0, 0, 0, \frac{1}{2}, \frac{1}{2}, \frac{1}{2})+$

Bi, O$_I$ and O$_{II}$ in 8(j): $\pm(x, y, 0; x, \bar{y}, \frac{1}{2})$

Li in 8(f): $\pm(x, 0, \frac{1}{4}; x, 0, \frac{3}{4})$

LiBiO$_2$: $a = 17.978$ Å, $b = 5.189$ Å, $c = 4.978$ Å [258]

Bi: $x = 0.0893$, $y = 0.2311$; O$_I$: $x = 0.211$, $y = 0.261$
O$_{II}$: $x = 0.088$, $y = -0.231$; Li: $x = 0.25$

LiSbO$_2$: ?

NaBiO$_2$ [259] as well as KBiO$_2$ and the isotypic RbBiO$_2$ and CsBiO$_2$ [260] crystallize in different monoclinic structures which represent some kind of three-dimensional chain structure with Bi nearly at right angles with 2 O neighbors (NaBiO$_2$: Bi—O = 2.04 (2×), 2.39 (2×) and 3.15 Å (2×); KBiO$_2$: Bi—O = 2.14 (2×), 2.42 (2×) and 3.45 Å (2×)).

The orthorhombic structure of NaAsO$_2$ is a layered chain structure [261] (Figure 59 and Table 39). Each As atom has the characteristic trigonal-pyramid coordination due to the p^3 bonds. [AsO$_3$] groups are linked together in zigzag chains by sharing two of the three O atoms. These chains run parallel with the c-axis and are bound together in two sets by the Na$^+$ ions. From the reported parameters, we calculate the following distances: As—O = 1.60 Å (to the lone O atom), 1.81 and 1.95 Å with an angle of 92.9° between the latter two and 97.6°

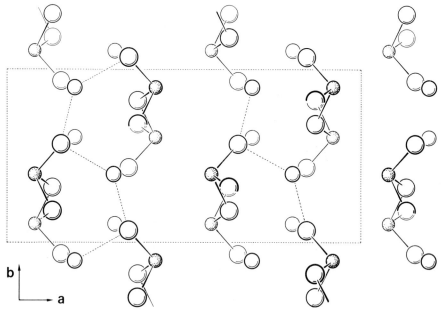

Fig. 59. The orthorhombic structure of NaAsO$_2$. The As—O bonds are all shown while only certain Na—O contacts are indicated by dotted lines. Stippled spheres: As, largest spheres: O.

TABLE 39

$NaAsO_2$ structure, orthorhombic, D_{2h}^{15}—Pbca (No. 61), $Z=8$.
All atoms in 8(c): $\pm(x, y, z; \frac{1}{2}+x, \frac{1}{2}-y, \bar{z}; \bar{x}, \frac{1}{2}+y, \frac{1}{2}-z; \frac{1}{2}-x, \bar{y}, \frac{1}{2}+z)$

$NaAsO_2$: $a = 14.314$ Å, $b = 6.779$ Å, $c = 5.086$ Å [261]

	x	y	z
Na	0.306	0.102	0.250
As	0.080	0.103	0.250
O_I	0.127	0.322	0.405
O_{II}	0.154	0.568	0.810

and 105.4° to the lone O atom. The resulting coordination of Na is strongly distorted tetrahedral (Na—O = 2.26, 2.32, 2.33, 2.48, 2.91, 3.07 and 3.15 Å) which is rather unusual for an Na^+ ion. The positional parameters obviously need refinement. The gross features of the structure, however, are believed to be correct since the layer character of the structure is consistent with the observed thin cleavage flakes.

Another possibility of s^2-diluting substitution in PbO (besides hypothetical $MPbO_2$ with M = Be, Mg, Ca, Sr, Ba, Eu, Zn, Cd, Hg) is realized in the Tl^+ phases $TlM^{3+}O_2$ with M^{3+} = Al, Ga, In, Cr, Fe, etc. Two hexagonal modifications are reported for $TlAlO_2$, $TlGaO_2$ and $TlFeO_2$ [262, 263, 264]. In the rhombohedral modification $TlAlO_2$(h) honeycomb layers of up- and down-pointing $[AlO_4]$ tetrahedra are connected via the free corners to a three-dimensional network with large holes. Six Tl ions are inserted in each of these holes in a layered fashion so that each Tl^+ gets 3 + 3 O neighbors (at 3.19 and 3.22 Å in β-$TlAlO_2$, at 3.24 and 3.29 Å in $TlFeO_2$).

Sulfides $TlM^{3+}S_2$ are reported for M = Al, Ga, Cr, Fe, Co [265] but none of them appears to crystallize in a true two-dimensional structure. The monoclinic $TlFeS_2$ (mineral raguinite) and $TlFeSe_2$ [266] are fibrous and might be similar to monoclinic $KFeS_2$ which is built up of chains of $(FeS_{2/2}S_{2/2})$ tetrahedra.

b. LANARKITE $Pb_2O(SO_4)$

The so-called basic lead sulfate $Pb_2O(SO_4)$ occurs in nature as the rare mineral lanarkite. The faint-yellow laminar crystals easily exfoliate along the planes $(\bar{2}01)$ indicated in Figure 60. Each Pb atom is located at the vertex of a distorted trigonal pyramid, characteristically with two short Pb—O distances of 2.30 Å and a longer one of 2.54 Å, reflecting the original $s^2p_x^1p_y^1p_z^0$ configuration of the cation. The coordination of Pb is completed by two additional O atoms at 2.80 Å from the neighboring layer. The sulfate complex is tetrahedral, $[SO_2O_{2/2}]$. There are three kinds of oxygen atoms. Those in 4(g) are tetrahedrally coordinated by Pb. The chains formed of these $[OPb_{4/2}]$ tetrahedra are linked to the $[SO_{2/2}O_2]$ tetrahedra by the O atoms in positions (4i). The O atoms in the eightfold positions make only one bond to the S atoms and thus carry formally one charge. It is these O atoms that are closest to the Pb atoms of the adjacent layers.

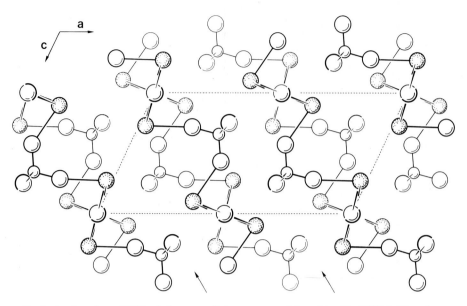

Fig. 60. Projection along [010] of the monoclinic structure of lanarkite Pb$_2$O(SO$_4$). The cleavage planes ($\bar{2}$01) are indicated by arrows. Dotted spheres: Pb, large spheres: O, small spheres: S.

Pb$_2$O(SO$_4$) retains this structure up to its melting point of 975 °C. The lanarkite structure is obviously adopted also by lead selenate, chromate and molybdate. Nothing is known of corresponding Sn^{2+} salts. Isoelectronic Bi$_2$O(SiO$_4$) and

TABLE 40
Lanarkite structure, monoclinic C$_{2h}^3$—C 2/m (No. 12), Z = 4.

$(0, 0, 0; \frac{1}{2}, \frac{1}{2}, 0) +$
O$_{IV}$ in 8(j): $\pm(x, y, z; x, \bar{y}, z)$
Pb$_I$, Pb$_{II}$, S, O$_{II}$ and O$_{III}$ in 4(i): $\pm(x, 0, z)$
O$_I$ in 4(g): $\pm(0, y, 0)$

Pb$_2$O(SO$_4$): $a = 13.769$ Å, $b = 5.698$ Å, $c = 7.079$ Å, $\beta = 115°56'$

	x	y	z	
Pb$_I$	0.146 6		0.105 0	
Pb$_{II}$	0.525 8		0.271 5	[266a]
S	0.829		0.344	
O$_I$		0.246		
O$_{II}$	0.111		0.429	
O$_{III}$	0.713		0.284	
O$_{IV}$	0.355	0.711	0.251	

Pb$_2$O(SeO$_4$): $a = 13.94$ Å, $b = 5.78$ Å, $c = 7.25$ Å, $\beta = 115.9°$ [1]

Pb$_2$O(CrO$_4$): $a = 13.80$ Å, $b = 5.70$ Å, $c = 7.10$ Å, $\beta = 114.1°$ [266b]

Pb$_2$O(MoO$_4$): $a = 14.225$ Å, $b = 5.789$ Å, $c = 7.336$ Å, $\beta = 114.0°$ [266c]

$Bi_2O(GeO_4)$ might be other representatives of this structure family, as well as $PbBiO(PO_4)$, $PbBiO(VO_4)$, $TlPbO(MnO_4)$, $TlBiO(SO_4)$, etc., where ordering of the cations may lead to superstructures.

$Pb_5O_4(SO_4)$ ($=4PbO \cdot PbSO_4$) appears to have a structure related to red PbO.

c. $Pb_3(PO_4)_2(r)$

The room-temperature structure of lead phosphate is a distorted version of the layered structure adopted by $Sr_3(PO_4)_2$, $Ba_3(PO_4)_2$, $Ba_3(VO_4)_2$, $Sr_3(CrO_4)_2$, etc. [267, 268]. $Pb_3(PO_4)_2$ itself adopts this more symmetrical structure above 172°C. The high-temperature form consists of sandwiches composed of [PO$_4$] tetrahedra which all point towards the interior. These sandwiches are stacked in a cubic-close-packed manner. The Pb atoms are inserted in the holes between these [PO$_4$] tetrahedra in such a way that we may describe a sandwich in terms of close-packed layers of composition (3 O + 1 Pb$_{II}$) ——(2 O + 1 Pb$_I$)——(3 O + 1 Pb$_{II}$) reminiscent of the mixed close packing of (CaO$_3$) or (KF$_3$) layers in CaTiO$_3$ and KNiF$_3$ and other perovskite-related compounds. The lead atoms themselves are in a close-packed arrangement *hch* or

```
    C A B        A B C        B C A        C A B
    :←—— monoclinic cell ——→:
    :←———————— hexagonal cell ————————→:
```

The low-temperature modification, which is a ferroelastic [269], is characterized by a lowering of the coordination number of the Pb atoms. We may describe the hexagonal high-temperature modification (at 200°C) [270] with the monoclinic cell of the low-temperature modification to show the relation between both

$$a'_{mon} = \left(\frac{a_h^2}{3} + \frac{4}{9}c_h^2\right)^{1/2} = 13.91 \text{ Å} \quad (13.816 \text{ Å})$$

$$b'_{mon} = a_h = 5.53 \text{ Å} \quad (5.692 \text{ Å})$$

$$c'_{mon} = \sqrt{3} a_h = 9.58 \text{ Å} \quad (9.429 \text{ Å})$$

$$\beta' = \text{arc tg}\left(\frac{2}{\sqrt{3}} \frac{c_h}{a_h}\right) = 103.3° \quad (102.36°)$$

The parameters of the room-temperature modification [270] are added in brackets for comparison. As can be seen, the cell of $Pb_3(PO_4)_2$ does not change significantly at the first-order transition [269a, 269b] but the equidistance of the layers gets lost. In the low-temperature form, the [PO$_4$] tetrahedra come closer together while the Pb^{II} atoms move out of the outer (O$_3$Pb) planes. This gives rise to an approximately trigonal pyramidal coordination of Pb$_{II}$ (Figure 61). Each

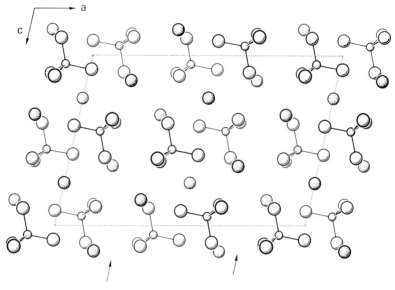

Fig. 61. Projection along [010] of the monoclinic structure of $Pb_3(PO_4)_2(r)$. The arrows indicate the cleavage planes. Stippled spheres: Pb, smallest spheres: P.

Pb_{II} atom has $1 O_{IV}$ neighbor at 2.33 Å, $1 O_{III}$ at 2.45 Å, $1 O_{II}$ at 2.50 Å, $1 O_I$ of the neighboring sandwich at 2.73 Å, $1 O_{II}$ at 2.81 Å, etc. The shortest Pb—O distance compares well with that in red PbO (2.31 Å). As is proved by the mica-like appearance of the crystals, the distance to the opposite oxygen layer is obviously too large to represent any significant bonding. The ionically bonded Pb_I within the O_6 octahedra has $2 O_I$ at 2.60 Å, $2 O_{III}$ at 2.63 Å and $2 O_{II}$ at 2.75 Å. It might well be replaced by, say, Ca or Sr ($\rightarrow CaPb_2(PO_4)_2$, etc., $SrBi_2(SiO_4)_2$, $Tl_2Sr(SO_4)_2 \ldots$,). These compositions correspond to palmierite $K_2Pb(SO_4)_2$

TABLE 41

$Pb_3(PO_4)_2(r)$ structure, monoclinic, C_{2h}^6—C 2/c (No. 15), $Z = 8$.
$(0, 0, 0; \frac{1}{2}, \frac{1}{2}, 0) +$

Pb in 4(e): $\pm(0, y, \frac{1}{4})$
other atoms in 8(f): $\pm(x, y, z; \bar{x}, y, \frac{1}{2} - z)$

$Pb_3(PO_4)_2(r)$ (24°C): $a = 13.816$ Å, $b = 5.692$ Å, $c = 9.429$ Å, $\beta = 102.36°$ [270]
(25°C): $a = 13.808$ Å, $b = 5.688$ Å, $c = 9.432$ Å, $\beta = 102.39°$ [269]

	x	y	z	
Pb_I		0.287 2		
Pb_{II}	0.318 4	0.310 4	0.351 7	
P	0.398 1	0.245 5	0.054 0	
O_I	0.350	0.029	0.110	[270]
O_{II}	0.359	0.466	0.123	
O_{III}	0.140	0.227	0.104	
O_{IV}	0.512	0.234	0.086	

whose structure represents an ordered version of the $Ba_3(PO_4)_2$ type [271]. Superstructures of the type $TlBiSr(PO_4)_2$ with Tl^+ and Bi^{3+} ordered in the outer layers might also be feasible. The distortion of the $Sr_3(PO_4)_2$ structure met in $Pb_3(PO_4)_2$ is obviously due to the s^2 lone electron pair of Pb^{2+} and we wonder whether the palmierite-type compounds $Tl_2Pb(SO_4)_2$, $Tl_2Pb(CrO_4)_2$, $Tl_2Sr(CrO_4)_2$ and $Tl_2Ba(CrO_4)_2$ [268] do not in fact (at least at low temperatures) crystallize in the $Pb_3(PO_4)_2(r)$ structure. This is not the case, however, for $Pb_3(VO_4)_2$ though its rhombohedral high-temperature structure [272] transforms at 100°C into a monoclinic structure [272a] with a unit cell ($a = 13.97$ Å, $b = 5.89$ Å, $c = 9.39$ Å, $\beta = 101°$) close to that of the phosphate. This phase is antiferroelectric and ferroelastic [272b] and the phase diagram of solid solutions $Pb_3(V_{1-x}P_xO_4)_2$ demonstrates that there is no continuity between the two end members [272c]. On both sides the phase-transition temperature decreases sharply. 25% of $Pb_3(VO_4)_2$ shift the transition to below 0°C and when x is between 0.35 and 0.60 the transition temperatures lie below the liquid-nitrogen temperature.

The corresponding arsenate $Pb_3(AsO_4)_2$ was reported to be tetragonal. Mixing of the phosphate or the arsenate with $BiPO_4$, $BiAsO_4$ or $BiVO_4$ leads to phases of the type $BiPb_3(PO_4)_3$, $BiPb_3(AsO_4)(PO_4)_2$, ..., $BiPb_3(AsO_4)_3$, etc., which are no longer layered but have the cubic eulytine structure of $BiPb_3(VO_4)_3$ and $Pb_4(PO_4)_2SO_4$.

From their tendency to cleave into thin sheets, we conclude that monoclinic $Pb_8P_2O_{13}$ [273] and triclinic $Pb_2P_2O_7$ [274] crystallize also in layer structures in contrast to $Pb_4P_2O_9$ and $Pb_5P_4O_{15}$ [275].

Three-dimensional apatites $Pb_{10}(MO_4)_6O$ or $Pb_5(MO_4)_3X$ form with $M = $ P, As, V and $X = $ F, Cl, Br, I, OH.

d. $SbPO_4$ AND RELATED PHASES

In the monoclinic structure of $SbPO_4$ [276] (Table 42) tetrahedra of the complex anion $(PO_4)^{3-}$ are linked by nearly rectangular O—Sb—O bonds to form chains along the c-axis. These chains are connected by almost linear O—Sb—O bonds parallel with the b-axis, thus building corrugated layers parallel with the bc plane. As in most other Sb^{III} oxygen compounds, antimony has a unilateral coordination [277] and forms the outermost part of the layers. The Sb—O distances correspond to a p^2 hybrid bond (1.98 and 2.04 Å at an angle of 88°) and roughly two p or (p, d) half bonds in the third direction (2.18 Å and 165°; 84° and 85° relative to the two single bonds). As is evident from Figure 62, the interlayer distances are significantly larger. Together with the lone electron pair, Sb has the characteristic trigonal-bipyramid coordination (Sb—O_4e) met also in other Sb^{3+} and Bi^{3+} compounds as e.g. in SbOF(r) and in tetragonal β-Bi_2O_3 [278].

$SbPO_4$ (Fig. 62) is the only representative of this structure type known up to now. $AsPO_4$ [279] and As_2O_4 which might be expected to adopt this structure, are reported to crystallize in the same orthorhombic unknown structure. A similar bonding could also be expected in the chemically related compounds Sb_2O_4

TABLE 42
$SbPO_4$ structure, monoclinic, C_{2h}^2—$P2_1/m$ (No. 11), $Z=2$.

Sb, P, O_I and O_{II} in 2(e): $\pm(x, \frac{1}{4}, z)$

O_{III} in 4(f): $\pm(x, y, z; x, \frac{1}{2}-y, z)$

$SbPO_4$: $a = 5.0868$ Å, $b = 6.7547$ Å, $c = 4.7247$ Å, $\beta = 94.66°$ [276]

	x	y	z
Sb	0.1802		0.2053
P	0.6110		0.7223
O_I	0.3346		0.8332
O_{II}	0.5546		0.3933
O_{III}	0.7692	0.0700	0.8183

($=Sb^{III}Sb^{V}O_4$), $SbVO_4$, $SbNbO_4$, $SbTaO_4$ and $BiPO_4$, $BiAsO_4$, $BiSbO_4$, $BiVO_4$, ... and perhaps even in $SnSO_4$, $PbSO_4$, ..., $SnCrO_4$, $TlMnO_4$, etc. The ferroelectric [281, 282] compounds α-Sb_2O_4 (servantite), $SbNbO_4$, $SbTaO_4$, $BiNbO_4(r)$ and $BiTaO_4(r)$ [283] crystallize in the orthorhombic stibiotantalite structure which is a framework structure with octahedral [NbO_6] and one-sided fourfold Sb—O coordination. The monoclinic structure of $BiSbO_4$ [283] and of isomorphous β-Sb_2O_4 [284, 285] is closely related to that of $SbTaO_4$. Both the $SbTaO_4$ and the $BiSbO_4$ structures can be described in terms of layers of oxygen octahedrons parallel with (001) with alternating orientations, representing (100)

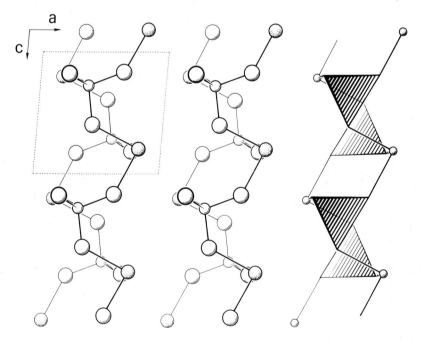

Fig. 62. The structure of $SbPO_4$.
Sb: dotted spheres.

sections of a hcp array. Half the octahedra are occupied by Sb^V or Ta^V in a chessboard-like manner. The fourfold coordination of Bi^{III} in $BiSbO_4$ and of Sb^{III} in $SbTaO_4$ is very similar to that of Sb^{III} in $SbPO_4$. However, in both $SbTaO_4$ and $BiSbO_4$, the s^2 cations are bound to oxygen atoms of both neighboring layers so that we end up with a layered three-dimensional structure.

In the triclinic antiferroelectric and ferroelastic high-temperature modifications β-$BiNbO_4$ and β-$BiTaO_4$ [286] puckered sheets arise from joining the ($MO_2O_{4/4}$) octahedra by the corners with the two free vertices adjacent (cis), as is illustrated in Figure 10c. This unusual feature distinguishes this structure from that of α-$BiNbO_4$ and β-Sb_2O_4 where the free vertices of the ($MO_2O_{4/4}$) octahedra are opposite (trans). Again the Bi atoms, which are inserted in the large holes between the zigzag sheets, extend bonds towards both sides and thus prevent the formation of a two-dimensional structure.

The room-temperature modification of $SnWO_4$ [287] crystallizes in the layered higher-symmetry structure of $SbNbO_4(h)$ and $SbTaO_4(h)$, stable in the latter compounds above their ferroelectric Curie point of 680 K. It is closely related to the tetragonal structure of $TlAlF_4$. The high-temperature modification $SnWO_4(h)$ is cubic with tungsten in tetrahedral coordination [286]. A tetrahedral complex anion is common also for the remaining compounds. $BiPO_4$ is known to have a monazite ($CePO_4$)-type low-temperature modification and a high-temperature form which is also monoclinic and similar to monazite [288]. Both structures are three-dimensional. $BiAsO_4$ too has two modifications, one with a monazite structure and a second with the tetragonal scheelite ($CaWO_4$) structure. In the monazite-type phases, Bi has a coordination number of (4+5), while in tetragonal $BiAsO_4$, Bi is in contact with eight separate $[PO_4]$ tetrahedra, Bi—O = 2.49 Å (4×) and 2.58 Å (4×). The monazite structure is met also with $PbSeO_4(r)$, $PbCrO_4$ (krokoite), $PbMoO_4$ and $PbWO_4$ while the scheelite structure, a modified version of the zircon ($ZrSiO_4$) structure, is adopted by $PbMoO_4$ (wulfenite), $PbWO_4$ (stolzite) and $TlReO_4$. Raspite represents a monoclinic modification of $PbWO_4$. It is pseudotetragonal and similar to stolzite and has perfect (100) cleavage. $PbSO_4$, $PbSeO_4(h)$, an unstable modification of $PbCrO_4$, $PbBeF_4$, $TlBF_4$ and $TlClO_4(r)$ crystallize in the orthorhombic baryte ($BaSO_4$) structure which is almost a layer structure with perfect cleavage on (001) and (210). $SnSO_4$ has a closely related structure [289] but with the trigonal-pyramidal environment of Sn^{2+} pronounced. Another case of a nearly two-dimensional structure is realized in orthorhombic $BiVO_4$ (pucherite) [290]. Zigzag rows of distorted $[VO_4]$ tetrahedra (V—O = 1.76 Å (2), 1.94 Å (2) and 2.69 Å (2)) run along the a-axis. The holes between these rows are filled with Bi^{3+} ions of coordination number 6 to 8 (Bi—O = 2.19 Å (2), 2.31 Å (2), 2.53 Å (2) and 2.73 Å (2)). This leads to a quasilayered structure with perfect (001) cleavage. For $BiVO_4$, two other structures have also been reported [291]. At low temperatures, a zircon-type phase was synthesized, which transformed above 400 °C into a β-fergusonite-type modification, a monoclinic variant of the zircon and scheelite structures.

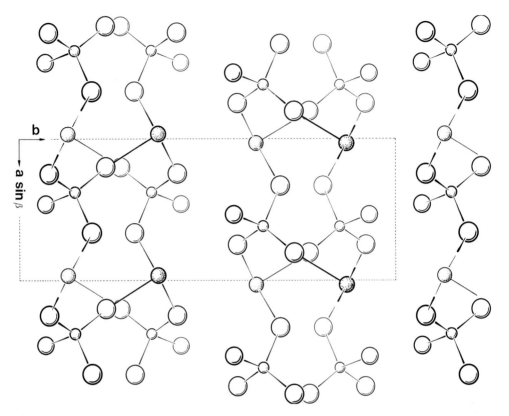

Fig. 63. The structure of SnPO$_3$F projected along c. Stippled spheres: smallest spheres: P, largest spheres: O.

The monoclinic structure of SnPO$_3$F [292] is reminiscent of the SbTaO$_4$ structure with the difference that the complex anion is now forming layers of isolated tetrahedra parallel with the (100) plane. Again these layers are linked by the cations, Sn being bound to 1 O at 2.15 Å on either side and to 2 O at 2.10 and 2.21 Å on the other side. A similar arrangement is found in SnHPO$_4$ (= SnPO$_3$OH) [293]. In SnHPO$_4$, the O neighbors of Sn^{2+} in one phosphate layer are at 2.12, 2.63 and 2.68 Å, those of the opposite layer at 2.27 and 2.38 Å. The bonds from Sn^{2+} are essentially directed towards the axial glide planes at $y = \pm\frac{1}{4}$. Therefore, double layers (SnPO$_3$F)$_\infty$ and (SnHPO$_4$)$_\infty$ can be recognized in Figures 63 and 64. The F$^-$ ions are not bound to Sn^{2+} and are located outside. In SnHPO$_4$, OH ions obviously substitute for F and these OH ions now give rise to H bonds to the oxygen neighbor of the opposite double layer at 2.50 Å. Nevertheless, the (001) planes midway between the glide planes are the cleavage planes found also in SnHPO$_4$.

It should be possible to synthesize the isoelectronic analogs In^{I}SO$_3$F, Sb^{III}SiO$_3$F and In^{I}HSO$_4$ with the same structure. On replacing Sn^{2+} by the larger Pb^{2+} ion, however, the structure will probably be modified though it may still be related.

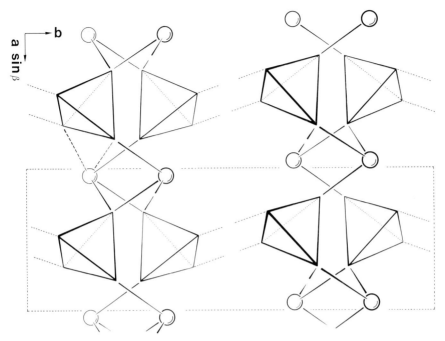

Fig. 64. The structure of SnHPO$_4$ projected along c. Sn spheres and [PO$_4$] tetrahedra are shown. Hydrogen bonds are indicated by dotted lines.

Thus, if we take twice the unit cells reported for PbHPO$_4$ and PbHAsO$_4$ (the mineral schultenite) the cells compare well with those of SnHPO$_4$ (see Table 43). A layering normal to the b-axis can be deduced from the extraordinarily good cleavage on (010). The lead salts, however, appear to be more dense, since the calculated value for the cell volume of PbHPO$_4$ is even 0.6% smaller than that of the Sn analog which points to a higher coordination number of the Pb atom.

TeSiO$_4$ might be another analog of the layer-type SbPO$_4$. Structure determinations on Te^{4+} salts are rather rare, however. An appropriate example is TeVO$_4$ which exists in two monoclinic modifications. Both black TeVO$_4$(r) and green TeVO$_4$(h) stable above 650°C, possess layered structures [294]. In TeVO$_4$(r) distorted (V^{IV}O$_2$O$_{2/2}$O$_{2/2}$) octahedra share edges to form isolated chains parallel with the b-axis. These chains are arranged in layers parallel with (001). The Te^{IV} atoms are inserted between these sheets of octahedron chains and connect them by forming trigonal-pyramid bonds to three different [VO$_6$] octahedra.

In TeVO$_4$(h), puckered layers of distorted corner-sharing [VO$_6$] octahedra parallel with (010) are again linked by Te—O bonds.

One might expect less distorted octahedra on replacing the unfavorable V^{4+} ion by a symmetrical ion such as Ti^{4+}. Maybe papers on TeTiO$_4$, TiZrO$_4$, etc. escaped our attention, but we are aware only of the cubic phases MTe$_3$O$_8$ with M = Ti, Zr, Hf, Sn [295].

TABLE 43
$SnPO_3F$ structure, monoclinic, C_{2h}^5—$P2_1/c$ (No. 14), $Z = 4$.
All atoms in 4(e): $\pm(x, y, z; x, \frac{1}{2}-y, \frac{1}{2}+z)$

$SnPO_3F$: $a = 4.621$ Å, $b = 12.644$ Å, $c = 6.194$ Å, $\beta = 99.3°$ [292]

	x	y	z
Sn	0.034	0.6299	0.2803
P	0.380	0.3517	0.336
O_I	0.250	0.421	0.484
O_{II}	0.205	0.270	0.202
O_{III}	0.665	0.306	0.450
F	0.461	0.431	0.159

and closely related:

$SnPO_3(OH)$: $a = 4.576$ Å, $b = 13.548$ Å, $c = 5.785$ Å, $\beta = 98°41'$ [293]

	x	y	z
Sn	0.0517	0.1571	0.1100
P	0.571	0.150	0.565
O_I	0.306	0.212	0.453
O_{II}	0.695	0.091	0.392
O_{III}	0.795	0.221	0.689
OH	0.472	0.077	0.748

for comparison:
$PbHPO_4$: $a' = 4.66$ Å, $b' = 2 \times 6.64$ Å, $c' = 5.77$ Å, $\beta = 97.2°$ [5]
$PbHAsO_4$: $a' = 4.85$ Å, $b' = 2 \times 6.76$ Å, $c' = 5.83$ Å, $\beta = 95.5°$ [5]

e. Tellurite β-TeO_2

The tellurium atom has two singly occupied p levels which can be used for the formation of two covalent p bonds at right angles to each other. The p^2 bonds will lie in a plane with the lone s^2 electron pair and depending upon the degree of admixture of sp^2 character, the X—Te—X angle may increase from 90° towards the 120° of pure sp^2 hybrid bonds. In tetravalent Te the additional two bonds will be either ionic bonds or covalent dp bonds normal to the plane, or a mixture of both. Such an asymmetric 4-coordination is indeed found in both modifications of TeO_2. In tetragonal paratellurite [296], the stable [297] rutile-derived α-TeO_2, the coordination figures are connected via corners only. In orthorhombic tellurite β-TeO_2 [298], however, these [TeO_4] units are joined to Te_2O_6 units and these in turn are connected to yield wavy layers parallel with the (b,c)-plane with the lone electron pairs pointing towards the neighboring layers (Figure 65). Each oxygen atom within the Te_2O_6 unit has 2Te neighbors at an angle of 102.0° while the oxygen atoms which connect the Te_2O_6 units have 2Te neighbors at an angle of 137.6°. The Te—O distances are 1.88 and 1.93 Å within the equatorial plane of the (TeO_4e) trigonal bipyramid (O—Te—O 101°) and 2.07 and 2.19 Å towards its apices. The difference between the latter two bonds reflects the different ionicity and the influence of the Te—Te repulsion (Te—Te = 3.17 Å). The low

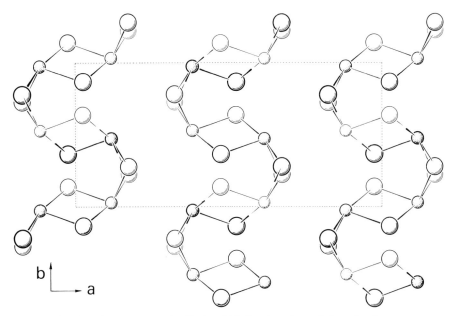

Fig. 65. The structure of tellurite β-TeO₂ viewed parallel to the layers.
Te: dotted spheres.

Te—Te distance is obviously also the reason for the lower energy gap of yellow tellurite compared with colorless paratellurite [296] where Te—Te = 3.74 Å.

TABLE 44
Tellurite β-TeO_2, orthorhombic, D_{2h}^{15}—Pbca (No. 61), $Z=8$.
All atoms in 8(c): $\pm(x, y, z; \frac{1}{2}+x, \frac{1}{2}-y, \bar{z}; \bar{x}, \frac{1}{2}+y, \frac{1}{2}-z; \frac{1}{2}-x, \bar{y}, \frac{1}{2}+z)$

β-TeO_2: $a = 12.035$ Å, $b = 5.464$ Å, $c = 5.607$ Å

	x	y	z	
Te	0.118 2	0.025 5	0.378 1	[298]
O$_I$	0.027	0.637	0.178	
O$_{II}$	0.174	0.222	0.086	

f. $2\,TeO_2\cdot HNO_3$

The orthorhombic structure of basic tellurium nitrate [299] is not simply a β-TeO_2 intercalation variant but contains different structural elements (Figure 66). Te_2O_6 units are linked to form a layered network with corrugated ($Te_6O_8O_{4/2}$) meshes normal to the a-axis. The nitrate groups are intercalated and attached alternately to one TeO₂ layer obviously by hydrogen bonds. All distances between atoms of adjacent layers are larger than the normal van der Waals distances. Again the coordination around Te is fourfold and together with the lone s^2 pair of Te^{IV} a trigonal bipyramid (TeO₄e) can be defined as coordination

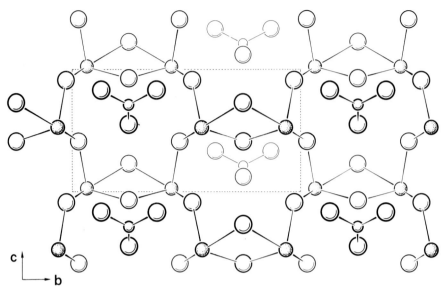

Fig. 66. The structure of 2TeO$_2$·HNO$_3$. Projection of the corrugated layer on to the (b, c) plane.
Te: dotted spheres
NO$_3^-$: isolated molecules.

polyhedron. In 2TeO$_2$·HNO$_3$, the bipyramids are symmetrically connected along a corresponding edge, whereas in β-TeO$_2$, the bipyramids are displaced parallel with the 'trigonal' axis so that Te at the apex of one bipyramid lies at an equatorial position in the other bipyramid. As in α-TeO$_2$ and β-TeO$_2$, the axial bonds are weaker than the equatorial bonds: Te—O distances are 1.88 and 1.95 Å versus 2.02 and 2.16 Å, the larger values of both sets occurring again in the Te—O$_2$—Te square containing the common pyramid edge. Therefore, it may be more appropriate to describe the TeO$_2$ layer in 2TeO$_2$·HNO$_3$ in terms of (TeO)$_\infty$ zigzag chains running along the c-axis, the chains being interlinked alternately with neighboring chains by oxygen pairs.

Corrugated layers are found also in orthorhombic Te$_2$O$_3$(SO$_4$) [1055].

g. Mixed Te^{IV} oxides: Te$_2$O$_5$, H$_2$Te$_2$O$_6$ and Te$_4$O$_9$

The characteristic trigonal-bipyramid coordination (Te^{IV}O$_4$e) met in β-TeO$_2$ and 2TeO$_2$·HNO$_3$ is found also in the monoclinic structure of Te$_2$O$_5$ = Te^{IV}Te^{VI}O$_5$ [300]. Te^{VI} has the usual octahedral coordination. The [Te^{VI}O$_2$O$_{4/2}$] octahedra share corners with four other octahedra to form infinite sheets of composition (Te^{VI}O$_4$)$_n^{2n-}$. These sheets are interlinked by the Te^{IV} bipyramids. Two of the bipyramidal O atoms are bridging atoms in a (Te^{VI}O$_4$)$_n$ sheet while each of the other two is shared with another bipyramid. Thus, the atoms between the (Te^{VI}O$_4$)$_n$ layers may be looked at as (Te^{IV}O)$_n^{2n+}$ zig-zag chains running along the b-axis. The structure of Te^{IV}O(Te^{VI}O$_4$) therefore is a layered three-dimensional network.

If we split this structure between the (TeO)$_n$ chains and rotate every second chip by 180°, we end up with the H$_2$Te$_2$O$_6$ structure [300] disregarding the H atoms. In H$_2$Te$_2$O$_6$, the Te^{VI} octahedra, still linked through corners, now form chains which are connected by the same —Te^{IV}—O—Te—O— chains to infinite sheets. These sheets are held together by H bonds and van der Waals forces only, resulting in perfect cleavage planes.

The rhombohedral structure of Te$_4$O$_9$ [298a] is built up of layers (TeO$_3$·3TeO$_2$)$_n$. Crystals show stacking disorder since these layers can be stacked also invertedly. The layers are made up of [Te^{VI}O$_{6/2}$] octahedra and [Te^{IV}O$_{4/2}$] units connected via corners. The [Te^{IV}O$_4$] unit conforms to the usual description as a trigonal bipyramid with one of the equatorial positions occupied by the lone electron pair. The distances are:

$$\text{Te}^{IV}\text{—1 O}_I \text{ at } 1.902 \text{ Å}$$
$$1 \text{ O}_{III} \text{ at } 1.883 \text{ Å}$$
$$1 \text{ O}_{II} \text{ at } 2.020 \text{ Å}$$
$$1 \text{ O}_{III} \text{ at } 2.144 \text{ Å}$$

TABLE 44a
Te$_4$O$_9$ structure, trigonal, C_{3i}^2—$R\bar{3}$ (No. 148), $Z = 2$ (6),
hexagonal axes: $(0, 0, 0; \frac{1}{3}, \frac{2}{3}, \frac{2}{3}; \frac{2}{3}, \frac{1}{3}, \frac{1}{3})$ +

Te^{IV}, O$_I$, O$_{II}$ and O$_{III}$ in 18(f): $\pm(x, y, z; \bar{y}, y-x, \bar{z}; x-y, \bar{y}, \bar{z})$
Te^{VI} in 6(c): $\pm(0, 0, z)$.

Te$_4$O$_9$: $a = 9.320$ Å, $c = 14.486$ Å ($a_{rh} = 7.230$ Å, $\alpha = 80.26°$) [298a]

	x	y	z
Te^{VI}	$\frac{1}{3}$	$\frac{2}{3}$	0.484 7
Te^{IV}	0.735 6	0.021 0	0.420 2
O$_I$	0.365 0	0.845 8	0.566 7
O$_{II}$	0.821 8	0.315 3	0.592 8
O$_{III}$	0.258 1	0.028 2	0.454 3

h. Li$_2$TeO$_3$

The monoclinic structure [301] of the highly refractive Li$_2$TeO$_3$ consists of strongly deformed double layers of close-packed oxygen atoms. Two thirds of all tetrahedral holes between these O sheets are occupied by Li atoms. If the second cation were Si or Ge and with a more polarisable anion, the other cations might well go into the remaining tetrahedral sites. It is because of the lone electron pair of Te^{IV} that the Te atoms do not sit inside these tetrahedra but just stick outside thus forming a trigonal pyramid with the lone electron pair pointing towards the neighboring double layer (Figure 67). As the Te atoms occupy only one third of the possible pyramidal sites outside each oxygen layer, there are two possibilities for stacking the double layers. A deviation from ideal symmetry, however, makes one position more favorable but nevertheless disordered crystals appear to form at high growth rates. A layer structure occurs also with monoclinic Li$_2$Te$_2$O$_5$ [1056].

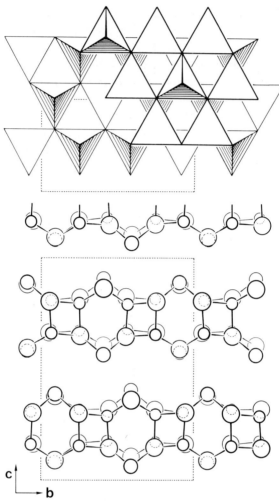

Fig. 67. The monoclinic structure of Li_2TeO_3 projected onto the (b, c)-plane (below). An idealized layer is given above. Large spheres: O, medium spheres: Te, small spheres: Li.

TABLE 45

Li_2TeO_3 structure, monoclinic, C_{2h}^6—C2/c (No. 15), $Z = 8$.
All atoms in 8(f): $(0, 0, 0; \frac{1}{2}, \frac{1}{2}, 0) \pm (x, y, z; x, \bar{y}, \frac{1}{2}+z)$

Li_2TeO_3: $a = 5.069$ Å, $b = 9.566$ Å, $c = 13.727$ Å, $\beta = 95.4°$

	x	y	z	
Li_I	0.798 1	−0.067 5	0.187 0	
Li_{II}	0.798 6	0.253 1	0.182 8	
Te	0.260 5	0.094 5	0.089 6	
O_I	0.641 4	0.092 3	0.114 0	[301]
O_{II}	0.184 1	−0.054 9	0.167 6	
O_{III}	0.183 2	0.244 8	0.169 4	

i. A LAYERED SULFATE, $(IO)_2SO_4$

The monoclinic structure of $(IO)_2SO_4$ is built up of sandwich-like layers parallel with (001). The layers consist of infinite $(IO)_n$ spiral chains along [010] held together by $[SO_4]$ tetrahedra (Figure 68). The coordination of each I comprises $2\,O_I$ at 1.97 and 1.98 Å, $1\,O_{II}$ at 2.35 Å and $1\,O_{III}$ at 2.42 Å in an almost planar arrangement. It is tempting to formulate the compound as a simple salt between IO^+ and SO_4^{2-} ions. From the intra-chain I—O_I bond lengths, Furuseth et al. [302] conclude that the bonding is predominantly covalent, although a certain degree of uneven charge distribution on I, S and O is caused by their different electronegativities, leading to the yellow color of these crystals. The proposed bonding scheme for I involves that two of its p-orbitals are engaged in σ-bonds to the two O_I atoms of the chain and the (originally empty) $d_{x^2-y^2}$ and d_{z^2} orbitals are used in dative σ-bonding with O_{II} and O_{III} of the $[SO_4]$ tetrahedra. The three remaining d-orbitals on each iodine atom are of appropriate symmetry for π-bond contribution to the I—O_I bonds.

The layer character of the $(IO)_2SO_4$ structure appears to be an accidental consequence of the Te-like $(IO)_n^+$ chains. The corresponding selenate $(IO)_2SeO_4$ may have the same or a closely related structure. For a hypothetical $(IS)_2SeO_4$ we would rather expect a three-dimensional structure. The somewhat related iodine compound $IAlCl_6 = (ICl_2)^+AlCl_4^-$ consists of chains in which ICl_2^+ and $AlCl_4^-$ ions alternate.

The monoclinic structure of gypsum $CaSO_4 \cdot 2H_2O$ [1, 4] is an example of an improper layer structure containing tetrahedral $[SO_4]^{2-}$ ions. Double layers $[Ca_2(SO_4)_2]_n$ are separated by H_2O double layers. Each Ca ion is bonded to 6 sulfate oxygen atoms and 2 oxygen atoms of water molecules. The $[CaSO_4]_n$

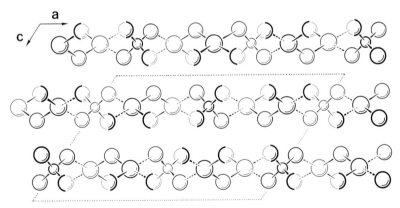

Fig. 68. The monoclinic structure of $(IO)_2 SO_4$ projected onto the (a, c)-plane. The weaker bonds between the $(IO)_n$ spiral chains and the $[SO_4]$ tetrahedra are indicated by broken arrows. Half the lower and half the upper atom is shown in the case of superposed O atoms.

I: large spheres
O: medium spheres
S: small spheres

TABLE 46
$(IO)_2SO_4$ structure, monoclinic, C_{2h}^6—C2/c (No. 15), $Z = 4$.
$(0, 0, 0; \frac{1}{2}, \frac{1}{2}, 0) +$

I, O_I, O_{II} and O_{III} in 8(f): $\pm(x, y, z; x, \bar{y}, \frac{1}{2}+z)$
S in 4(e): $\pm(0, y, \frac{1}{4})$

$(IO)_2SO_4$ (100 K): $a = 15.177$ Å, $b = 4.685\,4$ Å, $c = 9.810$ Å, $\beta = 125.17°$

	x	y	z	
I	0.163 5	0.706 6	0.251 9	
S		0.177 3		
O_I	0.283 5	0.986 1	0.357 6	[302]
O_{II}	0.096 0	−0.003 2	0.367 7	
O_{III}	0.021 2	0.355 0	0.147 1	

sheets are however not completely screened by the H_2O molecules since each water molecule is bonded to a Ca^{2+} ion and a sulfate O atom of one layer but also to a sulfate O atom of the adjacent layer, O—O = 2.82 Å. These O—H $\cdots$ O bonds are of course the weakest in the structure and account for the excellent cleavage and marked anisotropy of thermal expansion. The same structure is found in $CaSeO_4 \cdot 2H_2O$ and $YPO_4 \cdot 2H_2O$ and probably in other hydrated rare-earth phosphates and arsenates. The structure of $CaHPO_4 \cdot 2H_2O$ and $CaHAsO_4 \cdot 2H_2O$ is closely related.

j. $SnCl_2$

$SnCl_2$ crystallizes in the orthorhombic $PbCl_2$ structure which is adopted by quite a series of compounds such as $PbF_2(r)$, Pb(OH)Cl, Pb(OH)I, SbTeI (ordered version), $BaCl_2$, $BaBr_2$, BaI_2, ThS_2, $ThSe_2$, US_2, USe_2, CaH_2, SrH_2, BaH_2, EuH_2 and YbH_2. The idealized $PbCl_2$ structure is characterized by intergrown sawtooth-like layers of trigonal-prismatic columns. Therefore, the $PbCl_2$ structure is normally not a layer structure. The ideal coordination number of the cation is 6 + 3, that is, the cation has six neighbors forming a trigonal prism and three more in the equatorial plane outside the square faces of the prism. Nine bonds of equal length are impossible but this structure-type allows all kinds of distortions. Whereas the coordination is fairly regular in ThS_2, it is highly asymmetric in $SnCl_2$ due to the lone electron pair of Sn^{2+}. The trigonal-prismatic coordination is hardly recognizable (Figure 69). One equatorial bond is very strong (2.66 Å) and, moreover, the distances to the anions on one vertical edge are nearly as short (2.78 Å). Considering only these three shortest contacts, we may describe the structure in terms of zig-zag chains

$$\begin{array}{ccccc} & Cl & & Cl & \\ \diagdown & \diagup & \diagdown & \diagup & \diagdown \\ Sn & & Sn & & Sn \\ | & & | & & | \\ Cl & & Cl & & Cl \end{array}$$

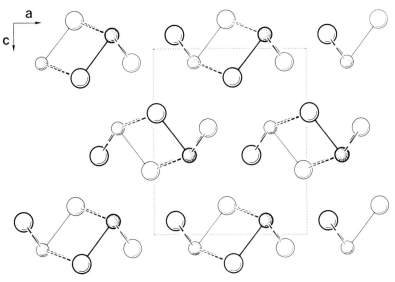

Fig. 69. The nearly layered structure of SnCl$_2$.
Sn: dotted spheres

In a first approximation we therefore have one single bond to the equatorial Cl and two half bonds within the chains (chain angle 105.6°). In the real structure each single Cl atom makes weak bonds (Sn—Cl = 3.06 Å) with two Sn atoms of the neighboring chain, thus linking the chains pairwise. These twin chains are arranged in a layered manner, with distances Sn—Cl = 3.22 Å to the next twin chain. The distance between these pseudolayers is Sn—Cl = 3.30 Å. The structure thus is intermediate between fibrous and layered.

In PbCl$_2$, the (1+2) coordination is much less pronounced: Pb—Cl = 2.80, 2.91 (2×), 3.04, 3.05 (2×) and 3.09 Å. The character of the PbCl$_2$-type compounds thus depends strongly upon the values of the positional parameters. Other lead-containing representatives are PbF$_2$(r), PbBr$_2$, Pb(OH)$_2$, Pb(OH)Cl, Pb(OH)Br, Pb(OH)I.

It would be interesting to know the parameters of SbTeI.

TABLE 47

PbCl$_2$ structure, orthorhombic, D_{2h}^{16}—Pnma (No. 62), Z = 4.
All atoms in 4(c): $\pm(x, \frac{1}{4}, z; \frac{1}{2}+x, \frac{1}{4}, \frac{1}{2}-z)$

SnCl$_2$: $a = 7.793$ Å, $b = 4.43$ Å, $c = 9.207$ Å [303]

	x	z
Sn	0.269	0.074
Cl$_I$	0.856	0.075
Cl$_{II}$	0.478	−0.155

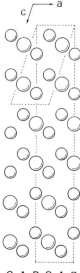

C A B C A B

Fig. 70. (110) section through the hexagonal cell of Tl$_2$O. The monoclinic cell is indicated in the upper part.

Large spheres: O.

k. Tl chalcogenides Tl$_2$X and Tl$_4$X$_3$

Tl$_2$O crystallizes in very thin black flakes. A monoclinic and a rhombohedral structure model, which differ only in minor details, have been reported [304]. We confine our discussion to the more reasonable rhombohedral model. According to both models, the Tl$_2$O structure can be described as an anti-CdI$_2$ polytype. Two anti-CdI$_2$ sandwiches are stacked in a cubic sequence which necessarily leads to a rhombohedral cell. The stacking of the Tl$^+$ ions thus is {chhc}$_3$. If we add the oxygen atoms, located at the octahedral sites, in lower-case letters then the hexagonal cell shown in Figure 70 can be described by

$$AcB \quad AcB \quad CbA \quad CbA \quad BaC \quad BaC$$

Within the hexagonal Tl layers, the Tl—Tl distance is equal to $a = 3.52$ Å. The distance from a Tl atoms to a neighboring Tl atom of the second layer of the same Tl$_2$O sandwich is 3.63 Å, the distance to a Tl atom of the adjacent sandwich is 3.88 Å. For comparison we note that in TlF(h) and TlI the metal-metal distances are 3.78 Å and 3.83 Å, respectively. Distances thus are relatively short, but it is

TABLE 48

Tl$_2$O structure, trigonal, D_{3d}^5—R$\bar{3}$m (No. 166), $Z = 2$.
All atoms in 2(c): $\pm(x, x, x)$

Tl$_2$O: $a = 12.776$ Å, $\alpha = 15.82°$ or $a = 3.516$ Å, $c = 37.84$ Å
x(Tl$_I$) = 0.123; x(Tl$_{II}$) = 0.290; x(O) = 0.417 [304]

TABLE 49
Tl_4O_3 structure, monoclinic, C_{2h}^2—$P2_1/m$ (No. 11), $Z = 2$.
All atoms in 2(e): $\pm(x, \frac{1}{4}, z)$

Tl_4O_3: $a = 10.89$ Å, $b = 3.473$ Å, $c = 7.622$ Å, $\beta = 109.4°$ [306]
$a = 10.88$ Å, $b = 3.45$ Å, $c = 7.61$ Å, $\beta = 109°35'$ [307a]

	x	z
Tl_I^+	0.132 7	0.226 2
Tl_{II}^+	0.146 2	0.694 1
Tl_{III}^+	0.633 5	0.930 1
Tl^{3+}	0.608 6	0.425 6 [307a]
O_I	0.384 2	0.355 7
O_{II}	0.823 1	0.529 3
O_{III}	0.403 3	0.761 9

hard to decide whether the dark color of the crystals is a consequence of the short Tl—Tl distance within a Tl plane or of the direct contact of two layers of cations with lone electron pairs. Nevertheless, we have doubts that the compound is intrinsically metallic (or semimetallic) though a metallic type of conductivity has been reported [305]. Tl_3O_4 ($= Tl_2O \cdot 5Tl_2O_3$) is also 'black' but it has a room-temperature resistivity of 1M $\Omega \cdot$cm and an energy gap $\geqslant 0.5$ eV [306]. Tl_2O exhibits a reversible phase transition at 354 °C [307]. All attempts to prepare In_2O seem to have been unsuccessful until now.

The monoclinic structure of the mixed oxide $Tl_4O_3 = Tl_3^+Tl^{3+}O_3$ shows the typical features of a compound with s^2 cations. The trivalent cations are at the center of a distorted octahedron. Each [TlO_6] octahedron is sharing four edges as part of a double-octahedron chain along the b-axis. The monovalent cations stick outside these octahedron chains in trigonal-pyramid coordination with the lone electron pairs thus screening the whole complex (Figure 71). The structure thus looks like a fibre structure. However, the fibres are linked in the c-direction by two relatively close Tl_{III}^+—O_{III} contacts whereas the bonding in the a-direction (Tl_{II}^+—O_{II}) is rather negligible, as follows from the bond distances:

Tl^{3+}——1 O_{II} at 2.20 Å Tl_{II}^+——2 O_{II} at 2.52 Å
                    2 O_{III} at 2.21 Å   1O_{III} at 2.67 Å
                    1 O_I at 2.32 Å   (1 O_{II} at 3.31 Å)
                    2 O_I at 2.38 Å Tl_{III}^+——1 O_{III} at 2.40 Å
Tl_I^+——2 O_{II} at 2.46 Å   2 O_I at 2.73 Å
                    1 O_I at 2.58 Å   2 O_{III} at 3.04 Å

This is an agreement with the observation that the black prismatic platelets are easily deformed [307a].

The congruently melting sulfide Tl_2S can be synthesized in the form of blue-black platy crystals with metallic luster and perfect (001) cleavage. Tl_2S, the

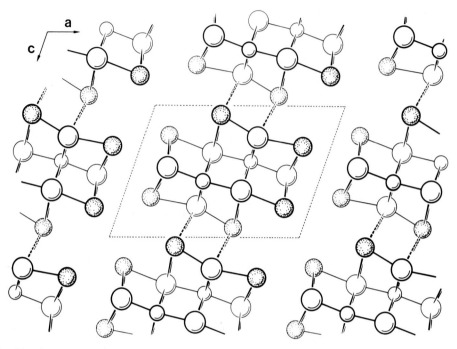

Fig. 71. The monoclinic structure of Tl_4O_3 projected along [010]. The Tl^+—O bonds that link the $(Tl_8O_6)_\infty$ columns are indicated by dashed lines.

Small spheres : Tl^{3+}
dotted spheres: Tl^+

TABLE 50

Tl_2S structure, trigonal, C_3^4—R3 (No. 146), $Z=9$ (27).
hexagonal axes: $(0, 0, 0; \frac{1}{3}, \frac{2}{3}, \frac{2}{3}; \frac{2}{3}, \frac{1}{3}, \frac{1}{3})+$
Tl_I—Tl_{VI}, S_{IV} and S_V in 9(b): x, y, z; $\bar{y}, x-y, z$; $y-x, \bar{x}, z$.
S_I, S_{II} and S_{III} in 3(a): $0, 0, z$.

Tl_2S: $a = 12.20$ Å, $c = 18.17$ Å ($a_{rh} = 9.29$ Å, $\alpha \approx 82°$) [1]
carlinite: $a = 12.12$ Å, $c = 18.175$ Å ($a_{rh} = 9.256$ Å, $\alpha = 81.80°$) [839]

	x	y	z	
Tl_I	0.119	0.191	0.957	
Tl_{II}	0.122	0.201	0.321	
Tl_{III}	0.118	0.209	0.656	
Tl_{IV}	0.233	0.103	0.140	
Tl_V	0.227	0.093	0.474	
Tl_{VI}	0.233	0.096	0.808	[985]
S_I			0.195	
S_{II}			0.519	
S_{III}			0.853	
S_{IV}	0.665	0.657	0.226	
S_V	0.333	0.337	0.195	

mineral carlinite, crystallizes in a distorted anti-CdI$_2$ structure with $a = 3a_0$, $c = 3c_0$. The packing of the cations differs from an ideal hexagonal close-packing in that it is expanded in the (001) layers: Tl—Tl = 3.52–4.35 Å and c/a = 1.49. While Tl$_{IV}$, Tl$_V$ and Tl$_{VI}$ lie in a plane the other cations, Tl$_I$, Tl$_{II}$ and Tl$_{III}$, form a corrugated layer. The same holds for the S layers. The [STl$_6$] octahedra are strongly distorted thus reflecting the influence of the lone electron pair of Tl$^+$. The S atoms are shifted from the centers of the Tl octahedra giving rise to markedly different Tl—S distances [985]: Tl$_I$—S$_{III}$ (2.78), S$_{IV}$ (2.92, 3.28), S$_V$ (2.91 Å); Tl$_{II}$—S$_I$ (3.14), S$_{IV}$ (3.28), S$_V$ (3.22 Å); Tl$_{III}$—S$_{II}$ (2.78), S$_{IV}$ (3.08), S$_V$ (3.23 Å); Tl$_{IV}$—S$_I$ (2.66), S$_{IV}$ (2.67), S$_V$ (2.68 Å); Tl$_V$—S$_{II}$ (2.55), S$_{IV}$ (2.51), S$_V$ (2.70 Å); Tl$_{VI}$—S$_{III}$ (2.61), S$_{IV}$ (2.51), S$_V$ (2.70 Å). The shortest Tl$^+$—Tl$^+$ distances between the sandwiches are Tl$_I$—Tl$_I$ 3.52 Å, Tl$_{II}$—Tl$_{II}$ 3.70 Å, Tl$_{III}$—Tl$_{III}$ 3.84 Å, Tl$_{IV}$—Tl$_V$ 3.73 Å and Tl$_{IV}$—Tl$_{VI}$ 3.63 Å which is close to the distances met in Tl$_2$O. The observed semiconductivity [308] with an energy gap of 0.6 eV proves that the Tl—Tl distance is by no means critical.

The structure of Tl$_4$S$_3$ (= Tl$_3^+$Tl^{3+}S$_4$) was claimed [309] to be a derivative of that of Tl$_2$S. However, its monoclinic structure [310] is not a layer structure, but is composed of chains of [TlS$_4$] tetrahedra held together by Tl$^+$ ions.

Tl$_2$Se has a somewhat layered three-dimensional structure. The tetragonal structure contains Tl double sheets alternating with similar single Se sheets. Additional Tl$_2$Se 'molecules' are inserted between every second Se—Tl—Tl—Se unit. Within the Tl double sheets, which are reminiscent of the Se layers of TlSe or of the Al layers of CuAl$_2$, extremely short Tl—Tl distances occur. Thus each Tl atom of a double layer has one Tl neighbor in the other sheet as close as 3.25 Å and a second Tl neighbor in the same sheet at 3.49 Å. Based on this coordination, we would conclude that Tl$_2$Se has a metallic character, which is unexpected from the semiconducting properties of Tl$_2$S and Tl$_2$Te.

4. MX and MX$_2$ Phases with Low Coordination Number

a. In AND Tl MONOHALIDES AND ALKALI HYDROXIDES

In and Tl monohalides are isoelectronic with the Sn and Pb monochalcogenides. The character of their structures, however, is different, although geometrically, the two structures are closely related. Common to both groups is, furthermore, the occurrence of NaCl-type phases and NaCl-type derivatives (see Table 51). CsCl-type modifications, however, are known only for the In and Tl halides, similar to the alkali halides and rare-earth monochalcogenides, but no such modifications are reported for the isoelectronic Sn and Pb chalcogenides.

TABLE 51
On the occurrence of TlI-type compounds

LiOH (r) anti-PbO structure	InF ?
(p)	InCl (r) def. NaCl structure, yellow
NaOH (r) anti-TlI structure [4]	(h) >120°C, ?, red
(h) TlI → NaCl structure [315]	InBr TlI structure
(p) unknown structure, CsCl ? [324]	InI (r) TlI structure
	(p) CsCl structure?
KOH (t_1) ~anti-TlI structure? [4]	
(t_2)	TlF (r) orthorh. def. NaCl struct. [317]
(r)	(h) >354 K, tetrag. NaCl str. [316]
(h) NaCl structure?	(>13 kb/rt) unknown structure [318]
RbOH (t) ~anti-TlI structure? [4]	TlCl (t) NaCl structure [319]
	(r) CsCl structure
	(film) TlI structure [314]
	TlBr (t) NaCl structure
	(r) CsCl structure
	(film) TlI structure [314]
	TlI (r) TlI structure
	(h/p) CsCl structure [320–323]
	(film) NaCl structure [319]

The TlI structure which has a slightly higher symmetry than the SnS structure can easily be described with a SnS-type cell if we use the following transformations:

$\mathbf{a}$(SnS) = $\mathbf{b}$(TlI)   M: x(SnS) = $\frac{1}{2}$ − y(TlI)
$\mathbf{b}$(SnS) = $\mathbf{c}$(TlI) z(SnS) = 0
$\mathbf{c}$(SnS) = −$\mathbf{a}$(TlI)   X: x(SnS) = −y(TlI)
                                           z(SnS) = $\frac{1}{2}$

This transformation yields for example for TlI:

a = 12.92 Å, b = 5.251 Å, c = 4.582 Å
Tl: x = 0.108, z = 0; I: x = −0.133, z = 0.500,

which is to be compared with the parameters listed in Table 33. Thus the TlI structure can be derived from a rocksalt structure by distortions similar to those occurring in the SnS structure, though the final coordination turns out to be different.

Alternatively, the TlI structure can be constructed by stacking double layers of the NaCl type with consecutive double layers being shifted relative to each other by $a/2$ (Figure 72). The cation sublattice then has to be displaced relative to the anion sublattice in order to form more or less pronounced molecular units. Each atom thus obtains 1+4 neighbors defining a deformed octahedron with its sixth corner missing. Bond distances are listed in Table 52. The structure is a compromise between a molecular structure with truly covalent single bonds and a three-dimensional ionic packing. If the positional parameters are correct then InBr is a purely molecular crystal in contrast with InI and TlI. The M—X

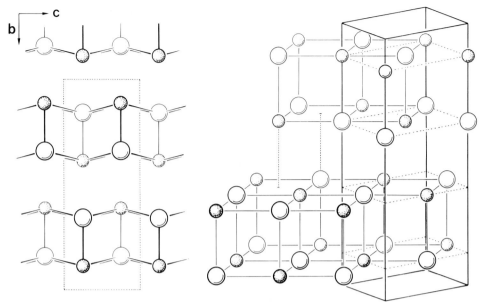

Fig. 72. The structure of TlI projected along [100] and its derivation from NaCl-type slabs. Dotted spheres: Tl.

distances between neighboring NaCl-type double layers are too large to correspond to any bonding, and the crystals will cleave between these double layers.

As in the CrB structure, the cations in the TlI structure are arranged in zig-zag chains. It is tempting therefore to assume covalent M—M bonds in order to engage the lone-pair electrons in bonding [311]. However, the M—M distances are much too long compared with Pauling's single-bond distances given in Table 52 so that we reject this interpretation. The decrease of the energy gap on going from the orthorhombic TlI-type to the cubic CsCl-type modifications is not due to the transition from trivalent to monovalent cations but is the consequence of a change from mainly single bonds to fractional bonds, as in the case of the SnS-type to NaCl-type transition of the Sn and Pb monochalcogenides.

It is worth noting that a partial substitution of the s^2 cation by Ag^+ completely changes the structure. Thus, $InAgI_2$ and $TlAgI_2$ crystallize in the tetragonal TlSe ($=Tl^+Tl^{3+}Se_2$) structure with Ag^+ at the tetrahedral sites.

TABLE 52
Interatomic distances in TlI-type halides.

	M—1X (Å)	M—4X (Å)	M—2X (Å)	M—M (Å)	M^{III}—M^{III} single bond distance (Å)
InBr	2.80	3.30	4.06	3.68	2.88
InI	3.23	3.46	3.95	3.58	2.88
TlI	3.35	3.50	3.87	3.83	2.92

TABLE 53
TlI structure, orthorhombic, D_{2h}^{17}—Cmcm (No. 63), $Z=4$.
All atoms in 4(c): $(0, 0, 0; \frac{1}{2}, \frac{1}{2}, 0) \pm (0, y, \frac{1}{4})$

Compound	$a(\text{Å})$	$b(\text{Å})$	$c(\text{Å})$	y_M	y_X	References
InBr	4.460	12.39	4.73	0.386	0.160	[4]
InI	4.75	12.76	4.91	0.398	0.145	[313]
TlCl	4.27	12.4	4.74			[314]
TlBr	4.39	12.5	4.96			[314]
TlI (25°C)	4.582	12.92	5.251	0.392	0.133 3	[4]

α-NaOH structure = anti-TlI structure with additionally H in 4(c)

Compound	$a(\text{Å})$	$b(\text{Å})$	$c(\text{Å})$	y_M	y_O	y_H	References
α-NaOH (24°C)	3.399 4	11.377	3.399 4	0.162 5	0.366 8	0.466 8	[315]
α-KOH	3.95	11.4	4.03	0.14	0.39		[4]
RbOH	4.15	12.2	4.30				[4]

β-NaOH structure, monoclinic, C_{2h}^2—$P2_1/m$ (No. 11), $Z=2$.
2nd setting, atoms in 2(e): $\pm(x, \frac{1}{4}, z)$ [315]
$a = 3.434$ Å, $b = 3.428$ Å, $c = 6.068$ Å, $\beta = 109°50'$ (>240°C)
Na: $x = 0.320$, $z = -0.171$; O: $x = 0.164$, $z = 0.277$; H: $x = 0.054$, $z = 0.386$ (?)
(there appears to be an error; compare Figure 5 of ref. [315])
The corresponding monoclinic cell for α-NaOH would be:
$a_{mon} = a_{orth} = 3.399$ Å   Na: $x = 0.162\ 5$, $z = -0.1750$
$b_{mon} = c_{orth} = 3.399$ Å   and   O: $x = 0.366\ 5$, $z = 0.233\ 6$
$c_{mon} = \frac{1}{2}(a_{orth}^2 + b_{orth}^2)^{1/2} = 5.937$ Å   H: $x = 0.446\ 8$, $z = 0.393\ 6$
$\text{tg}(\beta - 90°) = (a/c)_{orth} \rightarrow \beta = 106.64°$

If we derive the TlI structure from trigonal-prism layers (Figure 4b) we see that the cations are displaced towards one square prism face. As the cleavage plane runs amidst the prism layer the cations still form the outer layer, obviously with the lone electron pair pointing outwards. In the anti-TlI structure of NaOH(r) we find a hydrogen atom in place of the lone pair. Busing [312] deduced from the infrared spectra of NaOH and NaOD that hydrogen bonding between neighboring layers does not occur. At high temperatures, NaOH suffers a transition to a monoclinic phase caused by a small shift of the NaCl-type double layers. The monoclinic distortion of the cell (which increases the angle β of the monoclinic cell of α-NaOH by 3.2°) is such as to reduce the relative displacement of the double layers. The angle β increases monotonously with temperature but the crystal is melting before the NaCl structure is obtained. From this displacement it becomes evident that the proton does not bind an oxygen atom to a sodium atom of the neighboring double layer since Na, O and H shift as a rigid linear molecule.

TlF is one of the very few layer-type fluorides and obviously it owes its adherence to the layer-type family to the peculiarity of the s^2 cations. TlF is known to exist in an orthorhombic room-temperature and a tetragonal high-temperature modification [316]. The structure of both these phases derives from

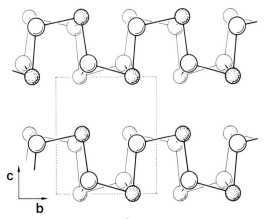

Fig. 73. The structure of TlF(r) projected on (100).
Dotted spheres: Tl.

the rocksalt structure. The low-temperature modification grows in the form of thin fragile flakes. Its orthorhombic structure [317] is not simply a squeezed NaCl structure with $(TlF)_n$ monolayers perpendicular to the longest axes and —Tl—F—Tl— chains along the shortest axis as was assumed earlier. Considerable distortions produce in fact NaCl-type double layers in which the F^- ions are drawn towards the interior. This leads to a stacking $\cdots$ Tl—F—F—Tl $\cdots$ Tl—F—F—Tl $\cdots$ with a Tl—F distance $\geqslant 3.50$ Å between the sheets and 2.6 Å within the sheets. The TlF structure, illustrated in Figure 73, is reminiscent of that of yellow PbO. The lone electron pairs on each Tl atom obviously screen each sheet. Short and long distances (e.g. 2.52 Å and 3.07 Å) alternate also along the shortest axis. The coordination of the Tl atoms is $1+1+2$.

In the tetragonal high-temperature modification stable above 82 °C [316], the asymmetry in the layer plane is removed and each atom has now four other neighbors at 2.68 Å. This tetragonal structure may transform into the NaCl structure at still higher temperatures.

TABLE 54
TlF(r) structure, orthorhombic, C_{2v}^4—Pma2 (No. 28), $Z = 4$.
All atoms in 2(c): $\frac{1}{4}$, y, z; $\frac{3}{4}$, $\bar{y}$, z.

TlF: $a = 5.1848$ Å, $b = 5.4916$ Å, $c = 6.0980$ Å.

	y	z
Tl_I	0.2412 (X+N)	0
Tl_{II}	0.7348 (X)	0.5129 (X+N)
	0.7450 (N)	
F_I	0.1943 (N)	0.4281 (N)
F_{II}	0.6713 (N)	0.1500? (N)

Positional parameters determined by X-ray diffraction (X) and neutron diffraction (N).

The z-parameter for F_{II} seems doubtful.

TABLE 55
Structural relationship among alkali halides and pseudohalides.

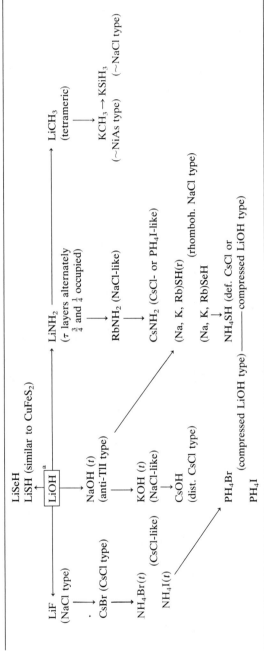

[a] No structural transition up to the melting point of 744 K [326].

Though a one-sided coordination favors the occurrence of layer-type structures, this is not a sufficient criterion, as is exemplified by InCl. This compound also crystallizes in a distorted NaCl structure showing the characteristic reduction of the coordination number, but the structure remains cubic.

b. LiOH AND RELATED STRUCTURES

If we neglect the H atom in LiOH then its structure corresponds to the anti-PbO type. Since the axial ratio and the free positional parameter of this tetragonal structure can vary in a certain range, different isopuntal structure [9] types are possible. Thus for $c/a = \sqrt{2}$ and $z(\text{anion}) = \frac{1}{4}$, a cubic close-packing of the anions results with the cations in tetrahedral holes. An axial ratio $c/a = 1/\sqrt{2}$ and an anion parameter $z = \frac{1}{2}$, on the other hand, correspond to the CsCl type with coordination number 8. As follows from Table 56, LiOH approximates a cubic close-packing with Li in deformed tetrahedral coordination. The position of the lone electron pair of PbO is here taken by H (corrected O—H distance 0.98 Å [325] similar to the lone pair – cation distance). The electron density corresponds to $Li^+O^{-0.7}H^{+0.7}$ and one electron smeared between the layers [1012]. In Table 55, LiOH is compared with chemically related compounds. Lithium amide has a closely related structure in which the layers of tetrahedral cation sites are alternately $\frac{3}{4}$ and $\frac{1}{4}$ occupied ($\frac{1}{2}\tau_1 + 1\tau_2$ and $\frac{1}{2}\tau_1$, respectively) instead of the completely occupied and completely empty layers of LiOH. This is obviously a consequence of the weaker dipole character of NH_2^-. LiF, with no dipole moment, crystallizes in the rocksalt structure. The structure of LiSH is similar to chalcopyrite whereas that of the hydrosulfides and hydroselenides of Na, K and Rb is a rhombohedrally deformed rocksalt type.

LiOH forms discrete phases with other alkali hydroxides such as $LiNa(OH)_2$, $LiNa_2(OH)_3$, $Li_2K(OH)_3$, $Li_3Rb(OH)_4$ and $Li_5Cs(OH)_6$. The larger cations, however, spoil the layer character though Li is still tetrahedrally coordinated [326a].

In Table 56, we have listed also a chemically quite different compound, InBi, which has an axial ratio midway between LiOH and CsCl. InBi is a rather nontrivial metal with a low melting point of 110°C. The other metallic III–V compounds, TlSb and TlBi, crystallize in a non-deformed CsCl structure, whereas the metallic high-pressure modification of InSb adopts the white-tin structure.

In the LiOH-like layer structure of InBi, each In is located at the center of a squeezed Bi tetrahedron with one of the Bi—In—Bi bond angles being 101.4° as compared with the ideal value of 109.5°. The bond distance In—Bi ($3.10 \cdots 3.13$ Å) is somewhat larger than the value expected for covalent bonds but corresponds fairly well to the sum of the metallic radii. The In—In distance within the square net is 3.54 Å which is not low enough to exclude nonmetallic properties. The Bi—Bi distance between the layers is 3.7 Å and thus is much larger than in arsenic-type bismuth. Whereas in sphalerite-type InSb, one set (τ_1) of tetrahedral holes in the anion ccp is occupied by the cations, in InBi, half τ_1

TABLE 56
LiOH type and related structures
LiOH, tetragonal D_{4h}^7—P4/nmm (No. 129), $Z=2$.
Li in 2(a): 0, 0, 0; $\frac{1}{2}$, $\frac{1}{2}$, 0.   or $\pm(\frac{1}{4}, \frac{1}{4}, 0)$
O and H in 2(c): 0, $\frac{1}{2}$, z; $\frac{1}{2}$, 0, $\bar{z}$ $\pm(\frac{1}{4}, \frac{3}{4}, z)$

Compound	a(Å)	c(Å)	c/a	z	References
LiOH	3.56	4.34	1.22	O: 0.194 H: 0.407	[325]
AgI (3kbar)	4.58	6.00	1.31	0.30	[327]
NH$_4$SH	6.011	4.009	0.667	0.34	[4]
NH$_4$Br (128 K)	5.697	4.046	0.710	0.53	[4]
PH$_4$Br	6.042	4.378	0.725	0.42	[4]
PH$_4$I	6.34	4.62	0.73	0.40	[4]
cubic CsCl	5.828	4.121	0.707	0.500	
ccp			1.414	0.250	
OPb	3.964	5.008	1.26	0.238 5	
InBi	5.015	4.781	0.953	0.38	[3]
	5.000	4.773	0.955	0.393	[328]
(87 K)	4.963	4.846	0.977		[329]
(351 K)	5.051	4.753	0.941		[329]
FeS	3.679	5.047	1.372	~0.25	[330]
	3.678	5.038	1.370		[331]
	3.684	5.041	1.368		[332]
FeSe (43.8 at.%Se)	3.773	5.529	1.466	0.26	[3,333]
FeTe (Fe$_{1.1}$Te)	3.820	6.281	1.644	0.285	[333]

and τ_2 are occupied in layers. The transition from the cubic structure of InSb to the tetragonal structure of InBi can be achieved by shifting the cations in every second (001) τ_1 layer along [001] by $a/2$ into the empty τ_2 in addition to a deformation. While In still can form sp^3 bonds as in InSb, the bonds on the anion now probably involve d functions and the overlapping of the energy bands originating mainly from Bi (s,p,d)-functions may account for the metallic properties. InBi is one of the few non-transition-element compounds for which the metallic properties are not already obvious from composition, though they are not unexpected on extrapolation from InP ($\Delta E = 1.41$ eV), InAs (0.41 eV) and InSb (0.24 eV). Under pressure, InBi may convert to the CsCl type. (p:[1046])

Nonmetallic properties have been reported [331] for the LiOH-type chalcogenide FeS. The unstable tetragonal FeS is known as the mineral mackinawite which is found with compositions ranging from M$_{1.023}$S to M$_{0.994}$S where M includes up to 10% Ni and some Co and Cu [334]. Pure FeS transforms into the hexagonal pyrrhotite Fe$_{0.9}$S at 160° [331] to 170°C [335]. The upper stability limit of mackinawite rises markedly with increasing substitution of Fe by Ni and Co.

FeS is an intriguing substance. Fe^{2+} with a d^6 configuration is rather unusual in tetrahedral coordination although it is met in stannite Cu$_2$FeSnS$_4$ and in two other metastable modifications of FeS, one with cubic sphalerite [335, 336] and one with hexagonal wurtzite structure [335]. It is funny that the transition proposed above for InSb → InBi really does occur in FeS: metastable cubic FeS transforms gradually to mackinawite at room temperature [335]. This phase transition is accompanied by a volume contraction of 14%. Whereas in zincblende-type FeS, each Fe has 12 Fe neighbors at 3.83 Å, it would have 4 Fe neighbors at 2.71 Å in a noncontracted LiOH-type cell. The new type of bonding due to the changed electronic configuration leads to a contraction of the Fe—Fe distance to 2.60 Å and of the Fe—S distance from 2.35 Å to 2.23 Å. From this short Fe—S distance and the absence of magnetic ordering down to 1.7 K [331], we conclude that Fe is in a low-spin state. Unfortunately, no magnetic-susceptibility measurements seem to be available which could confirm the assumed d-electron configuration. Whereas in octahedral low-spin Fe^{II}, an S state is achieved, two unpaired d electrons are left in tetrahedrally coordinated Fe^{II}. Now, in tetragonal FeS the closeness of the Fe neighbors will not only split the lowest doubly degenerate d levels but one of these two d orbitals will be involved in bonding. Thus we are left with three electrons that cannot be accommodated in a filled band. Furthermore, the Fe—Fe bonds themselves should cause metallic properties. However, Bertaut et al. [331] report definitely nonmetallic properties for this compound. The bond scheme offered by Kjekshus et al. [332] cannot explain semiconductivity either.

As mentioned above, the mackinawite-type FeS frequently forms with deviations from stoichiometry as is always the case for metallic [1013] $Fe_{1+\delta}Se$, the mineral achavalite, and the weakly ferro- or ferrimagnetic tellurides $Fe_{1.05}Te$—$Fe_{1.2}Te$. It was always assumed that the excess cations were inserted in the octahedral holes, in other words, that these phases were representatives of the defect Cu_2Sb type. From a structure refinement on mackinawite, Taylor and Finger [337] concluded, on the contrary, that there is a slight deficiency of sulfur in the structure and not a metal excess, so that the formula should be written as FeS_{1-x}. According to [1043] FeSe is metallic and Pauli paramagnetic. Compare [1044].

c. CuTe

The non-stoichiometric $Cu_{3-x}Te_2$ has an orthorhombic structure with only minor deviations from the tetragonal defective Cu_2Sb-type modification stable above 140°C, which is closely analogous to $Fe_{1+\delta}Te$. The additional metal atoms which are inserted in the 'octahedral' holes of the ccp anion array (the deviation from ideal close-packing reduces the coordination number in fact to 5, since the sixth corner in c-direction is too far away) connect the X—M_2—X sandwiches and thus destroy the layer character.

The stoichiometric monotelluride CuTe, the yellow-bronze colored mineral vulcanite, on the other hand, does crystallize in a true layer structure which, however, represents a fairly distorted version of the mackinawite type (Figure 74). The formerly planar square net of the tetrahedrally coordinated metal atoms is now a slightly puckered Cu plane. Moreover, the lattice is compressed along [100] which leads to the formation of straight Te chains along the a-axis with Te—Te =

TABLE 57

CuTe structure, orthorhombic, D_{2h}^{13}—Pmmn (No. 59), $Z=2$.

Cu in 2(b): $0, \frac{1}{2}, z; \frac{1}{2}, 0, \bar{z}$
Te in 2(a): $0, 0, z; \frac{1}{2}, \frac{1}{2}, \bar{z}$.

CuTe:	a(Å)	b(Å)	c(Å)	z(Cu)	z(Te)	
	3.16	4.08	6.94	0.46	0.22	[7]
	3.10	4.02	6.86	0.449	0.223	[338]
	3.15	4.09	6.95			[1]
	3.149	4.086	6.946			[339]

CuAgTe$_2$ structure, orthorhombic, C_{2v}^1—Pmm2 (No. 25), $Z=1$.

Cu in 1(c): $\frac{1}{2}, 0, z$
Ag in 1(b): $0, \frac{1}{2}, z$
Te$_I$ in 1(a): $0, 0, z$
Te$_{II}$ in 1(d): $\frac{1}{2}, \frac{1}{2}, z$

CuAgTe$_2$: $a=3.12$ Å, $b=4.05$ Å, $c=6.875$ Å [340]
z(Cu) $=0.539$, z(Ag) $=0.462$, z(Te$_I$) $=0.220$, z(Te$_{II}$) $=0.779$.

3.15 Å (to be compared with a Te—Te half-bond distance of 2.92 Å). In order to create a non-metallic modification of CuTe, the distortions should lead to Te pairs. In that case, the Cu atoms should not interact with each other in order to be monovalent. In the actual structure, d-electrons will be engaged in bonding. In spite of the puckering of the Cu plane, the Cu—Cu distances are still as low as 2.63 Å and thus Cu—Cu bonds, similar to those in tetragonal FeS, will help to reduce the number of unbalanced d-electrons.

Whereas AgTe has a three-dimensional structure, the mixed telluride CuAgTe$_2$ is claimed to crystallize in an ordered CuTe superstructure [340]. Structure and

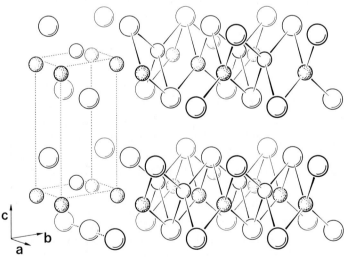

Fig. 74. The orthorhombic structure of CuTe and CuAgTe$_2$. The unit cell is shifted by $(0, \frac{1}{2}, \frac{1}{2})$ to show the relation with mackinawite FeS, rickardite Cu$_4$Te$_3$ and Cu$_2$Sb. The chain formation is indicated below the cell. Small spheres: Cu and Ag (dotted).

composition were determined from electron-diffraction patterns of films, which were prepared by evaporating a quenched CuAgTe$_2$ alloy. Since the reported lattice constants are identical with those of CuTe, we suspect that the effective Ag content is lower.

Only 3% of tellurium in CuTe can be replaced by selenium [341]. Unlike the iron monochalcogenides, CuS and CuSe seem to possess no true layer-type modification. The covellite structure of CuS and α-CuSe is layered in a certain sense as it is composed of Cu—X$_2$—Cu sandwiches which alternate with BN-like CuX layers, but the bonding is three-dimensional in spite of the perfect (001) cleavage. A quite unreasonable layer structure with Cu—Se = 1.63 Å has been proposed for CuSe [342].

Another 'layered' structure has been determined [343] on Ag$_3$TlTe$_2$ which, from its formula, might be a normal valence compound with monovalent cations. Its orthorhombic cell can be built up from layer packs Ag—Te—(Ag+Tl)—Te—Ag. These Ag$_3$TlTe$_2$ units are stacked in such a way that the atoms of the two contacting Ag layers form zig-zag chains in [100] direction with a rather short Ag—Ag distance of 3.05 Å which seems to indicate bonding between the layer units. For Ag$^+$ ions (ionic radius 1.26 Å) such an Ag—Ag distance would not be critical. However, if the outer Ag atoms of each layer unit really were ionized we would rather expect that adjacent units would shift by $b/2$ in order to increase the Ag—Ag distance. Anyway, on vapor-depositing the Ag$_3$TlTe$_2$ films onto NaCl crystals, plate-like textures always formed.

d. β-NiTe

The structure of the badly characterized phase β-NiTe represents the rhombohedral analog of the tetragonal FeS structure. In both β-NiTe and mackinawite, the idealized anion sublattice corresponds to a cubic close-packing. In β-NiTe, the layering of the occupied tetrahedral holes τ is hexagonal (Figure 75).

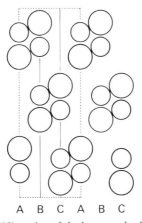

A B C A B C

Fig. 75. (110) section of the hexagonal cell of β-NiTe.

TABLE 58
Structure of β-NiTe, trigonal, D_{3d}^5—$R\bar{3}m$ (No. 166), $Z = 2(6)$.
All atoms in 2(c): $\pm(x, x, x)$ with rhombohedral axes or
in 6(c): $(0, 0, 0; \frac{1}{3}, \frac{2}{3}, \frac{2}{3}; \frac{2}{3}, \frac{1}{3}, \frac{1}{3}) \pm (0, 0, z)$ with hexagonal axes

---

β-NiTe: $a = 3.88$ Å, $c = 20.2$ Å
$z(\text{Ni}) = 0.129$, $z(\text{Te}) = 0.257$ [344]

The close-packing is elongated with respect to the ideal case, $c/3a = 1.74$ instead of the ideal value $(8/3)^{1/2} = 1.63$. The 'thickness' of a filled layer is 3.69 Å, and of an empty one, 3.05 Å. The Te—Te distance between the layers is 3.78 Å, which is distinctly larger than the corresponding value of 3.5 Å in $NiTe_2$. The Ni atoms in the tetrahedral holes are displaced from the center towards the base yielding Ni—Te = 2.48 Å (3×) and 2.63 Å (1×). This shift of the Ni atoms out of the Te-tetrahedron center increases somewhat the Ni—Ni distance, however, the nickel atoms are still as close as 2.71 Å which indicates some d—d-bonding in this metallic phase.

The layer structure of PtTe will be discussed together with Pt_3Te_4, Pt_2Te_3 and other transition-element chalcogenides.

e. TETRAHEDRAL MX_2 STRUCTURES (HgI_2, $AlPS_4$, $GaPS_4$, GeS_2)

MX_2 structures with tetrahedral coordination of the cations are to be expected with small (s, p) cations such as Be^{2+}, B^{3+}, Si^{4+} and P^{5+}. This is indeed observed but, in addition, surprisingly many phases exist also with Hg^{2+} and Zn^{2+} whereas the Cd compounds contain six-coordinated cations. Not many of the possible structures, however, are of the layer type. Quite a number of structures are known in which the $[MX_{4/2}]$ tetrahedra are connected via four corners. The three-dimensional orthorhombic structure of GeS_2 and $GeSe_2$ is one of the few tetrahedral structures which are not derived from an anion close-packing. Three-dimensional ε-$Zn(OH)_2$ and isostructural $Be(OH)_2$ are examples of the few hydroxides with tetrahedral cation coordination. As can be seen in Table 59, the cations in the structures of $InPS_4$ [345] and α-$ZnCl_2$ [346] uniformly occupy half of one set of tetrahedral holes in a cubic close-packing (ccp). The α-$ZnCl_2$ structure is found also in the high-pressure modifications of SiS_2 and GeS_2 [347]. The monoclinic modification [346] β-$ZnCl_2$ already shows a tendency towards layer formation. Its structure can be built up from a hcp anion array in which $\frac{1}{3}$ $(\tau_1 + \tau_2)$ and $\frac{1}{6}(\tau_1 + \tau_2)$ are alternately occupied by cations. The third modification [346] γ-$ZnCl_2$ finally crystallizes in a true layer structure, as also do γ-$ZnBr_2$, ZnI_2 and red HgI_2. The tetragonal HgI_2 structure, as illustrated in Figure 76a, can be derived from a ccp anion array with occupation sequence —τ_1—0—τ_1—0— (compare Table 59). For ideal ccp, the axial ratio $c/a = 2\sqrt{2}$ and the free parameter, $z = \frac{1}{8}$. For HgI_2, five modifications were detected in the range 0–

DISCUSSION OF SPECIAL COMPOUNDS

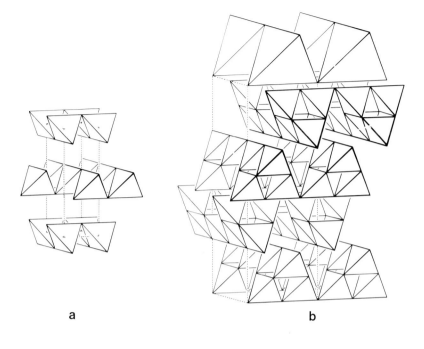

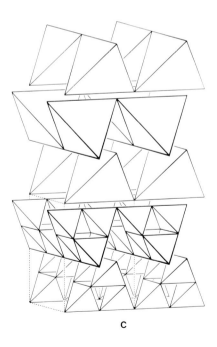

Fig. 76. The tetragonal structures of red and orange HgI_2. (a) red HgI_2 with the same orientation as orange four-layer HgI_2 for easy comparison; (b) orange four-layer HgI_2; (c) orange two-layer HgI_2; Part of the Hg_4I_{10} units are given as blocs.

500°C/20 kbar [350]. The yellow high-temperature form of HgI_2, which thus has a higher energy gap, crystallizes in the $HgBr_2$ structure with octahedrally coordinated Hg. At room temperature <20% of $HgBr_2$ is soluble in the tetragonal HgI_2, diminishing to zero near 130°C [349, 315]. With increasing $HgBr_2$ content, the energy gap of $Hg(I, Br)_2$ steeply increases. The tetragonal → orthorhombic transformation takes place at roughly that concentration where the curves ΔE vs. concn. for both modifications would cross [349]. From Raman-scattering studies [363] it was concluded that in contrast to yellow HgI_2, the interlayer force in red HgI_2 is comparable with the intralayer force relevant to low-frequency modes, which appears to be a rather non-trivial claim.

Metastable orange modifications of HgI_2 can be obtained from HgI_2 vapor or by recrystallization from organic solvents in the form of square platelets ranging in color from red-orange to yellow-orange. The yellow-orange crystals exhibit stacking disorder whereas the red-orange crystals have an ordered structure [352] consisting of corrugated sheets of P_4O_{10}-like $[Hg_4I_{10}]$ groups. These tetrahedral

TABLE 59

Occupation of tetrahedral voids in ccp MX_2 compounds as compared with a $CuFeS_2$ cell
ccp of X: $\frac{1}{4}, \frac{1}{4}, \frac{1}{8}; \frac{3}{4}, \frac{3}{4}, \frac{1}{8}; \frac{1}{4}, \frac{3}{4}, \frac{3}{8}; \frac{3}{4}, \frac{1}{4}, \frac{3}{8}; \frac{1}{4}, \frac{1}{4}, \frac{5}{8}; \frac{3}{4}, \frac{3}{4}, \frac{5}{8}; \frac{1}{4}, \frac{3}{4}, \frac{7}{8}; \frac{3}{4}, \frac{1}{4}, \frac{7}{8}$.
τ_1: ccp $-(\frac{1}{4}, \frac{1}{4}, \frac{1}{8})$, τ_2: ccp $+(\frac{1}{4}, \frac{1}{4}, \frac{1}{8})$, Ω: ccp $+(\frac{1}{2}, \frac{1}{2}, \frac{1}{4})$

τ_1	τ_2	$CuFeS_2$	$InPS_4$	α-$ZnCl_2$	OCu_2	HgI_2	SiS_2	BPS_4	$AlPS_4$	$MM'X_4$
0, 0, 0		Cu	In	Zn	O	Hg	Si	B	Al	M
$\frac{1}{2}, \frac{1}{2}, 0$		Fe				Hg				
	$\frac{1}{2}, 0, 0$									
	$0, \frac{1}{2}, 0$						Si	P	P	M'
$\frac{1}{2}, 0, \frac{1}{4}$		Fe	P							
$0, \frac{1}{2}, \frac{1}{4}$		Cu		Zn						
	$\frac{1}{2}, \frac{1}{2}, \frac{1}{4}$									
	$0, 0, \frac{1}{4}$				O					
$\frac{1}{2}, \frac{1}{2}, \frac{1}{2},$		Cu	In	Zn			Si	B		
$0, 0, \frac{1}{2}$		Fe			O				P	M'
	$0, \frac{1}{2}, \frac{1}{2}$					Hg				M'
	$\frac{1}{2}, 0, \frac{1}{2}$					Hg	Si	P	Al	
$0, \frac{1}{2}, \frac{3}{4}$		Fe	P							
$\frac{1}{2}, 0, \frac{3}{4}$		Cu		Zn						
	$0, 0, \frac{3}{4}$				O					
	$\frac{1}{2}, \frac{1}{2}, \frac{3}{4}$									

three-dimensional   layer   fibre   layered fibre

←——— 4 common corners ———→ ←— 2 common edges —→

TABLE 60

HgI_2 structure, (C13 type), tetragonal, D_{4h}^{15}—$P4_2/nmc$ (No. 137), $Z = 2$.
  I in 4(d): $0, \frac{1}{2}, z; \frac{1}{2}, 0, \bar{z}; 0, \frac{1}{2}, \frac{1}{2}+z; \frac{1}{2}, 0, \frac{1}{2}-z$.
  Hg in 2(a): $0, 0, 0; \frac{1}{2}, \frac{1}{2}, \frac{1}{2}$.

HgI_2: $a = 4.357$ Å, $c = 12.36$ Å, $z = 0.14$		[4]
$a = 4.361$ Å, $c = 12.450$ Å, $z = 0.1393$		[348]
$HgI_{2-x}Br_x$: $x < 0.4$ (25 °C)		[349]
γ-$ZnCl_2$: $a = 3.70$ Å, $c = 10.67$ Å, $z \approx \frac{1}{8}$		[346]
$ZnBr_2$: $a = 3.87$ Å, $c = 11.4$ Å		[4]
ZnI_2: $a = 4.27$ Å, $c = 11.80$ Å		[1]
(25 °C): $a = 4.338$ Å, $c = 11.788$ Å		[1]

$[Hg_4I_{10}]$ units, composed of four simple $[HgI_{4/2}]$ tetrahedra, are arranged much like the $[HgI_{4/2}]$ tetrahedra in red HgI_2 but without the empty layers. The anion sublattice is still cubic close-packed, but, in contrast to red HgI_2, there now exist two kinds of stacking as is illustrated in Figure 76. The disorder of the yellow-orange crystals is caused by an irregular succession of these two kinds of stacking. The superstructure of ZnI_2 [1010] might be similar to red-orange HgI_2.

It was argued [346] that the reported $CdCl_2$-type modification of $ZnCl_2$ was erroneous and would in fact correspond to a tetrahedral 'C 19'-type structure as shown in Figure 77. In this still hypothetical structure, the $[ZnClCl_{3/3}]$ tetrahedra share three corners with three neighboring tetrahedra. The bonding is extremely asymmetric since the anions in one layer are bonded solely by one single bond while the anions of the opposite layer are bonded to three cations. If this structure really exists it appears to be more appropriate for say SiPCl.

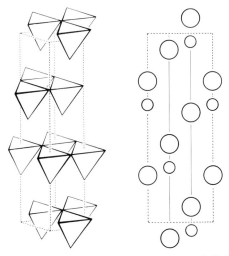

Fig. 77. The hypothetical C19τ structure of $ZnCl_2$.
Layers of $[ZnCl Cl_{3/3}]$ tetrahedra (left) and (110) section (right).

TABLE 61

C19 τ type, trigonal, C_{3v}^5—R3m (No. 160), $Z = 1(3)$.
hexagonal axes: all atoms in 3(a):
$(0, 0, 0; \frac{1}{3}, \frac{2}{3}, \frac{2}{3}; \frac{2}{3}, \frac{1}{3}, \frac{1}{3}) + (0, 0, z)$

$ZnCl_2$: $a = 3.77$ Å, $c = 17.80$ Å;
$z(Zn) = \frac{1}{8}$; $z(Cl_I) = 0$, $z(Cl_{II}) = \frac{1}{2}$ [1, 346]

A true layer structure with $[MX_{4/2}]$ tetrahedra sharing two corners and an opposite edge (compare Figure 22c) is realized in $GaPS_4$ [352a]. The monoclinic structure (Figure 78, Table 62) is based on a hexagonal close-packing. Modifications with different stacking might well be possible.

Two shared opposite edges of the $[MX_{4/2}]$ tetrahedra lead to fibre structures

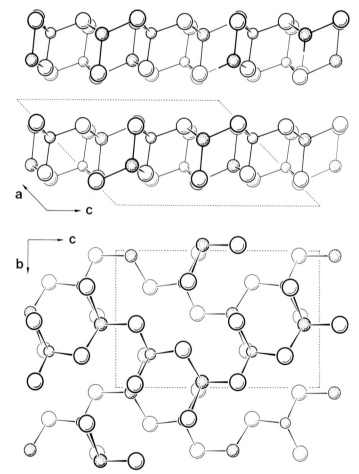

Fig. 78. Projection parallel and perpendicular to the layers of the $GaPS_4$ structure.
Smallest spheres: P
largest spheres : S
dotted spheres : Ga

TABLE 62

GaPS$_4$ structure, monoclinic, C_{2h}^5—P2$_1$/c (No. 14), Z = 4.
All atoms in 4(e): ±(x, y, z; x, $\frac{1}{2}$−y, $\frac{1}{2}$+z)

GaPS$_4$: a = 8.603 Å, b = 7.778 Å, c = 11.858 Å, β = 135.46°

	x	y	z	
Ga	0.624 0	0.536 9	0.235 4	
P	0.358 6	0.259 0	0.010 3	
S$_I$	0.224 1	0.040 1	0.010 3	[352a]
S$_{II}$	0.251 6	0.479 5	0.029 7	
S$_{III}$	0.697 6	0.268 3	0.204 8	
S$_{IV}$	0.748 4	0.773 2	0.207 8	

such as the SiS$_2$ (C42) type which is found in SiO$_2$, SiS$_2$, SiSe$_2$, BPS$_4$(r), BeCl$_2$, BeBr$_2$ and BeI$_2$. In the structure of AlPS$_4$ [353] the SiS$_2$-type fibre layers are alternately crossed, which leads to a fibrous layer structure (Figure 79; Table 63).

Other tetrahedral layer structures would be feasible with different packing of the [MX$_4$] tetrahedra or other stacking of fibres, etc. Thus, a less elegant HgI$_2$ modification may be created by filling the sites of one set of tetrahedral holes

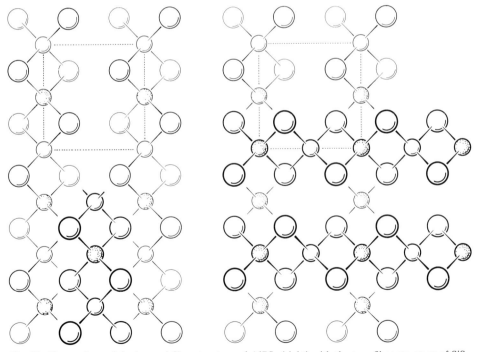

Fig. 79. Comparison of the layered fibre structure of AlPS$_4$ (right) with the true fibre structure of SiS$_2$ and BPS$_4$ (left). The structures are distinguished from each other solely by the stacking of the fibre layers.

Large spheres: S
dotted spheres: P

TABLE 63

$AlPS_4$ structure, orthorhombic, D_2^1—P222 (No. 16), $Z = 2$.
S_I and S_{II} in 4(u): x, y, z; $\bar{x}, \bar{y}, z$; $x, \bar{y}, \bar{z}$; $\bar{x}, y, \bar{z}$.
P in 1(c): $0, \frac{1}{2}, 0$ and 1(d): $0, 0, \frac{1}{2}$.
Al in 1(a): $0, 0, 0$ and 1(f): $\frac{1}{2}, 0, \frac{1}{2}$.

$AlPS_4$: $a = 5.61$ Å, $b = 5.67$ Å, $c = 9.05$ Å

	x	y	z
S_I	0.200	0.260	0.125
S_{II}	0.740	0.800	0.630	[353]

only: —τ_1—0—τ_1—0—. Table 59 may be useful for predicting chalcopyrite-related MX_2 structures. In fact, several tetrahedral layer structures still await determination. Thus, a layer-type high-temperature modification of BeI_2 is obviously based on a ccp of iodine with 32 molecular units in the orthorhombic cell [354]. A monoclinic layer structure is also reported for BPS_4(h) [355], and we suspected that the monoclinic high-temperature modification of GeS_2 [356] and the recently reported orthorhombic modification of $GeSe_2$ [356a] well might have been based on a hexagonal close-packing but with the Ge atoms in tetrahedral instead of the proposed [357] octahedral coordination (a distorted eight-layer

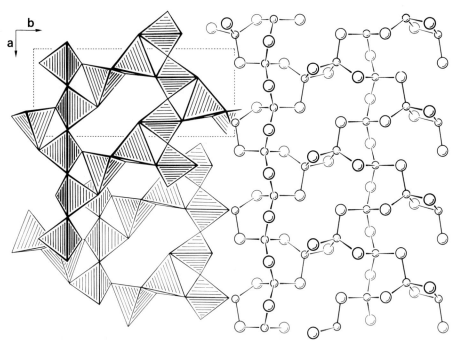

Fig. 79a. The monoclinic structure of GeS_2 (h) viewed normal to the layers. In the left part the [GeS_4] tetrahedra of both the upper layer (above) and the lower layer (below) are indicated. In the right part only the upper layer is represented.

Small spheres: Ge.

CdI$_2$ type was proposed for flaky modifications of GeS$_2$ and GeSe$_2$ with orthorhombic unit cells [357, 358] which obviously correspond to Zachariasen's three-dimensional modification). Meanwhile the monoclinic GeS$_2$(h) structure has been determined [356b]. It turned out to be somewhat more complicated since it contains elements of the SiS$_2$ structure. In GeS$_2$(h) chains formed by corner-sharing [GeS$_4$] tetrahedra run along the a-axis. Two such chains that are equivalent by the symmetry operation around the 2$_1$ axes are bridged via corners by two edge-sharing tetrahedra. These bridges repeat to both sides after every second tetrahedron as is illustrated in Figure 79a. There are two such layers within the identity period along the c-axis. The peculiarity of the structure thus explains the striking plastic flexibility of the crystals. Another GeS$_2$(h, p) phase with a tetragonal cell was prepared at 1–5 kbar and 575–840°C in the form of fine platelets [359, 360]. We presume therefore that this modification has a layer structure based on a cubic close-packing unless it contains similar structure elements as GeS$_2$(h).

Unfortunately, our knowledge on the isoelectronic boron chalcogenohalides is rather meagre though the existence of BSCl, BSBr, BSI, BSeCl, BSeBr and BSeI is certified. The boron selenohalides are described as white powders with an unctuous aspect, and a chain structure (BClS$_{2/2}$)$_\infty$ is proposed [361]. For the colorless needles of BSBr [362a], a molecular structure with planar B$_3$S$_3$ rings has been reported [1]. However, the (BSBr)$_3$ units are not arranged in a plane but are inclined towards each other and the crystals will probably cleave along (10$\bar{2}$) rather than along (010). The acicular crystals of (HBS$_2$)$_3$ = (BSSH)$_3$ [362] were postulated to have the same structure [1].

TABLE 63a

GeS$_2$(h) structure, monoclinic, C$_{2h}^5$–P2$_1$/c (No. 14), Z = 16.
all atoms in 4(e): ±(x, y, z; x, $\frac{1}{2}$ – y, $\frac{1}{2}$ + z)

GeS$_2$(h): a = 6.69 Å, b = 16.1 Å, c = 11.46 Å, β = 90.6° [356]
a = 6.720 Å, b = 16.101 Å, c = 11.436 Å, β = 90.88° [356b]

	x	y	z
Ge$_I$	0.343 0	0.153 1	0.221 3
Ge$_{II}$	0.171 4	0.151 4	0.779 8
Ge$_{III}$	0.839 6	0.002 6	0.705 7
Ge$_{IV}$	0.673 4	0.307 3	0.277 7
S$_I$	0.668 7	0.177 3	0.214 1
S$_{II}$	0.279 0	0.037 0	0.122 6
S$_{III}$	0.229 2	0.112 6	0.393 3
S$_{IV}$	0.172 6	0.256 4	0.136 9
S$_V$	0.427 2	0.331 9	0.400 0
S$_{VI}$	0.921 1	0.331 6	0.402 0
S$_{VII}$	0.676 7	0.390 9	0.123 6
S$_{VIII}$	0.166 1	0.474 5	0.201 1

GeSe$_2$(h): a = 7.037 Å, b = 16.821 Å, c = 11.826 Å, $\beta \approx$ 90° [356a]
a = 7.016 Å, b = 16.796 Å, c = 11.831 Å, β = 90.65° [1034]
for the atomic positions, see [1034]

5. Polycationic Tetrahedral Layer Structures

a. GeAs$_2$-TYPE COMPOUNDS

In its orthorhombic low-temperature modification GeAs$_2$ [371] contains two kinds of chemically and four kinds of crystallographically inequivalent anions. When viewed along the c-axis the structure consists of pentagonal tubes (Ge$_2$As$_3$)$_n$ bound together by the fourth As atoms (Figure 80). Each Ge atom is tetrahedrally coordinated by 4 As atoms. Half the As atoms are surrounded by 3 Ge at angles between 90° (as for p^3 bonds) and the tetrahedral bond angle (As$_I$: 90.4° (2×) and 99.1°; As$_{IV}$: 99.6° (2×) and 98.5°) as in normal valence compounds. The other two As atoms form zigzag chains along the c-axis and are bonded to one Ge atom only. The angles within the As chain are 96.4°, those towards the Ge neighbors are 98.4° and 100.8°. The As—As distances within the chain are 2.50 Å, slightly larger than expected for single bonds. The non-metallic character of GeAs$_2$ (an energy gap of 1.06 eV was reported [371a]) proves that the bonds within the As chains are single bonds.

The lone electron pairs on the As atoms separate the —(Ge$_2$As$_3$)$_n$—As$_n$—(Ge$_2$As$_3$)$_n$—As$_n$—units into discrete waved layers parallel with the cleavage plane (100).

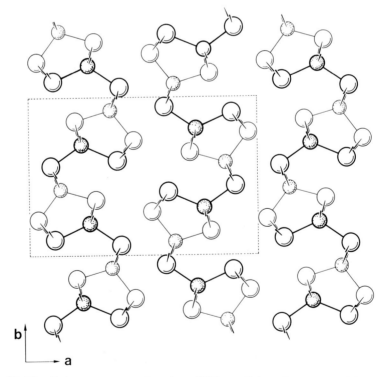

Fig. 80. The GeAs$_2$ structure projected on (001) parallel to the corrugated layers. Dotted spheres: Ge.

TABLE 64
GeAs$_2$ structure, orthorhombic, D_{2h}^9—Pbam (No. 55), $Z=8$.
  Ge$_I$, As$_I$ and As$_{II}$ in 4(g): $\pm(x, y, 0; \frac{1}{2}+x, \frac{1}{2}-y, 0)$
  Ge$_{II}$, As$_{III}$ and As$_{IV}$ in 4(h): $\pm(x, y, \frac{1}{2}; \frac{1}{2}+x, \frac{1}{2}-y, \frac{1}{2})$

GeAs$_2$: $a = 14.74$ Å, $b = 10.12$ Å, $c = 3.721$ Å [366]
     $a = 14.76$ Å, $b = 10.16$ Å, $c = 3.728$ Å [371]

	x	y		x	y	
Ge$_I$	0.137 8	0.419 1	Ge$_{II}$	0.266 7	0.202 4	
As$_I$	0.295 3	0.353 3	As$_{III}$	0.111 7	0.101 4	[4, 371]
As$_{II}$	0.038 4	0.226 0	As$_{IV}$	0.402 2	0.062 8	

SiP$_2$(r): $a = 13.97$ Å, $b = 10.08$ Å, $c = 3.436$ Å [369]
       $a = 13.99$ Å, $b = 10.09$ Å, $c = 3.43$ Å [389]
SiAs$_2$(r): $a = 14.53$ Å, $b = 10.37$ Å, $c = 3.636$ Å [369]
SiP$_2$(r'): $a = 13.64$ Å, $b = 20.06$ Å, $c = 3.51$ Å [390a]   (related)

Based on the chemical formula, one might first expect the structure of GeAs$_2$ to contain equivalent anions arranged in pairs. However, as far as we know, no such structure with the cation in tetrahedral coordination has been found yet. In the metallic high-temperature pyrite modification of SiP$_2$ and SiAs$_2$ the fourth bond orbital of the anion is engaged too, thus leading to a cation coordination number 6 [390].

We wonder whether it might be possible to synthesize isostructural phases such as AlPS or GaAsSe with the phosphorus and arsenic atoms in the zigzag chains. Will the As atoms form the chains in SiPAs?

The unit cell of SiP$_2$(r) was claimed to be in fact twice as large as the GeAs$_2$-type cell, requiring a doubling of the *b*-axis [390a]. The deviations of the other axis, however, are such that this cell might well belong to a second modification.

b. LAYER-TYPE GROUP III MONOCHALCOGENIDES AND GROUP IV MONOPNICTIDES

The smaller Group IIIB and IVB cations preferentially occur in tetrahedral coordination as is exemplified by the normal valence chalcogenides and the Group IIIB pnictides which are famous for their sp^3 bonds. Since most of the 1:1 compounds listed in Table 65 are semiconductors, they must crystallize with cation pairs unless half of the cations are monovalent as in the TlSe-type compounds. With the exception of InS (and possibly SiN) the cation pairs induce a layer structure. Three of these structures are related to the MoS$_2$ layer type. We even wonder why more modifications were not found since each MX$_2$ structure, which is based on layers of trigonal prisms or octahedra, can be used to generate a possible M$_2'$X$_2$ layer structure simply by replacing the cation M by a cation pair M'—M' pointing along the threefold axis. The trigonal-prismatic or octahedral (trigonal-antiprismatic) X$_3$—M—X$_3$ coordination thereby transforms into the

TABLE 65
Modifications of M^{III} monochalcogenides and M^{IV} monopnictides. Layer structures are underlined

AlS ?	SiN: Si_2N_2
GaS; hexag. GaS type [382]	SiP(r): <u>orthorhombic SiP type</u>
	(p): NaCl type
GaSe β: hexag. GaS type? [364, 382]	SiAs: <u>GaTe type</u> [366, 370, 371, 378]
γ: <u>rhombohedral</u> [364, 365]	GeP(r): <u>GaTe type</u> [369]
ε: <u>hexagonal ($P\bar{6}$)</u> [365]	(p_1): tetrag. SnP type [373]
polytypes [383a]	(p_2): NaCl type
GaTe: <u>monoclinic</u> [366]	GeAs(r): <u>GaTe type</u> [366]
metastable: GaS type [367]	(p): tetragonal SnP type [373]
InS: <u>orthorhombic</u>	GeSb (metastable): tetrag. 'NaCl' type [377]
InSe: <u>rhomb. γ-GaSe type</u> [366, 397]	SnP(r): <u>hexagonal layer structure</u> ($=SnP_3$?)
: <u>hexag. ε-GaSe type</u>? [381]	[375]
(p): <u>polytype transition</u> [368]	(p_1): tetragonal [374]
InTe: tetrag. TlSe type [365]	(p_2): NaCl type [374]
metastable: tetrag.	
NaCl type [1]	SnAs ($p \leqslant 150kb$): NaCl type [376]
>30kb: NaCl type [376, 368]	SnSb(r): rhombohedral NaCl type
TlS: tetragonal TlSe type	(p_1): NaCl type
TlSe: tetragonal $Tl^+(Tl^{3+}Se_2)^-$	(p_2): CsCl type [376]
TlTe: tetragonal [372]	

required tetrahedral X_3—M'—M' coordination. A selection of such real or hypothetical layer types is reproduced in Figure 81.

The accuracy of the reported positional parameters is such that we abstain from listing the bond distances. It is, however, readily seen that the interlayer bonding will depend on the kind of stacking. Thus, in the β-GaSe or GaS type, each cation has a remote anion neighbor in the adjacent layer and vice-versa. In the ε-GaSe and γ-GaSe structures, on the other hand, half the atoms have no opposite neighbor. The polar bonding across the layers will thus be strongest in the β-GaSe type. GaS, which, in this structure family, has the highest ionic contribution to the bonding, indeed does crystallize only with this stacking. We doubt that, using this interpretation, the metastable modification of GaTe really does crystallize in the GaS structure, and we would rather expect a ε-GaSe-type stacking. If we shift the GaS layers relative to each other, we end up with a stacking analogous to that found in NbS_2, where all the cations are in line. This stacking is obviously unfavorable in non-metallic compounds. As Basinski et al. [391] demonstrated, the stacking-fault energy is much higher in GaS than in ε-GaSe and γ-GaSe. According to Terhell and Lieth [383], ε-GaSe is partly disordered anyway.

In addition to the 2H and 3R modifications known for GaSe three new polytypes, 9R, 12R and 15R have recently been reported [383a]. The 9R polytype might represent a rhombohedral stacking of the 3R unit: *ABC BCA CAB* instead of *ABC ABC ABC* though the relative orientation of the sandwiches may be different.

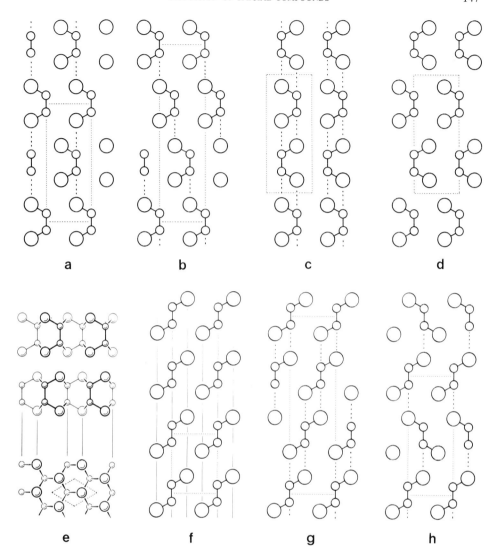

Fig. 81. Layer structures MX with cation pairs and tetrahedral cation coordination. (a) – (d) and (f) – (h): (110) sections through the hexagonal cells; (e): Projection onto the (110) plane (above) and onto the (001) plane of the GaS structure (c). This drawing should help imagination to link the rather abstract representation of Figures (a) – (d) and (f) – (h) with the three-dimensional picture of these structures.

(a): ε-GaSe type with stacking AββA CααC derived from the 2H—$Nb_{1+\delta}S_2$ type.
(b): γ-GaSe type with stacking AββA CααC BγγB derived from the rhombohedral MoS_2 type.
(c): β-GaSe or GaS type with stacking AββA BααB derived from the hexagonal MoS_2 type.
(d): hypothetical structure with stacking AββA CββC derived from the NbS_2 type.

(f, g, h): hypothetical MX layer structures AββC, AββC BγγA CααB and AββC AγγB derived from the CdI_2 (C6), $CdCl_2$ and CdI_2 (C27) type, respectively. The rhombohedral analog of Fig. (h), the type AββC AγγB CααB CββA BγγA BααC, has been omitted as well as the various combinations that are possible between the given types.

TABLE 66

The GaS, γ-GaSe and ε-GaSe structure types
GaS type, hexagonal, D_{6h}^4—P6$_3$/mmc (No. 194), $Z=4$.
All atoms in 4(f): $\pm(\frac{1}{3}, \frac{2}{3}, z; \frac{1}{3}, \frac{2}{3}, \frac{1}{2}-z)$

GaS: $a=3.585$ Å, $c=15.50$ Å; $z(Ga)=0.17$, $z(S)=0.60$ [382]
$a=3.586$ Å, $c=15.500$ Å [383]
$a=3.587$ Å, $c=15.492$ Å; $z(Ga)=0.1710$, $z(S)=0.6016$ [1004]
GaS$_{0.59}$Se$_{0.41}$: $a=3.660$ Å, $c=15.786$ Å; $z(Ga)=0.1725$, $z(S, Se)=0.6004$ [1030]
GaS$_{0.5}$Se$_{0.5}$: $a=3.672$ Å, $c=15.832$ Å [383]
GaS$_{0.25}$Se$_{0.75}$: $a=3.689$ Å, $c=15.871$ Å [383]
GaSe (β) ?: $a=3.742$ Å, $c=15.919$ Å; $z(Ga)=0.177$, $z(Se)=0.60$ [4]
$a=3.755$ Å, $c=15.94$ Å; $z(Ga)=0.180$, $z(Se)=0.590$ [3]
$a=3.759$ Å, $c=16.02$ Å [384]
GaTe: $a=4.06$ Å, $c=16.96$ Å; $z(Ga)=0.170$, $z(Te)=0.602$ [367]
Ga$_2$GeTe$_3$ (?): $a=4.05$ Å, $c=16.87$ Å [385]
GaGeTe$_2$ (?): $a=4.02$ Å, $c=16.82$ Å [385]
InSe: $a=4.05$ Å, $c=16.93$ Å; $z(In)=0.157$, $z(Se)=0.602$ [381]
Ga$_2$SeTe: $a=6.77$ Å, $c=15.62$ Å (16.52?) [387] probable space group C_{6h}^1—P6/m

γ-GaSe type, trigonal, C_{3v}^5—R3m (No. 160), $Z=2(6)$.
hexagonal axes $(0, 0, 0; \frac{1}{3}, \frac{2}{3}, \frac{2}{3}; \frac{2}{3}, \frac{1}{3}, \frac{1}{3})+$
all atoms in 3(a): 0, 0, z

γ-GaSe: $a=3.747$ Å, $c=23.910$ Å; $z(Ga_I)=0.050$, $z(Ga_{II})=0.950$,
$z(Se_I)=0.767$, $z(Se_{II})=0.567$ [365]
InSe: $a=4.015$ Å, $c=25.00$ Å [366]
$a=4.00$ Å, $c=24.85$ Å [397]
$a=4.00$ Å, $c=25.32$ Å ($a_{rh}=8.76$ Å, $\alpha=26.40°$)
$z(In_I)=0.0557=-z(In_{II})$, $z(Se_I)=0.7724$, $z(Se_{II})=0.5608$. [397a]
$a=4.0046$ Å, $c=24.960$ Å; $z(In)=\pm 0.0555$, $z(Se)=\frac{2}{3}\pm 0.106$ [403]
Ga$_2$InSe$_3$(?) [366a]

ε-GaSe type, hexagonal, C_{3h}^1—P6̄ (No. 174), $Z=4$.
Ga$_I$ in 2(g): $\pm(0, 0, z)$
Ga$_{II}$ and Se$_I$ in 2(i): $\frac{2}{3}, \frac{1}{3}, z; \frac{2}{3}, \frac{1}{3}, \bar{z}$
Se$_{II}$ in 2(h): $\frac{1}{3}, \frac{2}{3}, z; \frac{1}{3}, \frac{2}{3}, \bar{z}$

ε-GaSe: $a=3.743$ Å, $c=15.919$ Å: $z(Ga_I)=0.075$, $z(Ga_{II})=0.575$,
$z(Se_I)=0.150$, $z(Se_{II})=0.650$ [9]
$a=3.755$ Å, $c=15.946$ Å [383]
GaS$_{0.2}$Se$_{0.8}$: $a=3.719$ Å, $c=15.889$ Å [383]
GaSe polytypes: 9R $a=3.75$ Å, $c=71.73$ Å
12R $a=3.75$ Å, $c=95.64$ Å [383a]
15R $a=3.75$ Å, $c=119.55$ Å

4H-GaSe polytype, hexagonal, C_{6v}^4—P6$_3$mc (No. 186), $Z=8$.
Ga$_{III}$, Ga$_{IV}$, Se$_{III}$ and Se$_{IV}$ in 2(b): $\frac{1}{3}, \frac{2}{3}, z; \frac{2}{3}, \frac{1}{3}, \frac{1}{2}+z$
Ga$_I$, Ga$_{II}$, Se$_I$ and Se$_{II}$ in 2(a): 0, 0, z; 0, 0, $\frac{1}{2}+z$.

δ-GaSe: $a=3.755$ Å, $c=31.990$ Å

	z		z	
Ga$_I$	−0.0380	Se$_I$	0.1781	
Ga$_{II}$	0.0394	Se$_{II}$	0.3274	[999]
Ga$_{III}$	0.2119	Se$_{III}$	0.4255	
Ga$_{IV}$	0.2880	Se$_{IV}$	0.5752	

As expected for van der Waals layer structures, the thermal expansion and the isothermal compressibility are larger perpendicular to the layers than parallel with the layers. Measurements are reported for GaSe from 13 to 400 K and from 150 to 320 K, respectively [392].

From calculations of the pseudo-wavefunctions, Schlüter et al. [393, 394] have derived the charge densities of the valence electrons in GaSe. These densities are in fair agreement with what one would expect from a simple chemical picture though we were somewhat deceived to find no hint of the s^2 pair of the anion. The lowest energy bands are formed by Se 4s states only. They are obviously non-bonding and the electrons occupying them behave like core electrons.

It is remarkable that no stable Al analogs nor isoelectronic halides of the type ZnBr are known. For the 1:1 mixture of GaSe and GaTe, the compound Ga_2SeTe, a hexagonal structure with a $\sqrt{3}$ times larger unit cell has been reported [387]. Such a cell, however, would be appropriate for an ordered phase Ga_3Se_2Te, whereas for Ga_2SeTe, one would rather expect an orthorhombic cell with $a' = \sqrt{3}a_0$, $b' = a_0$, $c' = c_0$. Another paper [385] reports GaS-type cells for ternary tellurides Ga_2GeTe_3 and $GaGeTe_2$, which is doubtful. We think that any ternary phase with a GaSe-type related structure should have the composition $Ga_{1-x}Ge_{2x/3}Te$, as e.g. $Ga_3Ge_2Te_6$, since we find no motivation why a metallic phase with Ga—Ga bonds should form. $GaInS_2$ and $GaInSe_2$ could be candidates for ordered GaSe-type structures, whereas all phases containing Tl crystallize in the tetragonal or monoclinically deformed $Tl^+Tl^{3+}Se_2$ structure [395, 396].

As is evident from Table 65, a lower cation-to-anion radius ratio or a lower electronegativity difference changes the structures from the GaSe types to the

TABLE 67

GaTe structure, monoclinic, C_{2h}^3—C2/m (No. 12), $Z = 12$.
All atoms in 4(i): $(0, 0, 0; \frac{1}{2}, \frac{1}{2}, 0) \pm (x, 0, z)$

GaTe: $a = 17.37$ Å, $b = 4.074$ Å, $c = 10.44$ Å, $\beta = 104.2°$ [5, 360]

	x	z		x	z
Ga_I	0.0620	0.5817	Te_I	0.0399	0.8223
Ga_{II}	0.3621	0.0804	Te_{II}	0.3415	0.8236
Ga_{III}	0.2620	0.1993	Te_{III}	0.3426	0.4472

$Ga_{1-x}In_xTe$: $x < 0.05(?)$ [366a]
SiAs: $a = 15.98$ Å, $b = 3.668$ Å, $c = 9.529$ Å, $\beta = 106.0°$ [370]

	x	z		x	z
Si_I	0.0661	0.5884	As_I	0.0369	0.8239
Si_{II}	0.3697	0.0838	As_{II}	0.3368	0.8262
Si_{III}	0.2613	0.2076	As_{III}	0.3479	0.4543

or $a' = 21.23$ Å, $b = 3.667$ Å, $c = 9.530$ Å, $\beta' = 46.3°$ [378]
 with $\mathbf{a'} = \mathbf{a} + 2\mathbf{c}$.
GeP: $a = 15.14$ Å, $b = 3.638$ Å, $c = 9.19$ Å, $\beta = 101.1°$ [369]
GeAs: $a = 15.59$ Å, $b = 3.792$ Å, $c = 9.49$ Å, $\beta = 101.3°$ [9]

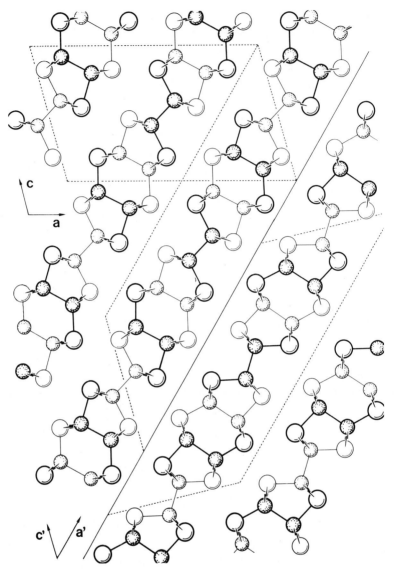

Fig. 82. The monoclinic structure of SiAs projected along [010]. Bryden's (a, c) and Schubert's cell (a', c') are indicated. Two (a', c') cells added along a glide plane (full line) define the orthorhombic cell of SiP. Dotted spheres: Si.

monoclinic GaTe or the orthorhombic SiP type. The coordination in these structures is essentially the same as in the former structures, except that they are more distorted, but instead of all being parallel, one third of the cation pairs is now orientated at nearly right angles. The structure of SiAs is illustrated in Figure 82. The straight line indicates a twin plane. The orthorhombic cell of SiP is composed of two SiAs cells, one on each side of such a twin plane. The SiP structure is thus a stacking variant of the SiAs type and may occur as a second

TABLE 68

SiP structure, orthorhombic, C_{2v}^{12}—Cmc2_1 (No. 36), $Z = 24$
(0, 0, 0; $\frac{1}{2}$, $\frac{1}{2}$, 0) +
all atoms in 4(a): 0, y, z; 0, $\bar{y}$, $\frac{1}{2}+z$.

SiP: $a = 3.54$ Å, $b = 20.84$ Å, $c = 13.96$ Å [388]
$a = 3.510$ Å, $b = 20.592$ Å, $c = 13.601$ Å [379]
$a = 3.5118$ Å, $b = 20.488$ Å, $c = 13.607$ Å [860]

	y	z		y	z	
Si$_\text{I}$	0.170 70	0.262 19	P$_\text{I}$	0.999 83	0.362 51	
Si$_\text{II}$	0.060 18	0.219 99	P$_\text{II}$	0.230 93	0.120 27	
Si$_\text{III}$	0.435 93	0.319 72	P$_\text{III}$	0.344 32	0.413 14	[860]
Si$_\text{IV}$	0.432 39	0.148 27	P$_\text{IV}$	0.541 26	0.117 96	
Si$_\text{V}$	0.795 00	0.163 11	P$_\text{V}$	0.689 59	0.364 73	
Si$_\text{VI}$	0.798 08	0.335 22	P$_\text{VI}$	0.885 98	0.069 94	

modification of the GaTe-type compounds. The GaTe structure can indeed be derived from the CdI$_2$ structure though the misorientation of one third of the cations causes severe distortions. The SiP structure then corresponds to the double-hexagonal C27-type CdI$_2$ stacking (*ABAC* of the anions). The cation pairs are arranged in stripes. Two stripes of the kind in Figure 81f with the pairs perpendicular to both stripe and layer, alternate with one stripe in which the pairs lie within the layer plane but still at a right angle to the stripe axis.

Since the GaTe and SiP structures contain one third of the cation pairs in different orientation we wonder whether these structures might be adopted by 2 : 1 GaSe-type mixtures such as In$_2$GaSe$_3$ or by GaSe-type–GaTe-type mixtures such as Ga$_2$SiPS$_2$ or Ga$_2$SiAsSe$_2$. The probability that two different cations develop cation-cation bonds in one phase, however, seems to be rather low. Whereas the system GeP—GeAs is continuous [399] we do not know of any reports on mixed-cation systems such as (Ga, In)Se or (Si, Ge)As.

We mentioned already that we do not know of any MX layer structure derived from hcp or ccp anion arrays as proposed in Figure 81f–h. In the cases of Si$_2$Te$_3$ and Ge$_2$Te$_3$, our expectations were not fulfilled although these phases would be candidates for a structure based on Figure 81f and Figure 7b. The red transparent Si$_2$Te$_3$ [400] was described earlier as CdI$_2$-type SiTe$_2$. The simplest explanation for the non-metallic character of this compound would be the occurrence of Si pairs in $\frac{2}{3}$ of the octahedral sites of a CdI$_2$-like structure. The axial ratio $c/a = 1.57$, however, is too low for Si—Si pairs parallel with the trigonal axis. If we want to keep the hcp Te stacking we have to place single Si atoms into octahedral and/or tetrahedral holes. As can be seen from Figure 27 the vertical pairs of tetrahedral sites α—α or γ—γ in a hcp $A\gamma b\alpha C\alpha b\gamma A$ are so close ($c/4$) whereas the octahedral holes are so remote ($c/2$) that Si—Si bond distances could be attained only by considerable distortions. Moreover, a hcp layer structure with octahedral pairs only (*AbCbA CbAbC*) would require a tripling of the CdI$_2$-type c-axis and would lead to a composition MX$_3$. Due to the threefold symmetry, a combination of adjacent tetrahedral and octahedral sites, such as γb or γa,

TABLE 69
Structure models for Si_2Te_3, trigonal, C_{3v}^4—P31c (No. 159), $Z = 4$.

(a) One quarter of the Si atoms in octahedral holes
  Si_{II}, Te_I and Te_{II} in 6(c): x, y, z; $y, x-y, z$; $y-x, \bar{x}, z$;
  $y, x, \frac{1}{2}+z$; $\bar{x}, y-x, \frac{1}{2}+z$; $x-y, \bar{y}, \frac{1}{2}+z$.
  Si_I in 2(b): $\frac{1}{3}, \frac{2}{3}, z$; $\frac{2}{3}, \frac{1}{3}, \frac{1}{2}+z$.
  Si_2Te_3: $a = 7.429$ Å, $c = 13.471$ Å [401]
  Si_I: $z = 0$; Si_{II}: $x = \frac{2}{3}, y = 0, z \approx 0.05$
  Te_I: $x = \frac{1}{3}, y = 0, z \approx 0.14$; Te_{II}: $x = \frac{2}{3}, y = 0, z \approx 0.38$

(b) All Si atoms in tetrahedral holes
  Si_{II} and Te_{IV} in 6(c): as above
  Si_I, Te_{II} and Te_{III} in 2(b): as above   [401]
  Te_I in 2(a): $0, 0, z$; $0, 0, \frac{1}{2}+z$.
  Si_I: $z = 0.5$; Si_{II}: $x = \frac{2}{3}, y = 0, z \approx 0.10$;
  Te_I: $z \approx 0.18$; Te_{II}: $z \approx 0.18$; Te_{III}: $z \approx 0.70$;
  Te_{IV}: $x = \frac{2}{3}, y = 0, z \approx 0.42$

necessarily leads to Si_4 trigonal pyramids. Several combinations of octahedral and/or tetrahedral Si sites were checked by Klein-Haneveld [401] and two possibilities (Table 69, Figure 83) were found to agree with the observed X-ray intensities. In both models clusters of 4Si form an obtuse trigonal pyramid with Si—Si distances of 2.5–2.6 Å depending on the flatness of the pyramid. In both cases, the Si clusters must be connected through $\frac{2}{3}$ bonds, the lower-symmetry Si atoms then showing a formal valence of $+3\frac{1}{3}$, while the high-symmetry Si has to be formally divalent relative to Te. The higher density of the lattice may favor this odd Si_4 cluster formation over the simple Si—Si pairs with coordination number 4 for all atoms. These clusters look like part of the Si sublattice met in SiCl and $CaSi_2$ which we discuss in Section 6. According to Dittmar [1034] the Si_2Te_3 structure is a distorted BiI_3 structure with Si—Si pairs in octahedral holes oriented in the three crystallographic directions (see Appendix, Table 202).

The hexagonal cell reported for Ge_2Te_3 [402] is related to that of Si_2Te_3:

$$a \approx \frac{1}{\sqrt{3}} a(Si_2Te_3), \qquad c = \frac{15}{4} c(Si_2Te_3).$$

The atomic arrangement proposed for this phase looks rather unlikely even if Ge_2Te_3 should prove to be metallic. All cations are assumed in octahedral coordination but instead of a rhombohedral cell with uniform stacking such as would be

  AbCbA BaCaB CaBaC AcBcA BcAcB CbAbC

or

  ABcAbCaBcABCaBcAbCaBCAbCaBcAbC

Chiragov and Talibov [402] proposed a rhombohedral cell

  AbCbA CbAbC BcAcB AcBcA CaBaC BaCaB

(capitals: Te atoms)

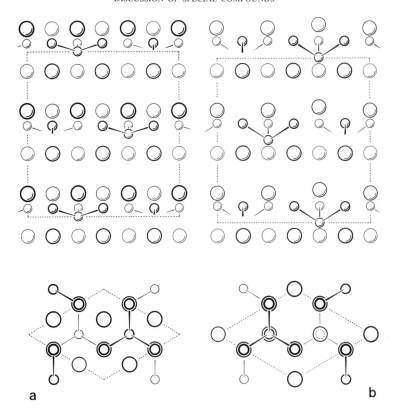

Fig. 83. (101) sections and projections along [001] of the two models proposed for Si_2Te_3. (a) One quarter of the Si atoms in octahedral holes; and (b) All Si atoms in tetrahedral holes. According to Dittmar [1034] Si pairs are located in octahedral holes.

which is rather unreasonable as it contains triple layers of Te atoms. Like Sn_2S_3 the possibly non-metallic Ge_2Te_3 also might contain divalent and tetravalent cations. We then would expect Ge^{2+} to be located in layers of octahedral holes of a Te close packing while Ge^{4+} should be found in tetrahedral holes similar to the $ZnIn_2S_4$ phases.

6. Compounds with Polycationic Layers

Quite a number of carbon and silicon compounds are known which contain the tetrahedrally coordinated cation in the form of puckered sheets.

Fluorination of graphite leads to a polycarbon monofluoride of ideal composition CF. Incompletely fluorinated samples with compositions $CF_{0.7\cdots0.8}$ are nearly black while those of compositions $CF_{0.8\cdots0.95}$ and $CF_{0.95\cdots1.13}$ are grey and white, respectively. Carbon monofluoride was found to be a superior lubricant especially at high temperatures and in oxidizing atmospheres [404]. CF is used

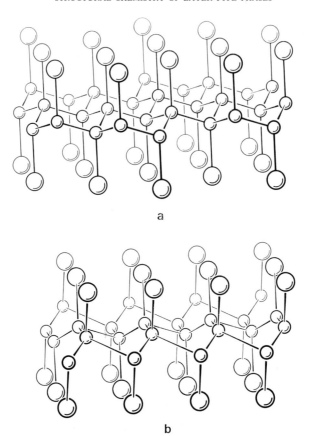

Fig. 84. Part of one layer of the structure of CF. (a) chair model realized in the Si monohalides; and (b) boat model after Ebert et al. [407].

also as the cathode of lithium high-energy-density batteries utilizing organic electrolytes [405]. Its structure is derived from graphite by insertion of three F atoms above and three below every hexagon in each layer. The layers become puckered since all valence electrons of the cations are now engaged in sp^3 σ-bonds. Originally, these layers have been assumed to be infinite arrays of trans-linked cyclohexane chairs [52], as illustrated in Figure 84a. This geometry has the highest symmetry and is met also in the Si monohalides. This structure model, however, is incompatible with nuclear-magnetic-resonance second-moment studies [407] that make an infinite array of cis-trans-linked cyclohexane boats (Figure 84b) more plausible. From the reported hexagonal unit cell $a = 2.54$ Å, $c = 5.80$ Å [406] as well as from the orthorhombic cell $a = 2.51$ Å, $b = 5.13$ Å, $c = 6.12$ Å [407] (the value of c depends on the fluorine content [52]) follows that the unit cell contains only one layer hence the stacking is AA, not AB or ABC as in graphite. Moreover, it is clear that the C—C distance is increased, reflecting the transition from $\frac{4}{3}$ bonds in graphite to single bonds in CF. The limiting stoichiometry of the graphite fluorides will depend upon the grain size of the reacting graphite powder. Carbon atoms at the edges of each layer will tend to bind both axially and equatorially disposed F atoms. The superstoichiometry of saturated CF_{1+x} thus will be determined by the relative amount x of marginal C atoms. The limiting composition of one single C hexagon will be CF_2.

The second graphite derivative, graphite oxide, is a very ill-defined compound, also known as graphite oxyhydroxide or graphitic acid. It appears to have no definite composition. Its rather high electrical resistance strongly varies with composition and can reach values of 10^7 Ω cm at a carbon-to-oxygen ratio of 3 [51]. The color depends on the carbon-to-oxygen ratio and on the water content. Dry samples are green or brown, but they become nearly colorless when hydrated, if the carbon-to-oxygen ratio is close to 2. Graphite oxide swells in water until the individual layers are independent.

In thoroughly dried samples the separation of the carbon layers is ~6.2 Å. This value increases continuously upon reaction with water and values as high as 11 Å have been reported [51]. We are not aware of a structure determination, but different models have been proposed with buckled layers similar to CF. The tetrahedral coordination is probable since most carbon atoms are attached to four distinct ligands and also in view of the nonmetallic properties. The oxygen is obviously bonded by other linkages to two carbon atoms in meta positions on the same layer plane. Carbon atoms not involved in C—O—C bridges may attach hydroxyl groups. The limiting composition for a pure graphite oxide would be C_2O, which however, was never observed. An ordered version of the structure proposed by Ruess is shown in Figure 85 for a composition $C_6O_2(OH)_2$. Hoffmann's model, on the other hand, predicts an ideal composition $C_8O_2(OH)_2$ and contains OH groups and C=C bonds.

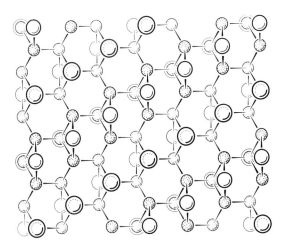

Fig. 85. One possibility for the atomic arrangement in a layer of a 'graphite oxide' $C_6O_2(OH)_2$.
OH: largest spheres
O: medium spheres
C: small dotted spheres.

Fluorination of graphite oxide leads to a light grey graphite oxyfluoride possibly by replacing the OH groups by fluorine [406].

Starting from $CaSi_2$, a whole series of silicon analogs to CF can be synthesized [408, 409]. In the rhombohedral structure of $CaSi_2$ the puckered Si layers are already present. The $CaSi_2$ structure can be described as a deformed close-packing

$$aBAbCBcABcACaBC$$

with the anions marked by capitals. By bleaching out the cations with $SbCl_3$ lepidoidal (scaly) silicon can be prepared as a dark brown powder. Other reactants lead to layered Si compounds. Thus the reaction with ICl and IBr will produce yellow SiCl and yellow-brown SiBr [410]. SiCl and SiBr can be converted to other $(SiX)_n$ layer compounds as for example by the reaction

$$(SiCl)_n + nLiCH_3 \rightarrow (SiCH_3)_n + nLiCl.$$

SiCl reacts with SbF_3 under benzene to form the light-yellow $(SiF)_n$ while $(SiH)_n$ is best prepared from $(SiBr)_n$ and $LiAlH_4$ in ether [411]. Of course, there is a steric condition for the formation of such layer compounds. Thus the reaction of SiCl with methanol CH_3OH does not progress further than to a mixed phase $Si(OCH_3)_{0.8}Cl_{0.2}$. Similarly with ethanol and propanol substitution stops at the concentrations $Si(OC_2H_5)_{0.75}Cl_{0.25}$ and $Si(OC_3H_7)_{0.67}Cl_{0.33}$. These mixed compounds are described as reddish-brown to brass-colored flakes in contrast to the fine powders of most other $(SiX)_n$ preparations.

Free valencies exist in the phases $Si(NHCH_3)_{0.8}$ and $Si(NHC_6H_5)_{0.33}$ [412] similar to the lepidoidal silicon $(Si)_n$. The reaction of $CaSi_2$ with SCl_2 leads to a mixed phase $Si(SCl)_{0.4}Cl_{0.6}$ while with S_2Cl_2 a polysulfide $(SiS)_n$ can be synthesized [413]. The polysulfide ion $\cdots$S—S$\cdots$ does not link different Si sheets but it really connects two Si atoms on the same layer. Taking the Si—Si distances from diamond-type Si, we have a short Si—Si distance of 2.35 Å between atoms of the two different layers of one puckered Si sheet. The Si—Si distance within one Si layer is then 3.84 Å. This close-packing distance has to be bridged by the S_2 polyanion which is easily possible since S—S ≈ 2.05 Å (Figure 86). Geometrically the corresponding selenide and telluride are possible as well. Replacement of S by Se and Te would decrease the X—X—Si angle to ~ 109 and $\sim 103°$, respectively, while it would be 90° for ideal tetrahedral angles at the Si atoms. To our knowledge, only a fibrous modification of SiSe, analogous to fibrous SiO and SiS, has been synthesized until now [414].

In the case of the nitride $(Si_6N_2)_n$ [415] the structural arrangement is less obvious. Based on the chemical formula one might expect N to be placed above discrete triangles of the Si layers. The distance to the center of such a triangle, however, is as large as 2.22 Å whereas the Si—N bond distance was found to be

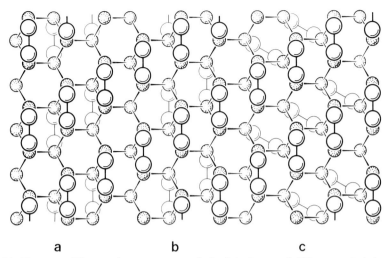

Fig. 86. Three possible atomic arrangements (a, b, c) in layers of SiS-type polychalcogenides.
S: large spheres
Si: small dotted spheres

1.74 Å in Si_3N_4. Even with P, the bonds would have to be plane which is not probable for phosphorus but might be adequate for a hypothetical BC_3. With the larger Sb atom, the Si—Sb—Si angles would be 97° which is between the sp^3 tetrahedral angle and the 90° for pure p^3 bonds.

If in $(SiH)_n$ we replace half the Si—Si bonds by Si—O—Si in such a way that finally Si_6 rings are linked via oxygen only, then we arrive at the layer compound siloxene $Si_6O_3H_6$ (Figure 87). Partial or total substitution of H by other groups is

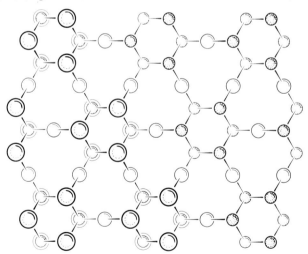

Fig. 87. One layer of the structure of bromosiloxene $Si_6O_3Br_6$. The Br atoms are omitted on the right part.

Br: large spheres
O: medium spheres
Si: small dotted spheres

possible as in the case of SiH or SiCl. Thus, $Si_6O_3H_{6-x}R_x$ with R = OH (x = 1, 2, 6(?)), OCH_3 (x = 1, 2, 2.8, 5.6), OC_2H_5 (x = 1, 2, 4, 6), NHC_2H_5 ($x \le 2$), Br ($x \le 6$) and I ($x \le 2$) have been obtained [416]. With divalent radicals, bridge formation is possible similar to the case of $(SiS)_n$. But since here the Si_6 rings are separated, only one bridge on each side of the Si_6 unit can form, so that for $x > 4$ substitution will be monomeric. Among the diols (1, 3)propanediol has the appropriate bridge length. As in alcohols, the Si—Br bonds only, but not the Si—H bonds, are solvolysed, the reaction of (1, 3) propanediol with $Si_6O_3H_{6-x}Br_x$ will strongly depend on the value of x. Thus, for $x = 2$ (one Br on each side of the Si_6 ring) the alcohol will react with one OH group only whereas for $x = 5$, two —O—$(CH_2)_3$—O— bridges form [417].

As in most of these phases the layers are not organized in the third dimension, all structural information is deduced from chemical reactions.

7. Compounds with d^8 Cations in Square-Planar Coordination

a. PdS_2 AND PdPS

In a strong crystal field, d^8 cations such as Pd^{2+} and Au^{3+} preferentially adopt a square-planar coordination. This two-dimensional coordination of the cation, however, does not necessarily lead to a layer structure. There might as well result a one-dimensional structure as in $PdCl_2$ or a three-dimensional structure as in PdS, PtS or PdP_2. In $PdCl_2$, PdS_2 and $Cs_2AgAuCl_6$ the square-planar coordination can be thought to evolve from an octahedral coordination by strongly elongating the octahedron of nearest anion neighbors along one diagonal. What was a layer structure 'before' now becomes a fibre structure ($PdCl_2$) while the three-dimensional pyrite structure transforms to the layer structure of PdS_2. The reduction of the cation coordination number from 6 to 4 is paralleled by a reduction of the anion coordination number from 4 (tetrahedral) in pyrite to three in PdS_2, the fourth S—Pd bond now becoming a lone electron pair of S pointing towards the adjacent layer. As follows from the atomic separations the distortion is more pronounced in the sulfide than in the selenide:

PdS_2: Pd—4 S at 2.30 Å $PdSe_2$: Pd—4 Se at 2.44 Å
         2 S at 3.28 Å   2 Se at 3.25 Å

TABLE 70

PdS_2 structure, orthorhombic, D_{2h}^{15}—Pbca (No. 61), $Z = 4$.
S in 8(c): ($x, y, z; \frac{1}{2}+x, \frac{1}{2}-y, \bar{z}; \bar{x}, \frac{1}{2}+y, \frac{1}{2}-z; \frac{1}{2}-x, \bar{y}, \frac{1}{2}+z$)
Pd in 4(a): $0, 0, 0; \frac{1}{2}, \frac{1}{2}, 0; 0, \frac{1}{2}, \frac{1}{2}; \frac{1}{2}, 0, \frac{1}{2}$.

PdS_2: $a = 5.460$ Å, $b = 5.541$ Å, $c = 7.531$ Å [4]
   $x = 0.107$, $y = 0.112$, $z = 0.425$

PdSSe: $a = 5.595$ Å, $b = 5.713$ Å, $c = 7.672$ Å [418]

$PdSe_2$: $a = 5.741$ Å, $b = 5.866$ Å, $c = 7.691$ Å [4]
   $x = 0.112$, $y = 0.117$, $z = 0.407$.

The anion–anion distances, S—S = 2.13 Å and Se—Se = 2.36 Å, are slightly larger than expected for covalent single bonds but the non-metallic character of these diamagnetic compounds is a proof that these are single bonds. When the distortion of the anion 'octahedra' in PdS_2 is reduced by pressure the semiconductor transforms into a metal before the pyrite structure is reached [419].

The isoelectronic compounds AgPS and AgPSe, known only as glasses [420], will probably contain monovalent silver as do the lautite-type analogs AgAsS and AgAsSe [421]. AuPS, however, might be a ternary PdS_2 analog.

The PdPS structure, recently determined by Jeitschko [422], is closely related to the PdS_2 structure. It may be interpreted as a combination of the PdS_2 and the PdP_2 structure. Pairs of PdS_2 layers are stacked in such a way that half the anions, the P atoms, come into close contact. As pointed out by Jeitschko [422] the puckered (3 + 4)-connected pentagonal net reproduced in Figure 88, occurs as a building element of most late transition-metal dipnictides and dichalcogenides, such as pyrite, marcasite, $NiAs_2$(r), PdP_2, PdS_2 and PdPS. The two kinds of stacking that generate the PdPS structure are illustrated in Figure 88a, b. A side view of the structures of both PdPS and PdS_2 is given in Figure 89 for comparison. In the upper part of the PdPS drawing the polyanionic unit S—P—P—S is emphasized.

The S atoms occupy the outer half of the anion sites of the double layer and thus retain the three-coordination met in PdS_2, whereas the P atoms mutually complement their coordination to a tetrahedron. Interatomic distances are as follows:

$$
\begin{array}{ll}
\text{Pd—2 P at 2.29 Å} & \text{P—1 S at 2.11 Å} \\
\phantom{\text{Pd—}}\text{2 S at 2.36 Å} & \phantom{\text{P—}}\text{1 P at 2.21 Å} \\
\phantom{\text{Pd—}}\text{1 Pd at 3.20 Å} & \phantom{\text{P—}}\text{2 Pd at 2.29 Å} \\
\phantom{\text{Pd—}}\text{1 S at 3.23 Å} & \text{S—1 P at 2.11 Å} \\
\phantom{\text{Pd—}}\text{1 P at 3.39 Å} & \phantom{\text{S—}}\text{2 Pd at 2.36 Å} \\
\end{array}
$$

As is to be expected from the ionic formula $Pd^{2+}(S-P-P-S)^{4-}$ the silvery blade-like crystals of PdPS and PdPSe are diamagnetic semiconductors with band gaps of 0.7 eV and 0.15 eV, respectively. PdPS and PdPSe are completely miscible [419]. PdAsS and PdSbS as well as PdAsSe and PdSbSe crystallize in a pyrite structure and are therefore metallic, whereas PdPTe is unknown. In this context it is of interest to note that substitution of P by As in the anion chains of PdP_2 is possible up to PdPAs whereas at the composition $PdAs_2$ only the pyrite structure is obtained. We wonder whether the structural PdPS-type → pyrite-type transition in Pd(P, As)S and Pd(P, As)Se solid solutions occurs near the semiconductor → metal transition point.

Whereas AgP_2 is a non-metallic polyphosphide containing monovalent silver, nothing is known of AuP_2 which might well contain Au^{3+} and represent a binary analog of PdPS. We think $Au^{III}P_2$ has a better chance to exist than the ternary PdPS analog $Au^{III}SiS$.

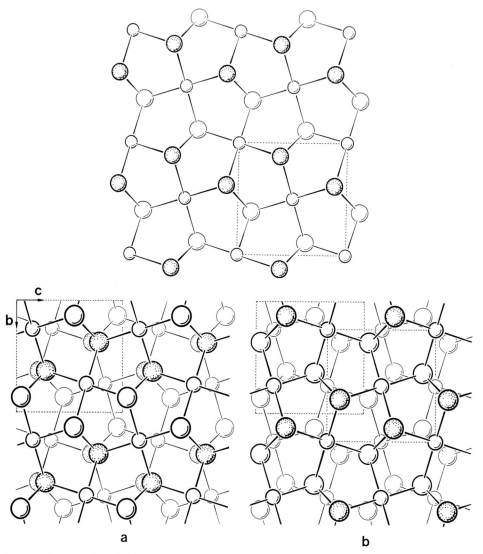

Fig. 88. Layer stacking in the PdPS structure as derived from the puckered (3+4)-connected net (above) according to Jeitschko [422].
(a) The double layer at $x \approx \pm 0.1$ held together by the nearly superposed P atoms.
(b) The PdS$_2$-type stacking of the layers at $x \approx 0.1$ and $x \approx 0.4$. The PdS$_2$ unit cell is also indicated.

Small spheres: Pd
dotted spheres: P

It is rather unexpected that under pressure (1000–1200 °C/65 kbar) the reaction of an atomic ratio of Pd:P:S = 1:1:3 yields silvery metallic phases with tetragonal cells ($a = 5.63$–5.67 Å, $c = 6.48$–6.43 Å) related to those of orthorhombic PdPS. Analysis of one product indicated a composition near PdP$_{0.33}$S$_{1.67}$ for these metallic phases [419]. It was not said whether these crystals appear also blade-like or rather three-dimensional. One might well imagine combinations of PdS$_2$ and

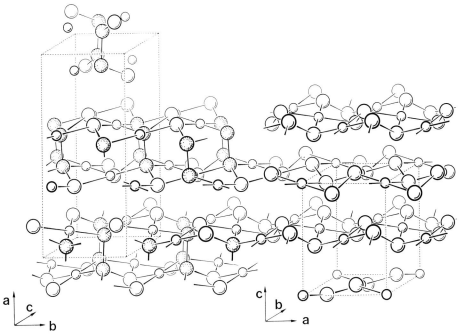

Fig. 89. The structures of PdPS (left) and PdS$_2$ (right) demonstrating the PdS$_2$-like superposition of the (Pd$_2$P$_2$S$_2$)$_\infty$ double layers. The (S—P—P—S) polyanion of PdPS is emphasized in the top layer. Small spheres: Pd, dotted spheres: P, large open spheres: S.

PdPS by an ordered stacking of PdS$_2$ and PdPS layers, m PdS$_2 \cdot n$ PdPS, such as PdP$_{1/2}$S$_{3/2}$ and PdP$_{1/3}$S$_{5/3}$. These phases, however, should be non-metallic and the corresponding unit cells should be a multiple of the PdPS cell rather than half of it. Metallic PdP$_{2/3}$S$_{4/3}$ prepared at high pressure turned out to have the pyrite structure [419].

If instead of the stacking sequence SPdP—PPdS defined by Figures 88a, b we repeat the stacking of Figure 88a (i.e. if we add two or more layers with translative displacement before we produce the PdS$_2$ gap with a 180° rotation of the next layer) we can create sheets SPdP—PPdP—PPdS = Pd$_3$P$_4$S$_2$ as well as Pd$_4$P$_6$S$_2$, etc.

TABLE 71

PdPS structure, orthorhombic, D_{2h}^{14}—Pbcn (No. 60), $Z = 8$.
All atoms in 8(d): $\pm(x, y, z; \frac{1}{2}-x, \frac{1}{2}-y, \frac{1}{2}+z; \frac{1}{2}+x, \frac{1}{2}-y, \bar{z}; \bar{x}, y, \frac{1}{2}-z)$

PdPS: $a = 13.304\,5$ Å, $b = 5.677\,7$Å, $c = 5.693\,2$ Å

	x	y	z	
Pd	0.113 7	0.252 9	0.159 1	
P	0.418 2	0.129 7	0.281 1	[422]
S	0.346 9	0.363 2	0.045 5	

PdPSe: $a = 13.569$ Å, $b = 5.824$ Å, $c = 5.856$ Å [419]

TABLE 72
$Pd_3(PS_4)_2$ structure, trigonal D_{3d}^3—P$\bar{3}$m1 (No. 164), $Z = 1$.
S_{II} in 6(i): $\pm(x, \bar{x}, z; x, 2x, z; 2x, x, \bar{z})$
Pd in 3(e): $\frac{1}{2}, 0, 0; 0, \frac{1}{2}, 0; \frac{1}{2}, \frac{1}{2}, 0$
P and S_I in 2(d): $\pm(\frac{1}{3}, \frac{2}{3}, z)$

$Pd_3(PS_4)_3: a = 6.836$ Å, $c = 7.239$ Å [419]
$S_I: z = 0.58$, $S_{II}: x = 0.174, z = 0.177$; P: $z = 0.30$

b. $Pd_3(PS_4)_2$

The structure of $Pd_3(PS_4)_2$ is transitional between layer type and three-dimensional. It crystallizes in the form of red-purple hexagonal platelets with pronounced (001) cleavage. If we write the chemical formula as $(Pd_3\square)$ $[S_6(PS)_2]$ the relation with the CdI_2 structure becomes obvious. However, the units P^+S^- protrude from the anion layers thus creating puckered layers which grip into each other. Figure 90 shows the main bonds corresponding to a description $Pd_3^{II}(P^+S_3S^-)_2$. The $[PdS_4]$ square appears to be fairly regular with Pd—S = 2.32 Å while the coordination polyhedron around the P atom is a trigonal pyramid rather than a tetrahedron with three distances P—S = 2.10 Å and the fourth P—S$^-$ = 2.00 Å. The covalently bound S atoms have 1P + 2Pd neighbors which together with the lone electron pair form a distorted tetrahedron. The layers are weakly held together electrostatically by the singular S^- ions (The S$^-$—Pd and the S$^-$—S distances to the next layer are 3.59 Å and 3.85 Å, resp.).

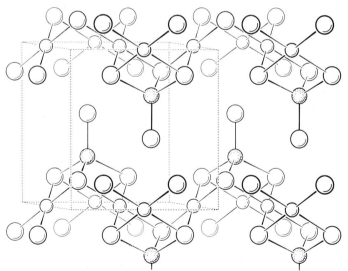

Fig. 90. The hexagonal structure of $Pd_3(PS_4)_2$.
small spheres: Pd
dotted spheres: P

The diamagnetic compound is a semiconductor with an optical energy gap of about 2.2 eV [419].

c. AuTe$_2$Cl

Trivalent gold usually occurs as diamagnetic d^8 ion in square-planar coordination. Examples are AuF$_3$, AuCl$_3$ with dimeric Au$_2$Cl$_6$ units, AuSeBr with infinite ribbons, AuTe$_2$Cl with weakly bound sheets. The latter two examples are polyanionic compounds containing Te—Te pairs. These Te pairs connect the [AuTe$_4$] squares to form corrugated layers as is illustrated in Figure 91a for AuTe$_2$Cl. In AuTe$_2$Cl, the channels in the AuTe$_2$ nets are filled with Cl atoms. As expected for a lone-pair ion, each Te is thus surrounded by 2Au + 1Te at 2.67 Å and 2.78 Å, respectively. The latter distance corresponds to the Te—Te single bond. In a covalent bond scheme there will be a backtransfer of charge from Te to Pd so that tellurium is formally Te$^+$. The Au—Cl distance is as large as 2.93 Å which is 0.6 Å larger than in AuCl$_3$. The weakened Cl—Au bond is obviously compensated by some ionic bonding between Cl and Te, the separation being 3.38 Å between Cl and the 4Te neighbors of the same layer and 3.22 Å to the

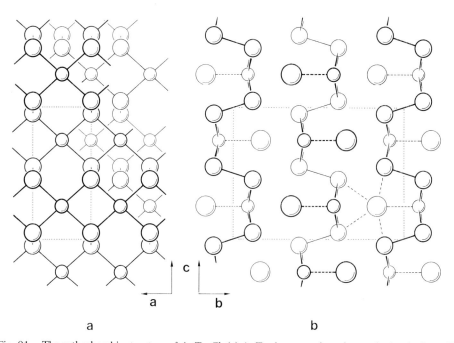

Fig. 91. The orthorhombic structure of AuTe$_2$Cl. (a) AuTe$_2$ layers projected onto the (a, c)-plane, Cl atoms omitted; and (b) Projection on (100). The Cl—Te contacts are indicated by broken lines for the Cl atom at the lower right corner of the unit cell.

Large spheres: Cl
small spheres: Au

TABLE 73
$AuTe_2Cl$ structure, orthorhombic, D_{2h}^{17}—Cmcm (No. 63), $Z=4$.
$(0, 0, 0; \frac{1}{2}, \frac{1}{2}, 0)+$
Te in 8(f): $\pm(0, y, z; 0, y, \frac{1}{2}-z)$
Au and Cl in 4(c): $\pm(0, y, \frac{1}{4})$

$AuTe_2Cl$: $a = 4.0199$ Å, $b = 11.8666$ Å, $c = 8.7728$ Å [423]
  Te: $y = 0.6108$, $z = 0.0516$; Au: $y = 0.0866$; Cl: $y = -0.1609$
$AuTe_2Br$: $a = 4.038$ Å, $b = 12.389$ Å, $c = 8.946$ Å [424]

2Te neighbors of the adjacent layer. As suggested by Haendler et al. [423] $AuTe_2Cl$, $AuTe_2Br$ and also $AuTe_2I$ may structurally be represented as

$$[Au^{III}(Te_2)_{4/4}]^+ X^-$$

In $AuTe_2I$, the $AuTe_2$ nets are symmetrically connected by the iodine atoms into a three-dimensional network. In $AuTe_2Cl$ and $AuTe_2Br$ ionic bonding to the neighboring layers also spoils the true layer character of these phases.

Based on the structural data one would expect non-metallic properties but the silvery white crystals are reported to show metallic conductivity [424].

d. $AuCl_3$ AND $AuBr_3$

Diamagnetic $AuCl_3$ is dimeric in the vapor as well as in the crystalline state. Single crystals can be obtained by sublimation at temperatures above 100°C. At 200°C thin 'cellophane-like' crystals parallel to (100) form whereas at 250–280° sublimation yields a lacy network of wine-red [001] needles within thin orange

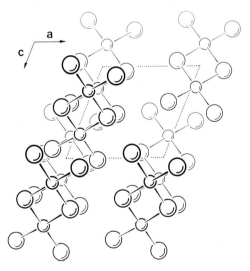

Fig. 91c. The monoclinic Au_2Cl_6 structure viewed in $[0\bar{1}0]$ direction.

TABLE 73a
$AuCl_3$ structure, monoclinic, C_{2h}^5—$P2_1/c$ (No. 14), $Z = 4$.
All atoms in 4(e): $\pm(x, y, z; x, \frac{1}{2}-y, \frac{1}{2}+z)$

$AuCl_3$: $a = 6.57$ Å, $b = 11.04$ Å, $c = 6.44$ Å, $\beta = 113.3°$ [931]

	x	y	z	
Au	0.041 5	0.086 8	0.233 7	
Cl_I	0.258	0.003	0.059	[931]
Cl_{II}	0.335	0.169	0.509	
Cl_{III}	0.820	0.162	0.395	

plates [931]. The monoclinic crystal structure of $AuCl_3$ is of a unique molecular layer type (Figure 91c). Planar Au_2Cl_6 units form zig-zag ribbons parallel to the b-axis and these ribbons are stacked parallel to the c-axis. Trivalent gold is in the usual square planar coordination with Au—Cl = 2.23 and 2.25 Å to the terminal Cl atoms and 2.33 and 2.34 Å to the bridging Cl atoms of the dimer. Cl—Cl distances between the dimers are 3.68 and 3.76 Å within the pseudolayers and 3.54 and 3.55 Å between them.

Diamagnetic AuF_3 and $AuBr_3$ have different structures though with a similar environment of Au^{III}. In $AuBr_3$ [942] the dimeric units all lie in approximately the same plane. The monoclinic cell (Figure 91d) contains two such layers. Distances Au—Br are 2.38, 2.40(2) and 2.41 Å to terminal bromine atoms and 2.46(3) and 2.47 Å to bridging bromine atoms. The shortest Br—Br distance is between the bridging anions, 3,37 Å, as compared with the distances between the terminal pairs, 3.39 and 3.44 Å.

The mixed halides $AuClBr_2$, $AuClI_2$, $AuBrCl_2$, $AuBrI_2$, $AuICl_2$ and $AuIBr_2$ can be synthesized by halogenation of the monohalides [986].

e. AuSe (SEE APPENDIX)

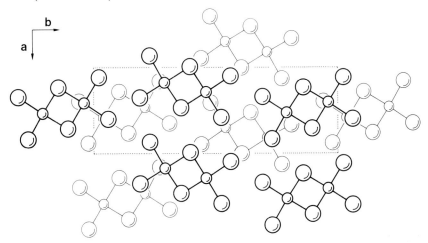

Fig. 91d. The monoclinic structure of $AuBr_3$ projected perpendicularly to the (001) plane. Deviations from planarity of the Au_2Br_6 dimers are neglected.

TABLE 73b
$AuBr_3$ structure, monoclinic, C_{2h}^5—$P2_1/c$ (No. 14), $Z = 8$.
all atoms in 4(e): $\pm(x, y, z; x, \frac{1}{2}-y, \frac{1}{2}+z)$

$AuBr_3$: $a = 6.831$ Å, $b = 20.41$ Å, $c = 8.105$ Å, $\beta = 119.74°$ [942]

	x	y	z	
Au_I	0.734 2	0.038 7	0.259 6	
Au_{II}	0.633 8	0.210 0	0.285 0	
Br_I	0.438 7	0.106 3	0.263 9	
Br_{II}	0.533 4	0.939 6	0.234 6	
Br_{III}	0.024 5	0.977 4	0.248 6	[942]
Br_{IV}	0.939 2	0.142 5	0.298 5	
Br_V	0.825 4	0.310 3	0.294 0	
Br_{VI}	0.333 1	0.269 8	0.289 2	

8. Oxides with Cation-Coordination Numbers 4 · · · 6

a. P_2O_5

Three crystalline and a glassy modification are known of P_2O_5. A metastable rhombohedral form crystallizes as fine needles. The structure of this volatile modification consists of discrete P_4O_{10} molecules. The stable orthorhombic form is built up of an infinite array of rings of ten $[PO_4]$ tetrahedra linked up to form a

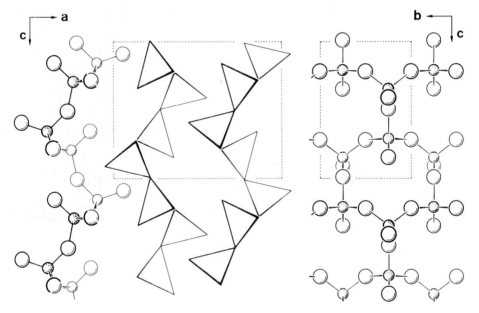

Fig. 92. The orthorhombic structure of metastable P_2O_5. In the projection on (010) on the left side the corrugated layers of $[PO_4]$ tetrahedra are emphasized. In the projection on (100) the six-membered rings are easily seen.

Small dotted spheres: P

TABLE 74

Structure of P_2O_5, orthorhombic, D_{2h}^{16}—Pnma (No. 62), $Z=4$.
O_{IV} in 8(d): $\pm(x, y, z; \frac{1}{2}+x, \frac{1}{2}-y, \frac{1}{2}-z; x, \frac{1}{2}-y, z; \frac{1}{2}+x, y, \frac{1}{2}-z)$
all other atoms in 4(c): $(x, \frac{1}{4}, z; \frac{1}{2}+x, \frac{1}{4}, \frac{1}{2}-z)$

P_2O_5: $a=9.23$ Å, $b=4.94$ Å, $c=7.18$ Å

	x	y	z	
P_I	0.244		−0.288	[425]
P_{II}	−0.098		0.156	
O_I	−0.219		0.011	
O_{II}	−0.142		−0.346	
O_{III}	0.055			
O_{IV}	0.136	0.000	−0.282	

three-dimensional network in which the oxygen atoms are approximately close-packed. In the third form which is also orthorhombic, there is a similar linking-up of the [PO_4] tetrahedra by sharing three O with neighboring P atoms as in V_2O_5. Rings of six [PO_4] tetrahedra form corrugated layers parallel with (100) as is illustrated in Figure 92. In all three forms the structural unit may be represented as [$P^+O_{3/2}O$] in order to allow for sp^3 bonds on the pentavalent phosphorus atoms.

b. THE d^0 TRANSITION-ELEMENT OXIDES AND RELATED HYDROXIDES

In an oxide $M^vO_{v/2}$ with cation valence v the cation coordination polyhedron is characterized by [$MO_{CN/n}$] where CN is the cation coordination number, the oxygen atoms being common to n polyhedra, $n=2CN/v$. On going from the left to the right part of Table 75, the valence v increases, and as a consequence, the packing of the polyhedra becomes looser. At the same time, the size of the cations decreases which is paralleled by a reduction of their coordination number. This can be checked in Table 75 where a region with network structures on the left

TABLE 75

Structural characterization of the d^0 transition-element oxides. Cation coordination numbers and coordination polyhedra.

Sc_2O_3	TiO_2	V_2O_5	CrO_3	Mn_2O_7	
CN 6	CN 6	CN 4–5	CN 4	CN 4	
[$MO_{6/4}$]	[$MO_{6/3}$]		[$MO_2O_{2/2}$]	[$MO_3O_{1/2}$]	
3-dim.	3-dim.	layers	chains	dimers	
Y_2O_3	ZrO_2	Nb_2O_5	MoO_3	Tc_2O_7	RuO_4
CN 6	CN 7+8	CN 6	CN 4–6	CN 4	CN 4
3-dim.	3-dim.	3-dim.	fibre-layers	dimers	[MO_4]
					monomers
Lu_2O_3	HfO_2	Ta_2O_5	WO_3	Re_2O_7	OsO_4
CN 6+7	CN 7	CN 6	CN 6	CN 4+6	CN 4
3-dim.	3-dim.	[$MO_{4/2}O_{2/4}$]	[$MO_{6/2}$]	[$MO_2O_{2/2}$]+[$MO_2O_{4/2}$]	monomers
		3-dim.	3-dim.	layers	

side is separated by three layer structures from the region with chain and molecular structures on the right side. Hydratization gives rise to an increase of the anion/cation ratio, thus reducing the value of n. The layer structure of V_2O_5 is thus transformed into a fibre structure in monoclinic $V_2O_5 \cdot 3H_2O$ [1], while the complex layers of Re_2O_7 are replaced by $Re_2O_7(OH_2)_2$ dimers in the layered molecular structure of $Re_2O_7 \cdot 2H_2O$ [426]. Incorporation of H_2O transforms the MoO_3 double layers into the single layers of metastable yellow β-$MoO_3 \cdot H_2O$ which can take up additional H-bonded H_2O molecules between these layers to form $MoO_3 \cdot 2H_2O$. Isolated double chains $[MoO_2O_{3/3}(OH_2)]_\infty$, on the other hand, are present in the stable 'white' modification of $MoO_3 \cdot H_2O$ which crystallizes as colorless needles [427, 428].

c. Tc_2O_7

Very thin pale-yellow platelets of Tc_2O_7 can be grown by sublimation in a temperature gradient. Platelet surfaces and cleavage planes are parallel with (100) which is well accounted for by the structure illustrated in Figure 93. It is a layered molecular structure with oxygen tetrahedra around each Tc atom, $[TcO_3O_{1/2}]$, connected by a linear M—O—M bridge to form a dimeric unit very much like the complex anion in the thortveitite ($Sc_2Si_2O_7$) structure. The Tc—O distance within the bridge is 1.84 Å, the distances to the other 3 oxygen atoms range from 1.66 Å to 1.71 Å. As pointed out by Krebs [429] these bonds are characterized by strong π-bonding contributions similar to those in the other oxides with high cation valence.

The structure of $Re_2O_7 \cdot 2H_2O$ bears similar features [426]. The dimeric unit is here made up of a $[ReO_3O_{1/2}]$ tetrahedron and a distorted $[ReO_3(OH_2)_2O_{2/2}]$ octahedron, which are connected by a practically linear Re—O—Re bond (179°).

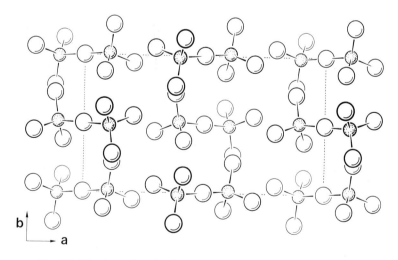

Fig. 93. The layered molecular structure of Tc_2O_7 projected along [001]. Dotted spheres: Tc

TABLE 76

Tc_2O_7 structure, orthorhombic, D_{2h}^{15}—Pbca (No. 61), $Z = 4$.

Tc, O_{II}, O_{III} and O_{IV} in 8(c): $\pm(x, y, z; \frac{1}{2}+x, \frac{1}{2}-y, \bar{z}; \bar{x}, \frac{1}{2}+y, \frac{1}{2}-z; \frac{1}{2}-x, \bar{y}, \frac{1}{2}+z)$
O_1 in 4(a): $0, 0, 0; \frac{1}{2}, \frac{1}{2}, 0; 0, \frac{1}{2}, \frac{1}{2}; \frac{1}{2}, 0, \frac{1}{2}$.

Tc_2O_7: $a = 13.756$ Å, $b = 7.439$ Å, $c = 5.617$ Å

	x	y	z	
Tc	0.106 2	0.013 6	0.198 5	[429]
O_{II}	0.120 9	0.221 4	0.294 4	
O_{III}	0.206 7	−0.058 5	0.060 7	
O_{IV}	0.085 0	−0.122 3	0.435 0	

d. Re_2O_7

Tetrahedral and octahedral cation coordinations are also met in the yellow Re_2O_7 [430] but here the Re_2O_9 units are connected to each other via 4 corners as is shown in Figure 94. The bond distances are similar to those in Tc_2O_7 and $KReO_4$ reflecting a considerable amount of π-bonding in the terminal as well as in the bridging bonds.

e. MoO_3 AND RELATED PHASES

The MoO_3 structure is usually described in terms of double layers of $[MoOO_{2/2}O_{3/3}]$ octahedra. These double layers can be generated from double

TABLE 77

Re_2O_7 structure, orthorhombic, D_2^4—$P2_12_12_1$ (No. 19), $Z = 8$.
All atoms in 4(a): $x, y, z; \frac{1}{2}-x, \bar{y}, \frac{1}{2}+z; \frac{1}{2}+x, \frac{1}{2}-y, \bar{z}; \bar{x}, \frac{1}{2}+y, \frac{1}{2}-z$.

Re_2O_7: $a = 12.508$ Å, $b = 15.196$ Å, $c = 5.448$ Å

	x	y	z	
Re_I	0.474 9	0.369 7	0.565 6	
Re_{II}	0.267 4	0.363 7	0.062 9	
Re_{III}	0.526 2	0.142 7	0.325 5	
Re_{IV}	0.219 7	0.136 5	0.808 0	
O_I	0.375 3	0.337 2	0.255	
O_{II}	0.583 7	0.368 7	0.283	[430]
O_{III}	0.500 6	0.234 9	0.513	
O_{IV}	0.580 9	0.382 2	0.766	
O_V	0.448 2	0.474 7	0.519	
O_{VI}	0.485 3	0.049 6	0.463	
O_{VII}	0.448 0	0.157 9	0.079	
O_{VIII}	0.373 9	0.337 5	0.762	
O_{IX}	0.661 4	0.137 8	0.231	
O_X	0.247 8	0.227 5	0.011	
O_{XI}	0.160 6	0.364 1	0.250	
O_{XII}	0.284 3	0.472 7	0.017	
O_{XIII}	0.262 8	0.043 7	0.967	
O_{XIV}	0.290 4	0.150 0	0.542	

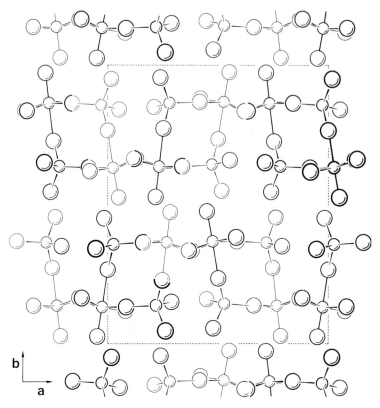

Fig. 94. The Re$_2$O$_7$ structure projected along [001], parallel to the cleavage plane (010). Small spheres: Re

chains of octahedra [MoO$_3$O$_{3/3}$] connected along two adjacent edges (*cis*) that extend parallel with the b-axis. These double chains have to be joined at two opposite corners to form the double layers. An examination of the Mo—O distances (1.67, 1.73, 1.95 (2×), 2.25 and 2.33 Å) leads to rather different conclusions. As expected the terminal bond is shortest, close to the value found for the corresponding M—O distance in Tc$_2$O$_7$ and Re$_2$O$_7$. The Mo—O distance to one of the shared octahedron corners is even shorter than expected from a comparison with the corresponding distances in Tc$_2$O$_7$ and Re$_2$O$_7$. The next two

TABLE 78

*MoO$_3$ structure, orthorhombic, D$_{2h}^{16}$—Pnma (No. 62), Z = 4.
All atoms in 4(c): ±(x, ¼, z; ½+x, ¼, ½−z)*

MoO$_3$: $a = 13.855$ Å, $b = 3.696\,4$ Å, $c = 3.962\,8$ Å

	x	z	
Mo	0.101 6	0.086 7	
O$_I$	0.435 1	0.499 4	[433]
O$_{II}$	0.088 6	0.521 2	
O$_{III}$	0.221 4	0.037 3	

distances, which complement the coordination polyhedron to a distorted tetrahedron, are 0.22 Å larger. Thus there is a clear tendency towards molybdenyl MoO_2^{2+} formation reminiscent of the uranyl ion UO_2^{2+}. This is confirmed by the IR absorption band at 360 nm. Considering only these 4 short distances as bonds, we can build up the structure from chains of tetrahedra $[MoO_2O_{2/2}]$ running in the direction of the b-axis, and indeed the MoO_3 crystals usually grow in the form of needles with the b-axis as the needle axis. This coordinative behaviour in the oxides distinguishes Mo from its 5d analog W for which even an undistorted

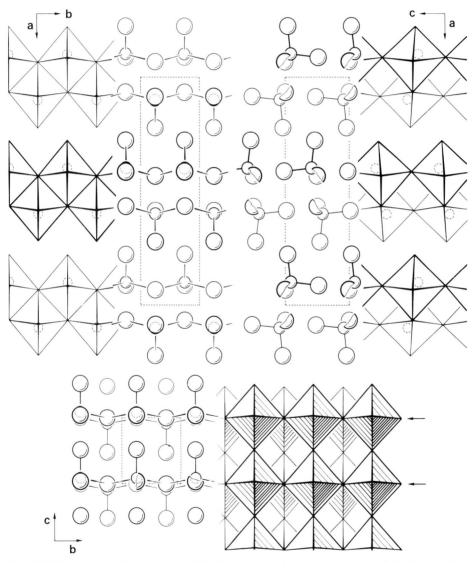

Fig. 95. The orthorhombic structure of MoO_3 projected along the three axes. Only the four short bonds are indicated while the completed coordination is given by the octahedra. The arrows point to the zigzag octahedron chains mentioned in the text.

ReO$_3$-type modification is known. This difference can be seen in the fact that MoO$_3$ and WO$_3$ show mutual solubilities of 3% only [431]. It has however certain 'geometrical' advantages to compose the MoO$_3$ structure from octahedra which fill the space to a higher degree than the tetrahedra. As pointed out already by Magnéli [432] partial reduction of the metal atoms is accompanied by an increase of their coordination number, i.e. by a lower degree of distortion of the [MoO$_6$] octahedra. This can be checked in the case of Mo$_2$O$_5$(OH) and MoO$_{2.8}$F$_{0.2}$, for which the oxidation number is 5.5 and 5.8, resp. Except for the additional H atoms, these two compounds have a structure closely related to that of MoO$_3$ but with a higher symmetry. The Mo atoms lie here exactly in the (100) middle section through the octahedra. The asymmetry of the bonds in the direction perpendicular to the layer plane, however, is conserved so that the coordination polyhedron is best defined as a square pyramid (Figure 95). This asymmetry is a consequence of the Mo—Mo repulsion across the shared octahedron edges. This Mo—Mo distance is still as long as in MoO$_3$ itself (~3.4 Å).

Mo$_2$O$_5$(OH) was described to crystallize in prismatic black crystals. A resistivity of 0.1 Ωcm was measured on powders [424]. Judging from the still highly asymmetric Mo coordination (Figure 96) we expect non-metallic properties as was reported for MoO$_{2.8}$F$_{0.2}$ [436]. Since half the Mo atoms then should be pentavalent one might expect either a paramagnetic localized d^1 state and cation ordering or else pair formation of half the Mo atoms.

If we shift the Mo atom towards the center of the [MoO$_6$] octahedron the Mo—Mo distance within the Mo zigzag chain is reduced to 3.07 or 2.56 Å if we use for the y-parameter of Mo the same value as for O$_{II}$ or O$_{I}$, resp. Thus, the structure is so flexible that it might well adapt for a compound such as hypothetical 'diamagnetic' MoOF$_2$ where two cation d-electrons could be bound in the Mo zigzag chains. In the case of the hydroxides, we would expect only O$_{III}$ to be replaceable by OH, however, the studies of Glemser et al. [437] point to close structural relationships for the whole family MoO$_{3-x}$(OH)$_x$ with $0.5 \leq x \leq 2$.

Both from its X-ray powder pattern and the method of preparation it appears that not only Mo$_2$O$_5$(OH) but also MoO$_2$(OH) crystallizes in a structure which is closely related to that of MoO$_3$ [434]. Reduction of MoO$_3$ with hydrogen always

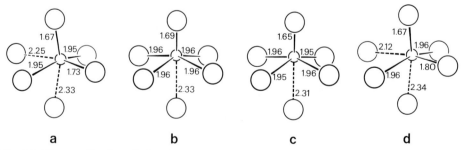

Fig. 96. The coordination of the Mo atoms in MoO$_3$ (a), Mo$_2$O$_5$(OH) (b), MoO$_{2.8}$F$_{0.2}$ (c) and Mo$_{18}$O$_{52}$ (d). Mo—O distances in Å.

TABLE 79

$Mo_2O_5(OH)$ *structure*, orthorhombic, D_{2h}^{17}—Cmcm (No. 63), $Z = 8$.
$(0, 0, 0, \frac{1}{2}, \frac{1}{2}, 0) +$
all atoms in 4(c): $\pm(0, y, \frac{1}{4})$

$Mo_2O_5(OH)(20°C)$: $a = 3.888$ Å, $b = 14.082$ Å, $c = 3.734$ Å

	y	
Mo	0.103 5	
O_I	0.939 7	[435]
O_{II}	0.586 5	
O_{III}	0.223 3	

$Mo_2O_{5.6}F_{0.4}(25°C)$: $a = 3.878$ Å, $b = 13.96$ Å, $c = 3.732$ Å

	y	
Mo	0.102 6	
O_I	0.937 1	[436]
O_{II}	0.580 5	
O_{III}	0.220 8	

for comparison: MoO_3 in Pbnm (No. 62), atoms in 4(c): $\pm(x, y, \frac{1}{4}; \frac{1}{2} - x, \frac{1}{2} + y, \frac{1}{4})$

	x	y
Mo	0.086 7	0.101 6
O_I	0.000 6	0.935 1
O_{II}		0.586 6
O_{III}	0.037 3	0.221 4

related (?):
$Mo_2O_4(OH)_2$, monoclinic
$a = 3.885$ Å, $b = 14.63$ Å, $c = 3.795$ Å, $\gamma = 96.9°$ [1]

leads to distinct blue hydroxides, first to $MoO_{2.5}(OH)_{0.5}$ and then to $MoO_2(OH)$ which probably contain Mo^{4+} and Mo^{6+}. Vice versa, oxidation transforms $MoO_2(OH)$ in a first stage back to $MoO_{2.5}OH$ and this latter phase is subsequently converted to MoO_3. A similar mechanism works in the case of the deep violet tungsten analogs. However, like WO_3 itself, these W phases have other structures. Tungsten blue $W_2O_5(OH)$ crystallizes in a tetragonal structure and the violet $H_{0.5}WO_3$ with the same stoichiometry is reported with the cubic ReO_3 structure [1], both closely related to the parent WO_3 structure. The cubic ReO_3 structure is also obtained on fluorinating MoO_3 up to $MoO_{2.4}F_{0.6}$ [436].

It is interesting that stronger hydrogen reduction of MoO_3 leads to red $Mo_5O_7(OH)_8$ and olive green $MoO(OH)_2$. It is not known however whether the structure of these MX_3 phases still bears a close relation to that of MoO_3 [437].

f. $MoO_3 \cdot 2H_2O$ AND $MoO_3 \cdot H_2O$

The dihydrates $MO_3 \cdot 2H_2O$ can be precipitated at room temperature from acidic aqueous molybdate and tungstate solutions. $MoO_3 \cdot 2H_2O$ grows in the form of lemon yellow transparent prisms from a HCl solution of MoO_3 [438]. In these modifications the water is present as H_2O molecules as follows from H-nuclear-resonance and infra-red studies. Half the water forms part of the cation coordination polyhedron which is a distorted octahedron [$MoOO_{4/2}(H_2O)$]. The structure consists of layers of these corner-linked octahedra with the remaining H_2O molecules intercalated between these ReO_3- or $Sc(OH)_3$-like sheets. As can be seen in Figure 96, the cation coordination polyhedra are strongly distorted. Three short Mo—O distances (1.69, 1.77 and 1.80 Å) are opposed to three long distances (2.05, 2.16 and 2.29 Å in the average). The largest distance corresponds to the Mo—OH_2 bond, while the smallest value belongs to the opposite terminal Mo—O bond, which comprises a considerable π-bonding contribution and a formal bond order >2 [438]. The octahedron layers are weakly held together by H bonds. Each coordinated water molecule forms two donating H bridges to two interlayer H_2O molecules with OH—O distances of 2.70 to 2.79 Å. Furthermore it acts as H acceptor in a bridge to a third interlayer H_2O molecule with O—HO distances of 3.06 to 3.15 Å. Thus each interlayer water molecule forms two bridges as H-acceptor to two coordinated H_2O molecules of different layers as well as a weaker bridge as H-donor to a third H_2O of the octahedron layer. The

TABLE 80
$MoO_3 \cdot 2H_2O$ structure, monoclinic, C_{2h}^5—$P2_1/n$ (No. 14′), $Z = 16$.
all atoms in 4(e): $\pm(x, y, z; \frac{1}{2}+x, \frac{1}{2}-y, \frac{1}{2}+z)$

$MoO_2 \cdot 2H_2O$ (20 °C): $a = 10.476$ Å, $b = 13.822$ Å, $c = 10.606$ Å, $\beta = 91.62°$.

	x	y	z
Mo_I	0.117 3	0.249 7	0.147 9
Mo_{II}	0.393 9	0.255 3	0.359 6
Mo_{III}	−0.137 2	0.251 9	0.398 0
Mo_{IV}	0.153 9	0.245 0	0.618 3
O_I	0.021 6	0.254 3	−0.025 5
O_{II}	0.231 3	0.208 5	0.266 1
O_{III}	0.263 5	0.208 7	0.019 2
O_{IV}	−0.026 6	0.243 8	0.228 7
O_V	0.481 2	0.262 3	0.219 0
O_{VI}	0.271 6	0.209 4	0.511 6
O_{VII}	−0.269 5	0.295 8	0.265 2
O_{VIII}	0.014 6	0.253 9	0.477 9
O_{IX}	0.152 0	0.368 5	0.143 1
O_X	0.342 3	0.372 7	0.377 0
O_{XI}	−0.172 3	0.133 4	0.389 7
O_{XII}	0.173 7	0.366 2	0.628 8
O_{XIII}	0.083 8	0.088 9	0.110 7
O_{XIV}	0.414 6	0.091 0	0.351 4
O_{XV}	−0.088 4	0.410 6	0.361 0
O_{XVI}	0.092 0	0.085 7	0.600 3
O_{XVII}	0.359 3	0.501 1	0.128 8
O_{XVIII}	0.361 2	0.511 9	0.624 3
O_{XIX}	0.142 6	0.503 5	0.376 2
O_{XX}	−0.133 9	0.489 6	0.123 4

[438]

$WO_3 \cdot 2H_2O$: $a = 10.53$ Å, $b = 13.84$ Å, $c = 10.53$ Å, $\beta = 90.08°$ [439]
related:
$MoO_3 \cdot H_2O$ structure, monoclinic, C_{2h}^5—$P2_1/c$ (No. 14), $Z = 8$.
$a = 7.55$ Å, $b = 10.69$ Å, $c = 7.28$ Å, $\beta = 91°$ [440]

second H atom of each interlayer H_2O molecule acts as a bridge (2.77 to 2.85 Å) to a terminal O atom of the adjacent octahedron layer, so that each terminal O atom is a H acceptor.

Dehydration of $MoO_3 \cdot 2H_2O$ in air proceeds in two steps both being topotactic in nature [440]. In the first step the intercalated water is removed. This reaction occurs between 60 and 80 °C and the product is the yellow form of $MoO_3 \cdot H_2O$. During dehydration the octahedron layers keep their gross feature and the relation between the two structures $MoO_3 \cdot 2H_2O$ (di, $P2_1/n$) and $MoO_3 \cdot H_2O$ (mono, $P2_1/c$) can be described as follows [440]:

$$(010)_{di} \parallel (010)_{mono}$$
$$[\bar{1}01]_{di} \parallel [001]_{mono}$$
$$[101]_{di} \parallel [100]_{mono}$$

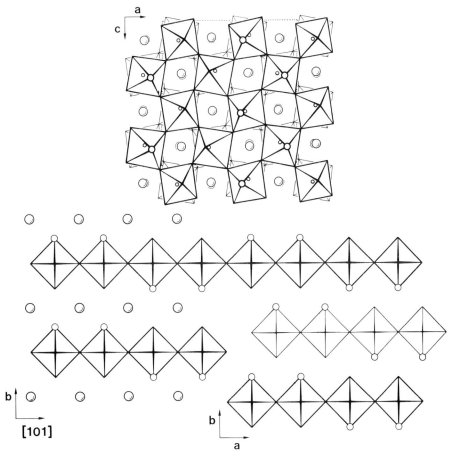

Fig. 97. The monoclinic structure of $MoO_3 \cdot 2H_2O$ projected along [010] (above). Only the coordinated and intercalated water molecules are shown while the positions of the other O atoms are defined by the octahedron corners. The Mo atoms are indicated for the upper layer only.

The lower part represents schematically a section through the layers of the structures of $MoO_3 \cdot 2H_2O$ (left) and its first (metastable) stage of dehydration, $MoO_3 \cdot H_2O$ (right).

In Figure 97 a section perpendicular to the layers is shown schematically. We have shifted the second layer by $a/4$ assuming a similar stabilization of this arrangement as in the case of MoO_3.

Finally, the dehydration leads to MoO_3. The transformation into the double octahedron layers of MoO_3 in the second step would be more straightforward if the H_2O molecules in the monohydrate were already located on the same side of an octahedron layer and adjacent in pairs of such layers. When starting dehydration with single crystals, the final product still has the appearance of the original crystals, disintegrated however by parallel cracks into a series of lamellae. $WO_3 \cdot H_2O$ appears to have the same structure as yellow $MoO_3 \cdot H_2O$ [6], so we wonder whether dehydration leads to a MoO_3-type modification of WO_3.

g. $Mo_{18}O_{52}$

Thermal decomposition of MoO_3 in vacuum leads to a number of almost black and flaky suboxides with ordered structures derived from the parent MoO_3 structure [441]. Quite a number of stepped layer structures can be constructed if we cut the MoO_3 double octahedron layers into strips and stick these strips together but shifted in height by an octahedron. Strips are mutually connected by edge sharing of the outer octahedra which implies an oxygen deficiency. Alternatively, these layers may be derived from an MoO_3 layer by a shear mechanism occurring along the border lines of the strips (Figure 98). Since the shear direction is inclined to the MoO_3 layer the resulting layer is regularly stepped. The strip axis makes a certain angle with the direction of the double octahedron chains of the MoO_3 structure (b-axis of the MoO_3 structure). The number of double octahedra along one chain is exactly 18 in $Mo_{18}O_{52}$ and 16 in $Mo_{16}O_{46}$. Along the shear zone ($z \approx \frac{1}{2}$ in Figure 98) the structure is disturbed in so far as the next outermost octahedron in each zigzag row is missing. Instead Mo atoms are found to occupy some tetrahedral holes. Thus, the composition of this family of compounds, defined by a variable band width, is Mo_nO_{3n-2} where $2n$ is the number of octahedra in a zigzag row.

Other families Mo_nO_{3n-m+1} can be generated by changing the degree to which the rows of adjacent strips overlap (defined as the number $2m$ of octahedra engaged in increased edge-sharing at each end of a row; $m = 6$ in $Mo_{18}O_{52}$) [441].

In $Mo_{18}O_{52}$ the cation polyhedra are similarly distorted as in MoO_3. The unit cell contains as many tetrahedra as Mo^{4+} atoms, which is equivalent to the number of missing O atoms. Tetrahedral sites instead of the ideal octahedral sites are favorable from an electrostatic point of view [441]. The shear operation would

TABLE 81

$Mo_{18}O_{52}$ structure, triclinic, C_i^1–$P\bar{1}$ (No. 2), $Z = 2$.
all atoms in 2(i): $\pm(x, y, z)$

$a = 8.145$ Å, $b = 11.89$ Å. $c = 21.23$ Å, $\alpha - 102.67°$, $\beta = 67.82°$, $\gamma = 109.97°$ [441] (for atomic positions see [441].)

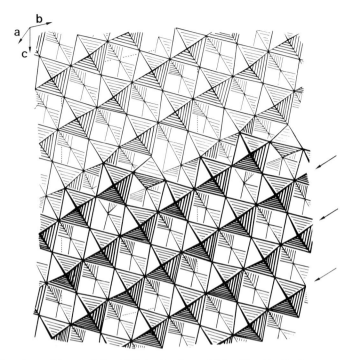

Fig. 98. One stepped layer of the triclinic structure of $Mo_{18}O_{52}$ idealized as a network of joined [MoO_6] octahedra and [MoO_4] tetrahedra. The a-axis points down below the plane of the paper. The arrows on the right side point to the zigzag rows each of which contains 18 octahedra at each level.

lead to 11 additional shared edges per unit cell, i.e. to relatively short Mo—Mo distances whereas in the actual structure the number of shared edges exceeds the corresponding number in MoO_3 by only one. Since the short Mo—Mo distances in $Mo_{18}O_{52}$ ($\geqslant 3.30$ Å) are still rather large and comparable with those in MoO_3 (3.44 Å) the excess d electrons are expected to be localized on Mo atoms in the crystallographically distinguished sites in the shear zone. It is tempting to locate the tetravalent molybdenum at the tetrahedral sites though from its electronic structure we would rather expect to find it in a less distorted octahedral coordination. There is one cluster of 4Mo atoms with 3 Mo—Mo contacts somewhat closer than in MoO_3 but there are no contacts comparable with the Mo—Mo bond distance as met in MoO_2 (2.50 Å).

We wonder whether it might be possible to synthesize isostructural insulating $Mo_{16}Ti_2O_{52}$ or $Mo_{16}Ge_2O_{52}$.

h. Layered vanadium oxyhydroxides

The structures of these phases can be built up of layers of oxygen octahedra that are connected via corners and edges [442]. The layers are however weakly connected by hydrogen bonds so that these structures are in fact not true layer

TABLE 82
Duttonite structure, monoclinic, C_{2h}^6—I2/c (No. 15'), $Z = 4$.

VO(OH)$_2$: $a = 8.80$ Å, $b = 3.95$ Å, $c = 5.96$ Å, $\beta = 90°40'$.
described as pseudo-orthorhombic in space group
D_{2h}^{28}—Imma (No. 74), $Z = 4$.

$(0, 0, 0; \frac{1}{2}, \frac{1}{2}, \frac{1}{2})+$
O$_I$ in 8(g): $\pm(\frac{1}{4}, y, \frac{1}{4}; \frac{3}{4}, y, \frac{1}{4})$ [442]
V and O$_{II}$ in 4(e): $\pm(0, \frac{1}{4}, z)$
$a = 5.96$ Å, $b = 8.80$ Å, $c = 3.95$ Å
$z(V) = 0.336$, $y(O_I) = 0.100$, $z(O_{II}) = 0.754$.

---

structures. Thus, the crystals of the simplest representative of this family, duttonite VO(OH)$_2$, are described as thin platelets with moderate (100) cleavage. The duttonite structure can be generated by corner linking along [010] of octahedron chains formed by trans-edge sharing parallel with the c-axis. The OH corners of the [VO$_{2/2}$(OH)$_{4/2}$] octahedra are linked by a hydrogen zigzag seam. The description of the coordination polyhedron by an octahedron is a geometrically useful but otherwise rather crude approximation. In fact, the V atom is strongly displaced from the center towards one corner of the octahedron giving rise to a short V—O distance of 1.65 Å and a large one of 2.30 Å. The coordination polyhedron is therefore better described as a square pyramid with the remaining four V—O distances being 2.02 Å. The VO(OH)$_2$ structure thus contains distinct vanadyl ions (VO)$^{2+}$. These polarized groups are arranged in alternate directions along the b-axis as in an antiferroelectric. The polarity of the H bonds in the zig-zag chains along the b-axis causes the symmetry to be monoclinic (pseudoorthorhombic), as is found in natural crystals. In synthetic preparations the polarity is disordered and thus the true symmetry of the structure is orthorhombic. Vanadium in duttonite is tetravalent and therefore should exhibit a localized magnetic moment corresponding to a d^1 state. VO(OH)$_2$ is expected to be non-metallic. It would be interesting to know the structure of diamagnetic insulating TiO(OH)$_2$ for comparison.

In the structure of the mineral häggite the chains which by corner linking form the layers are double octahedron chains as visualized in Figure 99c. The additional corner sharing reduces the mean valence of the cation to +3.5. The displacement of the cation out of the center of the oxygen octahedron is much less pronounced: V—O = 1.82 and 2.06 Å. The distances to the O atoms at the base of the pyramid are similar to those in VO(OH)$_2$, namely 1.97 and 2.01 Å.

One might assume that additional edge sharing of the cation coordination polyhedra is a means to bind excess d-electrons of the cations. However, while the V—V distance in the straight V chains along the b-axis is $b = 2.99$ Å, the additional contacts in the zigzag chains are even looser, 3.15 Å. The asymmetry of the cation coordination looks even as if it were caused by the V—V repulsion. These distances are definitely too large to represent cation-cation bonds, as

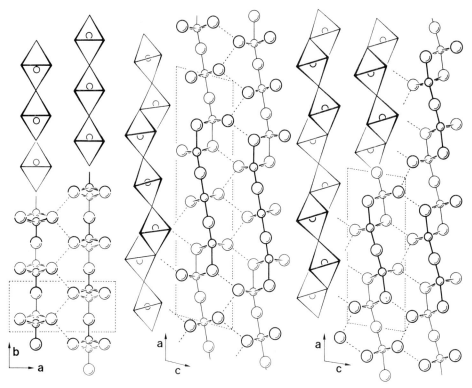

Fig. 99. The structures of vanadium oxyhydroxides. (a) duttonite VO(OH)$_2$; (b) protodoloresite V$_3$O$_3$(OH)$_5$; and (c) häggite V$_2$O$_2$(OH)$_3$. Large spheres: O, small spheres: V. Hydrogen bonds are indicated by dashed lines.

follows from a comparison with VO$_2$ where the cation displacement leads to distances of 2.65 and 3.12 Å. As there are no critical metal-metal bonds, we would expect the black häggite crystals to show non-metallic behavior. This implies localized d-electrons, hence V^{3+} and V^{4+} ions. Crystallographically, however, all cations are equivalent. The H bonds indicated in Figure 99 as dashed lines are based on the reported formula V$_2$O$_2$(OH)$_3$. The composition of one coordination

TABLE 83
Häggite structure, monoclinic, C_{2h}^3—C2/m (No. 12), $Z = 2$.
$(0, 0, 0; \frac{1}{2}, \frac{1}{2}, 0)+$
V, O$_{II}$ and O$_{III}$ in 4(i): $\pm(x, 0, z)$
O$_I$ in 2(c): $0, 0, \frac{1}{2}$

V$_2$O$_2$(OH)$_3$: $a = 12.17$ Å, $b = 2.99$ Å, $c = 4.83$ Å, $\beta = 98°15'$ [442]
(häggite)

	x	z
V	0.137 8	0.396 7
O$_{II}$	0.609 4	0.133
O$_{III}$	0.302 6	0.346

polyhedron therefore is assumed to be [VO$_{1/2}$(OH$_{2/2}$)$_{2/2}$(OH$_{1/2}$)$_{3/3}$] which leads to the formula V^{3+}V^{4+}O$_2$(OH)$_3$ for a double octahedron. If we remove the H bonds towards the three corners which are common to three octahedra then the coordination polyhedron can be described by the formula [VO$_{1/2}$(OH$_{2/2}$)$_{2/2}$O$_{3/3}$] corresponding to a composition V$_2$O$_3$(OH)$_2$ with V^{4+} only. Unfortunately no magnetic measurements on häggite are available that could plead in favor of this proposal.

A whole family of stepped layer structures can be created by combining the häggite and the duttonite structure. Designating with n the number of octahedron chains at the same height, the formula will be V$_n$O$_n$(OH)$_{2n-1}$, if we use the häggite formula corresponding to Figure 99. The representative with $n = 3$ was described as the black mineral protodoloresite [442]. The crystal structure contains two inequivalent cation sites which allows two possibilities for the cation oxidation numbers: V^{3+}V$_2^{4+}$ and V^{5+}V$_2^{3+}$. The middle octahedron is elongated but fairly regular: V—O = 1.93 Å (2×) and 1.98 Å (4×). The terminal octahedra show a similar but more pronounced shift of the cation out of the center towards the middle as does häggite. The V—O distances are 1.70 Å and 2.01 Å, 1.97 Å (2×) and 2.13 Å (2×). The V—V distance along the straight chains in b-direction is again $b = 2.99$ Å as in the case of häggite, but is even larger in the zig-zag chains, 3.23 Å. The distances lead to the assumption that the terminal V atoms are in the higher oxidation state, thus favoring the formula V^{3+}V$_2^{4+}$O$_3$(OH)$_5$. If this conclusion is correct then hypothetical CrTi$_2$O$_3$(OH)$_5$ well might be isostructural.

It is somewhat intriguing that the same positional parameters as for black protodoloresite have been suggested also for the related brown mineral doloresite which belongs to the family V$_n$O$_{n+1}$(OH)$_{2n-2}$ originating from a combination of 'oxidized häggite' and duttonite. Doloresite crystallizes in the form of fibrous flakes with the fibre axis b. As now all cations are tetravalent a crystallographic inequivalence is no longer required. A refinement of the crystal structure as well as magnetic measurements on well-defined crystals would be desirable.

TABLE 84

Doloresite structure, monoclinic, C$_{2h}^3$—C2/m (No. 12), $Z = 2$.
V$_I$ in 2(c), V$_{II}$, O$_I$—O$_{IV}$ in 4(i) (see häggite structure)

V$_3$O$_4$(OH)$_4$ (doloresite) and
V$_3$O$_3$(OH)$_5$ (protodoloresite):
$a = 19.64$ Å, $b = 2.99$ Å, $c = 4.83$ Å, $\beta = 103°55'$ [442]

	x	z
V$_I$	0.1766	0.3680
O$_I$	0.470	0.248
O$_{II}$	0.095	0.427
O$_{III}$	0.665	0.095
O$_{IV}$	0.286	0.362

TABLE 85
V_2O_5 structure, orthorhombic, D_{2h}^{13}—Pmmn (No. 59). $Z = 2$.
V, O_{II} and O_{III} in 4(e): 0, y, z; 0, ȳ, z; $\frac{1}{2}, \frac{1}{2} - y, \bar{z}$; $\frac{1}{2}, \frac{1}{2} + y, \bar{z}$.
O_I in 2(a): 0, 0, z; $\frac{1}{2}, \frac{1}{2}, \bar{z}$.

V_2O_5: $a = 3.563$ Å, $b = 11.510$ Å, $c = 4.369$ Å

	y	z	
V	0.148 7	0.108 6	
O_I		−0.003 1	[443]
O_{II}	0.146 0	0.471 3	
O_{III}	0.319 1	−0.002 6	

i. V_2O_5

The slate-like brown-violet crystals of V_2O_5 have a rather low melting point of ~670 °C. The orthorhombic structure of V_2O_5 [443] also is usually derived from an octahedron packing. An 'idealized' V_2O_5 structure arises from connecting MoO_3 double layers along their free corners. It then can be interpreted as a ReO_3-related one-dimensional shear structure of the family M_nX_{3n-1} and with a (001) shear plane [444]. In such a geometrically simplifying description the layer character of the V_2O_5 structure gets completely lost. In fact, however, the V atom is shifted out of the octahedron center thus acquiring a coordination somewhere between distorted tetrahedral and square pyramidal with distances V—O = 1.59, 1.78, 1.88 (2×), 2.02 and 2.79 Å. The difference between the largest and the shortest distance along the c-axis well accounts for the (001) cleavage plane. The second largest V—O distance (dashed lines in Figure 100) is only slightly larger

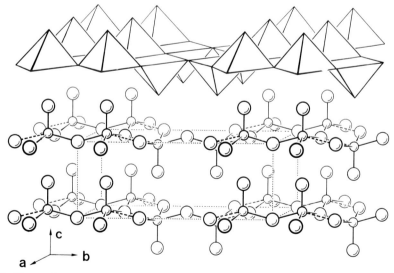

Fig. 100. The puckered layer structure of V_2O_5. The four short bonds are indicated by full lines whereas the fifths bond, that complements the coordination polyhedron to a trigonal bipyramid, is given as a dashed line. The bipyramids or distorted half-octahedra are emphasized above.
Small spheres: V

than the four tetrahedral bonds so that the fibre character due to the tetrahedron double chains along the a-axis is changed into a puckered-layer character. Intercalation of light alkali metals like Li and Na leads to vanadium bronzes $M_xV_2O_5$ with structures that still contain the V_2O_5 network in a more or less distorted version.

The dark brown mineral navajoite $V_2O_5 \cdot 3 H_2O = VO(OH)_3$ has a monoclinic yet unknown structure. The crystals are described as soft and fibrous with a silky luster [1].

9. Closed-Packed Structures with Group II and III Cations in Tetrahedral and Octahedral Coordination

Zn and Cd and to a lower degree Al and Ga form cations with a pronounced preference for tetrahedral coordination in chalcogenides and pnictides. The binary compounds all crystallize in three-dimensional structures under normal conditions. Indium is found in tetrahedral and octahedral coordination while the larger Tl^{3+} always adopts a coordination number of six. Most of the structures of the Group III M_2X_3 chalcogenides are based on close-packed anion arrays which would easily allow the formation of layer structures. However, as can be seen from Table 86 only very few layer-type modifications do in fact occur. There is some confusion about the modifications of In_2Se_3. Semiletov [446] described a room-temperature form which is essentially a wurtzite structure with every third cation layer missing

$$\dot{B}\beta C \quad B\beta C\alpha B \quad C\alpha\dot{B}$$

(pseudo-cell $a = 4.00$ Å, $c = 19.24$ Å, as originally reported by Hahn and Frank [447]). Superstructure lines are said to require $a' = 4a$ because $\frac{1}{16}$ of the In atoms have to be placed into octahedral holes. It is supposed that these atoms occupy octahedral holes in the layers whose tetrahedral holes are also taken up, i.e. this model corresponds to a true layer structure. The reliability factor, however, was only 0.22 and the bonding is not realistic. Semiletov [448] also determined a high-temperature modification stable above 200°C. This modification contains In atoms in tetrahedral sites only and is an ordered three-dimensional wurtzite-type superstructure. The phase transition is reflected also in resistivity, Hall-effect and Seebeck-effect measurements. The graphite-like modification revealed a surprisingly low and nearly constant resistivity of only 0.1Ω cm at room temperature but near 200°C it abruptly increased by several orders of magnitude. The Seebeck coefficient suddenly changed from 150–200 to 600–700 μV/°C at 200°C. The first-order phase transition was also detected in dilatometric studies [449, 450].

Likforman and Guittard [397], who have established the In–Se phase diagram, identify Semiletov's high-temperature modification but claim it to be the only

TABLE 86
The structures of the Group III M_2X_3 chalcogenides. Layer-type modifications are underlined.

Al_2S_3	(α): hexagonal tabular [1] (β): ZnO superstructure (γ): rhombohedral Cr_2O_3 type (p): tetrag., ordered defect spinel	Ga_2S_3	(r): defect ZnS type (α): monocl., ordered defect ZnO type	In_2S_3	(r, β): tetragonal ordered spinel (h, α): disordered defect spinel (h_2, γ): <u>NaCl-type deriv.</u> [451]
Al_2Se_3	(r): monocl., α-Ga_2S_3 type (p): defect spinel	Ga_2Se_3	(r): cubic, defect spinel (p): Bi_2Te_3 type [454]	In_2Se_3	(r, α): <u>rhombohedral</u> [446] (>200°C, β): ordered defect ZnO type (h, p?): <u>Bi_2Te_3 type</u> [450] (>650°C, γ): cubic (>750°, δ): monoclinic (h): <u>γ-In_2S_3(?)</u> [397]
Al_2Te_3	(r): hexagonal, defect ZnO type	Ga_2Te_3	(r): defect ZnS type	In_2Te_3	(r): ordered defect ZnS type (>600°C): disordered ZnS type (p): <u>Bi_2Te_3 type</u> [454] (film): hcp [455]

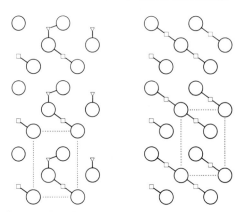

Fig. 101. (110) sections of one-pack M_2X_3 layer structures. (a) mixed octahedral/tetrahedral; and (b) γ-In_2S_3, a NaCl derivative.

phase stable below 550°C. Above this temperature In_2Se_3 was found to transform reversibly to the hexagonal form described by Hahn and Frank [447]. No further phase transition was detected up to the congruent melting point of 900°C, though single crystals of a third modification were found in samples obtained at high temperatures. As the cell of this trigonal modification ($a = 4.00$ Å, $c = 9.56$ Å) contains only one In_2Se_3 pack, this structure can be of a layer type only if the anion packing is ccc with either half the In atoms in tetrahedral and half in octahedral coordination:

$A\beta BaC$ or $Se\equiv In—Se\equiv In\equiv Se$ (lower case letters: octahedral cations; Greek letters: tetrahedral cations)

(dashes indicating the coordination number), or, what is more likely, with all cations in octahedral coordination as in the high-temperature modification of the sulfide, γ-In_2S_3 (Figure 101 and Table 87).

Osamura et al. [452] offer two more structure models for the layer-type In_2Se_3 phases. According to these authors the rhombohedral low-temperature modification is based on a close-packed Se arrangement in $(hhc)_3$ sequence. The idealized version is shown in Figure 102a. On checking the actual structure one recognizes that the reported atomic parameters cannot be accurate. The unit cell, which has a c-axis $\frac{3}{2}$ times that of Hahn and Frank [447] is, however, confirmed by Fitzgerald

TABLE 87
γ-In_2S_3 structure, trigonal, D_{3d}^3—$P\bar{3}m1$(No. 164), $Z = 1$.
In and S_{II} in 2(d): $\pm(\frac{1}{3}, \frac{2}{3}, z)$
S_I in 1(a): 0, 0, 0.

In_2S_3 (h_2): $a = 3.80$ Å, $c = 9.04$ Å $z(In) \approx \frac{1}{6}$, $z(S_{II}) \approx \frac{2}{3}$. [451]
$In_{2-x}As_xS_3$ ($x < 0.2$): $a = 3.806$ Å, $c = 9.044$ Å: $z(In) = 0.8097$, $z(S_{II}) = 0.3316$ similar with Sb substitution [451, 1020]
In_2Se_3 (h) probably: $a = 4.00$ Å, $c = 9.56$ Å [397]

TABLE 88

$In_2Se_3(r)$ structure, trigonal, C_{3v}^5—R3m (No. 160), $Z = 1$ (3).
hexagonal axes: $(0, 0, 0; \frac{1}{3}, \frac{2}{3}, \frac{2}{3}; \frac{2}{3}, \frac{1}{3}, \frac{1}{3})+$
all atoms in 3(a): 0, 0, z.

$In_2Se_3(r)$: $a = 4.05$ Å, $c = 28.77$ Å, $c/an = 0.789$
$z(In_\Omega) = 0.242$, $z(In_\tau) = 0.718$, $z(Se_\tau) = 0$, $z(Se_{II}) = 0.525$,
$z(Se_{III}) = 0.818$ [452]
proposal:
$z(In_\Omega) \approx \frac{1}{6}$, $z(In_\tau) \approx \frac{25}{36}$, $z(Se_I) = 0$, $z(Se_{II}) \approx \frac{5}{9}$, $z(Se_{III}) \approx \frac{7}{9}$.

[453]. We consider this model as improbable as Semiletov's [446] layer structure derived from wurtzite by removing every third cation layer, as in both structures the anions of one outer layer are bonded by a single bond only. If for rhombohedral α-In₂Se₃ the *chhchhchh* anion sequence is correct then we would favor a *(hch)₃* anion stacking and an occupation AβBaC BγCbA CαAcB as shown in Figure 102b. The high-temperature structure studied at 250°C [452] is of the Bi₂Te₃ type where all cations are octahedrally coordinated and the anion stacking is *(hch)₃*. This transition from mixed tetrahedral/octahedral to pure octahedral cation coordination seems to be inconsistent with the high resistivity of the high-temperature modification.

Electrical and structural studies on the same samples were performed by Popović et al. [456–458]. These authors find a hexagonal α-In₂Se₃(H) and a rhombohedral α-In₂Se₃(R) modification to be stable at room temperature. The

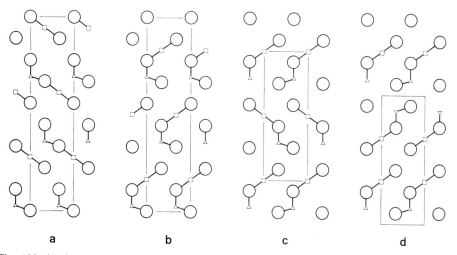

Fig. 102. (110) sections of possible layer-type indium chalcogenide In₂X₃ structures.
(a) Idealized version of the structure proposed by Osamura et al. [452] for In₂Se₃(r), where the anions of one boundary layer are attached by one single bond only.
(b) improved alternative using the same anion packing.
(c, d) possible models for the hexagonal structure of GaInS₃ with Ga at tetrahedral sites (triangles) and In at octahedral sites (squares).

former is obtained on cooling silica tubes containing the molten In$_2$Se$_3$ slowly from 1000°C with the sample at a homogeneous temperature, while crystals of the second modification grew when there was a temperature gradient. The hexagonal modification has lattice constants similar to those reported by Hahn and Frank [447], namely $a = 4.025$ Å, $c = 19.235$ Å, and the space group was found to be P6$_3$/mmc (No. 194). We are tempted to locate the atoms as follows:

Se$_{II}$ in 4(f): $\pm(\frac{1}{3}, \frac{2}{3}, z; \frac{1}{3}, \frac{2}{3}, \frac{1}{2}-z)$ with $z(\text{Se}) \approx \frac{1}{12}$

In in 4(e): $\pm(0, 0, z; 0, 0, \frac{1}{2}-z)$ with $z(\text{In}) \approx \frac{1}{3}$,

Se$_I$ in 2(d): $\pm(\frac{1}{3}, \frac{2}{3}, \frac{3}{4})$.

This corresponds to a $(hhh)_2$ stacking of the anions with all cations in octahedral holes, i.e. a NiAs structure with every third cation layer empty. Another possibility would be the Pt$_2$Sn$_3$ structure with $(chc)_2$ stacking. In both these structures In—In pairs would occur along the c-axis with relatively short distances of ~3.2 Å. However, if we put the cations into tetrahedral instead of octahedral holes we obtain a central Se layer whose atoms make two bonds in ±c-direction.

The rhombohedral α-In$_2$Se$_3$(R) phase with $a = 4.025$ Å, $c = 28.762$ Å has practically the same lattice constants as the Bi$_2$Te$_3$-type modification reported by Osamura et al. [452] and also the appropriate space group R$\bar{3}$m.

At 200°C a transformation to another rhombohedral modification takes place with a lattice contraction $\Delta c = -1.85\%$ and $\Delta a = -0.8\%$. At 205°C the cell of β-In$_2$Se$_3$ has $a = 4.00$ Å, $c = 28.33$ Å [456]. Thermal expansion coefficients are very similar for both modifications (see Figure 103). On cooling, single crystals transform back to the α-phase at 90 to 60°C. Powdered samples and pressed pellets, however, can be supercooled down to room temperature and this probably holds also for the films studied by Semiletov [446, 448] though this β-phase has not the same unit cell as Semiletov's.

According to van Landuyt et al. [459] the $\beta \to \alpha$ back transformation takes place via a superstructure due to the softening and freezing-in of a phonon mode. Moreover, these authors detected a new intrapolytypic transformation at −125°C.

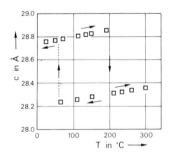

Fig. 103. Discontinuity of the hexagonal c-axis of In$_2$Se$_3$ at the α-β phase transition [456].

A periodic deformation of the α'-In_2Se_3 structure leads to a superstructure with a cell $\mathbf{a}' = \mathbf{a}_1 - \mathbf{a}_2$, $\mathbf{b}' = 2(\mathbf{a}_1 + \mathbf{a}_2)$, $\mathbf{c}' = \mathbf{c}$.

It can be expected that in mixed Ga—In chalcogenides, Ga will prefer the tetrahedral sites while In will occupy the octahedral sites. For the peritectically forming $GaInS_3$, a hexagonal cell $a = 3.88$, $c = 17.46$ Å, has been reported [460] without further specifications. A two-pack layer structure might be adequate as in In_2Se_3 or based on a hcp anion packing with the sequence

$B\gamma CaB$ $CaB\beta C$ (lower-case letters: octahedral cations, In; Greek letters: tetrahedral cations, Ga).

in space group $P\bar{3}m1$ with In in 2(c) and the remaining atoms in 2(d) and idealized z-values of $\frac{1}{3}$ (In), $\frac{1}{12}$ (S_I), $\frac{5}{12}$ (S_{II}), $\frac{3}{4}$ (S_{III}) and $\frac{1}{8}$ (Ga), as visualized in Figure 102c. A crystallographically more symmetric version is proposed in Figure 102d.

Another mixed phase, $Ga_2Se_3 \cdot 4\,In_2Se_3$, crystallizes in red hexagonal platelets ($a = 3.82$, $c = 100.54$ Å). From the cell shape and the easy cleavage parallel to (001) the authors [461] conclude that uniform layers of Ga, In and vacancies are stacked in c-direction. The rhombohedral cell contains 3×11 anion layers which is a rather queer number. The unit cells of the other phases in the system Ga_2S_3—In_2S_3 have larger a-axes which points to mixed layers of metals and vacancies, in agreement with the less pronounced cleavage.

Judging from the structure of the binary phases, one might expect layered ternary phases between ZnSe, CdSe and In_2Se_3. The 1:1 compounds, however, crystallize in three-dimensional structures, either in the Ag_2HgI_4(r) structure, a zincblende derivative with all cations in tetrahedral sites, or, under pressure, in the spinel or a defect NaCl structure. This behavior (eventually with the Ag_2HgI_4(h) structure for some tellurides) is met in all (Zn, Cd, Hg) (Al, Ga, In)$_2X_4$ chalcogenides with the exception of $ZnAl_2S_4$ and $ZnIn_2S_4$. Whereas the orthorhombic structure of $Zn\square Al_2S_4$(h) is a layered defective wurtzite derivative, the various polytypes of $ZnIn_2S_4$ crystallize in true layer structures [463–467]. Similar layer structures occur among the $M^{2+}Al_2S_4$ compounds with M = Mg, Mn, Fe, Ni [468, 469] and in high-pressure modifications of $MnGa_2S_4$ [470].

A great number of polytypes would be possible but analysis of the solved structures showed that the stacking of the outer anion layers of each pack is always of the h type. Moreover, in the outer anion sandwiches, the cations always occupy tetrahedral sites and specifically those tetrahedra which turn their bases towards the empty layer [466].

These mysterious rules may be understood by the following reasoning. In the covalent bond model, an octahedral cation site in these phases implies a higher charge transfer than does a tetrahedral one. It is therefore reasonable to locate the octahedrally coordinated cations near the center of the layer. For the tetrahedral sites of the outer anion double layer, however, we have no other equivalent choice. If we used the other τ set where all tetrahedra turn their bases towards the interior, then

TABLE 89

The ZnIn$_2$S$_4$ polytype structures

ZnIn$_2$S$_4$(I) structure, trigonal C$_{3v}^1$—P3m1 (No. 156), Z = 1.
  In$_\Omega$ in 1(c): $\frac{2}{3}, \frac{1}{3}, z$.
  In$_\tau$, S$_{III}$ and S$_{IV}$ in 1(b): $\frac{1}{3}, \frac{2}{3}, z$.
  S$_I$ and S$_{II}$ in 1(a): 0, 0, z.

ZnIn$_2$S$_4$: $a = 3.85$ Å, $c = 12.34$ Å, $c/an = 0.801$ [463]
  $z(\text{In}_\Omega) = 0.382$, $z(\text{In}_\tau) = 0.069$, $z(\text{Zn}) = 0.692$
  $z(S_I) = 0.007$, $z(S_{II}) = 0.495$, $z(S_{III}) = 0.269$, $z(S_{IV}) = 0.754$.

Inverse type:
FeAl$_2$S$_4$: Fe in 1(c), Al$_I$ in 1(a), Al$_{II}$ in 1(b) [464]

ZnIn$_2$S$_4$(IIa) structure, trigonal, D$_{3d}^3$—P$\bar{3}$m1 (No. 164), Z = 2.
  Zn, In$_\tau$, S$_I$ to S$_{IV}$ in 2(d): $\pm(\frac{1}{3}, \frac{2}{3}, z)$
  In$_\Omega$ in 2(c): $\pm(0, 0, z)$

ZnIn$_2$S$_4$: $a = 3.85$ Å, $c = 24.68$ Å, $c/an = 0.801$ [465]
  $z(\text{In}_\Omega) = 0.250$, $z(\text{In}_\tau) = 0.0941$, $z(\text{Zn}) = 0.594$
  $z(S_I) = 0.6935$, $z(S_{II}) = 0.4375$, $z(S_{III}) = 0.1938$, $z(S_{IV}) = 0.9375$.

ZnIn$_2$S$_4$(IIb) structure, hexagonal, C$_{6v}^4$—P6$_3$mc (No. 186), Z = 2.
  In$_\Omega$, In$_\tau$, S$_{III}$ and S$_{IV}$ in 2(b): $\frac{1}{3}, \frac{2}{3}, z$; $\frac{2}{3}, \frac{1}{3}, \frac{1}{2} + z$.
  Zn, S$_I$ and S$_{II}$ in 2(a): 0, 0, z; 0, 0, $\frac{1}{2} + z$.

ZnIn$_2$S$_4$: $a = 3.85$ Å, $c = 24.68$ Å, $c/an = 0.801$ [466]
  $z(\text{In}_\Omega) = 0.750$, $z(\text{In}_\tau) = 0.094$, $z(\text{Zn}) = 0.406$
  $z(S_I) = 0.0625$, $z(S_{II}) = 0.306$, $z(S_{III}) = 0.194$, $z(S_{IV}) = 0.938$.

ZnIn$_2$S$_4$(IIIa) structure, (inverse MgAl$_2$S$_4$ structure) trigonal,
C$_{3v}^5$—R3m (No. 160), Z = 1 (3).
  hexagonal cell: $(0, 0, 0; \frac{1}{3}, \frac{2}{3}, \frac{2}{3}; \frac{2}{3}, \frac{1}{3}, \frac{1}{3})+$
  all atoms in 3(a): 0, 0, z.

ZnIn$_2$S$_4$: $a = 3.850$ Å, $c = 37.06$ Å, $c/an = 0.802$ [467]
  $z(\text{In}_\Omega) = 0.166$, $z(\text{In}_\tau) = 0.937$, $z(\text{Zn}) = 0.396$.
  $z(S_I) = 0.040$, $z(S_{II}) = 0.294$, $z(S_{III}) = 0.459$, $z(S_{IV}) = 0.872$.
  $a = 3.85$ Å, $c = 37.02$ Å [465, 472]
  $z(\text{In}_\Omega) = 0.1665$, $z(\text{In}_\tau) = 0.9367$, $z(\text{Zn}) = 0.3959$,
  $z(S_I) = 0.0408$, $z(S_{II}) = 0.2926$, $z(S_{III}) = 0.4626$, $z(S_{IV}) = 0.8703$.

CdGaInS$_4$: $a \approx 4$ Å, $c \approx 40$ Å [462]

MgAl$_2$S$_4$ structure, (Mg in Ω, Al in τ)
proposal: trigonal, D$_{3d}^5$—R$\bar{3}$m (No. 166), Z = 1 (3)
  Mg in 3(a): 0, 0, 0
  remaining atoms in 6(c): $\pm(0, 0, z)$
  $z(\text{Al}) \approx \frac{11}{48}$, $z(S_I) \approx \frac{1}{8}$, $z(S_{II}) \approx \frac{7}{24}$.

MgAl$_2$S$_4$: $a = 3.674$ Å, $c = 36.10$ Å, $c/an = 0.819$; $a_{rh} = 12.21$ Å, $\alpha = 17°17'$ [468]
MnAl$_2$S$_4$: $a = 3.680$ Å, $c = 36.18$ Å, $c/an = 0.819$; $a_{rh} = 12.24$ Å, $\alpha = 17°17'$ [468]
  $a = 3.696$ Å, $c = 36.29$ Å, $c/an = 0.818$ [470]
FeAl$_2$S$_4$: $a = 3.639$ Å, $c = 35.70$ Å, $c/an = 0.818$; $a_{rh} = 12.09$ Å, $\alpha = 17°18'$ [468]
NiAl$_2$S$_4$: [469]

the outermost anion layer would be bonded to the sandwich by only one bond instead of the actual three bonds. If, moreover, the outer anion layers are in h stacking then all the cations on both sides are just opposite an anion of the adjacent pack, so that they will contribute to the interlayer bonding. With c-stacking of both boundary anion layers the cations would be opposite to each other and thus would weaken the attraction of the layers.

The two conditions strongly restrict the number of polytypic forms of these compounds. Thus, for $ZnIn_2S_4$-type modifications the following anion stacking is possible within one pack and in the unit cell:

$hhhh \rightarrow 1$ pack: $ZnIn_2S_4(I)$ type

$hchh \rightarrow 2$ packs: $ZnIn_2S_4(IIb)$ type

$hcch \rightarrow 3$ packs: $ZnIn_2S_4(IIIa)$ type

Different arrangements of the Zn and In_τ atoms is another means to create new polytypes. Thus, the $ZnIn_2S_4(IIa)$ type evolves from the type I by adding a second pack with interchanged tetrahedral cations (Figure 104).

$ZnIn_2S_4$ polytypes crystallize as soft, hexagonal platelets of light-red color, while the compounds richer in ZnS are bright yellow. It is remarkable that the

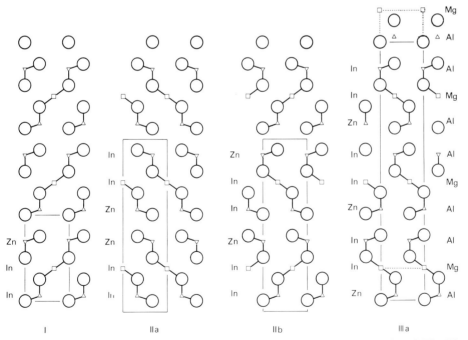

Fig. 104. (110) sections through the structures of the $ZnIn_2S_4$ polytypes I, IIa, IIb and IIIa. The structures are idealized as anion close packings. Triangles and squares represent tetrahedral and octahedral cation sites, respectively. $ZnIn_2S_4$ IIIa is the inverse $MgAl_2S_4$ structure whose cell is dotted.

TABLE 90

$Zn_2In_2S_5$(IIa) structure, hexagonal, C_{6v}^4—$P6_3mc$ (No. 186), $Z=2$.

Zn, In_τ and S in 2(b): $\frac{1}{3}, \frac{2}{3}, z; \frac{2}{3}, \frac{1}{3}, \frac{1}{2}+z$.
In_Ω in 2(a): $0, 0, z; 0, 0, \frac{1}{2}+z$.

$Zn_2In_2S_5$: $a=3.85$ Å, $c=30.85$ Å, $c/an=0.801$ [471]
$z(Zn_I)=0.070$, $z(Zn_{II})=0.314$, $z(In_\Omega)=0.193$, $z(In_\tau)=0.415$
$z(S_I)=0.040$, $z(S_{II})=0.149$, $z(S_{III})=0.237$, $z(S_{IV})=0.336$, $z(S_V)=0.445$.

$Zn_2In_2S_5$(IIIa) structure, trigonal, C_{3v}^5—$R3m$ (No. 160), $Z=1$ (3).
hexagonal axes $(0, 0, 0; \frac{1}{3}, \frac{2}{3}, \frac{2}{3}; \frac{2}{3}, \frac{1}{3}, \frac{1}{3})$+all atoms in 3(a): $0, 0, z$.

$Zn_2In_2S_5$: $a=3.85$ Å, $c=46.27$ Å, $c/an=0.801$ [472]
$z(Zn_I)=0.691$, $z(Zn_{II})=0.521$, $z(In_\Omega)=0.106$, $z(In_\tau)=0.921$,
$z(S_I)=0.005$, $z(S_{II})=0.469$, $z(S_{III})=0.607$, $z(S_{IV})=0.744$,
$z(S_V)=0.868$.

$c/a\cdot n$ values (see Tables 88–90) are rather close to the value $\sqrt{\frac{2}{3}}=0.816$ for ideal close packing of the anions so that it is easy to calculate the lattice constants of hypothetical polytypes. In the ZnS—In_2S_3 compounds half the In atoms are always found in octahedral sites and half in tetrahedral sites. It should be possible to replace the tetrahedrally coordinated In by Ga. Such a quaternary compound has indeed been synthesized: $CdGaInS_4$ [462] which we have tentatively added to the $ZnIn_2S_4$(IIIa)-type representatives.

A symmetric repartition of the cations is found in the $M^{2+}Al_2S_4$ compounds, as here the divalent cation occupies the central octahedron set. In analogy to the spinels, we might call the $ZnIn_2S_4$(I) structure the inverse $MgAl_2S_4$ type. However, it is not correct to term these structures hexagonal spinels since in the spinel, $\frac{2}{3}$ of the cations are located in octahedral sites which is just reciprocal to the situation in the $MgAl_2S_4$ structure.

The composition $Zn_2In_2S_5$ offers even more possibilities for layer structures. Thus the following anion stackings lead to cells with the given number of packs:

(hccch) → 1 pack

(hchch) → 2 packs

(hhcch) → 2 packs

(hhchh) → 6 packs (rhombohedral)

(hhhhh) → 2 packs

Two other two-pack polytypes can be constructed by combining two different packs: *hhchh hhhch* and *hchch hhhhh*. Furthermore, the sequency of Zn and In_τ can also be varied. The $M_2M_2'X_5$ structures would be more symmetric if all M cations were located in octahedral holes, as in hypothetical $Mg_2Al_2S_5$.

The observed structure of $Zn_3In_2S_6$ is somewhat exceptional as here the boundary τ set is made up of $\frac{1}{2}Zn+\frac{1}{2}In$. Such a structure might be more appropriate for symmetrical compositions like Zn_4TiS_6 or Cu_4ZnCl_6.

TABLE 91
$Zn_3In_2S_6$ structure, trigonal, D_{3d}^3—$P\bar{3}m1$ (No. 164), $Z=1$.
In_Ω in 1(b): $0, 0, \frac{1}{2}$
remaining atoms in 2(d): $\pm(\frac{1}{3}, \frac{2}{3}, z)$

$Zn_3In_2S_6$: $a = 3.85$ Å, $c = 18.5$ Å, $c/an = 0.80$ [473]
$z(Zn) = 0.299$, $z(Zn_{1/2}In_{1/2}) = 0.871$, $z(S_I) = 0.427$
$z(S_{II}) = 0.738$, $z(S_{III}) = 0.0795$.

The idealized polytypes are fully characterized by the following stacking sequences (the coordination or number of bonds is indicated by dashes; lower case letters: octahedral cations; Greek letters: tetrahedral cations)

$FeAl_2S_4$ type:

S≡Al—S≡Fe≡S—Al≡S
A β   B c   A α   B, (hhhh) anion stacking

$ZnIn_2S_4$(I) type:

S≡In—S≡In≡S—Zn≡S

$ZnIn_2S_4$(IIa) type:

S≡In—S≡In≡S—Zn≡S   S≡Zn—S≡In≡S—In≡S
B γ   C a   B β   C   B γ   C a   B β   C, $(hhhh)_2$

$ZnIn_2S_4$(IIb) type:

S≡In—S≡In≡S—Zn≡S   S≡In—S≡In≡S—Zn≡S
A β   C c   A α   C   A γ   C b   A α   B, $(hhch)_2$

$MgAl_2S_4$ type:

S≡Al—S≡Mg≡S—Al≡S   S≡Al—S≡Mg≡S—Al≡S
A β   B a   C γ   A   C α   A c   B β   C

S≡Al—S≡Mg≡S—Al≡S , $(hcch)_3$
B γ   C b   A α   B

$ZnIn_2S_4$(IIIa) type:

S≡Zn—S≡In≡S—In≡S   S≡Zn—S≡In≡S—In≡S
A β   B a   C γ   A   C α   A c   B β   C

S≡Zn—S≡In≡S—In≡S
B γ   C b   A α   B

hypothetical
$Mg_2Al_2S_5$ type:

S≡Al—S≡Mg≡S≡Mg≡S—Al≡S
A β   B a   C b   A α   B, (hccch) anion stacking

$Zn_2In_2S_5$(IIa) type:

$$S\equiv Zn-S\equiv In\equiv S-Zn\equiv S-In\equiv S$$
$$B\ \gamma\quad C\ a\quad B\ \beta\quad C\ \gamma\quad B$$
$$S\equiv Zn-S\equiv In\equiv S-Zn\equiv S-In\equiv S$$
$$C\ \beta\quad B\ a\quad C\ \gamma\quad B\ \beta\quad C,\ (h_{10})$$

$Zn_2In_2S_5$(IIIa) type:

$$S\equiv Zn-S\equiv In\equiv S-Zn\equiv S-In\equiv S$$
$$A\ \beta\quad B\ a\quad C\ \gamma\quad B\ \beta\quad C$$
$$S\equiv Zn-S\equiv In\equiv S-Zn\equiv S-In\equiv S$$
$$B\ \gamma\quad C\ b\quad A\ \alpha\quad C\ \gamma\quad A$$
$$S\equiv Zn-S\equiv In\equiv S-Zn\equiv S-In\equiv S$$
$$C\ \alpha\quad A\ c\quad B\ \beta\quad A\ \alpha\quad B,\ (hchhh)_3$$

$Zn_3In_2S_6$ type:

$$S\equiv(Zn, In)-S\equiv Zn-S\equiv In\equiv S-Zn\equiv S-(Zn, In)\equiv S$$
$$B\quad \gamma\quad C\ \beta\quad B\ a\quad C\ \gamma\quad B\quad \beta\quad C,\ (h_6)$$

These formulae also define a covalent bond state limit with single bonds if we add the formal charges as e.g. for $ZnIn_2S_4$:

$$S^+\equiv In^--S^{+2}\equiv In^{-3}\equiv S^{+2}-Zn^{-2}\equiv S^+$$

as opposed to the other extreme bond state, the ionic state

$$S^{-2}\vdots\vdots\ In^{3+}\cdots S^{-2}\vdots\vdots\ In^{3+}\vdots\vdots\ S^{-2}\cdots Zn^{2+}\vdots\vdots\ S^{-2}.$$

For an intermediate state with fractional resonating bonds we need not achieve the charge transfer except on the divalent cations at the boundary layer. Without charge transfer these cations would be bonded to the outer S layer only but not towards the interior because with their own two valence electrons, the cations are just able to bind the outer anion layer. This is the reason why we prefer to locate the trivalent tetrahedral cations in the outer layers.

These structures are visualized in Figure 104 and 105. Other polytypes were claimed to have been observed [463, 472, 474] but only c-values were reported that are exact multiples of those pertaining to the known polytypes.

Obviously, other compositions $mZnS\cdot nIn_2S_3$ are possible as well. If only integer values were allowed for m and n, then the c-values

$$c = p(m+3n)\times 3.1\ \text{Å}$$

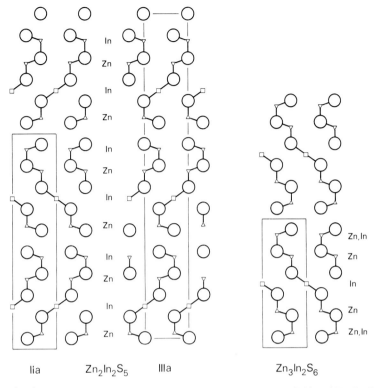

Fig. 105. (110) sections through the structures of $Zn_2In_2S_5$ IIa and IIIa (left) and $Zn_3In_2S_6$ (right).

TABLE 92

Phases in the system ZnS—In_2S_3. Experimental c-values ($a = 3.85$ Å for all phases) and ZnS concentrations determined by means of electron microprobe analysis [476]. Compositions derived from the number of S layers and from the experimental values of the ZnS concentration.

c(Å)	Number of S layers	Mole % ZnS	Close-lying compositions (mole % ZnS in parentheses)
37.39	12	83.4	$Zn_9In_2S_{12}(90)$; $Zn_{7\frac{1}{2}}In_3S_{12}$ (83.3)
68.56	22	83.4	$Zn_{13}In_6S_{22}(81.3)$; $Zn_{14}In_{5\frac{1}{3}}S_{22}(84.0)$
31.18	10	81.6	$Zn_7In_2S_{10}(87.5)$; $Zn_6In_{2\frac{2}{3}}S_{10}(81.8)$
56.04	18	79.3	$Zn_3In_2S_6(75)$; $Zn_{10}In_{5\frac{1}{3}}S_{18}(78.9)$
24.86	8	77.6	$Zn_5In_2S_8(83.3)$; $Zn_{4\frac{1}{4}}In_{2\frac{1}{2}}S_8(78.3)$
43.45	14	75.5	$Zn_4In_2S_7(80)$; $Zn_7In_{4\frac{2}{3}}S_{14}(75)$
18.62	6	70.1	$Zn_3In_2S_6(75)$; $Zn_{2\frac{1}{2}}In_{2\frac{1}{3}}S_6(68.2)$
55.84	18	66.5	$Zn_3In_2S_6(75)$; $Zn_7In_{7\frac{1}{3}}S_{18}(65.6)$
31.00	10	60.0	$Zn_2In_2S_5(66.7)$; $Zn_{3\frac{1}{2}}In_{4\frac{1}{3}}S_{10}(61.8)$
46.43	15	59.8	$Zn_2In_2S_5(66.7)$; $Zn_5In_{6\frac{2}{3}}S_{15}(60)$
55.59	18	55.1	$Zn_3In_4S_9(60)$; $Zn_5In_{8\frac{2}{3}}S_{18}(53.6)$
36.90	12	49.4	$ZnIn_2S_4(50)$

would give an indication about the composition. Thus, by means of microprobe analysis and a c-value of 24.8 Å, Gnehm et al. [475] identified the compound $Zn_5In_2S_8$. The results of Barnett et al. [476], however, demonstrate that partly disordered structures must exist, the metal atoms in certain layers being mutually exchanged as we have already met in the case of $Zn_3In_2S_6$. In Table 92, we compare the experimentally determined compositions with formulae conjectured from the number of S layers. The first formula corresponds to a structure with pure layers and the second to mixed layers.

10. Layer Compounds with B Metal Cations in Octahedral Coordination

Geometrically, any structure based on a close packing can be converted into a layer structure by completely removing the cations in certain layers. Bond considerations, however, show that this procedure is not applicable to the zincblende and wurtzite structures but that it works well in the case of the NaCl and the NiAs structures which both contain symmetrically bonded cations in octahedral coordination. In order to build up a layer structure we therefore simply have to squeeze a NaCl- or NiAs-type fragment between a sandwich of the form MX_2. Of course, additional conditions, e.g. regarding the electro-negativity difference obviously have to be fulfilled. Thus, electrostatic reasons probably prevent, say, $nNaCl\cdot CdCl_2$ or $nNaI\cdot NiI_2$ from crystallizing in such a layer structure. $SnS\cdot SnS_2 = Sn_2S_3$ and $PbS\cdot SnS_2 = PbSnS_3$ have no such layer structure either. The probability should be higher with compounds of the type $nPbTe\cdot PbI_2$ such as Pb_2TeI_2. The analogous pure chalcogenide is of the type γ-In_2S_3 or Bi_2Te_3 and it is in fact with compounds of the latter family in combination with rocksalt phases that a large number of compounds can be created, not however with the sulfides and only with the heaviest selenides. We note also that no binary chalcogenide is known with the $CdCl_2$ structure which is based on a ccp of the anions whereas the hexagonal analog, the CdI_2 structure, is widely spread.

On checking these chalcogenide layer structures, we notice one striking property common to most of these structures: With the exception of the γ-In_2S_3 structure they are all based on a ccp of all atoms. The stacking of the anions therefore is also cubic except at the boundaries. The stacking of the outermost layers of each pack is of the h type. By taking this stacking mode as a rule we can predict the structure of any composition provided we guess the repartition of the cations. In composite sandwich structures the anions of the boundary layers are inequivalent to the inner anions since they form only three bonds while the latter form six bonds. In other words, the bond strength of the three bonds of the boundary anions is twice that of the inner bonds. It appears therefore favorable to locate the higher valent cations next to the 'surface'. This means that in $Ge_2Bi_2Te_5$ the layer sequence will be

Te Bi Te Ge Te Ge Te Bi Te

but not Te Ge Te Bi Te Bi Te Ge Te though the latter sequence is crystallographically as symmetric as the former. In this overall cubic packing cations of adjacent layers face each other but their distance is far above the limit for interaction.

a. NON-TRANSITION-ELEMENT CdI_2-TYPE CHALCOGENIDES

This structure is made up of boundary layers only. In the simple trigonal C6 type structure the atoms are in ccc sequence which in turn corresponds to a *hh* stacking of the anions. This structure is frequent among the halides and the transition-element chalcogenides but rather seldom in the remaining part of the Periodic Table (compare Table 4). Structure refinements are rather rare for these simple structures though single crystals can easily be prepared by iodine-transport reactions. In the tin dichalcogenides the axial ratio is fairly close to the ideal value 1.633, so that the octahedral environment of the cation can be assumed to be at most slightly distorted. The tin and lead dichalcogenides are normal valence semiconductors. For the yellow SnS_2 and the grey-black $SnSe_2$ direct-transition band gaps of 2.88 and 1.62 eV have been determined from optical-absorption measurements [447] and an indirect-transition gap of 1.0 eV in $SnSe_2$ [529, 530]. Isostructural PbS_2 can be synthesized only under high pressure [478]. In Table 93 we have included two halide chalcogenides, BiTeBr and BiTeI. These compounds are semiconductors with energy gaps of ~0.6 eV [479, 480] and ~0.5 eV [481], respectively. Their homologs with Sb, Se and S crystallize in an orthorhombic fibre structure and some of them are famous for their ferro-electric and photoconductive properties.

It is noteworthy that the metallic high-pressure phase $Ag_2O(p)$ has an axial ratio much closer to the ideal value than the semiconducting phases BiTeBr and BiTeI.

While a large number of polytypes are known for CdI_2, it is only recently that such stacking variants have been detected in the case of SnS_2 [482, 483]. The next simple polytype 4H has been observed also in natural berndtite. It is based on a hchc stacking of the anions. Among iodine-transported single crystals, however, the most common structure encountered was the 18R type. Other crystals had a 8H (possible anion stacking *hchcchch*) and a 10H type [483].

b. TETRADYMITE (Bi_2Te_2S)-TYPE COMPOUNDS

As is evident from Table 94, this structure occurs under normal conditions in selenides and tellurides of the heavy cations Sb and Bi but under high-pressure conditions also in chalcogenides of Ga and In. Reduction of the M—X distances appears to separate the energies of p- and d-orbitals in Sb and Bi so that their chalcogenides tend to adopt structures with p^3 single bonds. Thus on application of pressure, Bi_2Se_3 transforms to the Sb_2S_3 structure, a transition quite similar to

TABLE 93
Non-transition-element CdI_2-type chalcogenides
CdI_2 (C6 type) structure, trigonal, D_{3d}^3—$P\bar{3}m1$ (No. 164), $Z=1$.
X in 2(d): $\pm(\frac{1}{3},\frac{2}{3},z)$ with $z \approx \frac{1}{4}$
M in 2(a): 0, 0, 0

	$a(\text{Å})$	$c(\text{Å})$	c/a	Ref.
$SnS_{2-x}O_x$				[508]
SnS_2	3.648 6	5.899 2	1.617	[483]
	3.645	5.901	1.619	[482]
	3.657	5.985	1.636	[484]
	3.645	5.891	1.616	[485a]
$SnS_{1.8}Se_{0.2}$	3.662	5.927	1.619	[485a]
$SnS_{1.6}Se_{0.4}$	3.678	5.961	1.621	[485a]
$SnS_{1.4}Se_{0.6}$	3.691	5.991	1.623	[485a]
$SnS_{1.2}Se_{0.8}$	3.706	6.019	1.624	[485a]
SnSSe	3.716	6.050	1.628	[485]
	3.723	6.047	1.624	[485a]
$SnS_{0.8}Se_{1.2}$	3.736	6.068	1.624	[485a]
$SnS_{0.6}Se_{1.4}$	3.753	6.085	1.621	[485a]
$SnS_{0.4}Se_{1.6}$	3.770	6.105	1.619	[485a]
$SnS_{0.2}Se_{1.8}$	3.782	6.116	1.617	[485a]
$SnSe_2$	3.811	6.137	1.610	[485]
	3.799	6.131	1.614	[485a]
	(77 to 370 K)			[530]
PbS_2	3.89	5.91	1.52	[478]
BiTeBr	4.23	6.47	1.53	[1]
	4.23	6.48		[479]
BiTeI	4.31	6.83	1.58	[1]
$Ag_2O(p)$	3.072	4.941	1.61	[1]

C27 polytype structure, hexagonal, C_{6v}^4—$P6_3mc$ (No. 186), $Z=1$.

Sn and S_I in 2(b): $\frac{1}{3},\frac{2}{3},z; \frac{2}{3},\frac{1}{3},\frac{1}{2}+z$
S_{II} in 2(a): $0,0,z; 0,0,\frac{1}{2}+z$.
$z(S_I) = 0$, $z(S_{II}) \approx \frac{3}{4}$, $z(Sn) \approx \frac{3}{8}$.

SnS_2: $a = 3.645$ Å, $c = 11.802$ Å, $c/a = 3.238$ [482]
$a = 3.648\,6$ Å, $c = 11.798\,4$ Å, $c/a = 3.234$ [483]

---

the NaCl → SnS type transition of PbS. On the other hand, Ga and In behave under pressure quite normally by increasing their coordination number.

In the tetradymite structure three Te Bi S Bi Te packs ($Te^{(1)}$ Bi $Te^{(2)}$ Bi $Te^{(1)}$ in Bi_2Te_3) are rhombohedrally stacked. The separation of the packs is of the order of the Van der Waals distance as can be seen from Table 95. Drabble and Goodman [515] have proposed a bond scheme for Bi_2Te_3 which is widely accepted. These authors also assume pure p-bonds of the $Te^{(1)}$ atoms but sp^3d^2 hybrid bonds of the central $Te^{(2)}$ (or S) atoms. Twenty-four valence electrons are needed for the twelve single bonds per unit cell or formula unit. In the covalent limit, the charge distribution is $Te^+Bi^-TeBi^-Te^+$ and only the $Te^{(1)}$ atoms keep

TABLE 94
Tetradymite (Bi_2Te_2S) structure, trigonal, D_{3d}^5—$R\bar{3}m$ (No. 166), $Z = 1(3)$.
rhombohedral axes: Bi and Te in 2(c): $\pm(x, x, x)$
S in 1(a): 0, 0, 0.
hexagonal axes: $(0, 0, 0; \frac{1}{3}, \frac{2}{3}, \frac{2}{3}; \frac{2}{3}, \frac{1}{3}, \frac{1}{3})+$
Bi and Te in 6(c): $\pm(0, 0, z)$
S in 3(a): 0, 0, 0

Compound	a_h (Å)	c_h (Å)	a_{rh} (Å)	$\alpha(°)$	$x_M = z_M$	$x_X = z_X$	Refs.
Sb_2TeSe_2	4.115	29.45	10.10	23.51			[486]
	4.105	29.503	10.116	23.41	0.3953	0.2146	[487]
Sb_2Te_2Se	4.188	29.937	10.268	23.53	0.3942	0.2146	[120a]
Sb_2Te_3	4.25	30.35	10.426	23.52	0.400	0.211	[4]
	4.264	30.458	10.447	23.55	0.3988	0.2128	[120a]
$Bi_8(Te_7S)S_4$	2×4.1968	2×29.452			0.3929	0.2133	[488, 488a]
Bi_2Te_2S	4.316	30.01	10.31	24.17	0.392	0.212	[4]
Bi_2Se_3	4.138	28.64	9.841	24.27	0.399	0.206	[4]
	4.143	28.636	9.840	24.30	0.4008	0.2117	[489]
	4.15	28.65			0.3957[a]	0.2197[a]	[489a, 518a]
Bi_2TeSe_2	4.218	29.240	10.046	24.24	0.3985	0.2115	[489]
Bi_2Te_2Se	4.28	29.86	10.255	24.08	0.3961	0.2117	[4]
	4.298	29.774	10.230	24.25	0.3958	0.2118	[489]
Bi_2Te_3	4.3835	30.487	10.473	24.17	0.400	0.212	[4]
	4.386	30.497	10.476	24.17	0.400	0.2095	[489]
					0.4004	0.2087	[490]
4 K	4.3717	30.3432	10.425	24.208			[518]
$Ga_2Se_3(p)$	3.99	27.8	9.55	24.1			[454]
In_2Se_3 (250°C)	4.05	29.41	10.1	23.2	0.401	0.222	[452]
(h)	4.033	29.44	10.09	23.07			[494]
$In_2Te_3(p)$	4.27	29.65	10.2	24.2			[495]
$Ta_2S_2C(h)$	3.276	25.62	8.747	21.59	0.38	0.22	[496]
$(Ta_2V)C_2$	3.045	21.81	7.481	23.45	$\sim\frac{7}{18}$	$\sim\frac{2}{9}$	[497]
Hf_3N_2	3.206	23.26	7.972	23.20			[498]
anion close packing					$\frac{7}{8}$ = 0.3889	$\frac{2}{9}$ = 0.2222	
ccp of all atoms					$\frac{2}{5}$	$\frac{1}{5}$	

$Sb_2(Te_{1-x}S_x)_3$	$x = ?$
$Sb_2(Te_{1-x}Se_x)_3$	$x \leq \frac{2}{3}$ [486]
	$x \sim 0.6$ [493]
$(Sb_{1-x}As_x)_2Te_3$	$x = ?$
$Bi_2(Se_{1-x}S_x)_3$	$x = ?$ ($x < 0.5$ [1])
$Bi_2(Te_{1-x}S_x)_3$	$x > 0.4$ [488, 499]
$Bi_2(Te, Se)_3$	continuous [489, 491, 492]
$(Bi_{1-x}Sb_x)_2Se_3$	$x \leq 0.7$ [486]
	$x < 0.28$ [493]
$(Bi, Sb)_2Te_3$	continuous at high temperatures [491, 500, 501, 502]
Bi_2Se_3—Sb_2Te_3	continuous [493]
$(1-x)Bi_2Te_3 \cdot xSb_2Se_3$	$x \leq 0.7$ [493]
$(Sb_{1-x}In_x)_2Te_3$	$x \leq 0.45$ [503]
$(Bi_{1-x}In_x)_2Se_3$	$x \leq 0.38$? (mixture of In_2Se_3 and Bi_2Te_3 structures?) continuous? [504]
$(Bi_{1-x}In_x)_2Te_3$	$x \leq 0.25$ [503]
	$x \leq 0.05$ [505]
$(1-x)Bi_2Te_3 \cdot xIn_2Se_3$	$x \approx 0.05$ at 400°C [506]
	($3In_2Se_3 \cdot 2Bi_2Te_3$: $a = 4.19$ Å, $c = 29.85$ Å [506])
$(1-x)Sb_2Te_3 \cdot xGeTe$	$x = 0.12$ at 500°C, 0.05 at 300°C [507]

[a] Calculated with Bi—Se distances of 2.81 Å for the outer $\frac{2}{3}$ bonds and 2.99 Å for the inner $\frac{1}{3}$ bonds [518a].

TABLE 95

Interatomic distances and separation of the layers in Bi_2Te_3-type compounds

Interatomic distances	Bi_2Te_2S	Bi_2Se_3	Sb_2Te_2Se	Bi_2Te_2Se	Sb_2Te_3	Bi_2Te_3
$X^{(1)}$—$X^{(1)}$ (Å)	3.69	3.52	3.75	3.66	3.74	3.64
M—$X^{(1)}$ (Å)	3.12	3.07	3.03	3.10	3.17	3.25
M—$X^{(2)}$ (Å)	3.05	2.85	2.98	3.04	2.98	3.07
Absolute and relative layer distances						
$X^{(1)}$—$X^{(1)}$ (Å)	2.72	2.58	2.87	2.69	2.81	2.61
$(2z_X-\frac{1}{3})c/a$	0.630	0.623	0.685	0.625	0.659	0.597
$X^{(1)}$—$X^{(2)}$ (Å)	3.64	3.48	3.55	3.62	3.67	3.78
$(\frac{1}{3}-z_X)c/a$	0.844	0.841	0.849	0.842	0.861	0.861
$X^{(1)}$—M (Å)	1.88	1.55	1.73	1.76	1.68	1.74
$(\frac{2}{3}-z_X-z_M)c/a$	0.436	0.374	0.414	0.409	0.393	0.397
$X^{(2)}$—M (Å)	1.76	1.93	1.82	1.86	1.99	2.03
$(z_M-\frac{1}{3})c/a$	0.408	0.466	0.435	0.433	0.468	0.464

Ideal close packing of the anions X:
$X^{(1)}$—$X^{(1)} = X^{(1)}$—$X^{(2)} = 0.8165a$, $X^{(1)}$—M = $X^{(2)}$—M = $0.4082a$, $z_M = \frac{7}{18}$, $z_X = \frac{2}{9}$
Ideal close packing of all atoms M+X:
$X^{(1)}$—$X^{(1)} = 0.8165a$, $X^{(1)}$—$X^{(2)} = 1.633a$,
$X^{(1)}$—M = $X^{(2)}$—M = $0.8165a$, $z_M = \frac{2}{5}$, $z_X = \frac{1}{5}$
M is at the center of the (deformed) [MX_6] octahedron if $z_X = 1-2z_M$.

their s^2 lone electron pair while $Te^{(2)}$ and Bi use all their s and p electrons for bonding. This model even explains why in alloys $Bi_2Te_{3-x}Se_x$ the more electronegative anion goes into the $Te^{(2)}$ sites. However, the model fails as soon as we add GeTe to form $GeBi_2Te_4$. In this case two electrons are missing. A single-bond limiting state cannot be realized in most compounds with high coordination number. In Sb and Bi the energy of the s-orbital is well separated from that of the p- and d-orbitals and in trivalent Sb and Bi compounds, the two s-electrons simply do not take part in the bonding, otherwise the anions would be pentavalent. Electrons would be missing also in the high-pressure modifications of Ga_2Se_3, In_2Se_3 and In_2Te_3. Without further speculating about the actual charge transfer we can say that the Te atoms contribute 2 electrons to the bonding and the Bi atoms contribute 3 electrons, according to a scheme

$$Te=Bi-Te-Bi=Te, \quad \text{with an ionic limit}$$
$$Te=Bi^+ \cdots Te^{2-} \cdots Bi^+=Te$$

where a dash now stands for a binding electron pair regardless of whether it corresponds to three $\frac{1}{3}$ bonds or to $1\frac{1}{2}$ $\frac{2}{3}$ bonds.

In the binary compounds, the difference between the two M—X distances corresponds surprisingly well to the expected difference of 0.18 Å between bonds differing in bond number by a factor of 2. In the mixed phases the central M—$X^{(2)}$ distance is considerably larger which reflects the ionic component of the bonding.

These compounds have attracted much attention as promising materials for thermoelectric refrigeration and power generation, since they are small-gap semiconductors with large Seebeck coefficients [491, 499]. Aiming at a reduction of the lattice thermal conductivity, several alloy systems have been studied, ternary [491, 499, 509] and quaternary ones [493, 510–512]. In mixed-anion systems the a-parameter of the hexagonal cell is linear in concentration while the c-parameter exhibits positive deviations from Végard's law [492, 493]. In mixed-cation alloys, on the other hand, it is the a-parameter which shows the main deviations, depending on thermal treatment. Careful investigations of the systems Bi_2Te_3—Bi_2Se_3 [492] and Sb_2Te_3—Sb_2Se_3 [487] reveal that on replacing Te by Se the deviations from a linear interpolation starts near M_2Te_2Se. This discontinuity results from the tendency of Se to first replace the $Te^{(2)}$ atoms. Once this position is filled up the rest of the Se atoms is randomly distributed among the $Te^{(1)}$ positions. The ordering near the concentration M_2X_2X' is reflected in the physical properties [491, 509]. In the system $Bi_2Te_{3-x}Se_x$ the optical-absorption edge reveals a linear shift from 0.15 eV at $x = 0$ to 0.31 eV at $x = 0.9$ and then a slow decrease of the gap to 0.276 eV for $x = 3$ [513].

In the mixed-cation alloys $(M_{1-x}M'_x)_2X_3$ cation ordering appears to occur at $x = \frac{1}{3}$ and $\frac{2}{3}$, i.e. ordering will be within the layers. In the system Bi_2Te_3—Sb_2Te_3 Testardi and Wiese [514] found a gentle peaking of the densities at compositions in the vicinity of $x = \frac{1}{3}$ and $\frac{2}{3}$ for zone-leveled samples but not for quenched (disordered) samples. This is reflected also in the concentration dependence of the a-values [500]. It is interesting that introduction of Sb_2Te_3 into Bi_2Te_3 first diminishes the thermal energy gap. The curve for rapidly crystallized alloys is approximately linear and extrapolates to zero for pure Sb_2Te_3. The energy-gap values for the slowly crystallized samples all lie above this straight line climbing to 0.3 eV for pure Sb_2Te_3 [516]. For Sb_2Te_3, however, metallic conductivity and a zero gap have also been reported. Preparation of clearly defined samples obviously poses certain problems. As the maximum melting point corresponds to a slightly Sb-rich composition Sb_2Te_3 will separate on cooling from a stoichiometric melt and the residual Te is segregated in the grain boundaries. The samples turn out to be always p-type and the intrinsic region is hard to reach in resistivity measurements. Weak doping with Pb increases the resistivity possibly because Pb improves the stoichiometry [517].

TABLE 96

$Bi_2Te_3(II)$ structure, trigonal, C_{3v}^5—R3m (No. 160), $Z = 1$.
All atoms in 3(a): $0, 0, z; \frac{1}{3}, \frac{2}{3}, \frac{2}{3}+z; \frac{2}{3}, \frac{1}{3}, \frac{1}{3}+z$.

$Bi_2Te_3(II)$: $a = 4.417$ Å, $c = 29.84$ Å [122a]
 $z(Bi_I) = 0.188$, $z(Bi_{II}) = 0.598$, $z(Te_I) = 0.389$, $z(Te_{II}) = 0.800$, $z(Te_{III}) = 0$.
$Sb_2Te_3(r)$? [121]

While most authors claim their samples to crystallize in the tetradymite structure, a piezoelectric effect has been detected on Sb_2Te_3, so that the structure of that modification cannot belong to the centrosymmetric group $R\bar{3}m$ but either to R3m or to R32 [121]. Possibly the tetradymite-type modification of Sb_2Te_3 is not an equilibrium phase [501]. It is argued [122a] that it has the same structure as the first high-pressure modification of Bi_2Te_3 which is claimed to be a cubic stacking of all atoms with the sequence

Te Bi Bi Te Te

Though now a Bi and a Te layer are interchanged, which drastically changes the bonding in this metallic phase, the distances are similar to those in non-metallic Bi_2Te_3:

$$Te_{I\ 3.45\ \text{Å}} Te_{II\ 3.03\ \text{Å}} Bi_{I\ 3.43\ \text{Å}} Bi_{II\ 3.27\ \text{Å}} Te_{III\ 3.04\ \text{Å}} Te_I$$

Under normal conditions tetradymite-type Bi_2Te_3 does not undergo any phase transition from 4 to 600 K as follows from X-ray measurements [518]. No anomalous temperature dependence of the (anisotropic) lattice expansion was observed as a function of charge-carrier density or carrier type.

At atmospheric pressure Bi_2Te_3 does not show superconductivity for carrier densities up to 1×10^{21} cm^{-3} and temperatures down to 0.02 K. However, superconductivity can be achieved in both n- and p-type Bi_2Te_3 by application of pressure. Starting near 50 kbar the transition temperature sharply increased with pressure up to a maximum $T_c \approx 2.8$ K near 67 kbar in a p-type sample with an initial carrier density of 2×10^{19} cm^{-3}. A similar behavior is shown by n-type $Bi_2Te_3(p_3)$ samples [517a]. It is noteworthy that in $Bi_2Te_3(p_1)$ superconductivity above 1.5 K was observed only on compressing p-type Bi_2Te_3 samples not however on n-type samples.

Considerations of bond-angle strain in the tetradymite structure led Pauling [518a] to the conclusion that the true composition of natural (= equilibrium) tetradymite is in fact $Bi_{14}Te_{13}S_8$ rather than Bi_2Te_2S. Already in 1934 when Harker determined the tetradymite structure he pointed out that the bond lengths Bi—S and Bi—Te are somewhat greater than the sum of Pauling's single-bond covalent radii and from this difference erroneously deduced metallic properties. In fact the crystallographically determined distances are Bi—S = 3.01 Å and Bi—Te = 3.08 Å as opposed to 2.86 Å and 3.04 Å calculated for Bi—S $\frac{1}{3}$ bonds and Bi—Te $\frac{2}{3}$ bonds, respectively. The values for the Bi—S distance differ distinctly by more than the experimental error. Moreover, bond angles calculated with the theoretical bond lengths deviate so much from the expected values as to make the composition Bi_2Te_2S and Harker's structure unstable relative to other phases. According to Glatz [518b] the composition Bi_2Te_2S does indeed not occur as equilibrium phase in the system Bi_2Te_3—Bi_2S_3. With samples annealed at 400°C,

DISCUSSION OF SPECIAL COMPOUNDS

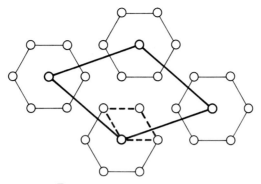

Fig. 105a. The $\sqrt{7}\,a_0$ supercell of tetradymite $Bi_{14}Te_{13}S_8$ [518a].

Glatz found an α-phase with 0–4% Bi_2S_3 ($a = 4.384$–4.379 Å, $c = 30.50$–30.45 Å), a β-phase with 25–30% Bi_2S_3 ($a = 4.255$, $b = 29.59$ Å) and a γ-phase with 34–50% Bi_2S_3 ($a = 4.255$–4.24 Å, $c = 29.60$–29.57 Å), but no phase with 33.3% Bi_2S_3.

Pauling argued that the strain brought into the structure by the different anion layers might be relieved by substitution of part of the atoms. Thus, an improvement of the bond angles is achieved in the α-phase by replacing every seventh Te atom of the central layer by an S atom. An ordered substitution leads to a $\sqrt{7}\,a_0$ superlattice shown in Figure 105a and a composition $Bi_{14}Te_{20}S$ which may represent the upper limit of the α-phase. The same supercell is obtained with the composition TeS_6 for the middle layer. This substitution which improves the bond angle of the middle layer corresponds to a formula $Bi_{14}Te_{15}S_6$ with 28.6% Bi_2S_3 well within the range of the β-phase. It improves the bond angle for the middle layer but that for the outer layers becomes too small. This value, in turn, can be increased by introducing S atoms into the outer layers as well. A layer stacking

$$Te_6S\text{——}Bi_7\text{——}TeS_6\text{——}Bi_7\text{——}Te_6S$$

corresponds to the formula $Bi_{14}Te_{13}S_8$ with 38.1% Bi_2S_3 and lies close to Soonpaa's [488a] formula $Bi_8Te_7S_5$ with 38.5% Bi_2S_3. It is this $Bi_{14}Te_{13}S_8$ structure that Pauling considers as the ideal composition for natural tetradymite.

An intriguing mineral with a composition $Bi_{2+x}Te_{3-x}$ near BiTe is wehrlite. From its perfect cleavage, its space group $R\bar{3}m$ and its unit cell ($a = 4.43$ Å, $c = 29.91$ Å [5]) one would expect a partially disordered variant of the tetradymite structure.

c. LAYER STRUCTURES BASED ON $NaCl$–CdI_2-TYPE MIXTURES

The Bi_2Te_2S structure discussed before does indeed already belong to this category as it is composed of one rocksalt layer intercalated between an MX_2 sandwich, though this is not evident from the chemical formula. However, this becomes clear immediately if we write down hypothetical mixed-anion analogs such as $Pb_2SeI_2 = PbSe \cdot PbI_2 = I\,Pb\,Se\,Pb\,I$. This phase does not exist under normal conditions and we were indeed disappointed by finding no layer-type compound of composition $nPbX \cdot PbI_2$ which would be analogous to the ternary chalcogenides that we are going to discuss below. In Table 97 we have compiled the phases that occur in tin and lead chalcogenohalide systems. All these compounds form peritectically. Where we expected to find layer-type compounds no compounds formed at all, i.e. under normal conditions most Pb-containing systems are of a eutectic type. Five compounds exist which from their composition could crystallize in a tetradymite-type structure. Two of them, Sn_2SBr_2 and Sn_2SI_2, are reported to have lamellar structures [533]. However, both α- and β-Sn_2SI_2 are in fact three-dimensional [1019]. Judging from the geometrical requirements only, each compound occurring in a MX—MX'_2 system has some chance to crystallize in a layer structure if MX has a NaCl-type modification and MX'_2 a CdI_2- or $CdCl_2$-type modification. Thus, for Sn_3SeI_4 a composite layer structure of the form $Sn_2SeI_2\,SnI_2\,Sn_2SeI_2\,SnI_2\ldots$ might be adequate while for $Sn_7S_3I_8$ a composite layer structure $Sn_4S_3I_2 \cdot 3\,SnI_2$ seems already less likely. Uniform layer structures based on close packing are possible only at compositions $M_nX_{n+1} = (n-1)MX \cdot M'X_2$ or $(n-2)MX \cdot M'_2X_3$. With less cations composite layer structures are compulsive. If again we apply the rule that all atoms should be stacked in a ccp manner then we can easily predict the structures provided they contain no mixed layers [550]. If $2n+1 = 6m+3$, where m = integer, then the structure will have a hexagonal cell, a rhombohedral cell in the other cases.

In the case of the (Ge, Sn, Pb)—(Sb, Bi)—(Se, Te) compounds, structure determinations partly have been made (or tried) on evaporated films with the

TABLE 97

Repartition of phases and character of phase diagrams for the systems (Sn, Pb)X—(Sn, Pb) (Br, I)$_2$. The possible layer-type compound is underlined.

SnS—SnBr$_2$	SnS—SnI$_2$	SnSe—SnBr$_2$	SnSe—SnI$_2$	SnTe—SnBr$_2$	SnTe—SnI$_2$
Sn$_2$SBr$_2$ hexagonal [533] Sn$_9$S$_2$Br$_{14}$ [534]	Sn$_2$SI$_2$ α monocl. β orthorh. [533, 1019] Sn$_7$S$_3$I$_8$ [534, 535] Sn$_3$SI$_4$ [534, 535]	eutectic type [534]	Sn$_2$SeI$_2$ [533] Sn$_3$SeI$_4$ [534]	eutectic type [534]	eutectic type [534]
PbS—PbBr$_2$	PbS—PbI$_2$	PbSe—PbBr$_2$	PbSe—PbI$_2$	PbTe—PbBr$_2$	PbTe—PbI$_2$
Pb$_7$S$_2$Br$_{10}$ Th$_7$S$_{12}$str. [536, 537] Pb$_2$SBr$_2$ [538]	Pb$_5$S$_2$I$_6$ monocl. [537, 539] Pb$_2$SI$_2$ tetrag. [538]	Pb$_4$SeBr$_6$ orthorh. [537]	eutectic type [536]	eutectic type [536]	eutectic type [536]

TABLE 98
$GeSb_2Te_4$ structure, trigonal, D_{3d}^5—R3m (No. 166), $Z = 1(3)$.
hexagonal axes: $(0, 0, 0; \frac{1}{3}, \frac{2}{3}, \frac{2}{3}; \frac{2}{3}, \frac{1}{3}, \frac{1}{3})+$
Sb, Te$_I$ and Te$_{II}$ in 6(c): $\pm(0, 0, z)$
Ge in 3(a): 0, 0, 0.

Compound	a_h (Å)	c_h (Å)	a_{rh} (Å)	α (°)	z_M	$z_{(X_I)}$	$z_{(X_{II})}$	Refs.
PbBi$_2$Se$_4$	4.16	39.20	13.29	18.01	0.428	0.139	0.286	[540]
GeSb$_2$Te$_4$	4.21	40.6	13.75	17.61	0.426	0.142	0.287	[541]
GeBi$_2$Te$_4$	4.28	39.2 (?)						[542]
SnSb$_2$Te$_4$	4.294	41.548	14.07	17.56	0.426 2	0.133	0.289	[543, 544]
SnBi$_2$Te$_4$	4.411	41.511	14.07	18.04	0.428 8	0.136	0.289	[543]
PbBi$_2$Te$_4$	4.452	41.531	14.08	18.19	0.428 5	0.136	0.289	[543]
In$_3$Te$_4$(p)	4.26	40.58	13.75	17.80	0.427 3	0.128 4	0.290 8	[495, 545]
Ge$_3$Te$_4$ (p)	4.11	38.68	13.11	17.93				[545]
anion close packing					$\frac{5}{12} =$ 0.416 7	$\frac{1}{8} =$ 0.125	$\frac{7}{24} =$ 0.291 7	
ccp of all atoms					$\frac{3}{7} =$ 0.428 6	$\frac{1}{7} =$ 0.142 9	$\frac{2}{7} =$ 0.285 7	

result that structures have been elaborated for phases to which a wrong composition had been assigned. From all the offered structural data and interpretations, we have selected those which agreed with our ideas, and in the case of Pb$_2$Bi$_2$Se$_5$, for instance, we have interchanged Pb and Bi.

The GeSb$_2$Te$_4$ structure is formally the anti-type of the laitakarite structure. However, the structure of the GeSb$_2$Te$_4$-type analogs consists of one kind of pack only. We have already explained above why we prefer a stacking Te=Sb—Te—Ge—Te—Sb=Te.

It is non-trivial that In$_3$Te$_4$(p) crystallizes in the GeSb$_2$Te$_4$ structure. In the system In$_{1-x}$Te there is a change in anion stacking near $x = 0.2$. InTe(p) crystallizes in the NaCl structure. It contains one excess valence electron per anion (or cation) and is a superconductor. On increasing x, T_c drops from ~ 3.5 K to 1.0 K near the phase boundary and it extrapolates to zero near $x = \frac{1}{3}$ where no excess valence electron would be left. Whereas in the disordered NaCl-type phases, the anion stacking is ccc, it changes to hcch and hch on ordering of the vacancies within discrete layers at $x = \frac{1}{4}$ and $x = \frac{1}{3}$, respectively. In$_3$Te$_4$(p) which has 0.25 excess electrons per anion was found to become superconducting below 1.2 K [495]. Ge$_3$Te$_4$(p) with the same structure and 1.0 excess (or 1.0 deficit) valence electrons has a higher transition temperature, $T_c \approx 1.7$ K. A valence distribution Ge$_2^{2+}$Ge^{4+}X$_4$ in a non-metallic phase would require a more electronegative anion and then would lead to another structure. According to the distances given [495] the layer packs in In$_3$Te$_4$(p) are clearly separated:

$$\cdots = \text{Te}^{(1)} \underset{3.98\,\text{Å}}{\quad} \text{Te}^{(1)} \underset{2.84\,\text{Å}}{=} \text{In}^{(2)} \underset{3.24\,\text{Å}}{\quad} \text{Te}^{(2)} \underset{3.02\,\text{Å}}{\cdots} \text{In}^{(1)} \underset{3.02\,\text{Å}}{\cdots} \text{Te}^{(2)} \underset{3.24\,\text{Å}}{\quad} \text{In}^{(2)} \underset{2.85\,\text{Å}}{=} \text{Te}^{(1)} \underset{3.98\,\text{Å}}{\quad} \text{Te}^{(1)} = \cdots$$

The asymmetric position of Te$^{(2)}$ with the rather long distance to the outer In atoms is remarkable. This situation is similar to that found in semiconducting PbBi$_2$Te$_4$ with

$$\cdots =\text{Te} \underset{3.62\,\text{Å}}{\quad} \text{Te}\underset{3.08\,\text{Å}}{=\!=\!=}\text{Bi}\underset{3.33\,\text{Å}}{\quad}\text{Te}\underset{3.16\,\text{Å}}{\quad}\text{Pb}\underset{3.16\,\text{Å}}{\quad}\text{Te}\underset{3.33\,\text{Å}}{\quad}\text{Bi}\underset{3.08\,\text{Å}}{=\!=\!=}\text{Te}\underset{3.62\,\text{Å}}{\quad}\text{Te}=\cdots$$

but contrasts with that in PbBi$_2$Se$_4$:

$$\cdots =\text{Se}\underset{3.24\,\text{Å}}{\quad}\text{Se}\underset{2.98\,\text{Å}}{=\!=\!=}\text{Bi}\underset{3.04\,\text{Å}}{\quad}\text{Se}\underset{3.04\,\text{Å}}{\quad}\text{Pb}\underset{3.04\,\text{Å}}{\quad}\text{Se}\underset{3.04\,\text{Å}}{\quad}\text{Bi}\underset{2.96\,\text{Å}}{=\!=\!=}\text{Se}\underset{3.24\,\text{Å}}{\quad}\text{Se}=\cdots$$

In the latter compound the separation of the packs is far below the van der Waals distance, so either PbBi$_2$Se$_4$ is metallic or the positional parameters are inaccurate. All three phases in the system PbSe—Bi$_2$Se$_3$ (3 PbSe·2 Bi$_2$Se$_3$, PbSe·Bi$_2$Se$_3$ and PbSe·2 Bi$_2$Se$_3$), however, are claimed to be semiconductors [540].

As in the case of Pb$_2$Bi$_2$Se$_5$, we have also interchanged Ge and Sb in Ge$_2$Sb$_2$Te$_5$ (a reliability factor of 23.6% had been achieved with the other occupation [547]).

We have renounced to list the structural data for Ge$_3$Bi$_2$Te$_6$ as we were not happy with the published results. In Figure 106, however, we present a model of what we would like to be the Ge$_3$Bi$_2$Te$_6$ structure. A rhombohedral 3×11 layers cell has been observed ($a = 4.21$ Å, $c = 61.0$ Å [551]). The proposed sequence of the layers (based on a reliability index of 18%) is

Te$_{Ge}$Te$_{Bi}$Te$_{Bi}$Te$_{Ge}$Te$_{Ge}$Te [551].

TABLE 99
$Pb_2Bi_2Se_5$ structure, trigonal, D_{3d}^3—P$\bar{3}$m1 (No. 164), $Z = 1$.
Bi, Se$_I$ and Se$_{II}$ in 2(d): $\pm(\frac{1}{3}, \frac{2}{3}, z)$
Pb in 2(c): $\pm(0, 0, z)$
Se$_{III}$ in 1(a): $\pm 0, 0, 0$.

Compound	a (Å)	c (Å)	$z(M^{2+})$	$z(M^{3+})$	$z(X_I)$	$z(X_{II})$	Refs.
Pb$_2$Bi$_2$Se$_5$	4.22	16.42	0.333	0.108	0.783	0.450	[546]
Ge$_2$Sb$_2$Te$_5$	4.20	16.96	0.317	0.106	0.788	0.421	[547]
Sn$_2$Sb$_2$Te$_5$	(4.37)	(17.35)					[550]
Ge$_2$Bi$_2$Te$_5$	(4.32)	(17.07)					[550]
Pb$_2$Bi$_2$Te$_5$	4.46	17.5	0.329	0.109	0.782	0.440	[548]
Sn$_4$Sb$_5$	4.32	15.75	0.335^a	0.107^a	0.775^a	0.441^a	[549]
ccp of identical spheres			$\frac{1}{3}$	$\frac{1}{9}$	$\frac{7}{9}$	$\frac{4}{9}$	
anion ccp			$\frac{3}{10}$	$\frac{1}{10}$	$\frac{4}{5}$	$\frac{2}{5}$	

^{a} averaged from values given in P3m1
Values in parentheses are calculated only.

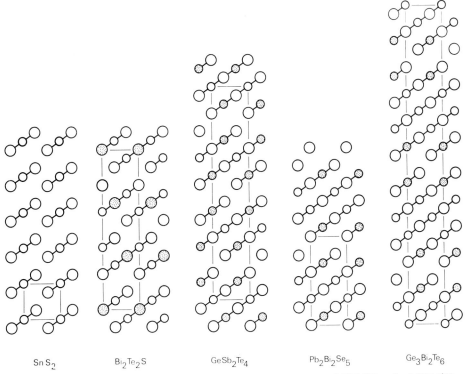

Fig. 106. (110) sections of the multiple-sandwich structures of the type $n\mathrm{MX}\cdot\mathrm{MX}_2'$ and $m\mathrm{MX}\cdot\mathrm{M}_2'\mathrm{X}_3$.
Large spheres: anions
CdI$_2$-type structure of SnS$_2$
Bi$_2$Te$_2$S tetradymite structure, S atoms dotted
GeSb$_2$Te$_4$ structure, Sb atoms dotted
Pb$_2$Bi$_2$Se$_5$ structure, Bi atoms dotted
Ge$_3$Bi$_2$Te$_6$ structure, Bi atoms dotted.

A sequence Te$_{Ge}$Te$_{Bi}$Te$_{Ge}$Te$_{Bi}$Te$_{Ge}$Te would at least be symmetric but would give rise to ionic charges whereas we propose in Figure 106 a neutral symmetric stacking with the Bi atoms in the outermost layers.

Odd structures have been reported for the PbBi$_4$Te$_7$-type phases. For PbBi$_4$Te$_7$ a statistical distribution of Pb and Bi has been proposed [553] and even layer sequences of the form Bi—Te—Bi—(Te, Pb)—(Te, Pb)—(Te, Pb)—Bi—Te—··· with (Te, Pb) = Te$_{13/16}$Pb$_{3/16}$ [555, 556]. Obviously, this was an erroneous interpretation of a Pb$_2$Bi$_2$Te$_5$ phase [554]. The only reasonable model appears to us the one showing GeBi$_2$Te$_4$·Bi$_2$Te$_3$ composite packs [543, 547, 552] as illustrated in idealized form as an anion close packing in Figure 107. In Table 100 we have added the positional parameters for this anion close packing as well as for a cubic close-packing of identical spheres. For GeBi$_4$Te$_7$ and PbBi$_4$Te$_7$ the reported values coincide strikingly with the ccp model which was the guiding model for the structure determination. In the case of GeSb$_4$Te$_7$ one feels some

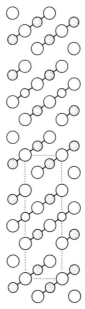

Fig. 107. (110) section of the composite layer structure of GeBi$_4$Te$_7$ = GeBi$_2$Te$_4$·Bi$_2$Te$_3$.
Large circles: Te; dotted circles: Bi

tendency towards the anion cp model which is to be expected for non-metallic phases.

For the semiconducting GeBi$_4$Te$_7$ an energy gap of 0.21 eV has been determined from electrical-resistivity measurements [552] while PbBi$_4$Te$_7$ behaves like Bi$_2$Te$_3$ with a high impurity content [557].

TABLE 100
GeSb$_4$Te$_7$ structure, trigonal, D_{3d}^3—$P\bar{3}m1$ (No. 164), $Z = 1$.
Te$_I$, Te$_{II}$, Sb$_I$ and Sb$_{II}$ in 2(d): $\pm(\frac{1}{3}, \frac{2}{3}, z)$
Te$_{III}$ in 2(c): $\pm(0, 0, z)$
Ge in 1(b): $0, 0, \frac{1}{2}$
Te$_{IV}$ in 1(a): $0, 0, 0$.

Compound	a (Å)	c (Å)	$z(M_I)$	$z(M_{II})$	$z(X_I)$	$z(X_{II})$	$z(X_{III})$	Refs.
PbBi$_4$Se$_7$	(4.25)	(22.68)						[550]
GeSb$_4$Te$_7$	4.21	23.65	0.079	0.346	0.575	0.844	0.268	[547, 554]
SnSb$_4$Te$_7$	(4.37)	(23.79)						[550]
GeBi$_4$Te$_7$	4.36	24.11	0.087	0.335	0.58	0.832	0.250	[552]
	4.352	23.925						[543]
PbBi$_4$Te$_7$	4.42	23.6(?)	0.083	0.335	0.582	0.834	0.252	[548]
	4.44	3×23.9						[553]
ccp of identical spheres			$\frac{1}{12}$ (0.083 3)	$\frac{1}{2}$ (0.333 3)	$\frac{7}{12}$ (0.583 3)	$\frac{5}{6}$ (0.833 3)	$\frac{1}{4}$ (0.25)	
anion close packing			$\frac{1}{14}$ (0.071 4)	$\frac{5}{14}$ (0.357 1)	$\frac{4}{7}$ (0.571 4)	$\frac{6}{7}$ (0.857 1)	$\frac{2}{7}$ (0.285 7)	

Values in parentheses are calculated only.

d. Tetradymite-bismuth composites

The stoichiometric phase BiTe has an ordered structure that can be described as a stacking of Bi_2Te_3 and Bi_2 layers. BiTe is only one member of a whole family of phases, which might be considered as a special kind of ordered solid solutions. It was probably Strunz [519] who first recognized that the Bi chalcogenide minerals with odd stoichiometry are built up of layers the number of which is defined by the chemical formula. Structure determinations [521–525] finally led to the conclusion that all these particular phases are created by an ordered stacking of Bi_2Te_3 (tetradymite)-type units and Bi_2 bismuth-type double layers [522, 525]. The interatomic distances confirm this view:

BiSe:

$=Se\underset{3.28\,Å}{}Bi\equiv\underset{3.05\,Å}{}Bi\underset{3.28\,Å}{}Se=\underset{2.91\,Å}{}Bi\underset{3.03\,Å}{}Se\underset{3.10\,Å}{}Bi\equiv\underset{2.89\,Å}{}Se\underset{3.52\,Å}{}Se\equiv\underset{2.89\,Å}{}Bi$

Bi_4Se_3:

$Bi\equiv\underset{3.04\,Å}{}Bi\underset{3.46\,Å}{}Se=\underset{2.90\,Å}{}Bi\underset{3.08\,Å}{}Se\underset{3.08\,Å}{}Bi=\underset{2.90\,Å}{}Se\underset{3.46\,Å}{}Bi\equiv\underset{3.04\,Å}{}Bi$

Within the Bi_2 unit the Bi—Bi distances are slightly smaller than in elemental bismuth as expected for a non-metallic Bi double layer. The distances within the Bi_2Se_3 unit compare well with those in the tetradymite-type modification. The interlayer Bi—Se distances, however, show remarkable differences from compound to compound. Thus, from the short distance one might expect semimetallic properties for BiSe rather than the reported semiconductivity [520, 527, 528]. The resistivity of all these phases is in fact extremely low.

The accuracy of the experimental values for the atomic positions is not very satisfactory in most cases. This shows up, for instance, in an unmotivated scatter of equivalent M—X distances. As these multi-layer structures consist of identical packs, there exist relations between the site parameters z, which are not particularly well fulfilled in the case of Bi_8Se_9, as an example. There might be a printing error in $z(Bi_{III})$ for which we have added an alternative in Table 102.

If we use the distances observed in M_2X_3 phases and reasonable values for the M—X interlayer separations, we can calculate the structural data for any combination between m Sb_2 or Bi_2 double layers and nM_2X_3 units. Such a catalog of lattice parameters for observed and hypothetical $M_{2(m+n)}X_{3n}$ phases is offered in Table 104 and some of the structures are illustrated in Figure 108. Of course the atomic positions could be given as well. The large cells might at a first glance look unreasonable but Brebrick [526] proved that reality can be even worse. Near a nominal composition $Bi_{45}Te_{55}$, six definite phases were detected. Of these only the one with the smallest cell is included in Table 104. The remaining phases consist of between 3×43 layers ($c = 258.69$ Å) and 3×82 layers ($c = 493.29$ Å).

TABLE 101

(a) *BiSe* structure, trigonal, D_{3d}^3—P$\bar{3}$m1 (No. 164), $Z = 3$.
  Bi_I, Bi_{II}, Se_I and Se_{II} in 2(d): $\pm(\frac{1}{3}, \frac{2}{3}, z)$
  Bi_{III} and Se_{III} in 2(c): $\pm(0, 0, z)$

	BiSe	SbTe	BiTe
a(Å)	4.18 [525]	4.26	4.40 [525]
c(Å)	22.84	23.9	23.97
$z(M_I)$	0.291	0.292	
$z(M_{II})$	0.541 [523]	0.533 [523]	
$z(M_{III})$	0.126	0.126	
$z(X_I)$	0.056	0.055	
$z(X_{II})$	0.789	0.789	
$z(X_{III})$	0.362	0.364	

(b) *Bi$_8$Se$_7$* structure, trigonal, D_{3d}^3—P$\bar{3}$m1 (No. 164), $Z = 3$.
  $Bi_I \cdots Bi_{VIII}$, $Se_I \cdots Se_{VII}$ in 2(d), $Bi_{IX} \cdots Bi_{XII}$, $Se_{VIII} \cdots Se_{XI}$ in 2(c), Se_{XII} in 1(a).
Bi$_8$Se$_7$: $a = 4.22$ Å, $c = 85.65$ Å [525]
  Bi: $z_I = 0.022\,2$, $z_{II} = 0.081\,1$, $z_{III} = 0.284\,4$, $z_{IV} = 0.353\,3$, $z_V = 0.488\,8$,
  $z_{VI} = 0.557\,8$, $z_{VII} = 0.755\,6$, $z_{VIII} = 0.820\,0$, $z_{IX} = 0.068\,0$, $z_X = 0.134\,4$,
  $z_{XI} = 0.332\,2$, $z_{XII} = 0.400\,0$.
  Se: $z_I = 0.157\,8$, $z_{II} = 0.221\,4$, $z_{III} = 0.422\,2$, $z_{IV} = 0.620\,0$, $z_V = 0.693\,3$,
  $z_{VI} = 0.886\,7$, $z_{VII} = 0.957\,8$, $z_{VIII} = 0.200\,0$, $z_{IX} = 0.264\,4$, $z_X = 0.464\,4$.

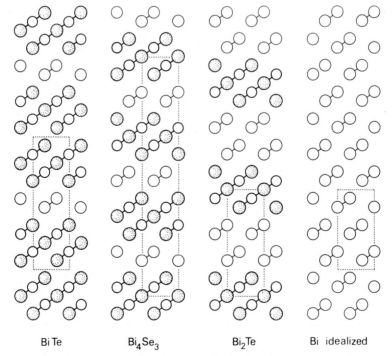

Fig. 108. Composite layer structures of the type $m\text{Bi}_2 \cdot n\text{Bi}_2\text{Te}_3$. (110) sections of BiTe = Bi$_2 \cdot$2Bi$_2$Te$_3$, Bi$_4$Se$_3$ = Bi$_2 \cdot$Bi$_2$Se$_3$, Bi$_2$Te = 2Bi$_2 \cdot$Bi$_2$Te$_3$, and Bi$_2$ idealized. Anions stippled.

TABLE 102
Bi_8Se_9 structure, trigonal, D_{3d}^5—R3m (No. 166), $Z=1(3)$
hexagonal axes: $(0, 0, 0; \frac{1}{3}, \frac{2}{3}, \frac{2}{3}; \frac{2}{3}, \frac{1}{3}, \frac{1}{3})+$
Bi_I to Bi_{IV} and Se_I to Se_{IV} in 6(c): $\pm(0, 0, z)$
Se_V in 3(a): $0, 0, 0$.

Bi_8Se_9: $a = 4.16$ Å, $c = 97.1$ Å [525]

	Bi_I	Bi_{II}	Bi_{III}	Bi_{IV}	Se_I	Se_{II}	Se_{III}	Se_{IV}
z:	0.1764	0.1167	0.3539 (0.3530?)	0.4127	0.0618	0.2343	0.2970	0.4676

At the composition $Bi_{43}Te_{57}$ Brebrick detected even one phase with 3×85 layers ($c = 513.08$ Å) and a second one with 3×86 layers ($c = 519.20$ Å). These phases represent a rather unusual type of ordered solid solutions and remind one of the valleriite-related minerals $CuFeS_2 \cdot Mg(OH)_2$, $FeS \cdot nMg(OH)_2$, etc.

In our list of representatives of the laitakarite structure, we have added some truly metallic phases, Sn_4P_3, Hf_4N_3, V_4C_3, etc. In these compounds the bonding is quite different from that in the $Bi_4Se_3 = Bi_2 \cdot Bi_2Se_3$ analogs. In Sn_4P_3 and Sn_4As_3 the Sn—Sn distance within the Sn double layers is 3.24 Å [490] which is larger than in metallic tin (4×3.02 Å and 2×3.17 Å). Whereas the Bi_4Se_3 analogs

TABLE 103
Laitakarite (Bi_4Se_2S) structure, trigonal, D_{3d}^5—R$\bar{3}$m (No. 166), $Z = 1(3)$.
hexagonal cell $(0, 0, 0; \frac{1}{3}, \frac{2}{3}, \frac{2}{3}; \frac{2}{3}, \frac{1}{3}, \frac{1}{3})+$
Bi_I, Bi_{II} and Se in 6(c): $\pm(0, 0, z)$
S in 3(a): $0, 0, 0$.

Compound	a_h (Å)	c_h (Å)	a_{rh} (Å)	$\alpha(°)$	$z(X)$	$z(M_I)$	$z(M_{II})$	Refs.
Sb_4Te_3	4.26	41.55	14.07	17.42				[525]
$Bi_4(S_{0.9}Se_{0.1})_3$ (ikunolite)	4.15	39.19	13.28	18.00	0.426	0.142	0.287	[519, 3]
$Bi_{4+x}S_2Se_{1-x}$ (selenojoseite)	4.21	39.74	13.47	17.98				[1]
Bi_4Se_2S (laitakarite)	4.225	39.93	13.530	17.97				[1, 519]
Bi_4Se_3	4.27	40.0	13.6	18.1	0.417	0.1445	0.2870	[524]
Bi_4TeS_2 (joseite A)	4.25	39.77	13.48	18.13				[519]
Bi_4Te_2S (joseite B)	4.34	40.83	13.84	18.03				[5, 519]
$Bi_{4-x}(Te, Se, S)_{3-x}$ (joseite)	4.41	42.09	14.26	17.80				[1]
Bi_4Te_3	4.43	41.87	14.19	17.96				[525]
Sn_4P_3	3.968	35.34	12.001	19.03	0.4295	0.1342	0.2896	[490]
	3.9677	35.331	11.998	19.036	0.4289	0.1341	0.2895	[531]
Sn_4As_3	4.090	36.06	12.25	19.22	0.4299	0.1358	0.2889	[490]
Hf_4N_{3-x}	3.214	31.12	10.54	17.53				[498]
V_4C_3	2.917	27.83	9.428	17.80	0.417	0.1265	0.291	[532]
Nb_4C_3	3.14	30.1	10.2	17.7				[532]
Ta_4C_3	3.116	30.00	10.16	17.64				[532]

TABLE 104
Composite layer structures $mM_2 \cdot nM_2X_3 = M_{2(m+n)}X_{3n}$.
Calculated values for hypothetical phases are in parentheses

m/n	$M_{2(m+n)}X_{3n}$	Sb—Te a (Å), c (Å)	Bi—Se a (Å), c (Å)	Bi—Te a (Å), c (Å)	Space group	Number of layers
0	M_2X_3	4.26, 30.44	4.14, 28.64	4.38, 30.57	$R\bar{3}m$	15
$\frac{1}{8}$	M_3X_4	(4.26, 84.92)	(4.18, 80.32)	(4.40, 85.25)	$P\bar{3}m1$	42
$\frac{1}{6}$	M_7X_9	(4.27, 193.89)	(4.20, 183.68)	(4.41, 194.78)	$R\bar{3}m$	96
$\frac{1}{5}$	M_4X_5	(4.27, 53.79 [525])	(4.21, 51.54 [525])	4.411, 54.33 [526]	$P\bar{3}m1$	27
$\frac{1}{4}$	M_5X_6	(4.27, 133.00)	(4.22, 126.40)	(4.42, 133.80)	$R\bar{3}m$	66
$\frac{2}{7}$	$M_{18}X_{21}$	(4.27, 78.53)	(4.23, 74.72)	4.421, 78.20 [526]	$P\bar{3}m1$	39
$\frac{1}{3}$	M_8X_9	(4.27, 101.33 [525])	4.16, 97.1 [525]	(4.42, 103.57 [525])	$R\bar{3}m$	51
$\frac{1}{2}$	MX	4.26, 23.90 [523]	4.15, 22.84 [520]	4.40, 23.97 [521]	$P\bar{3}m1$	12
$\frac{2}{3}$	$M_{10}X_9$	(4.28, 113.82)	(4.30, 109.60)	(4.45, 115.15)	$R\bar{3}m$	57
$\frac{5}{7}$	M_8X_7	(4.28, 88.83 [525])	4.22, 85.65 [525]	(4.45, 91.08 [525])	$P\bar{3}m1$	45
$\frac{4}{5}$	M_6X_5	(4.28, 65.06 [525])	4.22, 62.6 [519]	(4.45, 66.75 [525])	$P\bar{3}m1$	33
1	M_4X_3	4.26, 41.55 [523]	4.27, 39.97 [524]	4.43, 41.87 [525]	$R\bar{3}m$	21
2	M_2X	4.25, 17.58 [522]	(4.41, 17.42 [525])	(4.49, 18.09 [525])	$P\bar{3}m1$	9
$\frac{5}{2}$	M_7X_3	(4.29, 116.39 [525])	(4.43, 116.4 [525])	4.47, 119.0 [5]	$R\bar{3}m$	60
3	M_8X_3	(4.29, 64.19)	(4.44, 64.15)	(4.50, 66.00)	$R\bar{3}m$	33

cleave between the M_2 and the M_2X_3 units, these units are no longer defined in Sn_4P_3 and Sn_4As_3, as the bonds between them are the strongest. Actually the distances in Sn_4P_3 are the following:

$$\cdots Sn\, 3.24\,\text{Å}\, Sn\, 2.66\,\text{Å}\, P\, 2.95\,\text{Å}\, Sn\, 2.76\,\text{Å}\, P\, 2.76\,\text{Å}\, Sn\, 2.95\,\text{Å}\, P\, 2.66\,\text{Å}\, Sn\, 3.24\,\text{Å}\, Sn \cdots$$

For Sn_4P_3 the true composition was given as $Sn_4P_{2.65}$ [490]. Were there not the inappropriate Sn—P distances, it would be fascinating to speculate about a semiconducting layer compound with divalent tin separated by cationic lone electron pairs:

$$\cdots Sn^{2+}Sn^{2+}\!=\!P\!-\!Sn^{2+}\!-\!P_{2/3}\!-\!Sn^{2+}\!-\!P\!=\!Sn^{2+}Sn^{2+}\!-\!P \cdots$$

and in order to round off the structure we would try to replace the central incomplete P layer by a complete S layer, but we doubt that Sn_4P_2S really would crystallize in an anti-$GeSb_2Te_4$ structure. Sn_4As_3 (and we expect the same for Sn_4P_3) in the actual structure is a superconductor ($T_c \approx 1.2$ K) [545].

11. Transition-Element Compounds with Octahedral Cation Coordination

a. SMYTHITE Fe_3S_4

The mineral smythite, stable below 75 °C, is found, usually, together with pyrrhotite, as flaky inclusions in calcite crystals. The two iron minerals are hard to distinguish. Both are ferromagnetic and possess a layered hexagonal structure. However, only the structure reported for smythite [558] corresponds to a true layer type (Figure 109). NiAs-type layer packs $Fe_3\square S_4$ are rhombohedrally stacked so that the anion stacking is $(chhc)_3$, just the reverse of the stacking met in $PbBi_2Te_4$ and also different from the $hhhh$ stacking in Fe_3Se_4. As in Fe_3Se_4, 3 Fe are in line along the c-axis with relatively long Fe—Fe distances of 2.84 Å. While the distance between the S layers within the Fe_3S_4 units is 2.85—2.74—2.85 Å, that between different packs is as large as 3.07 Å with S—S = 3.67 Å which accounts for the observed perfect cleavage on (001). If it is not due to metal–metal bonds the unequal thickness of the S sandwiches might suggest a Fe^{3+}—Fe^{II}—Fe^{3+} charge distribution and possibly non-metallic properties. However, this nice picture is an idealization as in fact, electron-microprobe analyses of several smythite specimens revealed compositions $(Fe, Ni)_{3+x}S_4$, where $x = 0.25$–0.30 [559]. All smythite samples contained nickel in concentration from 0.4 up to

TABLE 105

Fe_3S_4 (smythite) structure, trigonal, D_{3d}^5—$R\bar{3}m$ (No. 166), $Z = 1(3)$.
hexagonal axes: $(0, 0, 0; \frac{1}{3}, \frac{2}{3}, \frac{2}{3}; \frac{2}{3}, \frac{1}{3}, \frac{1}{3})+$
S_I, S_{II} and Fe_{II} in 6(c): $\pm(0, 0, z)$
Fe_I in 3(b): $0, 0, \frac{1}{2}$.

$(Fe, Ni)_{3+x}S_4$ (smythite): $a = 3.47$ Å, $c = 34.5$ Å ($a_{rh} = 11.66$ Å, $\alpha = 17°7'$) [558]
$z(S_I) = 0.2888$, $z(S_{II}) = 0.1270$, $z(Fe_{II}) = 0.4176$
$(Fe, Ni)_{3.3}S_4$: $a = 3.47$ Å, $c = 3.44$ Å, possibly monoclinic [559]

7%. According to Taylor and Williams [559] smythite is possibly monoclinic. It is further maintained that smythite is not a phase in the Fe—S system nor does it have a polymorphic relationship with greigite, the cubic spinel-type Fe_3S_4 with pure c-type anion stacking (a spinel-type $FeNi_2S_4$ also exists). With the latter point, these authors are at variance with Yamaguchi and Wada [560] who claim (not very conclusively) to have observed the greigite → smythite transition in an electrostatic field. However, they define smythite roughly as a greigite elongated along [111] which is rather approximate.

b. $(Fe, Ni)_3Te_2(h)$

An odd layer structure has been reported [562] for the rhombohedral high-temperature modification of the metallic iron-nickel tellurides with the composition ranges 17–32 at.% Fe, 24–41 at.% Ni and 41–45 at.% Te. The structure can be described as a combination of the β-NiTe and the $CdCl_2$ structure, i.e. within one Te double layer all tetrahedral and octahedral interstices are equally occupied by metal atoms (Figure 109). For $Fe_{0.28}Ni_{0.28}Te_{0.44} = Fe_{1.27}Ni_{1.27}\square_{0.46}Te_2$ the average occupancy of the metal positions is $\sim 85\%$. Such an arrangement is rather uncommon as it leads to extremely small metal–metal distances. The filled Te double layers are 3.78 Å thick, while the sandwiches are only 3.06 Å apart. But this latter distance still corresponds to a Te—Te distance of 3.83 Å and

TABLE 106
$Fe_{1.5}Ni_{1.5}Te_2$ structure, trigonal, C_{3v}^5—R3m (No. 160). Z = 3.
$(0, 0, 0; \frac{1}{3}, \frac{2}{3}, \frac{2}{3}; \frac{2}{3}, \frac{1}{3}, \frac{1}{3})+$
all atoms in 3(a): 0, 0, z.

	$a_h(Å)$	$c_h(Å)$	$a_{rh}(Å)$	$\alpha(°)$	
$Fe_{1.11}Ni_{1.33}Te_2$	3.969	20.37	7.166	32.16	
$Fe_{1.5}Ni_{1.05}Te_2$	3.976	20.39	7.174	32.18	
$Fe_{1.27}Ni_{1.27}Te_2$[a]	3.972	20.34	7.157	32.22	[561]
$Fe_{0.95}Ni_{1.59}Te_2$	3.970	20.44	7.189	32.06	
$Fe_{1.21}Ni_{1.44}Te_2$	3.976	20.52	7.214	32.00	
$Fe_{0.88}Ni_{1.77}Te_2$	3.956	20.52	7.213	31.84	
$Fe_{1.38}Ni_{1.38}Te_2$	3.978	20.54	7.152	31.96	

[a] | Te_I | Te_{II} | M_Ω | M_{τ_1} | M_{τ_2} | [562]
z = 0 | 0.482 5 | 0.231 | 0.868 | 0.607 |

hence Van der Waals contacts in agreement with the graphite-like aspect of the iodine-transported single crystals. The distance between the tetrahedral metal positions is 2.74 Å compared with 2.71 Å in β-NiTe. Assuming the asymmetric Ω position to be correct, the extremely short distances between tetrahedral and octahedral sites are τ_1—Ω = 2.38 Å and τ_2—Ω = 2.45 Å (with a symmetric Ω position τ—Ω = 2.41 Å). Therefore, we think it more realistic to assume 100%

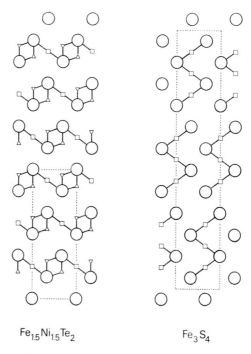

Fig. 109. (110) sections of the idealized structures of $(Fe, Ni)_3Te_2(h)$ and Fe_3S_4 (smythite). Squares and triangles symbolize octahedral and tetrahedral cation sites, respectively.

TABLE 107
Pt_2Sn_3 structure, hexagonal, D_{6h}^4—$P6_3/mmc$ (No. 194), $Z = 2$.
Pt and Sn_{II} in 4(f): $\pm(\frac{1}{3}, \frac{2}{3}, z; \frac{1}{3}, \frac{2}{3}, \frac{1}{2}-z)$
Sn_I in 2(b): $\pm(0, 0, \frac{1}{4})$

Pt_2Sn_3: $a = 4.334$ Å, $c = 12.960$ Å
$\qquad z(Pt) = 0.143$. $z(Sn_{II}) = 0.930$ [7]
$Au_4In_3Sn_3$: $a = 4.50$ Å, $c = 13.11$ Å [7]

---

occupancy for the tetrahedral sites and 50% occupancy for the octahedral sites though we are still not quite satisfied with this structure. In the low-temperature modification, stable below 140°C, there are no more empty layers, as it is then only the other half of the octahedral sites that are engaged.

c. Pt_2Sn_3

This metallic compound has only formally a layer structure based on a *chc chc* 'anion stacking' (Figure 110). This structure recalls to mind that of smythite and it might occur also with chalcogenides, say Fe_2S_3. In Pt_2Sn_3, the anion close packing is strongly squeezed: $c/6a = 0.498$ instead of 0.816 for ideal close packing. It is, however, the interlayer bonding, not the Pt—Pt pair formation that is responsible for the main contraction. The thickness of one SnPtSn sheet is 2.33 Å while the separation of the composite sandwiches is 1.81 Å. The tin atoms of the central layer, Sn_I, have a trigonal-prismatic coordination by 6Pt at 2.86 Å. The Sn atoms in the boundary layer, however, possess not only the 3Pt neighbors of the sandwich at 2.68 Å, but one more Pt neighbor at 2.76 Å (completing an $SnPt_4$ tetrahedron) as well as $3Sn_{II}$ at 3.09 Å which both belong to the adjacent sandwich. Pt in turn has 3+3 Sn neighbors which form a compressed octahedron as well as a seventh Sn neighbor of the adjacent sheet and another Pt neighbor at 2.77 Å. The Pt—Pt distance is considerably larger than the Sn sandwich thickness of 2.33 Å reflecting the attraction by the seventh Sn in the next pack. The bonding thus is three-dimensional.

The second representative known of this structure type is neither Pd_2Sn_3 nor the isoelectronic Au_2In_2Sn but $Au_2In_{1.5}Sn_{1.5}$ which has a slightly higher valence-electron concentration. We wonder whether a mixed-cation analog $IrAuSn_3$ would exist. The structure of Pt_2Ge_3, on the other hand, is a NiAs derivative and contains Pt atoms between all Ge layers:

$$\overset{|}{P}tGePt_{1/2}\square_{1/2}Ge\square_{1/2}Pt_{1/2}GePtGePt_{1/2}\square_{1/2}Ge\square_{1/2}Pt_{1/2}Ge\overset{|}{P}t.$$

d. The structures of PtTe (ZrCl), Pt_3Te_4 and Pt_2Te_3

In the phase diagram of the Pt—Te system four binary compounds have been detected, the congruently melting $PtTe_2$ with CdI_2-type structure and the peritectic phases PtTe, Pt_3Te_4 and Pt_2Te_3 [563] which crystallize in closely related

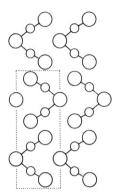

Fig. 110. (110) section of the structure of Pt$_2$Sn$_3$.
Large circles: Sn.

structures. While NiTe and PdTe adopt the NiAs structure, PtTe develops a new structure which is a true layer type. The stacking of all atoms is of the c-type but instead of the NaCl-type stacking the sequence is Te—Pt—Pt—Te (Figure 111). The sandwich thus contains a puckered central metal layer where each Pt is bonded to three Pt neighbors (only one Pt—Pt bond can be seen in the sections given in Figure 111 but one should keep in mind that Pt practically lies on a threefold axis). These Pt—Pt distances are as short as 2.72 Å. The separation of the boundary Te layers of adjacent sandwiches is 2.74 Å with a Te—Te distance of 3.57 Å, which is larger than in PtTe$_2$. PtTe is a diamagnetic metal that becomes superconducting below 0.59 K [23, 564].

The structures of Pt$_3$Te$_4$ (= Pt$_2$Te$_2$ + PtTe$_2$) and Pt$_2$Te$_3$ (= Pt$_2$Te$_2$ + 2 PtTe$_2$) are composite layer structures. For Pt$_3$Te$_4$ this thickness of the intercalated PtTe$_2$ layer is calculated as 5.26 Å as compared with 5.22 Å found in pure PtTe$_2$. We wonder what happens with the superconductivity. Unfortunately PtTe$_2$ itself is tested only down to 1.2 K but PdTe$_2$ has a transition temperature of ~1.5 K. It might be possible to intercalate PdTe$_2$ instead of PtTe$_2$ (→ PdPt$_2$Te$_4$ and PdPtTe$_3$) or even NiTe$_2$ or IrTe$_2$.

One might further speculate that hypothetical isoelectronic AuSb has a better chance to crystallize in the PtTe structure than metallic IrI. However, isomorphous halides do indeed exist, but their structure is described in the more symmetrical space group R$\bar{3}$m. In fact, the deviations of the monoclinic PtTe structure from rhombohedral symmetry are probably within the experimental accuracy and we add the data for PtTe in space group R$\bar{3}$m in Table 109. The corresponding hexagonal cell is also indicated in Figure 111.

These group IV monohalides are described as extremely soft and of graphite character. HfCl was obtained in the form of large black lustrous scale-like crystals [567]. ZrBr crystals had the form of hexagonal plates with perfect cleavage [568]. Due to their thermal stability, the chlorides may be used as high temperature lubricants. Disproportionation of ZrCl and HfCl into the tetrachloride and metal begins only near 870 and 790°C, respectively [567].

TABLE 108
Structures of PtTe and PtTe—PtTe$_2$ composites

(a) *PtTe* structure, monoclinic, C_{2h}^3—C2/m (No. 12), $Z=4$.
$(0, 0, 0; \frac{1}{2}, \frac{1}{2}, 0)+$
all atoms in 4(i): $\pm(x, 0, z)$

PtTe: $a=6.865$ Å, $b=3.962$ Å, $c=7.044$ Å, $\beta=108.98°$ [563]
Pt: $x=0.870$, $z=0.111$; Te: $x=0.598$, $z=0.294$.
(orthorhombic cell with $Z=10$:
$a=6.6144$ Å, $b=11.865$ Å, $c=5.6360$ Å [1])

(b) *Pt$_3$Te$_4$* structure, monoclinic, C_{2h}^3—C2/m (No. 12), $Z=2$.
Pt$_I$ in 2(d): $0, \frac{1}{2}, \frac{1}{2}$
remaining atoms in 4(i): as above

Pt$_3$Te$_4$: $a=6.906$ Å, $b=3.991$ Å, $c=12.019$ Å, $\beta=101.03°$ [563]

	x	z
Pt$_{II}$	0.363	0.061
Te$_I$	0.045	0.170
Te$_{II}$	0.802	0.384

(c) *Pt$_2$Te$_3$* structure, monoclinic, C_{2h}^3—C2/m (No. 12), $Z=4$.
all atoms in 4(i): as above.

Pt$_2$Te$_3$: $a=6.933$ Å, $b=4.002$ Å, $c=17.119$ Å, $\beta=97.75°$ [563]

	x	z
Pt$_I$	0.365	0.040
Pt$_{II}$	0.463	0.349
Te$_I$	0.048	0.126
Te$_{II}$	0.779	0.278
Te$_{III}$	0.148	0.436

As is to be expected the cells of these halides are much more elongated than that of PtTe. In this structure the coordination for ZrCl is as follows:

Zr—3 Zr at 2.87 Å
3 Cl at 3.11 Å(?)
Cl—3 Zr at 3.11 Å(?)
3 Cl at 2.81 Å(?)

We have however serious doubts about the correctness of the parameter value given for the Cl position in ZrCl. With the proposed z_X value (Table 109) the distances appear to be more reasonable:

Zr—Cl = 2.73 Å
Cl—Cl = 3.61 Å   (compare [1039])

With three covalent Zr—Zr single bonds, ZrCl as well as the other isomorphous halides might well be diamagnetic semiconductors. Struss and Corbett [569] found for HfCl a net paramagnetism of 53×10^{-6} emu/mol^{-1} (after a diamagnetic correction of 42×10^{-6}) without a detectable temperature dependence down to at

TABLE 109
ZrCl structure, trigonal, D_{3d}^5—$R\bar{3}m$ (No. 166), $Z = 2(6)$.
hexagonal axes: $(0, 0, 0; \frac{1}{3}, \frac{2}{3}, \frac{2}{3}; \frac{2}{3}, \frac{1}{3}, \frac{1}{3})+$
all atoms in 6(c): $\pm(0, 0, z)$.

	a_{rh}(Å)	α(°)	a_h(Å)	c_h(Å)	z_M	z_X	Refs.
ZrCl	9.12	21.62	3.41	26.66	0.372 4	0.129 1	[565, 566]
						0.11[a]	
HfCl	9.095	21.40	3.377	26.65			[567]
ZrBr	9.570	21.12	3.507	28.06			[568]
HfBr							
PtTe	7.043	32.68	3.963	19.983	0.370	0.098	[563]

[a] proposal, confirmed by [1039]

least $-150\,°C$. Evidently, these halides have a certain range of homogeneity, as is demonstrated for example by the variation of the *c*-parameter (28.0–28.4 Å) of ZrBr [568]. It is therefore not surprising that the experimental results on the electrical properties are contradictory. Struss and Corbett [569] deduce a metallic character from the few-ohm dc resistance of a 1 mm HfCl layer. On a ZrCl single crystal of $4 \times 4 \times 0.1$ mm^3 Troyanov [565] measured a resistivity of 500–1000 Ω cm along the *c*-axis but only 0.02 Ω cm parallel to the layers. In the case of

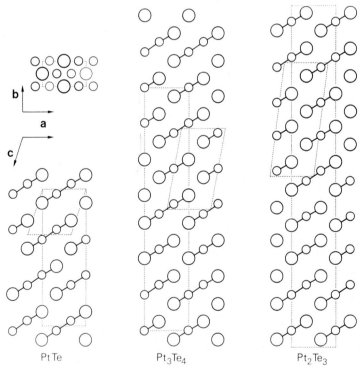

Fig. 111. The monoclinic structures of PtTe, Pt$_3$Te$_4$ and Pt$_2$Te$_3$. Left upper corner: Projection of one PtTe layer onto the (a, b) plane. Below: Section of the PtTe structure at $b/2$ which corresponds to the (110) section of the hexagonal (rhombohedral) cell of the ZrCl structure. This hexagonal cell is indicated also for Pt$_3$Te$_4$ and Pt$_2$Te$_3$ (right). Small circles: Pt.

a ZrBr single crystal the corresponding values are 100–500 and 0.08 Ω cm. No data on the temperature dependence nor any optical data are available. A different stacking of the AbcA-type XMMX sandwiches is reported for ScCl and ZrBr [1033].

It is worth noting that the isoelectronic NbS is not isomorphous but has a filled-up $2H_a$—NbS_2 structure.

e. THE $FePS_3$ AND THE $FePSe_3$ STRUCTURES

In order to emphasize the structural features we should designate these two types as $Fe_2(P_2)S_6$ and $Fe_2(P_2)Se_6$. Thus $FePS_3$ is in fact what chemists call a hexathiohypodiphosphate. However, there is no clear separation into an ionic cation and a complex anion $[(P_2)S_6]^{4-}$. Instead, the structures of $Fe_2(P_2)S_6$ and $Fe_2(P_2)Se_6$ are best described as slightly distorted $CdCl_2$- and CdI_2-type structures, respectively, with ordered occupation of the octahedral holes by Fe and (P_2). In this way, a tetrahedral coordination is achieved for each P atom:

$$P—3S+1P, \quad S—1P+2Fe, \quad Fe—6S.$$

The P—P distances range from 2.19–2.24 Å corresponding to a single bond. The cation therefore is divalent and phosphorus uses all its valence electrons, i.e. in

TABLE 110

$FePS_3$ structure, monoclinic, C_{2h}^3—C2/m (No. 12), Z = 4.
(0, 0, 0; $\frac{1}{2}$, $\frac{1}{2}$, 0) +
S_I in 8(j): $\pm(x, y, z; x, \bar{y}, z)$
P and S_{II} in 4(i): $\pm(x, 0, z)$
Fe in 4(g): $\pm(0, y, z)$

Compound	a(Å)	b(Å)	c(Å)	$\beta(°)$	Refs.
$MgPS_3$	6.07	10.53	6.80	107.1	[571]
VPS_3	5.85	10.13	6.66	107.1	[571]
$MnPS_3$	6.07	10.55	6.80	107.1	[571]
	6.087	10.530	6.800	107.31	[572]
$FePS_3$[a]	5.934	10.28	6.72	107.1	[571]
	5.949	10.288	6.720	107.17	[572]
$CoPS_3$	5.91	10.24	6.68	107.1	[571]
$NiPS_3$	5.83	10.10	6.63	107.1	[571]
	5.811	10.076	6.628	106.96	[572]
$PdPS_3$	5.97	10.32	6.73	107.1	[571]
$ZnPS_3$	5.96	10.28	6.73	107.1	[571]
$CdPS_3$	6.17	10.67	6.82	107.1	[571]
$SnPS_3$	5.99	10.36	6.80	107.1	[571]
$CrPSe_3$	6.13	10.66	6.67	107.1	[1016]
$NiPSe_3$	6.16	10.66	6.86	107.1	[571]

[a] $FePS_3$: $x(S_I) = 0.248\,1$, $y(S_I) = 0.165\,9$, $z(S_I) = 0.247\,9$; $x(S_{II}) = 0.750\,8$, $z(S_{II}) = 0.247\,4$; $x(P) = 0.057\,1$, $z(P) = 0.170\,8$; $y(Fe) = 0.332\,8$ [570]

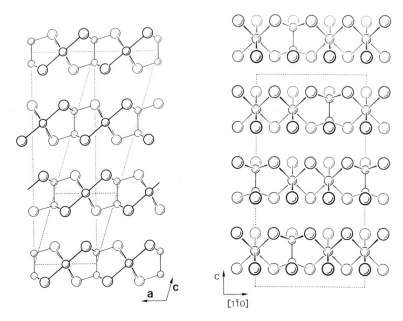

Fig. 112. The structures of monoclinic FePS$_3$ (left) and rhombohedral FePSe$_3$ (right). For FePS$_3$ a section at $b=\frac{1}{2}$ is shown and the hexagonal cell of the corresponding CdCl$_2$ structure is also indicated (for which the figure represents a (110) section as for FePSe$_3$ (right)).
Large spheres: anions, smallest spheres (vertical pairs): P, medium spheres: Fe.

TABLE 111

FePSe$_3$ structure, trigonal, C_3^4—R3 (No. 146), $Z=2(6)$.
hexagonal axes: $(0, 0, 0; \frac{1}{3}, \frac{2}{3}, \frac{2}{3}; \frac{2}{3}, \frac{1}{3}, \frac{1}{3})+$
Se$_I$ and Se$_{II}$ in 9(b): x, y, z; $\bar{y}, x-y, z$; $y-x, \bar{x}, z$;
Fe$_I$, Fe$_{II}$, P$_I$ and P$_{II}$ in 3(a): 0, 0, z.

Compound	a_h(Å)	c_h(Å)	a_{rh}(Å)	$\alpha(°)$	Refs.
CdPS$_3$	6.211	19.61	7.456	49.23	[573]
SnPS$_3$	5.98	19.32	7.31	48.3	[571]
MgPSe$_3$	6.39	20.12	7.65	49.3	[571]
MgCaP$_2$Se$_6$	6.39	20.12	7.65	49.3	[571]
VPSe$_3$	5.84	18.44	7.01	49.2	[1016]
MnPSe$_3$	6.38	19.19	7.38	51.2	[571]
FePSe$_3$[a]	6.265	19.80	7.526	49.2	[570]
	6.262	19.810	7.528	49.15	[574]
CdPSe$_3$	6.49	19.97	7.64	50.3	[571]

[a]FePSe$_3$:		x	y	z	
	Fe$_I$			0.165 8	
	Fe$_{II}$			−0.168 7	
	P$_I$			0.443 7	[570]
	P$_{II}$			−0.443 1	
	S$_I$	0.331 5	−0.003 0	0.080 0	
	S$_{II}$	−0.344 9	−0.027 2	−0.085 4	

the covalent limit is P^+. These compounds therefore are semiconductors, black opaque with V, Fe, Co, Ni and Pd, but transparent colorless to yellow with Mg, Zn and Cd and red with Sn [571]. $MnPS_3$, $FePS_3$, $NiPS_3$ and $FePSe_3$ ($T_N = 123$ K) were found to order antiferromagnetically [572, 574]. $VPSe_3$ is metallic [1016].

The two structures represent filled-up $MX_3 = M_2\square X_6$ layer types. Thus the $Fe_2(P_2)Se_6$ structure corresponds to a filled-up BiI_3 type while $Fe_2(P_2)S_6$ is an $AlCl_3$ type completed to give a $CdCl_2$-type structure. It is interesting that with increasing metal–anion electronegativity difference, the structures change from CdI_2-type to $CdCl_2$-type and finally to three-dimensional structures with Sn [575], Pb [576] and Ca [571] similar to the binary chalcogenides. For $HgPS_3$ a triclinic and for $HgPSe_3$ a monoclinic cell have been reported [571], both structures obviously representing similar layer types. Various polytypes might exist as well and for $SnPS_3$ a stacking variant has indeed been found [575].

Up to now only sulfides and selenides have been synthesized, tellurides have not been synthesized. We wonder whether isomorphous $M^{2+}AsTe_3$ might exist as well, perhaps with larger cations. Other possible candidates for these structures are compounds of the type $M_2^{3+}(Si_2)X_6$ such as $AlSiS_3$ and $CrSiSe_3$ or even $Mn^{2+}Ti^{4+}(Si_2)S_6$. The structure is stable also with holes on the cation sites, e.g. in $In_{2/3}PS_3$ [1015] ($a = 3a_0$). $Cr_{2/3}^{III}PS_3$ and $Fe_{2/3}^{III}PS_3$ might exist as well. Intercalation of $\frac{1}{3}$ pyridine in $MnPSe_3$ ($a = 6.38$ Å, $c = 3 \times 12.56$ Å) reduces its resistivity from $>10^8$ to 10^5 Ω cm [1016].

f. CdI_2-TYPE TRANSITION-ELEMENT CHALCOGENIDES

The CdI_2 structure is one of the simplest structures, based on a hcp anion stacking with every second layer of octahedral sites occupied by cations. In many transition-element—chalcogen systems it represents the last member of a continuous series of subtractive NiAs-type solid solutions $T_{2-x}X_2$. In certain systems such as $Zr_{2-x}Te_2$, $V_{2-x}S_2$, $V_{2-x}Te_2$, $Cr_{2-x}S_2$, ... $Fe_{2-x}Te_2$, the 1:2 stoichiometry is not reached under normal conditions. The CdI_2 structure is thus an ordered superstructure of a defective NiAs structure and it appears that in certain transition-element chalcogenides it is difficult to reach complete order. This means that part of the cations, the 'disordered cations', are found within the 'empty' layers. It was established that in $HfTe_{1.97}$ the anion sublattices also contain vacancies [586] and for $Zr_{1+\delta}Te_2$ Zr atoms were even found to substitute for Te atoms [602].

On checking through the list of representatives, one is both surprised and disappointed to find only rare, and in most cases, estimated values for the free parameter z of the anion positions. Only in a very recent study on $Ti_{1.023}S_2$ [579] was an effort made to determine the atomic positions accurately. The results are remarkable in two respects. First, the free parameter is virtually equal to the ideal value $z = \frac{1}{4}$ although the axial ratio 1.67 is larger than the ideal value $c/a = 1.633$ for close packing. Second, the CdI_2-type cation positions are completely filled and only the excess cations occupy octahedral positions in the 'empty layer'. In other words, $Ti_{1.023}S_2$ is an ideal NiAs—CdI_2 mixed crystal in contrast to $ZrTe_{2-x}$ and $HfTe_{2-x}$. A value $z = \frac{1}{4}$ means that the anion layers are equidistant and that the anion octahedra around the Ti atoms are elongated to a trigonal antiprism. The thickness of the X—M—X sandwiches is generally $2zc$ and the empty space is generally $(1-2z)c$. For axial ratios $c/a > 1.633$ an ideal octahedral coordination around the cation would still be possible if $z = 1/\sqrt{6}(c/a)$. This would imply a separation of the empty layers larger than the thickness of the sandwiches. For $c/a < \sqrt{\frac{8}{3}}$ one might expect z values $> \frac{1}{4}$ since the anions are now no longer in contact within the (001) planes. The only experimental values we found refer to $NiTe_2$, $PdTe_2$, PtS_2, $PtSe_2$ and $PtTe_2$ and these values are also rather close to $\frac{1}{4}$. This structural puzzle is only one of the intriguing features of the CdI_2-type phases. Another puzzle is the energy band structure. As there are no bonds between like atoms, tetravalent cations are needed to make a CdI_2-type chalcogenide non-metallic. From Table 112 we see that this structure occurs with d^0, d^1, d^5, d^6 and d^8 cations. We would expect an undistorted structure with $c/a = 1.63$ and non-metallic properties for the compounds with configurations d^0 and d^6. Our expectations, however, are only in part substantiated. In $NiTe_2$ and $PdTe_2$, which is a superconductor, the values for c/a are extremely low and the energy gap has vanished [603]. As expected, the gap is largest in the 5d compounds despite the low c/a, but this increased 'non-metallicity' does not prevent $PtTe_2$ from being metallic. The transition from non-metallic to metallic behavior appears to occur between $PtSe_2$ and PtSeTe [418].

$TiTe_2$ is obviously a semimetal due to an overlap of the cation d-states with the p-states of the anion. Optical [578, 608], electrical [605, 606], specific-heat [606] and X-ray-photoemission data give evidence that in $TiSe_2$ and possibly even in TiS_2 a similar overlap takes place. We hesitate, however, to believe that stoichiometric CdI_2-type $ZrTe_2$ and $HfTe_2$ should also be metallic [578]. For HfS_2 and $HfSe_2$, an indirect gap of 1.96 and 1.13 eV was determined from optical-absorption measurements [604]. On the other hand, the observed non-stoichiometry and disorder may be caused by the change in conductivity character (or vice-versa?). The study of the systems $Hf(S, Te)_2$ and $Hf(Se, Te)_2$, as well as the corresponding Zr systems might be decisive. Nitsche [508] prepared $Zr_{1-x}Ti_xS_2$ single crystals. These and even more $Hf_{1-x}Ti_xS_2$ may supply further information about the electronic structure of TiS_2.

It is worth noting that in such mixed systems the lattice constant a is a linear function of concentration, whereas in both mixed-anion and mixed-cation systems (Figure 113) c is slightly curved.

The band structures of ZrS_2, $ZrSe_2$, HfS_2 and $HfSe_2$, based on a semiempirical tight-binding method, have been presented by Murray et al. [609].

In the d^1 compounds the axial ratio c/a is distinctly larger than the ideal value. The anion octahedron therefore is trigonally elongated. The contrary holds for the d^5 compounds which contain a hole in the t_{2g} shell instead of the single electron. One might thus assume that the distortions are due to a Jahn-Teller effect

TABLE 112

Transition-element chalcogenides crystallizing in the CdI_2-type structure
CdI_2 structure, trigonal, D_{3d}^3—$P\bar{3}m1$ (No. 164), $Z=1$.
  anion in 2(d): $\pm(\frac{1}{3}, \frac{2}{3}, z)$
  cation in 1(a): 0, 0, 0.

MX_2	a(Å)	c(Å)	c/a	z	Refs.
TiS_2	3.408	5.701	1.673	$\frac{1}{4}$	[1]
	3.404 9	5.691 2	1.671 5		[1]
	3.412	5.695	1.669		[4]
$(TiS_{2.00})$	3.407	5.689	1.667		[577]
	3.404	5.699	1.674		[578]
	3.407 3	5.695 3	1.671 5	0.250 1	[989]
$Ti_{1.023}S_2$	3.409	5.694	1.670	0.249 3	[579]
$Ti(S, Se)_2$					[585]
$TiS_{1.3}Se_{0.7}$	3.443	5.818	1.690		[577]
$TiSSe$	3.457	5.868	1.697		[577]
$TiS_{0.8}Se_{1.2}$	3.477	5.900	1.697		[577]
$TiS_{0.3}Se_{1.7}$	3.508	5.967	1.701		[577]
$TiSe_2$	3.536	6.003	1.698		[577]
	3.542	6.015	1.698		[578]
	3.541	5.986	1.690		[4]
	3.548	5.998	1.691		[1]
	3.533	5.995	1.697		[1]
	3.535	6.004	1.698		[580]
	3.540	6.008	1.697	0.255 04	[991]
$TiSe_{1.5}Te_{0.5}$	3.587	6.159	1.717		[577]
$TiSeTe$	3.643	6.291	1.727		[577]
$TiSe_{0.5}Te_{1.5}$	3.702	6.425	1.736		[577]
$TiS_{1.8}Te_{0.2}$	3.442	5.808	1.687		[577]
$TiS_{1.5}Te_{0.5}$	3.501	5.961	1.703		[577]
$TiSTe$	3.586	6.192	1.727		[577]
$TiS_{0.8}Te_{1.2}$	3.623	6.285	1.735		[577]
$TiS_{0.5}Te_{1.5}$	3.674	6.387	1.738		[577]
$TiTe_2$	3.763	6.529	1.735		[577]
	3.757	6.513	1.734		[4]
	3.76	6.48	1.72		[580]
	3.778	6.493	1.719		[991]
ZrS_2	3.662	5.813	1.587		[580]
	3.661	5.829	1.592		[578]
	3.661 7	5.827 5	1.591 5		[581]
	3.667	5.817	1.586		[582]
$(Zr, Sn)S_2$					[591]
$ZrSSe$	3.715	6.013	1.615		[582]
	(See Figure 113)				
$ZrSe_2$	3.774	6.131	1.624		[582]
	3.770	6.137	1.628		[580]
	3.773	6.133	1.625		[578]
	3.770	6.129	1.626	0.255	[988]
$Zr(Se, Te)$					[583]
$ZrTe_2$	3.950	6.630	1.679		[1, 4]
	3.952	6.660	1.685		[580]
$Zr_{1.05}Te_2$	3.952 4	6.625	1.676		[581]

Table 112 (Continued)

MX_2	$a(Å)$	$c(Å)$	c/a	z	Refs.
HfS_2	3.635	5.837	1.606		[580]
	3.623	5.841	1.612		[584]
	3.632	5.850	1.611		[581]
	3.630	5.854	1.612		[578]
$HfS_{1.5}Se_{0.5}$	3.653	5.927	1.623		[584]
HfSSe	3.682	6.006	1.631		[584]
	3.694	6.061	1.641		[578]
$HfS_{0.5}Se_{1.5}$	3.711	6.076	1.637		[584]
$HfSe_2$	3.741	6.143	1.642		[584]
	3.748	6.159	1.643		[580]
	3.747	6.158	1.643		[578]
	3.744 0	6.155	1.644		[581]
	3.733	6.146	1.646		[604]
$HfTe_2$	3.951	6.659	1.685		[578]
$(Hf_{1.25}Te_2)$	3.950 9	6.651	1.683		[581]
$HfTe_{1.94}$ disordered	3.949 2	6.651 4	1.684		[586]
$VS_2(?)$					[508]
VSe_2	3.348	6.122	1.827		[1]
$V_{1-x}Nb_xSe_2$, $x \leqslant 0.7$					[1041]
$NbSe_2$ (1050°C)	3.53	6.29	1.78	$\approx \frac{1}{4}$	[587]
$NbTe_2$(h)					[611]
TaS_2	3.36	5.90	1.76	$\approx \frac{1}{4}$	[30]
	3.395	5.902	1.738		[1]
	3.346	5.860	1.751		[1]
	3.365	5.853	1.739		[588]
	3.360	5.900	1.755		[589]
	3.365	5.897	1.752		[590]
$Ta_{0.9}W_{0.1}S_2$	3.331	5.944	1.784		[589]
$TaSe_2$	3.476 9	6.272 2	1.804		[592]
	3.480	6.254	1.797		[593]
$TaTe_2$(h)					[611]
$Fe_{2/3}Ni_{1/3}Te_2$	3.794 0	5.569	1.468		[361]
$Fe_{1/2}Ni_{1/2}Te_2$	3.813 0	5.491	1.440		[361]
$Fe_{0.31}Ni_{0.69}Te_2$	3.836 4	5.402	1.410		[361]
CoSeTe (h?)	3.70	5.09	1.38		[594]
$CoTe_2$	3.784	5.403	1.428		[4]
$RhTe_2$ (h)	3.92	5.41	1.38	0.25	[1]
$IrTe_2$	3.93	5.393	1.37		[4]
NiSeTe	3.70	5.15	1.39		[418]
(kitkaite)	3.716	5.126	1.379		[1]
$NiTe_2$	3.854 6	5.259 4	1.364	0.252	[595]
	3.861	5.297	1.372	$\frac{1}{4}$	[4]
	3.854 7	5.261	1.365		[561]
	3.854 2	5.260 4	1.365		[600a]
$NiPo_2$	~3.95	~5.68	~1.44		[596]
PdSeTe	3.90	4.98	1.28		[418]
$PdTe_2$ (20°C)	4.036 5	5.126 3	1.270	0.247	[597, 600a]
(284°C)	4.049 5	5.168 5	1.276		[597]
(574°C)	4.063 2	5.230 3	1.287		[597]
PtO_2	3.10	4.29–4.41			[598]
	3.100	4.161	1.342		[599]

Table 112 (Continued)

MX$_2$	a(Å)	c(Å)	c/a	z	Refs.
PtS$_2$	3.537	5.019	1.419		[4]
20°C	3.543 1	5.038 9	1.422	0.250	[597, 600]
250°C	3.544 9	5.082 2	1.434		[597]
562°C	3.547 4	5.100 0	1.438		[597]
	3.543 2	5.038 8	1.422	0.227	[600a]
PtSe$_2$	3.732	5.072	1.359	0.250	[1]
20°C	3.727 8	5.081 3	1.363	0.255	[597, 600a]
503°C	3.737 7	5.123 5	1.371		[597]
946°C	3.748 2	5.140 5	1.371		[597]
PtSeTe	3.89	5.11	1.31		[418]
PtTe$_2$	4.010	5.201	1.297		[4]
20°C	4.025 9	5.220 9	1.297	0.254	[597, 600a]
591°C	4.038 6	5.302 6	1.313		[597]
870°C	4.045 2	5.327 9	1.317		[597]
(Pd, Pt)(Te, Bi)$_2$ (merenskyite)	3.98	5.125	1.288		[5]
(Pt, Pd)(Te, Bi)$_2$ (moncheite)	4.05	5.29	1.31		[5]
CuTe$_2$ (p)	3.95^a	5.49^a	1.39		[601a]
ZnTe$_2$ (p)	3.98	5.25	1.32		[601]
Polytypes					
PtO$_2$ (II) (C_{6v}^4—P6$_3$mc) hchc	3.10	8.32	2 × 1.34		[599]
PtO$_2$ (III) hcc hcc (?)	3.11	12.60	2 × 1.35		[599]

^{a} Tentative, not isolated as pure phase [601a]

necessary to produce non-metallic properties. However, the distortions are obviously insufficient to split the bands completely. In VSe$_2$ the magnetic moment corresponds to one localized d-electron but the temperature dependence of the electrical resistivity demonstrates the metallic character of the compound. In diamagnetic TaS$_2$ and TaSe$_2$, however, the resistivity points to semiconductivity. In a non-metallic d^1 compound without metal-metal bonds paramagnetism due to the localized d electron is expected. The observed diamagnetism has been explained by spin-orbit coupling which should even create a 'diamagnetic ferromagnet' at low temperatures [610]. Meanwhile, it became evident that this pseudo-semiconductivity is connected with changes in the Fermi surfaces due to minor lattice distortions and superstructure formation connected with charge-density waves.

The CdI$_2$-type structure is the high-temperature structure for NbSe$_2$, NbTe$_2$, TaS$_2$, TaSe$_2$ and TaTe$_2$, stable above 800–900 °C. The CdI$_2$-type modifications can be metastably retained by rapid quenching. On heating the metastable phases up to near 200 °C they transform to the stable low-temperature modifications with trigonal-prismatic cation coordination. Within the metastable low-temperature

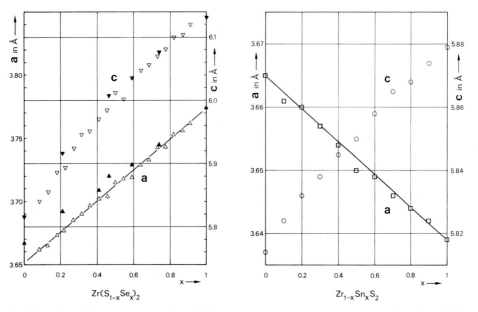

Fig. 113. The lattice constants a and c of CdI$_2$-type mixed crystals of the systems ZrS$_2$—ZrSe$_2$ [582] and ZrS$_2$—SnS$_2$ [591].

range of existence the CdI$_2$-type phases reveal additional transitions. Thus 1T—TaS$_2$ suffers first-order transitions at 352 K and 200 K, while a single first-order transition was observed in 1T—TaSe$_2$ at 473 K. The virtually temperature-independent magnetic susceptibility jumps at the corresponding temperatures. The basal-plane resistivity, which is some 10^{-4} Ω cm above the transition in both 1T—TaS$_2$ and 1T—TaSe$_2$, increases discontinuously and exhibits a semiconductor-like behavior in TaS$_2$ down to the lower transition with an apparent activation energy of 0.02 eV. At the lower transition in TaS$_2$ the resistivity increases by a further order of magnitude and at 4 K, ρ remains ~ 0.05 Ω cm [612]. The Seebeck coefficient of 1T-TaS$_2$ changes sign from negative to positive on warming through the 352 K transition, whereas the Hall coefficient remains negative, only a break occurs. Above the upper transition the charge carriers suffer heavy scattering. In the near-infra-red reflectivity spectrum little change is observed upon passing through this transition. Only after replacing more than 70% of Ta by Ti does the free-carrier reflectivity begin to appear normal [606].

Electron-diffraction studies revealed that these phenomena are coupled with the occurrence of superstructures which are caused by charge-density waves. Theoretical work suggests that these two-dimensional compounds are likely to be susceptible to Fermi-surface-driven instabilities. The two-dimensional character of the Fermi surface implies planar surfaces normal to the layers. Large parallel sections of Fermi surface, spanned by a vector $\mathbf{q}_0$, lead, in real-space potential, to an oscillatory component of wavelength $1/q_0$. This must introduce a periodic structural distortion which is in general incommensurate with the lattice. In these d^1 compounds the charge maxima in the charge-density waves will center upon the cations. On decreasing the temperature the increasing amplitude of the structural distortion will favor lock-in to some commensurate period. This is what happens with 1T—TaSe$_2$ (and partially with 1T—TaS$_2$) at the (upper) transition. Whereas above this temperature an incommensurate superlattice with a reciprocal vector $\mathbf{a}^* = \mathbf{a}_0^*/3.6$ exists, there is a sudden jump to a rotated superlattice with the $\mathbf{a}$-axis ($a = a_0\sqrt{13} = 3.6056 a_0$) pointing in [$3\bar{1}0$] direction of the CdI$_2$-type subcell.

The change of the physical properties at the superlattice transition suggests the formation of a gap in the electronic spectrum around the Fermi level over a large fraction of the Fermi surface. In TaSe$_2$ it is estimated that roughly 90% of the Fermi surface is destroyed by a gap.

In Figure 114 the cation displacements are shown for the $\sqrt{13}a_0$ superlattice assuming their motion to be simply directed to the charge maxima formed by the charge-density waves. The 13 cations per supercell then fall into three types in the ratio 1:6:6.

The temperature at which this superlattice is formed in these systems is strongly affected by substitutional doping. Thus the transition drops to room temperature in Ta$_{1-x}$Ti$_x$S$_2$ with $x = 0.06$, in Ta$_{1-x}$Ti$_x$Se$_2$ with $x = 0.1$, in Ta$_{1-x}$Nb$_x$S$_2$ with $x = 0.05$ and in Ta$_{1-x}$V$_x$Se$_2$ with $x = 0.05$ [612]. This indicates that the main factor in reducing the lock-in temperature is disorder and not electron concentration. In fact, for both TaS$_2$ and TaSe$_2$, the Fermi-surface size is better matched to the superlattice geometry under 2–3% Ti doping than it is for the pure materials [612].

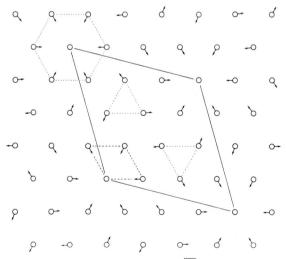

Fig. 114. Basal-plane displacements of the cations in the $\sqrt{13}\ a_0$ superlattice of 1T—TaSe$_2$ assuming that the cation motion is simply directed to the charge maxima formed by the charge-density waves [612]. Equivalent cation positions are emphasized by two triangles and a hexagon, respectively.

The situation in TaS$_2$ [612, 613, 812] is slightly more complicated as there appears to be a minor mismatch of the $\sqrt{13}a_0$ superstructure which gives rise to the second transition at 200 K. A commensurate $\sqrt{13}a_0$ superstructure has however been found by Brouwer and Jellinek [590] by means of X-ray diffraction ($a = a_0\sqrt{13} = 12.13$ Å, $c = 3c_0 = 17.69$ Å, C_3^1—P3).

It is worth noting that 1T—TaS$_2$ remains normal down to 0.07 K whereas the modifications with trigonal-prismatic cation coordination become superconducting [612].

Structural distortions have been discovered also in other CdI$_2$-type phases. For VSe$_2$ a continuous distortion has been observed starting near 55 K whereas in TiSe$_2$ a discontinuous distortion takes place at 200 K [613a]. These transitions may be taken as an indication of the metallic character of these phases. Thompson [613a] pointed out that the distortion-transition temperatures are correlated with ionicity and therefore with the axial ratio c/a independent of the symmetry of the cation coordination, i.e. for octahedral as well as trigonal-prismatic coordination. He found a linear relationship between T_0 and c/a. For the discontinuous transitions $T_0 = 0$ when $c/a = 1.608$ while for the continuous transitions $T_0 = 0$ when $c/a = 1.793$. If ZrTe$_2$ and HfTe$_2$ really are metallic then a discontinuous transition is to be expected at rather low temperatures.

g. The NbTe$_2$ structure

On cooling NbTe$_2$ and TaTe$_2$ to room temperature, their high-temperature CdI$_2$ structure obviously suffers similar distortions as 1T—TaS$_2$ and 1T—TaSe$_2$. Incommensurate and commensurate structures have been observed [611], the latter with $a = \sqrt{19}a_0$, i.e. $\mathbf{a} = [320]$, in contrast to the $\sqrt{13}a_0$ superlattice of 1T—TaSe$_2$. The monoclinic structure stable at room temperature, however, contains 6 formula units per cell or 18 in a pseudo-hexagonal cell with $c = 3c_0$, compared with $Z = 19n$ ($n = 1$ or 3) for the superstructure detected by electron diffraction [611]. The monoclinic cell is related to the original CdI$_2$-type cell by the relations:

$$a \approx 3\sqrt{3}a_0$$

$$b \approx a_0$$

$$c \approx c_0/\sin \beta$$

$$\mathrm{tg}(\pi-\beta) \approx \frac{c_0}{\sqrt{3}a_0}$$

The distortions are much more pronounced than in the superstructures of 1T—TaS$_2$, the cation shifts being almost an order of magnitude larger. In TaTc$_2$, the originally equivalent distances become Ta$_\mathrm{I}$—Ta$_\mathrm{I}$ = b = 3.65 Å, Ta$_\mathrm{I}$—Ta$_\mathrm{II}$ = 3.32 Å, Ta$_\mathrm{II}$—Ta$_\mathrm{II}$ = 4.51 Å. The shortened metal–metal distances as emphasized

TABLE 113
$NbTe_2$ structure, monoclinic, C_{2h}^3—C2/m (No. 12), Z = 6.
(0, 0, 0; $\frac{1}{2}$, $\frac{1}{2}$, 0)+
Nb_{II} and all Te atoms in 4(i): ±(x, 0, z)
Nb_I in 2(a): 0, 0, 0.

$NbSe_xTe_{2-x}$: $x \leqslant 1.4$ [614]
$NbTe_2$: a = 19.39 Å, b = 3.642 Å, c = 9.375 Å, β = 134°35'

	x	z	
Nb_{II}	0.639 7	0.988 2	
Te_I	0.649 7	0.289 8	[615]
Te_{II}	0.297 0	0.214 8	
Te_{III}	0.996 1	0.302 0	

a = 19.12 Å, b ≈ 3.68 Å, c ≈ 9.18 Å, β ≈ 134.0°
(calculated from CdI_2-like cell a = 3.68 Å, c = 6.61 Å [616])
TaSeTe: a ≈ 18.42 Å, b ≈ 3.53 Å, c ≈ 8.97 Å, β ≈ 133.2°
(calculated from the orthorhombic cell with
a = 6.09 Å, b = 10.66 Å, c = 6.54 Å [617])
$TaTe_2$: a = 19.31 Å, b = 3.651 Å, c = 9.377 Å, β = 134°13'.

	x	z	
Ta_{II}	0.639 6	0.988 5	
Te_I	0.648 3	0.285 1	[615]
Te_{II}	0.297 2	0.217 9	
Te_{III}	0.994 4	0.297 5	

a ≈ 19.01 Å, b ≈ 3.694 Å, c ≈ 9.170 Å, β ≈ 133.6°
calculated from the orthorhombic cell a = 6.415 kX, b = 10.90 kX,
c = 6.64 kX [618] using the relations
$a_{mon} \approx \frac{3}{2}(a_{orth}^2 + b_{orth}^2)^{1/2}$
$b_{mon} \approx \frac{1}{6}(9a_{orth}^2 + b_{orth}^2)^{1/2}$
$c_{mon} \approx (\frac{1}{4}(a_{orth}^2 + b_{orth}^2) + c_{orth}^2)^{1/2}$
$tg(\pi - \beta) \approx 2c_{orth}(a_{orth}^2 + b_{orth}^2)^{-1/2}$

in Figure 115 cannot give rise to non-metallic properties which would require single pair bonds. In $NbTe_2$ and $TaTe_2$ there will be a certain overlap of the $d(t_{2g})$ band with the anion p-band. Wilson et al. [612] speculated that the ribboned structure arises from a charge-density wave for which λ has fallen from $3.1a_0$ in the sulfide and selenide to around $2.6a_0$ in the telluride. The $3a_0$ superstructure comprises 1.5 $NbTe_2$-type cells [612].

It is noteworthy that in the mixed system $NbSe_xTe_{2-x}$ the $NbTe_2$ structure can be retained at room temperature up to x = 1.4 [614], in $TaSe_xTe_{2-x}$ at least up to x = 1 (Table 113). A study of the systems $Zr_xNb_{1-x}Te_2$ and $Hf_xTa_{1-x}Te_2$ might supply interesting information on the electronic structure of both end members.

While $NbTe_2$ undergoes a transition to the superconducting state at $T_c = 0.74$ K, $TaTe_2$ remains normal down to 0.05 K [619].

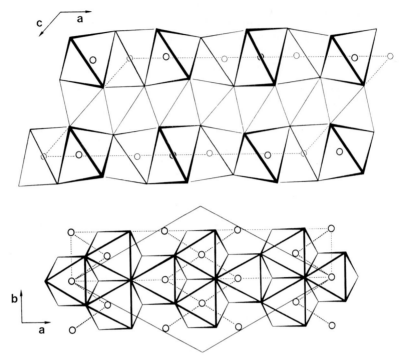

Fig. 115. The monoclinic structure of NbTe$_2$ visualized as stacking of [MX$_6$] octahedra. Above: Section perpendicular to the layers. Below: View of one layer. The shortened metal—metal distances are emphasized by dotted lines. The cell of the $3a_0$ superstructure is indicated by full lines.

h. Calaverite, Krennerite and Sylvanite

Calaverite (AuTe$_2$ or Au$_{1-x}$Ag$_x$Te$_2$ with $x \leqslant 0.13$), krennerite (Au$_{1-x}$Ag$_x$Te$_2$ with $x < 0.3$), sylvanite (AgAuTe$_4$) and kostovite (CuAuTe$_4$) are crystallographically related minerals. The structure of calaverite may be interpreted as a strongly squeezed CdI$_2$-type structure with $c/a \approx 1.22$. Its monoclinic cell corresponds to the (distorted) orthohexagonal cell with $a = \sqrt{3}a_0$, $b = a_0$, $c = c_0$. The sylvanite structure of AgAuTe$_4$ and CuAuTe$_4$ is an ordered version of the calaverite structure while the krennerite structure represents an internally twinned calaverite type. So one might expect at least calaverite and sylvanite to be layer structures. However, calaverite was said to have no cleavage [620], while krennerite has perfect cleavage on (001) and sylvanite on (010).

In both calaverite and krennerite two of the six Au—Te distances are distinctly shorter (2.68 vs. 2.97 Å in calaverite), so that both structures contain parallel and nearly linear Te—Au—Te units. The shortened bonds lie in the (110) plane of the original CdI$_2$ structure and are identical with those indicated in Figure 106. In the krennerite structure the linear molecules point in the [010] direction and in calaverite they are parallel with [10$\bar{1}$]. It is thus plausible that (001) is the cleavage plane in krennerite, but we see no reason why (101) should not be an equivalent cleavage plane in calaverite. Indeed, Te atoms of adjacent TeAuTe sandwiches are considerably closer than the Van der Waals distance. Tellurium in calaverite has two interlayer Te neighbors at 3.19 Å and a third one at 3.47 Å. This may account for the three-dimensional character of the bonding system. However, in krennerite, the compression of the TeAuTe—TeAuTe stacking is similar though less symmetric:

Te_I, Te_{II}——3 Te at 3.08, 3.34 and 3.51/3.54 Å
Te_{III}, Te_V——3 Te at 3.02, 3.28 and 3.54 Å.

The interlayer bonding should even be stronger. In sylvanite, finally, the Te—Te interlayer distances are 2.87, 3.55 and 3.65 Å. The closest Te—Te contacts thus are as short as in the spiral chains of elemental tellurium (2.86 Å). Since the strongest Te—Te bonds run across the (010) plane, these bonds should break on cleaving the crystal along (010).

Though these structures can be directly derived from the CdI_2 type, they no longer display a two-dimensional character.

i. $PtBi_2(h_2)$

Whereas the semiconducting platinum pnictides PtP_2, $PtAs_2$ and $PtSb_2$ crystallize in the pyrite structure with tetravalent Pt, the vanishing energy gap in $PtBi_2$ probably gives rise to the occurrence of a distorted pyrite variant stable at room temperature and a distorted CdI_2 variant stable below the melting point [621]. In this high-temperature modification, one third of the cations (i.e. half the Pt_{II} atoms) is shifted towards one anion layer and one third towards the opposite

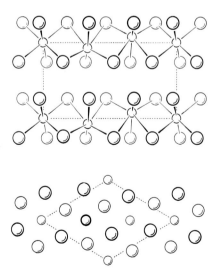

Fig. 116. Basal-plane projection (below) and (110) section (above) of the hexagonal structure of $PtBi_2$ (h_2). Large spheres: Bi

TABLE 114

$PtBi_2(h_2)$ structure, trigonal, C_{3i}^1—$P\bar{3}$ (No. 147), $Z = 3$.
Bi in 6(g): $\pm(x, y, z;\ \bar{y}, x-y, z;\ y-x, \bar{x}, z)$
Pt_{II} in 2(d): $\pm(\frac{1}{3}, \frac{2}{3}, z)$
Pt_I in 1(a): $0, 0, 0$.

$PtBi_2(h_2)$: $a = 6.57$ Å, $c = 6.16$ Å
Pt: $z = 0.920$; Bi: $x = 0.050$, $y = 0.365$, $z = 0.260$ [621]

---

anion layer. This displacement leads to an irregular coordination for Pt_{II}:

Pt_I—6 Bi at 2.76 Å versus Pt_{II}—3 Bi at 2.66 Å
3 Bi at 2.84 Å.

The geometry of this structure is illustrated in Figure 116.

It is noteworthy that an increase of the electron concentration as met in metallic PtSbTe and PtBiTe does not induce a structural change. Only in $PtTe_2$, which with the pyrite structure would contain formally divalent Pt, does this structural change occur, although the CdI_2-type telluride is nevertheless metallic and structural distortions have not been reported.

We wonder whether the pyrite → $PtBi_2(h_2)$ transition can be simulated in a sequence PtGeSe → PtSnTe, PtPbTe. A minor reduction of the electron concentration as exemplified by Pt_4PbBi_7, leads to a new structure.

j. Pt_4PbBi_7, $MoTe_2(h)$ AND WTe_2

The minor reduction of the electron concentration in $PtBi_2$ by partial substitution of Bi with Pb transforms the h stacking of the anion layers into a hc stacking. The anion layers are strongly distorted and the cations are shifted out of the centers of the anion octahedra giving rise to Pt zig-zag chains with Pt—Pt distances of 2.96 Å compared to 4.41 Å between the chains.

In hypothetical nonmetallic CdI_2-type $PtBi_2$ the hexavalent cation might give rise to two holes in the t_{2g} band. These holes could be eliminated by the formation of two Pt—Pt single bonds. The actual Pt zigzag chains in Pt_4PbBi_7, however, are not sufficient to reach this goal but the distortions may reduce the free Fermi surface.

A similar situation is met in $MoTe_2(h)$ and WTe_2. Except for the monoclinic deformation of the cell, the structures of $MoTe_2(h)$ and Pt_4PbBi_7 are identical. We

TABLE 115

Pt_4PbBi_7 structure, orthorhombic, D_{2h}^{16}—Pnma (No. 62), $Z = 1$.
All atoms in 4(c): $\pm(x, \frac{1}{4}, z;\ \frac{1}{2}+x, \frac{1}{4}, \frac{1}{2}-z)$

$Pt_{33}Pb_{10}Bi_{57}$: $a = 6.84$ Å, $b = 4.12$ Å, $c = 12.55$ Å

	x	z
Pt	0.328 1	0.513 9
$(Pb_{0.15}Bi_{0.85})_I$	0.124 4	0.168 4	[621]
$(Pb_{0.15}Bi_{0.85})_{II}$	0.883 6	0.601 3

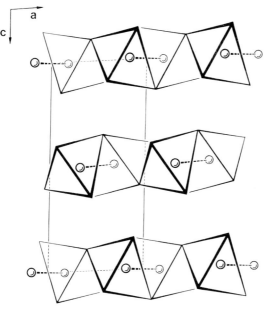

Fig. 117. The hc stacking of the [MoTe$_{6/3}$] octahedra in the monoclinic structure of MoTe$_2$(h). The zig-zag chains formed by the metal atoms are indicated by dashed arrows.

have therefore abstained from reproducing the Pt$_4$PbBi$_7$ structure. We have, instead, added in Table 116 the structural data transformed to the space group of MoTe$_2$(h) for comparison. Whereas in Pt$_4$PbBi$_7$ the mean $c/2a$ value is 1.7, it is between 1.9 and 2.0 in MoTe$_2$(h) and WTe$_2$. Below 850–900°C MoTe$_2$ is

TABLE 116
MoTe$_2$(h) structure, monoclinic, C_{2h}^2—$P2_1/m$ (No. 11), $Z = 4$.
All atoms in 2(e): $\pm(x, \frac{1}{4}, z)$

MoTe$_2$(h): $a = 6.33$ Å, $b = 3.469$ Å, $c = 13.86$ Å, $\beta = 93°55'$

	x	z
Mo$_I$	0.183 3	0.007 4
Mo$_{II}$	0.681 9	0.493 4 [622]
Te$_I$	0.588 8	0.104 5
Te$_{II}$	0.903 0	0.851 4
Te$_{III}$	0.443 2	0.648 4
Te$_{IV}$	0.056 3	0.395 5

for comparison: Pt(Bi, Pb)$_2$ of Table 115 rewritten in space group $P2_1/m$

	x	z
Pt$_I$	0.171 9	0.013 9
Pt$_{II}$	0.671 9	0.486 1
Bi$_I$	0.616 4	0.101 3
Bi$_{II}$	0.875 6	0.831 6
Bi$_{III}$	0.375 6	0.668 4
Bi$_{IV}$	0.116 4	0.398 7

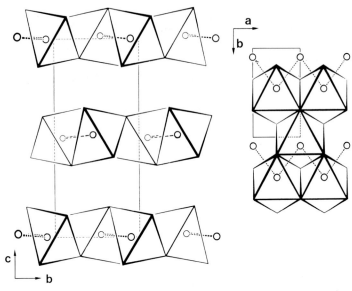

Fig. 118. Stacking of the cation-coordination polyhedra in the orthorhombic structure of WTe$_2$. The zig-zag chains formed by the cations are indicated.

non-metallic and crystallizes in the 2H$_b$—MoS$_2$ structure. The energy gap diminishes gradually with increasing temperature. As is to be expected the structural change does not restore the non-metallic character. In MoTe$_2$(h) the Mo—Mo distance within the zig-zag chain is still as large as 2.90 Å and in WTe$_2$, W—W = 2.86 Å. Thus, the distortions are much too small to secure non-metallic properties but they may generate a $2a_0$ superstructure with a largely reduced Fermi surface. Both MoTe$_2$(h) and WTe$_2$ are diamagnetic. MoTe$_2$(h) becomes superconducting near 0.3 K while WTe$_2$ remains normal down to this temperature [23] but an eventual transition temperature is expected to lie lower anyway. The semiconductor-to-metal transition in MoTe$_2$ was studied by Haas et al. [652].

TABLE 117

WTe$_2$ structure, orthorhombic, C_{2v}^7—Pmn2$_1$ (No. 31), $Z = 8$.
All atoms in 4(b): x, y, z; $\bar{x}, y, z$; $\frac{1}{2}-x, \bar{y}, \frac{1}{2}+z$; $\frac{1}{2}+x, \bar{y}, \frac{1}{2}+z$.

WTe$_2$: $a = 6.282$ Å, $b = 3.496$ Å, $c = 14.073$ Å

	x	y	z	
W$_I$	$\frac{1}{2}$	0.900 5	0	
W$_{II}$	0	0.541 4	0.014 9	[622]
Te$_I$	$\frac{1}{2}$	0.294 1	0.903 5	
Te$_{II}$	0	0.800 2	0.860 0	
Te$_{III}$	0	0.355 9	0.655 1	
Te$_{IV}$	$\frac{1}{2}$	0.851 7	0.610 7	

$a = 6.270$ Å, $b = 3.405$ Å, $c = 14.028$ Å [623]
$a = 6.278$ Å, $b = 3.483$ Å, $c = 14.054$ Å [624]

WS$_x$Te$_{2-x}$, $x < 0.2$ [1042]
WSe$_x$Te$_{2-x}$, $x < 0.52$ [1042]

TABLE 118
ReSe$_2$ structure, triclinic, C_i^1—P$\bar{1}$ (No. 2), Z = 4.
All atoms in 2(i): ±(x, y, z)

	a(Å)	b(Å)	c(Å)	α(°)	β(°)	γ(°)	Refs.
ReS$_2$	6.401	6.362	6.455	118.97	91.60	105.04	[625]
ReSSe	6.584	6.481	6.591	118.93	91.59	104.78	[625]
ReSe$_2$	6.727 2	6.606 5	6.719 6	118.94	91.83	104.93	[626]
	6.728	6.602	6.716	118.94	91.82	104.90	[625]

	x	y	z	
ReSe$_2$: Re$_I$	0.493 7 ($\frac{1}{2}$)	0.303 8 ($\frac{1}{4}$)	0.302 0 ($\frac{1}{4}$)	
Re$_{II}$	0.494 3 ($\frac{1}{2}$)	0.747 4 ($\frac{3}{4}$)	0.302 7 ($\frac{1}{4}$)	
Se$_I$	0.222 8 ($\frac{1}{4}$)	0.487 5 ($\frac{1}{2}$)	0.360 0 ($\frac{3}{8}$)	[626]
Se$_{II}$	0.222 0 ($\frac{1}{4}$)	0.983 0 (0)	0.362 0 ($\frac{3}{8}$)	
Se$_{III}$	0.708 9 ($\frac{3}{4}$)	0.489 3 ($\frac{1}{2}$)	0.122 1 ($\frac{1}{8}$)	
Se$_{IV}$	0.708 9 ($\frac{3}{4}$)	0.989 3 (0)	0.122 1 ($\frac{1}{8}$)	

(Figures in parentheses refer to an ideal CdCl$_2$-type packing)
Related structures:
TcS$_2$: a = 6.659 Å, b = 6.375 Å, c = 6.465 Å, α = 118.96°, β = 62.97° γ = 103.61° [625]
TcSe$_2$: probably isostructural with TcS$_2$
TcTe$_2$, monoclinic, Cc or C2/c
  a = 12.522 Å, b = 7.023 Å, c = 13.828 Å, β = 101.26° [625]

k. ReSe$_2$ and TcS$_2$

In the Re and Tc dichalcogenides there are three excess d electrons that must be localized in order to produce non-metallic properties. These compounds are diamagnetic semiconductors with energy gaps near 1 eV. Thus, the cation clustering is evidently sufficient to secure non-metallic properties.

Only the ReSe$_2$ structure is known in detail [626]. It represents a distorted CdCl$_2$ type, i.e. the anions form a ccp array in which the cations occupy every second layer of octahedral holes. The distortions are due to the Re clusters which

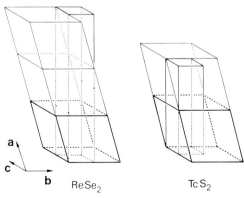

Fig. 119. The relation between the hexagonal CdCl$_2$-type cell and the triclinic superstructure of ReSe$_2$ (left) and the relation between the hexagonal CdI$_2$-type cell (two cells are indicated) and the triclinic superstructure of TcS$_2$ (right).

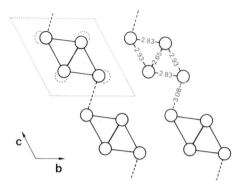

Fig. 120. Cation layer of a Se—Re—Se sandwich of the triclinic ReSe$_2$ structure. The cation clustering and the corresponding Re—Re distances are indicated.

are shown in Figure 120. Each Re atoms has 3 Re neighbors but the distances are rather irregular, ranging from 2.65 to 3.08 Å. If we neglect the bond corresponding to the largest distance we may assume within the remaining Re$_4$ clusters one double bond instead and the rest single bonds which is equivalent to half d^4 and half d^2 cations.

The TcS$_2$ structure appears to be derived from the CdI$_2$ structure [625]. The relations between the unit cells of the triclinic ReSe$_2$ and TcS$_2$ structures and their basic structures are illustrated in Figure 119.

12. Group V and VI Dichalcogenides with Trigonal-Prismatic Cation Coordination

In a trigonal-prismatic surrounding the fivefold orbital degeneracy of the d levels is lifted. Whether the singlet $a'_1(d_{z^2})$ or the doublet $e''(d_{xy}, d_{x^2-y^2})$ lie lowest depends upon the axial ratio of the anion trigonal prism. For a packing of hard spheres and lower axial ratios the a_1 state is lowest. For 4d^2 and 5d^2 cations a diamagnetic and nonmetallic state thus is possible. In d^1 compounds, nonmetallic properties would require spin-pairing energies larger than the band width. This requires a larger ionicity which in turn favors octahedral coordination.

All compounds with d^2 cations in trigonal-prismatic coordination are non-metallic. There is no smooth transition from non-metallic to metallic properties within the same structure as it occurs in CdI$_2$-type d^0 compounds. Semiconductor-to-metal transitions are accompanied here by a structural change (e.g. in MoTe$_2$ and WTe$_2$).

In the metallic d^1 TX$_2$ phases, both trigonal-prismatic and octahedral coordinations can occur alternately within the same structure. In the low-temperature modifications, the cation coordination is purely trigonal prismatic. It changes to mixed trigonal prismatic/octahedral at intermediate temperatures and in TaS$_2$ and TaSe$_2$ finally becomes purely octahedral at high temperatures.

TABLE 119

$2H_a$–NbS_2 structure, hexagonal, D_{6h}^4—$P6_3/mmc$ (No. 194), $Z = 2$.

S in 4(f): $\pm(\frac{1}{3}, \frac{2}{3}, z; \frac{1}{3}, \frac{2}{3}, \frac{1}{2}-z)$
Nb in 2(b): $\pm(0, 0, \frac{1}{4})$

Compound	a(Å)	c(Å)	c/2a	z	Refs.
NbS_2	3.31	11.89	1.80	$\sim\frac{1}{8}$	[627]
NbSSe	3.382	12.321	1.822		[23]
$NbSe_2$ 25°C	3.444 6	12.544 4	1.821	0.117 2	[628]
20°C	3.442	12.54	1.821		[587, 629]
	3.444	12.552	1.822		[633]
$NbSe_{0.81}Te_{0.19}$	3.53	12.86	1.82		[614]
NbSeTe	3.568	13.053	1.829		[23]
TaS_2	3.315	12.10	1.825	$\sim\frac{1}{8}$	[30]
	3.316	12.070	1.820		[588]
$TaS_{0.4}Se_{1.6}$	3.414	12.635	1.850		[630]
$Ta_{0.98}Mo_{0.02}Se_2$	3.430	12.760	1.860		[631]
$TaSe_2$	3.437	12.72	1.851	0.118	[593, 629]
	3.436 0	12.696	1.847	0.125	[592]
$Nb_{1-x}W_xSe_2$					
(x < 0.25)					[632]
(x < 0.15)					[633]
$Nb_{0.9}W_{0.1}Se_2$	3.426	12.582	1.836		[633]
$Nb_{0.75}Mo_{0.25}Se_2$	3.398 5	12.644	1.860		[631]
$Nb_{0.9}Mo_{0.1}Se_2$	3.426 2	12.550	1.831		[631]
$Nb_{0.98}Mo_{0.02}Se_2$	3.436 3	12.531	1.823		[631]
$Ta_{0.98}Mo_{0.02}Se_2$	3.430 0	12.760	1.860		[631]
Anti-type:					
Hf_2S	3.37	11.79	1.75	0.100 7	[634]
Hf_2Se	3.44	12.32	1.79		[635]
Ideal close packing				0.112 4	

The NbX_2 and TaX_2 phases are all metallic. If the mixed structures would occur at the lowest temperatures, one might expect an electronic reordering into d^0 (octahedral) and d^2 (trigonal-prismatic) cations leading to semiconductivity. Such mixed structures would also be adequate for hypothetical $ZrMoS_4$, $HfWSe_4$, etc. Papers on theoretical calculations and experimental studies of the electronic energy bands of these dichalcogenides are cited in [612], others are [813–824].

The geometry of the various polytypes has already been discussed in Chapter 2. In Figure 121, the known polytypes are represented by (110) sections. Bearing in

TABLE 120

$2H$–$NbSe_2(t)$, hexagonal, D_{6h}^4—$P6_3/mmc$ (No. 194), $Z = 8$.

Se_{II} in 12(k): $\pm(x, 2x, z; 2\bar{x}, \bar{x}, z; x, \bar{x}, z; x, 2x, \frac{1}{2}-z; 2x, x, \frac{1}{2}+z; \bar{x}, x, \frac{1}{2}+z)$
Nb_{II} in 6(h): $\pm(x, 2x, \frac{1}{4}; 2\bar{x}, \bar{x}, \frac{1}{4}; x, \bar{x}, \frac{1}{4})$
Se_I in 4(f): $\pm(\frac{1}{3}, \frac{2}{3}, z; \frac{1}{3}, \frac{2}{3}, \frac{1}{2}-z)$
Nb_I in 2(b): $\pm(0, 0, \frac{1}{4})$

$2H$—$NbSe_2(t)$ at 15 K: $a = 6.880$ Å, $c = 12.482$ Å [628]
$x(Nb_{II}) = 0.496\,7$; $z(Se_I) = 0.117\,8$; $x(Se_{II}) = 0.167\,8$; $z(Se_{II}) = 0.116\,0$.

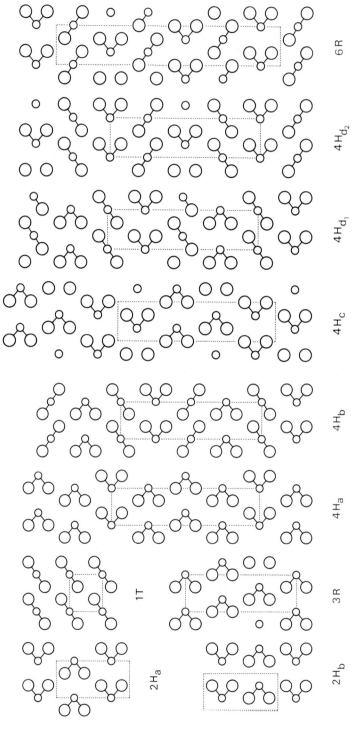

Fig. 121. (110) sections of the various polytypes observed in layer-type MX_2 compounds with octahedral and trigonal-prismatic cation coordination. Large circles: anions; small circles: cations.

TABLE 121

$2H_b$-MoS_2 structure, hexagonal, D_{6h}^4—$P6_3/mmc$ (No. 194), $Z = 2$.
S in 4(f): $\pm(\frac{1}{3}, \frac{2}{3}, z; \frac{1}{3}, \frac{2}{3}, \frac{1}{2}-z)$
Mo in 2(c): $\pm(\frac{1}{3}, \frac{2}{3}, \frac{1}{4})$

Compound	$a(\text{Å})$	$c(\text{Å})$	$c/2a$	z	Refs.
MoS_2	3.156	12.275	1.945		[1]
26 °C	3.160 4	12.295	1.945	0.629	[1, 4]
	3.160 2	12.294	1.945		[636]
	3.162	12.29	1.943		[578]
$MoSe_2$	3.291	12.91	1.961		[1]
25 °C	3.288	12.900	1.962	$\sim\frac{5}{8}$	[1, 3]
	3.288	12.92	1.965		[637]
	3.285 4	12.918	1.966		[631]
$Nb_{0.6}Mo_{0.4}Se_2$	3.376 6	12.648	1.873		[631]
$Ta_{0.75}Mo_{0.25}Se_2$	3.398 0	12.756	1.877		[631]
$Mo_{0.75}W_{0.25}Se_2$	3.288	12.941	1.968		[638]
$Mo_{0.5}W_{0.5}Se_2$	3.288	12.95	1.969		[637]
$Mo_{0.25}W_{0.75}Se_2$	3.287	12.965	1.972		[638]
$MoSe_{1.5}Te_{0.5}$	3.331	13.34	2.002		[637]
$MoSeTe$	3.401	13.66	2.008		[637]
$MoSe_{0.5}Te_{1.5}$	3.454	13.84	2.003		[637]
$MoTe_2$	3.518 2	13.974	1.986	0.621	[640]
25 °C	3.517	13.949	1.983		[639]
	3.521	13.96	1.982		[637]
	3.519	13.964	1.984	0.620	[641]
$Mo_{0.75}W_{0.25}Te_2$	3.521	13.96	1.982		[637]
WS_2	3.171	12.36	1.949		[1]
	3.176	12.397	1.952		[1]
25 °C	3.154	12.362	1.960		[1, 3]
	3.155	12.35	1.957		[636]
$Nb_{0.5}W_{0.5}S_2$	3.242 2	12.124	1.870		[589]
$Ta_{0.75}W_{0.25}S_2$	3.263 0	12.134	1.859		[589]
$W(S,Se)_2$					[1042]
$WS_{2-x}Te_x$					
($x \leq 0.3$)					[1042]
WSe_2 25 °C	3.280	12.950	1.974		[1, 3]
	3.285	12.96	1.973		[637]
	3.282	12.937	1.971		[642]
	3.284	12.984	1.977		[633]
$W_{1-x}Nb_xSe_2$					
($x \leq 0.46$)					[633]
$W_{0.74}Nb_{0.26}Se_2$	3.329	12.810	1.924		[633]
$W_{0.54}Nb_{0.46}Se_2$	3.360	12.784	1.902		[633]
$Ta_xW_{1-x}Se_2$	a, c linear				[642]
$Ta_{0.4}W_{0.6}Se_2$	3.347 0	12.829	1.916		[631]
$Ta_{0.75}W_{0.25}Se_2$	3.397 2	12.777	1.881		[631]
$WSe_{1.5}Te_{0.5}$	3.342	13.43	2.009		[637]
$WSe_{2-x}Te_x$					
($x \leq 0.62$)					[1042]
$Mo_{0.5}W_{0.5}Se_{1.5}Te_{0.5}$	3.331	13.37	2.007		[637]
$Mo_{0.5}W_{0.5}SeTe$	3.395	13.70	2.018		[637]

mind the threefold symmetry around the c-axis, one can easily deduce the coordination of the atoms. In the $2H_b$—MoS_2 structure each cation has another anion neighbor above and below, which may help to hold the layers together. Only semiconductors are known with this structure, whereas with the rhombohedral $3R$—MoS_2 structure both semiconducting d^2 phases as well as metallic d^1 phases occur. The remaining polytypes are known only with metallic phases. In Tables 119, 121–128 we have indicated the reduced axial ratio c/na per sandwich. For a trigonal-prismatic packing of spheres this value is 1.816 5, for mixed coordination it is 1.724 7. The high values observed for the non-metallic compounds are rather striking. The d^1 phases with trigonal-prismatic cation coordination all become superconducting.

In the recent past, diverse opinions have been put forward to explain the occurrence of the various structural types [21, 22, 652–654] Madhukar [654] has shown that in these mainly covalent compounds the small variation in the ionicity of the bonds competes against the polarization contribution of the bonding orbitals in determining the relative stability of octahedral versus trigonal-prismatic cation coordination. In a plot of the ratio of the single-bond covalent radii of the cation and anion versus bond ionicity, all compounds with trigonal-prismatic

TABLE 122

$3R$-MoS_2 structure, trigonal, C_{3v}^5—$R3m$ (No. 160), $Z = 1(3)$
hexagonal axes: $(0, 0, 0; \frac{1}{3}, \frac{2}{3}, \frac{2}{3}; \frac{2}{3}, \frac{1}{3}, \frac{1}{3})+$
all atoms in 3(a): 0, 0, z
$z(Mo) = 0$

Compound	a_h(Å)	c_h(Å)	$c/3a$	a_{rh}(Å)	α(°)	$z(X_I)$	$z(X_{II})$	Refs.
MoS_2	3.166	18.41	1.938	6.403	28.63	0.247 7	0.419 0	[26]
	3.17	18.38	1.93			$\frac{1}{4}$	$\frac{5}{12}$	[627]
	3.163	18.37	1.936	6.390	28.66			[636]
	3.16	18.45	1.95			0.252	0.414	[643]
$MoSe_2$ (h, p)	3.292	19.392	1.964	6.738	28.28			[644]
WS_2	3.162	18.50	1.950	6.431	28.46			[636]
NbS_2	3.330 3	17.918	1.793	6.275	30.78	0.246 4	0.420 1	[645]
	3.33	17.81	1.78					[627]
$Nb_{0.4}W_{0.6}S_2$	3.220 2	18.207	1.885	6.347	29.39			[589]
$NbSe_2$	3.45	18.88	1.82			0.243	0.421	[27]
	3.459	18.77	1.809	6.568	30.54			[629]
TaS_2	3.32	17.9	1.80			$\sim\frac{1}{4}$	$\sim\frac{5}{12}$	[30]
	3.32	18.29	1.84	6.39	30.18			[1]
$Ta(S, Se)_2$	a and c linear							[630]
$TaSe_2$	3.434 8	19.177	1.861	6.693	29.74			[592]
	3.436 9	19.213	1.863	6.705	29.70	$\sim\frac{1}{4}$	$\sim\frac{5}{12}$	[593, 629]
25 °C	3.446	19.101	1.848	6.671	29.94			[646]
$Ta_{0.98}W_{0.02}Se_2$	3.429 0	19.111	1.858	6.671	29.79			[631]
$Ta_{0.75}W_{0.25}Se_2$	3.397 2	19.166	1.881	6.683	29.45			[631]
	3.408	19.132	1.871	6.674	29.59			[646]
$Ta_{0.5}W_{0.5}Se_2$	3.367	19.186	1.899	6.684	29.18			[646]
$Ta_{0.4}W_{0.6}Se_2$	3.353	19.194	1.908	6.684	29.05			[646]
$ZrCl_2$	3.382	19.38	1.91	6.75	29.0	0.246	0.421	[647]

TABLE 123

$4H_a$—$NbSe_2$ structure, hexagonal, D_{3h}^1—$P\bar{6}m2$ (No. 187), $Z=4$.

Se_{III} and Se_{IV} in 2(i): $\frac{1}{3}, \frac{1}{3}, z; \frac{2}{3}, \frac{1}{3}, \bar{z}$

Se_{II} in 2(h): $\frac{1}{3}, \frac{2}{3}, z; \frac{1}{3}, \frac{2}{3}, \bar{z}$

Nb_{III} and Se_I in 2(g): $\pm(0, 0, z)$

Nb_{II} in 1(d): $\frac{1}{3}, \frac{2}{3}, \frac{1}{2}$

Nb_I in 1(a): $0, 0, 0$

Compound	a(Å)	c(Å)	$c/4a$	z_M	$z(Se_I)$	$z(Se_{II})$	$z(Se_{III})$	$z(Se_{IV})$	Refs.
$NbSe_2$	3.44	25.24	1.83	0.250	0.432	0.066	0.185	0.318	[27]
20°C	3.444	25.23	1.832	$\sim\frac{1}{4}$	$\sim\frac{7}{16}$	$\sim\frac{1}{16}$	$\sim\frac{3}{16}$	$\frac{5}{16}$	[587]
					(0.437 5)	(0.062 5)	(0.187 5)	(0.312 5)	[629]
$Ta_{0.98}W_{0.02}Se_2$	3.429 0	25.480	1.858						[631]
$TaSe_2$	3.436 2	25.399	1.848						[592]
	3.43	25.5	1.86						[629]
Ideal close packing			1.816 5	$\frac{1}{4}$	0.431 2	0.068 8	0.181 2	0.318 8	
general formula				$2(z_\pi + z'_\Omega) = \frac{1}{2} - 2(z_\pi + z''_\Omega)$	$\frac{1}{2} - z_\pi$	z_π	$z_\pi + 2z'_\Omega$	$3z_\pi + 2z'_\Omega$	

$2z'_\Omega c = 2H_a$-NbS_2-type anion-layer separation between adjacent XMX sandwiches

$2z''_\Omega c = 3R$-MoS_2-type anion-layer separation between adjacent XMX sandwiches

$2z_\pi c$ = trigonal-prism layer height.

ideal close-packing: $z'_\Omega = z''_\Omega$ $c = 4a(1 + \sqrt{\frac{2}{3}})$

$2z'_\Omega c = \sqrt{\frac{2}{3}}a$ $z_\pi = 0.068\,82$

$2z_\pi c = a$ $z'_\Omega = 0.056\,19$

coordination are separated from those with CdI_2 structure by a curve defined by the values of TaS_2, $TaSe_2$ and $MoTe_2$. This plot reflects the fact that the favorable conditions for trigonal-prismatic cation coordination are a small ionicity of the bonds and a large cation/anion radius ratio. Since d^2sp^3 hybrid bonds require less hybridisation of the d-orbitals they are preferred whenever the d levels lie higher in energy, though the hybridized d^4sp orbitals have a higher bond strength than the octahedral d^2sp^3 orbitals. The charge transfer due to a high ionicity will also favor octahedral coordination since in that case the anion—anion separation is greater than in trigonal-prismatic coordination.

TABLE 124

$4H_b$-$TaSe_2$ structure, hexagonal, D_{6h}^4—$P6_3/mmc$ (No. 194), $Z=4$.

Se_I and Se_{II} in 4(f): $\pm(\frac{1}{3}, \frac{2}{3}, z; \frac{1}{3}, \frac{2}{3}, \frac{1}{2} - z)$

Ta_{II} in 2(b): $\pm(0, 0, \frac{1}{4})$

Ta_I in 2(a): $0, 0, 0; 0, 0, \frac{1}{2}$.

Compound	a(Å)	c(Å)	$c/4a$	$z(Se_I)$	$z(Se_{II})$	Refs.
$Nb_{1-x}V_xSe_2$ ($x = 0.11\ldots 0.20$)						
($x = 0.11\ldots 0.20$)						[1041]
TaS_2	3.332	23.62	1.772			[648]
$TaSe_2$	3.46	25.18	1.82	0.065	−0.184	[27]
	3.457 5	25.143	1.818			[592]
	3.455 4	25.148	1.820			[593, 649]
ideal close packing			$\frac{1}{2}(1+\sqrt{6}) =$ 1.724 7	$z_\Omega =$ 0.059 2	$\frac{1}{4} - z_\pi =$ 0.177 5	

$2z_\Omega c$ = height of octahedral XMX layers

$2z_\pi c$ = height of trigonal-prismatic XMX layers

TABLE 125

$4H_c$-$TaSe_2$ structure, hexagonal, C_{6v}^4—$P6_3mc$ (No. 186), $Z = 4$.
Ta_I, Ta_{II}, Se_{III} and Se_{IV} in 2(b): $\frac{1}{3}, \frac{2}{3}, z$; $\frac{2}{3}, \frac{1}{3}, \frac{1}{2}+z$.
Se_I and Se_{II} in 2(a): $0, 0, z$; $0, 0, \frac{1}{2}+z$.

$TaSe_2$: $a = 3.4340$ Å, $c = 25.501$ Å, $c/4a = 1.857$ [631]
$a = 3.436$ Å, $c = 25.532$ Å, $c/4a = 1.858$ [649]

	$z_{obs.}$	$z_{id.cp.}$ [649]	$z_{general}$
Se_I	0.050	0.0562	z_Ω''
Se_{II}	0.178	0.1938	$2z_\pi + z_\Omega''$
Se_{III}	0.307	0.3062	$\frac{1}{2}-(2z_\pi + z_\Omega'')$
Se_{IV}	0.442	0.4438	$\frac{1}{2}-z_\Omega''$
Ta_I	0.628	0.6250	$\frac{1}{2}+z_\pi + z_\Omega''$
Ta_{II}	0.881	0.8750	$-(z_\pi + z_\Omega'')$

$z_{id.cp.}$: calculated assuming each Se to be equidistant from its 10 Se neighbors (and each Ta to be equidistant from its 6 Se neighbors) [649] which corresponds to ideal close packing.

TABLE 126

$4H_{d_1}$-$NbSe_2$ structure, (model I), hexagonal, C_{3v}^1—$P3m1$ (No. 156), $Z = 4$.
Se_V to Se_{VIII} in 1(c): $\frac{2}{3}, \frac{1}{3}, z$
Nb_{IV}, Se_{II} to Se_{IV} in 1(b): $\frac{1}{3}, \frac{2}{3}, z$
Nb_I to Nb_{III} and Se_I in 1(a): $0, 0, z$

Nb: $z_I = 0$, $z_{II} = \frac{1}{4}$, $z_{III} = \frac{3}{4}$, $z_{IV} = \frac{1}{2}$
$NbSe_2$ (950°C): $a = 3.48$ Å, $c = 25.45$ Å, $c/4a = 1.828$ [687]

	$z(Se)$	$z(Se)$ for ideal close packing	$z(Se)$ calculated with z_Ω and z_π from $4H_b$-$TaSe_2$
Se_I	$\frac{1}{2}-z_\Omega$	0.4408	0.435
Se_{II}	z_Ω	0.0592	0.065
Se_{III}	$\frac{3}{4}-z_\pi$	0.6775	0.684
Se_{IV}	$\frac{3}{4}+z_\pi$	0.8225	0.816
Se_V	$\frac{1}{4}-z_\pi$	0.1775	0.184
Se_{VI}	$\frac{1}{4}+z_\pi$	0.3225	0.316
Se_{VII}	$\frac{1}{2}+z_\Omega$	0.5592	0.565
Se_{VIII}	$-z_\Omega$	0.9408	0.935

$z = \dfrac{h_\Omega}{2c}$, $z = \dfrac{h_\pi}{2c}$

ideal close-packing: $h_\Omega = \sqrt{\frac{2}{3}}a$, $h_\pi = a$, $c = 2(1+\sqrt{6})a$,

$$z_\Omega = \frac{1}{12+2\sqrt{6}} = 0.0592$$

$$z_\pi = \frac{1}{4(1+\sqrt{6})} = 0.0725$$

$4H_b$-$TaSe_2$: $z_\Omega = 0.065$, $z_\pi = 0.066$ [27]

TABLE 127

$4H_{d_2}$-$NbSe_2$ structure (model II), hexagonal, D_{3h}^1—$P\bar{6}m2$ (No. 187), $Z=4$.
Se$_{III}$ and Se$_{IV}$ in 2(i): $\frac{2}{3}, \frac{1}{3}, z; \frac{2}{3}, \frac{1}{3}, \bar{z}$
Se$_I$ and Se$_{II}$ in 2(h): $\frac{1}{3}, \frac{2}{3}, z; \frac{1}{3}, \frac{2}{3}, \bar{z}$
Nb$_{III}$ in 2(g): $\pm(0, 0, z)$ with $z = \frac{1}{4}$
Nb$_{II}$ in 1(d): $\frac{1}{3}, \frac{2}{3}, \frac{1}{2}$
Nb$_I$ in 1(a): $0, 0, 0$

NbSe$_2$ (950°C): $a = 3.48$ Å, $c = 25.45$ Å, $c/4a = 1.828$ [587]

	z(Se)	z(Se) for ideal close packing	ideal z(Se) calculated with z_Ω and z_π from $4H_b$-TaSe$_2$
Se$_I$	z_π	0.072 5	0.066
Se$_{II}$	$\frac{1}{4} + z_\Omega$	0.309 2	0.315
Se$_{III}$	$\frac{1}{4} - z_\Omega$	0.190 8	0.185
Se$_{IV}$	$\frac{1}{2} - z_\pi$	0.427 5	0.434

At low temperatures, the structures with trigonal-prismatic cation coordination also become distorted via charge-density waves in order to reduce the free Fermi surface in analogy to the case of the CdI$_2$-type modifications [612]. These distortions appear to develop in two steps. The first step is a second-order transition to an incommensurate superstructure which is followed by a first- (or second-) order transition to a commensurate superstructure (e.g. a $3a_0$ superstructure at 90 K in TaSe$_2$ [655]). The transition temperatures are distinctly lower than in the CdI$_2$-type phases but likewise they decrease with increasing ionicity of the bonds [613a]: TaS$_2$ $T_0 = 75$ K [656], TaSe$_2$ $T_0 = 122$ K [655], NbSe$_2$ $T_0 = 33.5$ K [655, 938]. There are some discrepancies regarding the superlattice of NbSe$_2$.

TABLE 128

$6R$-TaS_2 structure, trigonal, C_{3v}^5—$R3m$ (No. 160), $Z = 1$ (3).
hexagonal axes: $(0, 0, 0; \frac{1}{3}, \frac{2}{3}, \frac{2}{3}; \frac{2}{3}, \frac{1}{3}, \frac{1}{3})+$
all atoms in 3(a): $0, 0, z$.

Compound	a(Å)	c(Å)	$c/6a$	z(M$_I$)	z(M$_{II}$)	z(X$_I$)	z(X$_{II}$)	z(X$_{III}$)	z(X$_{IV}$)	Ref.
TaS$_2^b$	3.34	35.94	1.79							[1]
	3.335	35.85	1.792	$\sim\frac{1}{12}$	$\sim\frac{11}{12}$	$\sim\frac{5}{24}$	$\sim\frac{3}{8}$	$\sim\frac{11}{24}$	$\sim\frac{5}{8}$	[30]
TaSe$_2$	3.46	37.9	1.83							[651]
	3.455	37.77	1.822							[629]
	3.455 8	37.826	1.824							[592]
^{a}				0.084	−0.083	0.207	0.373	0.461	0.627	
Ideal close packing			1.724 7	0.087 8	−0.078 9	0.215 0	0.372 8	0.469 4	0.627 2	
general formula				$z'_\Omega + z_\pi$	$-z_\Omega - z'_\Omega$	$3z'_\Omega + 2z_\pi$	$\frac{1}{3} + z'_\Omega$	$\frac{1}{3} + z'_\Omega + 2z_\pi$	$\frac{2}{3} - z'_\Omega$	

^{a} z-values calculated from those reported for $4H_b$-TaSe$_2$ [27]
$2z'_\Omega c$ = interlayer spacing = height of 'empty octahedral sandwich'
$2z_\Omega c$ = height of occupied octahedral sandwich
$2z_\pi c$ = height of trigonal-prism sandwich

ideal close packing: $c = 3(1+\sqrt{6})a$; $z'_\Omega = z_\Omega = \dfrac{1}{18+3\sqrt{6}} = 0.039\,45$; $z_\pi = \dfrac{1}{6+6\sqrt{6}} = 0.048\,32$

^{b} A 12-layer superstructure ($a = 3.335$ Å, $c = 71.24$ Å at room temperature) has been observed between the two intrapolytypic-transition temperatures 311 and 15 K [937].

While, according to Moncton et al. [655], the superlattice remains incommensurate down to 5 K, Marezio et al. [628] determined at 15 K a commensurate $2a_0$ superlattice (Table 120). A pressure of 1 kbar is sufficient to remove the distortions in NbSe$_2$ [65]. It is noteworthy that the $2H_a$-type modification exhibits a superconductive-transition temperature $T_c(0) = 7.35$ K while in the distorted modification $T_c(0) = 7.16$ K [657]. As the structural distortions occur within the layers, sharp discontinuities are observed in the resistivity within the layers but only gradual changes in the much higher resistivity perpendicular to the layers [656]. A consequence of the electronic rearrangement is the reversal of the Hall effect observed in NbSe$_2$ at 26 K [658].

The linear dependence of the transition temperatures on ionicity and axial ratio worked out by Thompson [613a] permits one to predict transition temperatures. One noteworthy prediction is that none of the polytypes of NbS$_2$ would be expected to undergo distortions [613a].

In the mixed octahedral/trigonal-prismatic phases two independent transitions occur: a high-temperature transition in octahedral layers and a low-temperature transition in trigonal-prismatic layers [613a, 648]:

$4H_b$—TaS$_2$: 320 K and 15 K
$4H_b$—TaSe$_2$: 410 K and (25 K?)
$6R$—TaS$_2$: 311 K and 15 K.

Similar to the case of graphite, intercalation compounds readily form above all with the metallic d^0 and d^1 TX$_2$ compounds. Hydrogen [648a], alkali and alkaline-earth metals as well as 3d transitions elements can be inserted between the XTX sandwiches. NH$_3$ and H$_2$O may additionally be intercalated. Furthermore, dozens of intercalation compounds with organic radicals have been synthesized. The large organic molecules greatly increase the distance between the cation layers. Surprisingly, the superconductive-transition temperature of TaSe$_2$, for example, is thereby shifted to considerably higher temperature [648b].

13. Compounds with Cation Coordination Number ⩾ 8

a. Layer-Type Rare-Earth Polytellurides

A mysterious group of compounds is formed by the layer-type rare-earth polytellurides. Two structures have been determined, NdTe$_3$ and Nd$_2$Te$_5$, but there appear to exist various combinations of these two prototypes. These are all based on the Cu$_2$Sb- or Fe$_2$As-type LnTe$_2$ structure. A projection of these structures is given in Figure 122. The tetragonal LnTe$_2$ structure is layered but the bonding in it is three-dimensional. The generating unit of the LnTe$_3$ structure possesses an

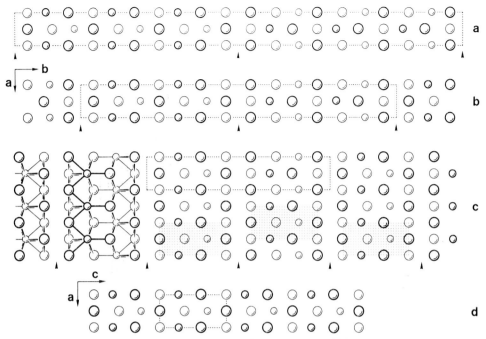

Fig. 122. Projections of the rare-earth polytelluride structures parallel to the layers.
(a): hypothetical Ln_3Te_7 structure ($=LnTe_3\cdot 2LnTe_2$)
(b): Nd_2Te_5 structure ($=NdTe_3\cdot NdTe_2$)
(c): $NdTe_3$ structure. Three generating units are dotted.
(d): $LnTe_2$ structure for comparison.
Large spheres: Te. The Van der Waals gaps are indicated by arrows.

additional Te layer, i.e. it contains both the top and the bottom Te layer of the $LnTe_2$ cell. In the unit cell of orthorhombic pseudotetragonal $LnTe_3$, two such units are stacked as indicated in Figure 122c by the dotted areas. The cations thereby remain nine-coordinated as in $LnTe_2$ but the boundary Te atoms loose two cation neighbors. The coordination in the $NdTe_3$ structure is in detail:

Nd——4 Te_{III} at 3.21 Å, flat square pyramid ⎱ 2 electrons involved
      1 Te_{III} at 3.25 Å, above the Te_{III} square ⎰ in bonding
      2 Te_I + 2 Te_{II} at 3.35 Å, square pyramid ⎱ 1 electron
               in opposite direction ⎰

Te_I (or Te_{II})——4 Te_{II} at 3.08 Å, square planar   ($1\frac{1}{2}$ electrons?)
              2 Nd at 3.35 Å $\frac{1}{2}$ electron
              2 Te_{III} at 4.09 Å
              2 Te_I + 2 Te_{II} at 4.24 Å, of adjacent sheet

Te_{III}——4 Nd at 3.21 Å, flat square pyramid ⎱ 2 electrons
       1 Nd at 3.25 Å, opposite to the Nd square ⎰
       4 Te_{III} at 3.86 Å

TABLE 129

$NdTe_3$ structure, orthorhombic, C_{2v}^{12}—Cmcm (No. 63), $Z = 4$.
$(0, 0, 0; \frac{1}{2}, \frac{1}{2}, 0) +$
all atoms in 4(c): $\pm(0, y, \frac{1}{4})$
$NdTe_3$: $y(Nd) = 0.830\,6$, $y(Te_I) = 0.070\,5$, $y(Te_{II}) = 0.429\,4$,
$y(Te_{III}) = 0.704\,7$ [659]

Compound	$a(\text{Å})$	$b(\text{Å})$	$c(\text{Å})$	Ref.
YTe_3	4.303	25.49	4.303	[660]
$LaTe_3$	4.422	26.09	4.422	[660]
	4.41	26.1	4.41	[670]
$CeTe_3$	4.398	25.99	4.398	[660]
	4.38	26.0	4.38	[661]
$PrTe_3$	4.376	25.89	4.376	[660]
	4.461	25.86	4.461	[662]
	4.374	25.57	4.374	[663]
$NdTe_3$	4.364	25.80	4.364	[660]
	4.35	25.80	4.35	[659]
$SmTe_3$	4.335	25.65	4.335	[660]
$GdTe_3$	4.326	25.58	4.326	[660]
$TbTe_3$	4.310	25.52	4.310	[660]
$DyTe_3$	4.296	25.45	4.296	[660]
$HoTe_3$	4.290	25.40	4.290	[660]
	4.273	24.87	4.273	[664]
$ErTe_3$	4.282	25.36	4.282	[660]
	4.31	25.45	4.31	[669]
$TmTe_3$	4.274	25.34	4.274	[660]
$LuTe_3$	4.277	25.137	4.278	[665]
$NpTe_3$	4.355	25.40	4.355	[666]
$PuTe_3$	4.343	25.55	4.343	[667]
$AmTe_3$	4.339	25.57	4.339	[668]
$CmTe_3$	4.34	25.7	4.34	[828]
$LaSb_2Sn(?)$	4.228	22.99	4.478	[2]

The structure of Ln_2Te_5 is a combination of $LnTe_2$ and $LnTe_3$, i.e. the additional Te layer follows only after two $LnTe_2$ units: $Ln_2Te_5 = LnTe_2 \cdot LnTe_3$. Geometrically many other combinations $mLnTe_2.nLnTe_3$ are possible, such as Ln_4Te_9 (3:1), Ln_3Te_7 (2:1), Ln_3Te_8 (1:2), Ln_4Te_{11} (1:3), etc. For compositions with $n = 1$, the unit cell consists of two $\{nLnTe_2 \cdot LnTe_3\}$ units shifted relative to each other by $a/2$. The pseudotetragonal axis thus becomes

$$b = 2n\, b_{LnTe_2} + b_{LnTe_3}$$

A similar formula holds for other compositions except for those with an even number of Te per formula, such as Ln_3Te_8 and Ln_5Te_{14}. In these cases the b-axis may be half as large (different polytypes, of course, are possible).

Knowing the atomic positions of the basic structures $LnTe_2$, $LnTe_3$ and Ln_2Te_5, the positions in polytypes of any $mLnTe_2 \cdot nLnTe_3$ phase can be calculated. Two examples, hypothetical Ln_3Te_7 and Ln_4Te_{11} structures, are offered in Table 131.

TABLE 130

Nd_2Te_5 structure, orthorhombic C_{2v}^{22}—Cmcm (No. 63), $Z=4$.

$(0, 0, 0; \frac{1}{2}, \frac{1}{2}, 0)+$

all atoms in 4(c): $\pm(0, y, \frac{1}{4})$

Nd_2Te_5: $y(Nd_I) = 0.901\ 0$, $y(Nd_{II}) = 0.691\ 0$, $y(Te_I) = 0.400$, $y(Te_{II}) = 0.460\ 0$
$y(Te_{III}) = 0.250\ 0$, $y(Te_{IV}) = 0.826\ 0$, $y(Te_V) = 0.618\ 0$ [660]

Compound	a(Å)	b(Å)	c(Å)	Ref.
Y_2Te_5				[660]
La_2Te_5	4.465	44.7	4.465	[660]
Ce_2Te_5	4.444	44.5	4.444	[660]
	4.42	44.35	4.42	[661]
Pr_2Te_5	4.426	44.3	4.426	[660]
	4.366		4.366	[663]
Nd_2Te_5	4.409	44.1	4.409	[660]
	4.380	44.0	4.380	[671]
Sm_2Te_5	4.362	43.8	4.362	[660]
Gd_2Te_5	4.336	43.6	4.336	[660]
Tb_2Te_5				[660]
Dy_2Te_5	4.299	43.3	4.299	[660]
Ho_2Te_5	4.411 (?)	45.03 (?)	4.411 (?)	[664]

TABLE 131

(a) Hypothetical Nd_3Te_7 structure, orthorhombic, C_{2v}^{12}—Cmcm (No. 63), $Z=4$.
All atoms in 4(c)

Nd_3Te_7: $a \approx 4.46$ Å, $b \approx 62.6$ Å, $c \approx 4.46$ Å

	y		y		y
Nd_I	0.636	Te_I	0.029	Te_V	0.584
Nd_{II}	0.783	Te_{II}	0.177	Te_{VI}	0.731
Nd_{III}	0.930	Te_{III}	0.323	Te_{VII}	0.878
		Te_{IV}	0.471		

calculated with the data of $NdTe_2$ and $NdTe_3$ [659]

(b) Hypothetical Nd_4Te_{11} structure, orthorhombic, C_{2v}^{12}—Cmcm (No. 63), $Z=4$
All atoms in 4(c).

Nd_4Te_{11}: $a \approx 4.38$ Å, $b \approx 95.65$ Å, $c \approx 4.38$ Å

	y		y		y
Nd_I	0.180 5	Te_{II}	0.124 1	Te_{VII}	0.579 6
Nd_{II}	0.277 2	Te_{III}	0.215 0	Te_{VIII}	0.653 3
Nd_{III}	0.545 7	Te_{IV}	0.310 8	Te_{IX}	0.750 0
Nd_{IV}	0.910 8	Te_V	0.384 1	Te_X	0.846 7
Te_I	0.010 8	Te_{VI}	0.481 0	Te_{XI}	0.944 8

calculated with the values of $NdTe_3$ [659] and Nd_2Te_5 [660].

With the cerium-group elements La through Sm, the following compositions have been detected: (Ln_4Te_9), Ln_3Se_7, Ln_3Te_7, Ln_4Te_{11} [672] and U_3Te_7 [683].

The $LnTe_3$ phases decompose peritectically at temperatures between 550 and 830°C [660]. $LuTe_3$ can only be prepared under pressure. The Ln_2Te_5 phases have a restricted stability range, being metastable at room temperature (La_2Te_5 exists between ~670 and ~810°C, Nd_2Te_5 between ~500 and ~900°C, and Ho_2Te_5 between ~700 and ~800°C [660]). The Ln_3Se_7, Ln_3Te_7, U_3Te_7 [683a] and Ln_4Te_{11} phases also form peritectically [672]. Pr_3Se_7 [662], Pr_3Te_7 and Pr_4Te_{11} [672] were reported to be tetragonal.

The $LnTe_3$ and Ln_2Te_5 phases crystallize as golden or red golden (pink) flakes. Pr_4Te_{11} too was said to be red golden while Pr_3Se_7 and Pr_3Te_7 were described as light grey-greenish and bright silvery, respectively [673]. The most mysterious property of all these phases, however, is the reported semiconductivity [662, 672, 673]. In the rare-earth compounds the cation is certainly trivalent so that in the two Te layers of the $LnTe_3$ unit, $\frac{3}{4}$ of the valence electrons of all Te_I and Te_{II} atoms should be localized in Te—Te bonds. In $LnTe_2$, on the other hand, $\frac{1}{2}$ of the valence electrons of the planar Te layers have to be fitted in Te—Te bonds. But there is no significant difference in the a-axis which defines the Te—Te distance of the various phases. The anisotropy of the electrical resistivity may be pronounced since the Te—Te distances across the layers are close to the Van der Waals distance of 4.40 Å. We did not find any detailed resistivity data. For Pr_3Se_7 a room-temperature resistivity of $2 \cdot 10^4$ Ω cm and a Seebeck coefficient of 200–250 μV deg^{-1} (p-type) was reported while for $PrTe_3$ the corresponding values were given as 10^{-4} Ω cm and 20 μV deg^{-1} (p-type) [672]. Bucher et al. [673] however, stated that the resistivity of their $PrTe_3$ sample was relatively high and they declared $PrTe_3$, Pr_2Te_5, Pr_4Te_{11} and $TmTe_3$ explicitly as semiconductors [662, 673]. In the system Sm—Te Yarembash et al. [690] observed semiconductive properties only on SmTe, Sm_3Te_4 and Sm_2Te_3 and not on Sm_3Te_7, Sm_2Te_5 and $SmTe_3$.

Another perplexing question is the occurrence of Np^{4+} in $NpTe_3$ as deduced from the lattice constants [666]. If $LaTe_3$ really is intrinsically nonmetallic then $NpTe_3$ with tetravalent Np is expected to be metallic. But why then does it not like UTe_3 and $NpSe_3$ adopt the $ZrSe_3$ structure with which it would have a chance to be semiconducting? Would it be metallic anyway?

b. $PdBi_2(h)$

While Pd is divalent and tetravalent in PdP_2 and $PdAs_2$, respectively, the formal valence in metallic $PdBi_2$ appears to be even higher. $PdBi_2$ crystallizes in two modifications both being superconductors. The room-temperature modification is monoclinic. It represents some kind of puckered layer structure derived from the CoGe structure by removal of all the octahedrally coordinated metal atoms. The puckered BiPdBi sheets are linked through Bi—Bi pairs at half-bond distance.

TABLE 132
$PdBi_2(h)$ structure, tetragonal, D_{4h}^{17}—I4/mmm (No. 139), $Z = 2$.
$(0, 0, 0; \frac{1}{2}, \frac{1}{2}, \frac{1}{2})$ + Bi in 4(e): $\pm(0, 0, z)$
Pd in 2(a): 0, 0, 0.

$PdBi_2(h)$: $a = 3.362$ Å, $c = 12.983$ Å, $z = 0.363$ [674]
LaI_2; $a = 3.922$ Å, $c = 13.97$ Å, $z = 0.365$ [1018]
CeI_2: $a = 3.888$ Å, $c = 13.95$ Å [1018]
$PrI_2(t)$: $a = 3.864$ Å, $c = 13.94$ Å [1018]
$NdI_2(p)$: $a \approx 3.83$ Å, $c \approx 13.90$ Å [983]

The high-temperature modification represents a strongly elongated $MoSi_2$ type of structure with a pronounced layer character (for comparison: $MoSi_2$ $c/a = 2.46$, $AlCr_2$ $c/a = 2.88$, $PdBi_2(h)$ $c/a = 3.86$). Unlike in $MoSi_2$, the anions in $PdBi_2$ no longer form pairs across the layers (along the c-axis), this distance being 3.56 Å, which is even larger than the Bi—Bi distance $a = 3.36$ Å within the layers. The cations are surrounded by eight anions at the corners of an elongated cube as is illustrated in Figure 123. With a high electronegativity difference between anion and cation and no anion-anion contacts, nonmetallic properties would be imaginable with diamagnetic d^4 cations. However, the metallic bonds within the Bi layers spoil the nonmetallic picture and the dt_{2g} band will be partly occupied.

c. THE $ZrSe_3$ STRUCTURE

In the monoclinic $ZrSe_3$ structure, infinite trigonal anion prisms are mutually connected to form layers parallel to the (001) plane. These layers are closely related to those met in the $PuBr_3$ structure, the most striking difference being the

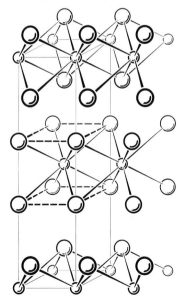

Fig. 123. The elongated tetragonal $MoSi_2$-type structure of $PdBi_2(h)$
Small spheres: Pd.

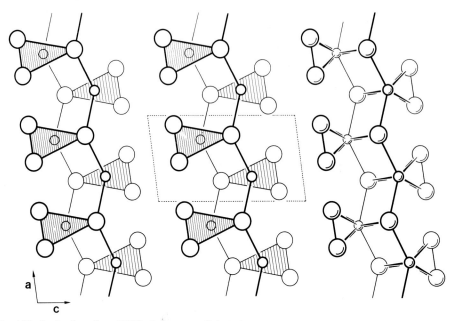

Fig. 124. Projection along [010] of the monoclinic ZrSe$_3$ structure. The squeezed trigonal prisms are emphasized in the left part. Small spheres: cations.

slender form of the [ZrSe$_{6/2}$] prisms (Figure 124). Among the representatives of the ZrSe$_3$ structure listed in Table 133 only tetravalent cations are found and at least the sulfides and selenides all are non-metallic [676a]. As a consequence they must be polyanionic compounds since two anion valence electrons per formula have to be saturated by anion-anion bonds. This anion pairing is responsible for the slender form of the trigonal prisms. The short Se—Se distance in ZrSe$_3$ exactly corresponds to a single bond. The coordination is characterized as follows:

$$\begin{array}{l} \text{Zr} \text{---} 2 \text{ Se}_\text{I} \text{ at } 2.72 \text{ Å} \\ \phantom{\text{Zr} \text{---}} 2 \text{ Se}_\text{II} \text{ at } 2.73 \text{ Å} \\ \phantom{\text{Zr} \text{---}} 2 \text{ Se}_\text{III} \text{ at } 2.74 \text{ Å} \end{array} \right\} \text{ forming the corners of the trigonal prism}$$

$$\begin{array}{l} 1 \text{ Se}_\text{I}' \text{ at } 2.86 \text{ Å} \\ 1 \text{ Se}_\text{I} \text{ at } 2.89 \text{ Å} \end{array} \right\} \text{ in the equatorial plane}$$

$$\text{Se}_\text{II}\text{---}\text{Se}_\text{III} = 2.35 \text{ Å}$$
$$\text{Se}_\text{II}\text{---}\text{Se}_\text{III}' = 3.06 \text{ Å}$$

The cation—anion bonds within the equatorial plane are distinctly weaker than those within the prism and this may account for the needle-like habit of the crystals that can be grown by halogen-vapor transport reactions.

Phase transitions have been detected in USe$_3$ at 1120 °C and in UTe$_3$ at 935 °C [683].

TABLE 133
$ZrSe_3$ structure, monoclinic, C_{2h}^2—$P\,2_1/m$ (No. 11), $Z = 2$.
All atoms in 2(e): $\pm(x, \frac{1}{4}, z)$.

Compound	a(Å)	b(Å)	c(Å)	β(°)	Ref.
TiS_3	4.973	3.433	8.714	97.74	[676]
	4.958	3.400 6	8.778	97.32	[677]
ZrS_3	5.123	3.627	8.986	97.15	[676]
	5.124 3	3.624 4	8.980	97.28	[677]
$Zr(S, Se)_3$					[677]
$ZrSe_3$	5.410 9	3.748 8	9.444	97.48	[677]
$ZrTe_3$	5.893 9	3.925 9	10.100	97.82	[677]
HfS_3	5.100	3.594	8.992	98.16	[676]
	5.092 3	3.595 2	8.967	97.38	[677]
$Hf(S, Se)_3$					[677]
$HfSe_3$	5.388	3.721 6	9.428	97.78	[677]
$HfTe_3$	5.879	3.902 2	10.056	97.98	[677]
$ThTe_3$	6.14	4.31	10.44	98.4	[678]
US_3	5.45	3.92	9.11(18.22?)	99.5	[679]
	5.40	3.90	9.13(18.26?)	99.5	[680]
	5.37	3.96	9.06	97.20	[682]
USe_3	5.68	4.06	9.63(19.26?)	99.60	[681]
	5.65	4.06	9.55	97.5	(682)
	5.68	4.06	9.60(19.20?)	99.33	[683]
UTe_3	6.090	4.226	10.302	98.0	[684]
NpS_3	5.36	3.87	9.05(18.10?)	99.5	[685]
$NpSe_3$	5.64	4.01	9.53(19.06?)	100.40	[686]

Compound	x(M)	z(M)	x(X_I)	z(X_I)	x(X_{II})	z(X_{II})	x(X_{III})	z(X_{III})	Ref.
Type A									
ZrS_3	0.716 3	0.344 7	0.236 9	0.445 7	0.527 5	0.828 4	0.120 1	0.830 1	[943]
$ZrSe_3$	0.715	0.343	0.236	0.447	0.545	0.825	0.112	0.831	[675]
HfS_3	0.716 1	0.345 2	0.238 9	0.445 4	0.535 8	0.829 8	0.123 2	0.830 3	[943]
Type B									
TiS_3	0.284 8	0.347 2	0.760 8	0.449 5	0.468 0	0.823 8	0.879 5	0.826 3	[943]
$HfSe_3$	0.285 5	0.343 7	0.763 0	0.446 7	0.454 5	0.826 7	0.888 7	0.832 7	[943]
$ZrTe_3$	0.293 1	0.334 0	0.760 9	0.443 9	0.433 9	0.832 4	0.903 7	0.838 4	[943]

$x_B = -x_A$, $y_B = y_A$, $z_B = z_A$.

A recent study [943] has revealed the existence of a variant of the $ZrSe_3$ type. The two types differ only in the values of the x-parameters: $x_B = -x_A$. The two arrangements thus are a kind of mirror image of each other. The deviation of β from 90° prevents a perfect right-/left-hand identity between variant A and B. A detailed discussion of the electronic structure of ZrS_3 and $ZrSe_3$ based on X-ray photoemission spectra was presented by Jellinek et al. [988].

d. THE HfTe₅ STRUCTURE

In the orthorhombic HfTe$_5$ structure [687] the cation has virtually the same coordination as in the ZrSe$_3$ structure. The trigonal prisms are linked via tellurium zigzag chains (Te$_{III}$) instead of being connected directly as in the ZrSe$_3$ structure. The coordination is as follows:

$$\begin{array}{ll} \text{Hf} \quad 2\,\text{Te}_I \text{ at } 2.95\,\text{Å} & \text{Te}_{II}-\text{Te}_{II} = 2.76\,\text{Å} \\ \phantom{\text{Hf}\quad} 4\,\text{Te}_{II} \text{ at } 2.94\,\text{Å} & \text{Te}_{III}-\text{Te}_{III} = 2.91\,\text{Å} \\ \phantom{\text{Hf}\quad} 2\,\text{Te}_{III} \text{ at } 2.96\,\text{Å} & \text{Te}_{III}-\text{Te}_{III}-\text{Te}_{III}\ 86.2° \end{array}$$

The Hf atoms thus form eight bonds of equal strength. Te$_I$ has only 2 Hf neighbors (Hf—Te$_I$—Hf 84.6°), therefore it could form two single bonds. The Te$_{II}$—Te$_{II}$ distance nicely corresponds to a single bond so that only one electron is left for binding the 2 Hf neighbors. The valence electrons available from Te$_I$ and Te$_{II}$ thus would be sufficient to bind the four valence electrons of the cation. However, the Te$_{III}$—Te$_{III}$ distance within the chains roughly corresponds to a half bond so that almost one electron per Te$_{III}$ is free for cation—anion bonding. As a consequence, in the actual structure there is no balance between the four cation valence electrons and the anion valence electrons available for cation—cation bonding. If on the other hand, the Te$_{III}$ chains were independent of the [HfTe$_{6/2}$] prisms, saturation of the bonds would well be possible and these compounds then

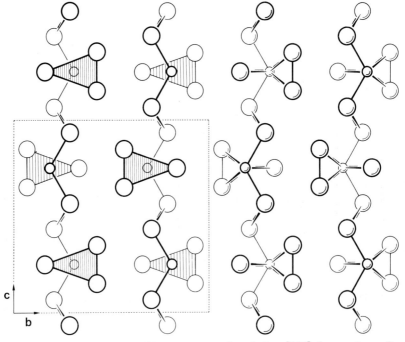

Fig. 125. The orthorhombic HfTe$_5$ structure projected along [100]. Large spheres: Te.

TABLE 134

$HfTe_5$ structure, orthorhombic, D_{2h}^{17}—Cmcm (No. 63), $Z=4$.

$(0, 0, 0; \frac{1}{2}, \frac{1}{2}, 0) +$

Te$_{II}$ and Te$_{III}$ in 8(f): $\pm(0, y, z; 0, y, \frac{1}{2}-z)$
Hf and Te$_I$ in 4(c): $\pm(0, y, \frac{1}{4})$.

ZrTe$_5$: $a = 3.9876$ Å, $b = 14.502$ Å, $c = 13.727$ Å [687]
HfTe$_5$: $a = 3.9743$ Å, $b = 14.492$ Å, $c = 13.730$ Å [687]

	y	z	
Hf	0.3143		
Te$_I$	0.6635		[687]
Te$_{II}$	0.9299	0.1494	
Te$_{III}$	0.2099	0.4353	

ZrSe$_3$Te$_2$?
HfSe$_3$Te$_2$?
UTe$_5$?

---

would be nonmetallic. Both ZrTe$_5$ and HfTe$_5$ are diamagnetic with temperature-dependent susceptibilities [687]. No electrical-resistivity data are available but diffuse-reflectance spectra seem to point to metallic properties. Single crystals of these peritectic phases, stable below 500°C, can be grown by the iodine transport technique. For a comparison of the bonding it would be interesting to know the interatomic distances in ZrTe$_5$ and HfTe$_5$. We wonder whether UTe$_5$ [683, 683a] which decomposes peritectically at 490°C also adopts the HfTe$_5$ structure.

e. TaSe$_3$

At the time when only the ZrSe$_3$ structure was known in detail, it was assumed that NbS$_3$, TaS$_3$, NbSe$_3$ and TaSe$_3$ all belonged to the same type. Since Nb and Ta possess one more valence electron their trichalcogenides then would necessarily be metallic. In order to restore the nonmetallic character by slight modifications of the ZrSe$_3$ structure half the anion pairs have to be loosened since the formation of cation pairs is geometrically impossible. This is roughly what happens in the TaSe$_3$ structure if one disregards the fact that the now undistorted prisms are shifted against each other so that the manner of connecting the prisms is different. Nevertheless, TaSe$_3$ shows a metallic behavior [688]. If we check the distances we detect indeed that our expectation outlined above is not fully substantiated:

$$\begin{array}{ll}
\text{Ta}_I \text{---} 2\,\text{Se}_I \text{ at } 2.65 \text{ Å} & \text{Ta}_{II} \text{---} 2\,\text{Se}_{IV} \text{ at } 2.64 \text{ Å} \\
\phantom{\text{Ta}_I \text{---}} 2\,\text{Se}_{II} \text{ at } 2.64 \text{ Å} & \phantom{\text{Ta}_{II} \text{---}} 2\,\text{Se}_V \text{ at } 2.64 \text{ Å} \\
\phantom{\text{Ta}_I \text{---}} 2\,\text{Se}_{III} \text{ at } 2.64 \text{ Å} & \phantom{\text{Ta}_{II} \text{---}} 2\,\text{Se}_{VI} \text{ at } 2.60 \text{ Å} \\
\hline
\phantom{\text{Ta}_I \text{---}} 1\,\text{Se}_{II}' \text{ at } 2.80 \text{ Å} & \phantom{\text{Ta}_{II} \text{---}} 1\,\text{Se}_I \text{ at } 2.72 \text{ Å} \\
\phantom{\text{Ta}_I \text{---}} 1\,\text{Se}_{IV} \text{ at } 2.82 \text{ Å} & \phantom{\text{Ta}_{II} \text{---}} 1\,\text{Se}_{IV} \text{ at } 2.74 \text{ Å} \\
\text{Se}_{II}\text{---}\text{Se}_{III} = 2.90 \text{ Å} & \text{Se}_V\text{---}\text{Se}_{VI} = 2.58 \text{ Å} \\
\multicolumn{2}{c}{\text{Se}_{III}\text{---}\text{Se}_V = 2.65 \text{ Å}}
\end{array}$$

TABLE 135
TaSe$_3$ structure, monoclinic, C_{2h}^2—P$2_1/m$ (No. 11), $Z=4$.
All atoms in 2(e): $\pm(x, \frac{1}{4}, z)$.

TaSe$_3$: $a = 10.402$ Å, $b = 3.495$ Å, $c = 9.829$ Å, $\beta = 106.26°$

	x	y	
Ta$_I$	0.193 7	0.625 0	
Ta$_{II}$	0.672 3	0.951 8	
Se$_I$	0.869 1	0.201 1	
Se$_{II}$	0.915 4	0.576 0	[688]
Se$_{III}$	0.633 8	0.422 2	
Se$_{IV}$	0.400 6	0.882 0	
Se$_V$	0.383 2	0.258 2	
Se$_{VI}$	0.151 2	0.075 0	

related (?):
TaS$_3$: $a = 36.791$ Å, $b = 15.177$ Å, $c = 3.340$ Å, $Z = 24$,
probable space group C222$_1$ [688]

The normal prisms are built up around Ta$_I$ by Se$_I$, Se$_{II}$ and Se$_{III}$ and the slender ZrSe$_3$-like prisms around Ta$_{II}$ by Se$_{IV}$, Se$_V$ and Se$_{VI}$. However, not only is the Se$_{II}$—Se$_{III}$ bond removed in the first kind of prism but in addition the Se$_V$—Se$_{VI}$ bond is also weakened to less than a half bond. Moreover, one Se of the adjacent normal prism is almost as close. This bond is indicated in Figure 126 by a broken line. Thus, a saturation of the bonds can no longer be achieved. It may be due to the low electronegativity difference that it is energetically favorable for the compound to be metallic though the structure reveals a clear tendency to the geometrically possible saturation of the bonds. The sulfide TaS$_3$, on the other

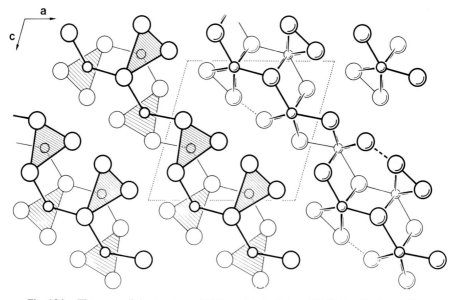

Fig. 126. The monoclinic structure of TaSe$_3$ projected along [010]. Small spheres: Ta.

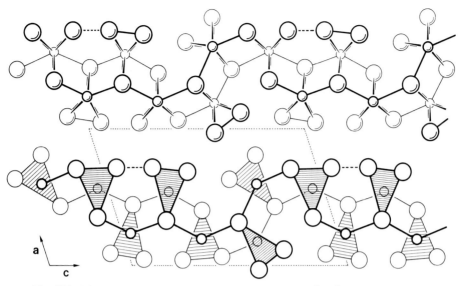

Fig. 127. The monoclinic structure of NbSe$_3$ projected along [010]. Large spheres: Se.

hand, is non-metallic [688]. Its orthorhombic cell is six times as large, but the atomic positions have not yet been determined.

f. NbSe$_3$

The monoclinic cell of NbSe$_3$ [689] is $1\frac{1}{2}$ times as large as that of TaSe$_3$ but it bears no direct relation to the latter. The NbSe$_3$ structure is reminiscent of both the ZrSe$_3$ structure and the TaSe$_3$ structure as is verified by Figure 127. Four and

TABLE 136

NbSe$_3$ structure, monoclinic, C_s^1—Pm (No. 6), $Z = 6$.
Nb$_{IV}$ to Nb$_{VI}$ and Se$_X$ to Se$_{XVIII}$ in 1(b): $x, \frac{1}{2}, z$
Nb$_I$ to Nb$_{III}$ and Se$_I$ to Se$_{IX}$ in 1(a): $x, 0, z$.
NbSe$_3$: $a = 10.006$ Å, $b = 3.478$ Å, $c = 15.626$ Å, $\beta = 109.50°$ [689]

Atoms in position 1(a)	x	z	Atoms in position 1(b)	x	z
Nb$_I$	0.5378	0.2776	Nb$_{IV}$	0.2315	0.0400
Nb$_{II}$	0.1809	0.6199	Nb$_V$	0.1910	0.3610
Nb$_{III}$	0.5595	0.9494	Nb$_{VI}$	0.5915	0.6977
Se$_I$	0.4399	0.0934	Se$_X$	0.3240	0.2262
Se$_{II}$	0.0413	0.0632	Se$_{XI}$	0.7119	0.2404
Se$_{III}$	0.0807	0.9177	Se$_{XII}$	0.6852	0.3959
Se$_{IV}$	0.4018	0.4067	Se$_{XIII}$	0.3140	0.5605
Se$_V$	0.0579	0.4293	Se$_{XIV}$	0.1133	0.7272
Se$_{VI}$	0.0354	0.2414	Se$_{XV}$	0.9697	0.5845
Se$_{VII}$	0.4275	0.7504	Se$_{XVI}$	0.3462	0.9093
Se$_{VIII}$	0.6253	0.5766	Se$_{XVII}$	0.7000	0.8776
Se$_{IX}$	0.8038	0.7354	Se$_{XVIII}$	0.7087	0.0684

two adjacent prisms are linked as in ZrSe$_3$ but the relative orientation of these two units induces a lower symmetry than in TaSe$_3$. Moreover, of the six [NbSe$_6$] prisms per unit cell only two (those around Nb$_{III}$ and Nb$_V$) show no Se—Se bonding. The shortest anion—anion contact occurs in the prism around Nb$_{II}$, Se$_{XIV}$—Se$_{XV}$ = 2.22 Å; but the distance is as large as 2.53 Å between Se$_{XI}$ and Se$_{XII}$ in the prism around Nb$_I$. Similar to the case of TaSe$_3$, the distance from Se$_{XI}$ to Se$_{XVIII}$ of the adjacent prism is nearly as short. The coordination around each Nb atom is in detail:

Nb$_I$—	2 Se$_X$ at 2.66 Å	Nb$_{II}$—	2 Se$_{XIII}$ at 2.55 Å
	2 Se$_{XI}$ at 2.66 Å		2 Se$_{XIV}$ at 2.66 Å
	2 Se$_{XII}$ at 2.61 Å		2 Se$_{XV}$ at 2.65 Å
	1 Se$_I$ at 2.71 Å		1 Se$_V$ at 2.82 Å
	1 Se$_{IV}$ at 2.78 Å		1 Se$_{VII}$ at 2.62 Å
	Se$_{XI}$—Se$_{XII}$ = 2.53 Å		
	Se$_{XI}$—Se$_{XVIII}$ = 2.68 Å		Se$_{XIV}$—Se$_{XV}$ = 2.22 Å(?)
Nb$_{III}$—	2 Se$_{XVI}$ at 2.66 Å	Nb$_{IV}$—	2 Se$_I$ at 2.63 Å
	2 Se$_{XVII}$ at 2.70 Å		2 Se$_{II}$ at 2.69 Å
	2 Se$_{XVIII}$ at 2.62 Å		2 Se$_{III}$ at 2.65 Å
	1 Se$_I$ at 2.88 Å		1 Se$_X$ at 2.74 Å
	1 Se$_{VII}$ at 2.94 Å		1 Se$_{XVI}$ at 2.66 Å
	Se$_{XVII}$—Se$_{XVIII}$ = 2.95 Å		Se$_{II}$—Se$_{III}$ = 2.43 Å
			Se$_{II}$—Se$_{VI}$ = 2.80 Å
Nb$_V$—	2 Se$_{IV}$ at 2.64 Å	Nb$_{VI}$—	2 Se$_{VII}$ at 2.70 Å
	2 Se$_V$ at 2.62 Å		2 Se$_{VIII}$ at 2.67 Å
	2 Se$_{VI}$ at 2.65 Å		2 Se$_{IX}$ at 2.65 Å
	1 Se$_X$ at 2.84 Å		1 Se$_{XIII}$ at 2.89 Å
	1 Se$_{XIII}$ at 2.94 Å		1 Se$_{XVII}$ at 2.65 Å
	Se$_V$—Se$_{VI}$ = 2.87 Å		Se$_{XVII}$—Se$_{XVIII}$ = 2.95 Å

The distance between Se$_{XIV}$ and Se$_{XV}$ is shorter than expected for a single bond and therefore may be inaccurate. The above list reveals that the bonding between the trigonal prisms varies considerably. The distance is largest between prism II and prism V. This appears to be in agreement with the fibrous aspect of the NbSe$_3$ crystals. Charge-density-wave driven phase transitions at 145 K and 59 K have recently been detected in this metallic phase [1051].

14. Mixed-Anion Compounds MXY

Due to the different electronegativity of the two kinds of anions, these phases adopt in most cases structures which are not simply ordered superstructures of binary phases MX$_2$, though this indeed does occur as can be checked by examining Tables 137 and 138. Among the rare-earth and transition-element

TABLE 137

The structure types occurring with rare-earth and transition-element mixed-anion compounds MXY. Potential layer types are printed in italics without differentiating between 3-dimensional and 2-dimensional PbFCl-type representatives.

	MOF	MOCl	MOBr	MOI	MSF	MSCl	MSBr	MSI	MSeF	MTeF
Sc	monocl. ZrO$_2$ CaF$_2$	FeOCl?	FeOCl?	FeOCl?						
Y	PbFCl YOF (<560°) CaF$_2$	PbFCl	PbFCl	PbFCl	α: *PbFCl* β: hex.YSF	LuSBr?	β: LuSBr α:	hex.GdSI	YSeF polytypes	
La	*PbFCl* YOF CaF$_2$	*PbFCl*	*PbFCl*	*PbFCl*	*PbFCl*	CeSI	CeSI	CeSI	hex.α-LaSeF β: *PbFCl* (>750°C)	*PbFCl*
Ce	*PbFCl* YOF CaF$_2$	*PbFCl*	*PbFCl*	*PbFCl*	*PbFCl*	CeSI	CeSI	α: CeSI (SrI$_2$) β: *SmSI*	*PbFCl*	*PbFCl*
Pr	*PbFCl* YOF CaF$_2$	*PbFCl*	*PbFCl*	*PbFCl*	*PbFCl*	CeSI	α: CeSI β: NdSBr γ:	*SmSI*	*PbFCl*	*PbFCl*
Nd	*PbFCl* YOF CaF$_2$	*PbFCl*	*PbFCl*	*PbFCl*	*PbFCl*		NdSBr	*SmSI*	*PbFCl*	*PbFCl*
Sm	YOF CaF$_2$	*PbFCl*	*PbFCl*	*PbFCl*	*PbFCl*		NdSBr	*SmSI*	*PbFCl*	
Eu	YOF CaF$_2$	*PbFCl*	*PbFCl*	*PbFCl*	*PbFCl*		NdSBr		*PbFCl*	
Gd	YOF CaF$_2$	*PbFCl*	*PbFCl*	*PbFCl*	*PbFCl*		NdSBr	hex.GdSI	*PbFCl*	
Tb	YOF CaF$_2$	*PbFCl*	*PbFCl*	*PbFCl*	*PbFCl*		NdSBr	hex.GdSI		
Dy	YOF CaF$_2$	*PbFCl*	*PbFCl*	*PbFCl*	*PbFCl*	FeOCl	FeOCl	hex.GdSI	YSeF polytypes	

Table 137 (Continued).

	MOF	MOCl	MOBr	MOI	MSF	MSCl	MSBr	MSI	MSeF	MTeF
Ho	YOF, CaF$_2$	PbFCl	PbFCl	PbFCl	α: PbFCl, β:hex.YSF	FeOCl	FeOCl	hex.GdSI	YSeF polytypes	
Er	YOF, CaF$_2$	PbFCl, SmSI	PbFCl	PbFCl	α: PbFCl, β: hex.YSF	FeOCl	FeOCl	hex.GdSI	YSeF polytypes	
Tm	YOF, CaF$_2$	SmSI	PbFCl	PbFCl	hex.YSF	FeOCl	FeOCl	hex.GdSI	YSeF polytypes	
Yb	YOF, CaF$_2$	SmSI	PbFCl	PbFCl		FeOCl	FeOCl	hex.GdSI	YSeF polytypes	
Lu	YOF, CaF$_2$	SmSI	PbFCl		hex.YSF	FeOCl	FeOCl	hex.GdSI	YSeF polytypes	
Ac	CaF$_2$	PbFCl	PbFCl							
Th	CaF$_2$	PbFCl								
U	tetrag.									
Np										
Pu	PbFCl, CaF$_2$	PbFCl	PbFCl	PbFCl						
Am		PbFCl	PbFCl	PbFCl						
Cm		PbFCl	PbFCl	PbFCl						
Bk		PbFCl	PbFCl	PbFCl						
Cf	CaF$_2$	PbFCl								
Es		PbFCl								
Ti	rutile	FeOCl	FeOCl							
V	rutile	FeOCl	FeOCl							
Nb										
Ta										
Cr	rutile?	FeOCl	FeOCl				FeOCl? cubic	hex.CrSI		
Mo						MoSBr	MoSBr	MoSBr		
W										
Mn	rutile?									
Fe	rutile	FeOCl								

TABLE 138

The structure types of mixed-anion compounds MXY with B-element cations. Italics are used for layer structures and Roman letters for three-dimensional structures; dashed underlining indicates an intermediate case.

Compound	M = B	Al	Ga	In	Tl	Sb	Bi
MOF			GaOF	ortho.CaF$_2$-deriv. (LaOF)	cubic, ~CaF$_2$	*t*: orth.chain *r*: orth. ~layer *h*: cubic	*PbFCl*
MOOH		α: diaspore β: γ: boehmite	diaspore boehmite orthorh.	orthorh. InOOH (def. rutile)			
MOCl		*AlOCl*	*AlOCl*	*FeOCl*	*PbFCl?*	monocl.SbOCl	*PbFCl*
MOBr		*AlOCl*	*AlOCl*	*FeOCl*			*PbFCl*
MOI		*AlOCl*	GaOI (needle)	*FeOCl*			*PbFCl*
MSF			GaSF (3-dim.)	InSF			
MSCl	B$_3$S$_3$Cl$_3$	orth. *AlSCl*	*AlSCl* (?)	*CdCl$_2$*	*PbFCl*	SbSI	SbSI
MSBr		*AlSCl*	*AlSCl* (?)	*CdCl$_2$*	*PbFCl*	SbSI	SbSI
MSI	BSI	*AlSCl*	*AlSCl* (?)	*InSeI*	*PbFCl*	SbSI	SbSI
MSeF				*InSeF*			
MSeCl	BSeCl(?)	*AlSeCl*	*AlSCl*(?)	*CdCl$_2$*		SbSI	orthorh.BiSeCl
MSeBr	BSeBr	*AlSeCl*	*AlSCl*(?)	*CdCl$_2$*		SbSI	SbSI
MSeI	BSeI	*AlSeCl*	*AlSCl*(?)	tetragonal InSeI		SbSI	SbSI
MTeF							
MTeCl		AlTeCl (needles)	GaTeCl	mon.*InTeCl*			BiTeCl
MTeBr		AlTeBr (needles)	GaTeBr	mon.InTeBr			CdI$_2$
MTeI		AlTeI (needles)	AlSCl?	*InTeBr*		orth.(PbCl$_2$?) SbSI	CdI$_2$

compounds there are three main layer types, the PbFCl type, the SmSI type and the FeOCl type. The YOF structure observed in the low-temperature modifications of the LnOF compounds is a rhombohedrally distorted ordered derivative of the fluorite CaF_2 structure with distinct layering - - -FYO OYF FYO- - -. In the PbFCl structure the layer character strongly depends upon the arrangement of the Cl double layers. The two Cl layers may coalesce to a puckered single layer, and this is what happens in the PbFCl-related structures, in the orthorhombic CeSI type [714] (or SrI_2 type [738]) and the monoclinic NdSBr type [715]. All these structures are built up by stacking sheets of the type $(PbF)_n$ and double sheets of n Cl atoms. While the $(PbF)_n$ sheets are analogous to the tetrahedron layers of the PbO structure, the corresponding $(SmS)_n$ sheets in the rhombohedral SmSI structure are distorted versions of the double-tetrahedron layer illustrated in Figure 24a.

a. THE PbFCl (MATLOCKITE) STRUCTURE

In the PbFCl structure the smallest and more electronegative anions form a planar square layer framed by the cations which themselves span a planar layer of tetrahedra around the anions. The PbO-type $(PbF)^+$ layer is neutralized by a layer of the second anion. Thus, sheets $Cl_n Pb_n F_{2n} Pb_n Cl_n$ are superposed in such a way that each Cl is opposed to a cation of the adjacent sheet. Each cation thus acquires a neighborhood Pb—4F + 4Cl + 1Cl.

The four F atoms span a square pyramid with the base length $a/\sqrt{2}$, while the 4 Cl define a different square pyramid in the opposite direction with base length a. The fifth Cl is located above this Cl square in the neighboring layer. It is this M—X distance which determines the layer character of these compounds. Flahaut [691] has proposed the axial ratio together with the metal position parameter as characteristics to separate the PbFCl-type compounds from the isopuntal anti-Fe_2As-type or ZrSiS-type phases (which, together with BaMgSi, BaMgGe, are omitted in Table 139):

for PbFCl type: $c/a < 2$–2.3, $z_M < 0.23$
for ZrSiS type: $c/a > 2$, $z_M > 0.25$

In the true PbFCl-type compounds, the $[M_{4/4}F]_n$ or $[M_{4/4}O]_n$ sheets are clearly distinguished from the double layers of the second anion whereas in the ZrSiS-type phases the metal atoms form part of the double layer of the larger anions. The structure of most of the metallic ZrSiS-type phases thus is only layered and the bonding is three-dimensional. Crystals of the ZrSiS, NbSiAs and UPS families grow as square platelets but without the micalike (001) cleavage as observed on BiOCl etc. In Table 139 we have added the distances M—4X and M—1X for comparison. Although in many cases the accuracy of the atomic parameters is rather low, we see that in surprisingly many compounds the interlayer M—X distance is even shorter than the four interlayer M—X distances. Only with small

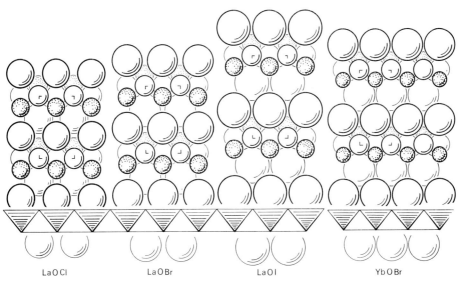

Fig. 128. The tetragonal PbFCl structure: Transition from a three-dimensional lattice to a layer structure due to a variation of the anion Cl → Br → I and of the cation La → Yb, respectively. Largest spheres: Halogen atoms. The cations are stippled. The tetrahedral $[OLn_{4/4}]_n$ skeleton, common to the LaOX compounds, is emphasized below. For regular tetrahedra and spherical ions $a = \sqrt{8/3}(r_{M^{3+}}+r_{O^{2-}})$, $z_M = a/2\sqrt{2}c$ and $a \geq 2r_{X_{VdW}}$.

cations and large anions X do true layer structures result as for example in CaHI, BiOBr, BiOI, LnOI, UNBr and UNI, as well as in ZrSiTe and HfSiTe.

If we compare a series of compounds with the same tetrahedron layer, such as ThNCl, ThNBr and ThNI, we notice that the lattice constant a remains practically constant, while the value of c increases and the layer character becomes more pronounced. It follows from Table 139 that the layer character becomes more pronounced in rare-earth series, LnOCl or LnOBr, on going from La to a heavier rare-earth atom. For MFX phases Beck [983] derived from electrostatic calculations that with smaller cations M as well as with larger anions X c/a will increase. The limiting phase with $r(X) \approx r(F)$ has $c/a \approx 1.5$ which roughly corresponds to the $FeSi_2$ type. Symmetry reduction transforms the $FeSi_2$ type to the PbFCl type with $c/a = \sqrt{2}$.

It is remarkable that in the LnOCl series the PbFCl structure becomes unstable at ErOCl with respect to the SmSI structure. The O—O distances are equal to $a/\sqrt{2}$ and Cl—Cl $= a$ within the same layer, but these contacts are not decisive since the LnOBr series exists down to LuOBr with distinctly smaller a values.

In view of the structural analogy between fluorides and hydrides we have tentatively included in Table 139 three phases LnHTe which were reported as tetragonal only. These compounds may correspond to BaHI. The existence of isostructural phases LnHSe similar to BaHBr may be possible as well. One might even speculate about the existence of analogous $Th^{4+}HN$ and UHN. With the

TABLE 139

PbFCl structure, tetragonal, D_{4h}^7—P4/nmm (No. 129), $Z = 2$.
Pb and Cl in 2(c): $0, \frac{1}{2}, z; \frac{1}{2}, 0, \bar{z}$
F in 2(a): $0, 0, 0; \frac{1}{2}, \frac{1}{2}, 0$.

Compound	a(Å)	c(Å)	c/a	z_M	z_X	M−4X (Å)	M−1X (Å)	Ref.
CaHCl	3.851	6.861	1.782	0.146	0.695	2.93	3.77	[1, 4]
SrHCl	4.100	6.961	1.698	0.199	0.66	3.06	3.21	[1, 4]
BaHCl	4.408	7.202	1.634	0.215	0.65	3.27	3.13	[1, 4]
EuHCl	4.074	6.896	1.693					[692]
CaHBr	3.858	7.911	2.051	0.140	0.67	3.11	4.19	[4]
SrHBr	4.254	7.290	1.714	0.155	0.68	3.24	3.83	[4]
BaHBr	4.564	7.418	1.625	0.175	0.67	3.43	3.67	[4]
EuHBr								
CaHI	4.071	8.941	2.196	0.16	0.675	3.23	4.61	[4]
SrHI	4.371	8.450	1.933	0.20	0.70	3.20	4.23	[4]
BaHI	4.828	7.867	1.629	0.19	0.68	3.56	3.85	[4]
CaFCl	3.90	6.84	1.76					[2]
	3.894	6.809	1.749					[983]
SrFCl	4.125 9	6.957 9	1.686	0.201 5	0.642 9	3.11	3.07	[693]
	4.129	6.966	1.687					[983]
BaFCl	4.393 9	7.224 8	1.644	0.204 9	0.647 2	3.29	3.20	[693]
	4.391	7.226	1.646					[983]
SmFCl	4.135	6.992	1.691					[983]
EuFCl	4.098	6.945	1.695	~0.20	~0.64			[692]
	4.127	6.984	1.692					[694]
	4.118	6.971	1.693					[983]
TmFCl	3.956	6.849	1.731					[983]
YbFCl	3.940	6.825	1.732					[983]
PbFCl	4.106	7.23	1.76	0.20	0.65	3.10	3.25	[4]
CaFBr	3.883	8.051	2.073					[983]
SrFBr	4.218	7.337	1.739					[983]
BaFBr	4.503	7.435	1.651					[983]
SmFBr	4.235	7.316	1.728					[983]
EuFBr	4.219	7.312	1.733					[983]
YbFBr	3.983	7.546	1.895					[983]
PbFBr	4.18	7.59	1.82	0.195	0.65	3.18	3.45	[2, 4]
CaFI	4.29	8.70	2.03					[983]
SrFI	4.253	8.833	2.077					[983]
BaFI	4.66	7.96	1.71					[764]
	4.654	7.977	1.714					[983]
SmFI	4.282	8.604	2.009					[983]
EuFI	4.249	8.732	2.055					[983]
YbFI	4.050	8.998	2.222					[983]
PbFI	4.235	8.81	2.08					[695]
BaClI								[770]
LaHTe	4.50	9.15	2.03					[1]
CeHTe	4.45	9.10	2.04					[1]
PrHTe	4.42	9.05	2.05					[1]
γ-YOF	3.938	5.47	1.39					[14]
γ-LaOF	4.091	5.852	1.430					[14]
CeOF								
PrOF								
NdOF								[696]
PuOF	4.05	5.72	1.41					[14]
BiOF	3.748	6.224	1.661	0.208	0.65	2.79	2.75	[4]
YOCl	3.903	6.597	1.690	0.18	0.64	3.00	3.03	[4]
LaOCl	4.119	6.883	1.671	0.178	0.635	3.18	3.15	[4]
25°	4.120	6.882	1.670					[1]
CeOCl	4.080	6.831	1.674					[4]

Table 139 (Continued)

Compound	a(Å)	c(Å)	c/a	z_M	z_X	M−4X(Å)	M−1X(Å)	Ref.
PrOCl	4.051	6.810	1.681	0.18	0.64	3.12	3.13	[4]
25°	4.051	6.802	1.679					[1]
NdOCl	4.018	6.782	1.688	0.18	0.64			[4]
25°	4.025	6.775	1.683					[1]
PmOCl	4.020	6.740	1.677					[14]
SmOCl	3.982	6.721	1.688	0.17	0.63	3.11	3.09	[1, 4]
EuOCl	3.965	6.695	1.689					[1, 4]
	3.9646	6.695	1.689	0.170	0.630			[1, 772]
GdOCl	3.950	6.672	1.689					[4]
TbOCl	3.927	6.645	1.692					[4]
	3.921	6.628	1.690	0.166	0.63	3.08	3.08	[712]
DyOCl	3.911	6.620	1.693					[4]
	3.920	6.602	1.684	0.168	0.629	3.08	3.04	[712]
HoOCl	3.893	6.602	1.696	0.17	0.63	3.05	3.04	[4]
α-ErOCl	3.88	6.58	1.70					[4]
AcOCl	4.25	7.08 (7.07?)	1.66					[4]
UOCl	4.00	6.85	1.71					[1, 4]
PuOCl	4.012	6.792	1.693	0.18	0.64			[4]
	4.004	6.779	1.693					[14]
AmOCl	4.00	6.78	1.70	0.18				[4]
CmOCl	3.985	6.752	1.694					[698]
BkOCl	3.966	6.710	1.692					[14]
CfOCl	3.956	6.662	1.684					[14, 699]
EsOCl								
430°	3.97	6.75	1.70					[700]
20°	3.948	6.702	1.698					[701]
BiOCl	3.891	7.369	1.894	0.170	0.645	3.07	3.50	[4]
	3.883	7.347	1.892					[1]
U(NH)Cl	3.972	5.810	1.715	0.168	0.630	3.13	3.15	[761]
YOBr	3.838	8.241	2.147					[713]
LaOBr	4.145	7.359	1.775	0.164	0.635	3.28	3.47	[4]
	4.159	7.392	1.777					[713]
CeOBr	4.138	7.487	1.809					[713]
PrOBr	4.071	7.487	1.839					[713]
NdOBr	4.017	7.619	1.897	0.16	0.64	3.22	3.66	[4]
	4.024	7.597	1.888	0.16	0.64			[713]
PmOBr	3.98	7.56	1.90					[14]
SmOBr	3.950	7.909	2.002					[713]
	3.952	7.914	2.003					[1, 772]
EuOBr	3.908	7.973	2.040					[713]
	3.924	8.015	2.043	0.145	0.660	3.18	4.46	[1, 772]
GdOBr	3.895	8.116	2.084					[713]
TbOBr	3.891	8.219	2.112					[713]
DyOBr	3.867	8.219	2.125					[713]
HoOBr	3.832	8.241	2.151					[713]
ErOBr	3.821	8.264	2.163	0.16	0.64			[713]
TmOBr	3.806	8.288	2.178					[713]
YbOBr	3.780	8.362	2.212					[713]
	3.784,7	8.309	2.195	0.133	0.670	3.14	4.46	[1, 772]
LuOBr	3.770	8.387	2.225					[713]
AcOBr	4.29	7.42	1.73					[4]
UOBr								
NpOBr								
PuOBr	4.022	7.571	1.882	0.16	0.64	3.22	3.63	[4]
AmOBr								
CmOBr								[844]

Table 139 (Continued)

Compound	a(Å)	c(Å)	c/a	z_M	z_X	M−4X(Å)	M−1X(Å)	Ref.
BkOBr	3.95	8.1	2.05					[844]
CfOBr	3.900	8.110	2.079					[701]
BiOBr	3.916	8.077	2.063	0.154	0.653	3.18	4.03	[4]
25°C	3.926	8.103	2.064					[1]
YOI								
LaOI	4.144	9.126	2.202	0.135	0.660	3.48	4.79	[4]
CeOI								
PrOI	4.085 3	9.162 4	2.243					[784]
NdOI	4.056	9.183	2.264					[784]
PmOI	4.00	9.18	2.30					[14]
SmOI	4.008	9.192	2.293					[735]
	4.012	9.176	2.287					[784]
EuOI	3.993	9.186	2.301	0.120	0.675	3.39	5.10	[14, 736]
GdOI								
TbOI								
DyOI	3.936	9.183	2.333					[737]
HoOI								
ErOI								
TmOI	3.887	9.166	2.358	0.125	0.680	3.28	5.09	[735]
YbOI	3.870	9.161	2.367					[735]
LuOI								
AcOI								
UOI								
NpOI								
PuOI	4.042	9.169	2.268	0.13	0.67	3.40	4.95	[4]
AmOI	4.011	9.204	2.295					[734]
CmOI								
BkOI	4.0	7.5	1.9					[844]
CfOI	3.97	9.13	2.30					[701]
BiOI	3.985	9.129	2.291	0.132	0.668	3.36	4.89	[4]
25°C	3.994	9.149	2.291					[1]
α-YSF	3.77	6.80	1.80					[702]
α-LaSF	4.04	6.97	1.73					[702]
	4.024	6.979	1.734	0.180	0.626	3.15	3.11	[703]
	4.01	6.97	1.74					[696]
CeSF	4.00	6.94	1.74					[702]
	4.010	6.951	1.733	0.180	0.615	3.17	3.02	[703]
	3.99	6.92	1.73					[696]
PrSF	3.96	6.92	1.75					[702]
NdSF	3.93	6.91	1.76					[702]
	3.90	6.85	1.76					[696]
SmSF	3.87	6.88	1.78					[702]
	3.85	6.83	1.77					[697]
EuSF	3.874	6.735	1.739					[703]
GdSF	3.83	6.85	1.79					[702]
	3.82	6.81	1.78					[697]
TbSF	3.81	6.84	1.80					[702]
DySF	3.78	6.82	1.80					[702]
	3.77	6.79	1.80					[697]
α-HoSF	3.76	6.79	1.81					[702]
α-ErSF	3.74	6.78	1.81					[702]
β-LaSeF	4.14	7.17	1.73					[704]
CeSeF	4.09	7.15	1.75					[704]
PrSeF	4.05	7.13	1.76					[704]
NdSeF	4.020	7.050	1.75					[705]
	4.02	7.12	1.77					[704]
SmSeF	3.97	7.10	1.79					[704]
EuSeF								

Table 139 (Continued)

Compound	a(Å)	c(Å)	c/a	z_M	z_X	M−4X(Å)	M−1X(Å)	Ref.
GdSeF	3.93	7.08	1.80					[704]
NdTeF	4.487	7.345	1.637					[705]
TlSCl(?)	4.27	6.43	1.51					[706]
TlSBr(?)	4.30	8.96	2.08					[706]
TlSI(?)	4.72	9.79	2.07					[706]
ThNCl	4.097	6.895	1.683	0.165	0.635	3.21	3.24	[707]
UNCl	3.979	6.811	1.712	0.169	0.616	3.17	3.04	[708]
	3.979	6.811	1.712	0.160	0.66	3.07	3.41	[709]
ThNBr	4.110	7.468	1.817	0.151	0.637	3.31	3.63	[707]
UNBr	3.944	7.950	2.016	0.144	0.650	3.23	4.02	[708]
ThNI	4.107	9.242	2.250	0.124	0.668	3.48	5.03	[707]
UNI	3.990	9.206	2.307	0.121	0.669	3.42	5.04	[708]
ThOS	3.963	6.746	1.702	0.200	0.647	2.99	3.02	[4]
PaOS	3.832	6.704	1.749					[4]
UOS	3.843	6.694	1.742	0.200	0.638	2.93	2.93	[4]
				0.199	0.643			[710]
NpOS	3.825	6.654	1.740	0.200	0.638	2.91	2.91	[4]
	3.815	6.623	1.736					[685]
PuOS	3.80	6.59	1.73					[711]
ThOSe	4.038	7.03	1.74	0.18	0.63	3.15	3.16	[4]
PaOSe								
UOSe	3.908	6.996	1.790					[4]
	3.901	6.978	1.789	0.192	0.627	3.04	3.04	[1]
NpOSe								
PuOSe	4.151	8.369	2.016					[1, 4]
ThOTe	4.118	7.549	1.833	0.18	0.63	3.25	3.40	[4]
UOTe	4.004	7.491	1.871	0.173	0.629	3.20	3.42	[691]

exception of the Tl sulfohalides, whose adherence to this structure type has still to be confirmed, there is as yet only one compound which contains an anion other than H, F, O or N in the planar central layer, namely BaClI [770].

The compounds MgHCl, MgHBr and MgHI listed in [2] have been shown to be physical mixtures of the binary phases [763].

SmFCl undergoes an isostructural electronic transition at 20 kbar similar to SmS [983].

PbFCl-type ThN(NH$_2$) is unknown, but the yellow imide Th(NH)$_2$ has a layer structure derived from α-Th$_3$N$_4$ (Table 200, Appendix).

b. THE SmSI STRUCTURE

The rhombohedral SmSI structure represents a trigonal analog of the tetragonal PbFCl structure. With respect to the latter structure, the cation coordination number is reduced from 8–9 to 7:

Sm——1 S at 2.84 Å   along the c-axis

3 S at 2.73 Å ⎫
3 I at 3.29 Å ⎭   forming a distorted octahedron.

The square antiprism of the PbFCl structure is replaced in the SmSI structure by a trigonal antiprism. Here, however, the smaller anions form a strongly corrugated layer or a double layer and the additional neighbor above one antiprism base belongs to the inner anion double layer. Thus it does not represent a possible bond that may or may not exist depending upon the ratio of the atomic radii. On

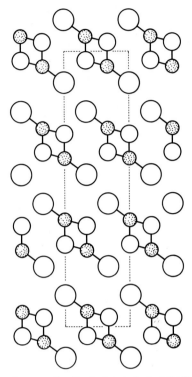

Fig. 129. (110) section of the hexagonal cell of rhombohedral SmSI. Cations are represented by dotted spheres, while the largest spheres symbolize the iodine atoms.

the contrary there are distinct sheets —I(SmSSSm)I—. The structure can be derived from a strongly deformed anion close packing with stacking sequence (hcch)$_3$. The distortions arise from the formation of the central tetrahedron layer [Sm$_{4/4}$S]$_n$. The S—S distance is 3.34 Å which is less than the distance of 3.6 Å observed in normal sulfides.

The occurrence of this structure is obviously connected with the relative size of cation and anions. We wonder whether this structure may occur also with say hypothetical ThPI or UPI.

TABLE 140

SmSI structure, trigonal, D_{3d}^5—$R\bar{3}m$ (No. 166), $Z = 2(6)$.
hexagonal axes: $(0, 0, 0; \frac{1}{3}, \frac{2}{3}, \frac{2}{3}; \frac{2}{3}, \frac{1}{3}, \frac{1}{3})+$
all atoms in 6(c): $\pm(0, 0, z)$

Compound	a_{rh}(Å)	α(°)	a_h(Å)	c_h(Å)	z_M	z_X	z_{Hal}	Ref.
β-ErOCl								[716]
TmOCl								[716]
YbOCl	9.523	22.56	3.726	27.380	0.384 1	0.301 4	0.112 1	[716]
LuOCl								[716]
β-CeSI	11.26	23.48	4.58	32.83				[702]
PrSI	11.24	23.77	4.63	32.75				[702]
NdSI	11.22	23.67	4.60	32.70				[702]
SmSI	11.21	23.37	4.54	32.69	0.388 5	0.301 7	0.115 9	[717]

c. The β-ZrNCl random structure

Whereas at room temperature both ZrNCl and ZrNBr crystallize in the PbFCl structure with a cation coordination number of 8–9, high-temperature modifications have been found [830] with a reduced coordination number of 6. The structure of these phases is built up of CdI_2-type double sandwiches XZrN NZrX (Table 140a) which are stacked in an irregular manner. This random stacking reduces the apparent unit cell to one third. A similar disorder is known for $CdBr_2$ where an irregular mixture of the CdI_2 and $CdCl_2$ type can be obtained.

TABLE 140a

Crystallographic data for the hexagonal double sandwiches of the β-ZrNCl structure
Zr in $\pm(0, 0, z)$; N in $\pm(\frac{1}{3}, \frac{2}{3}, z)$, X in $\pm(\frac{2}{3}, \frac{1}{3}, z)$
(refering to a unit cell (a_0, c_0) of one double sandwich;
$a_0 = a'\sqrt{3}$, where a' refers to the apparent subcell given below)

Compound	a'(Å)	c(Å)	z(Zr)	z(N)	z(X)	Ref.
ZrNCl(h)	2.081	9.234	0.147	0.10	0.332	[830]
ZrNBr(h)	2.100	9.751	0.137	0.095	0.331	[830]

d. The FeOCl and the γ-FeO(OH) structure

The orthorhombic FeOCl structure is a distorted version of the PbFCl structure. It contains a central sheet $(FeO)_n^+$ analogous to the $(PbF)_n^+$ sheet in PbFCl but as in the SmSI structure the central anion layer is puckered and, moreover, the Cl

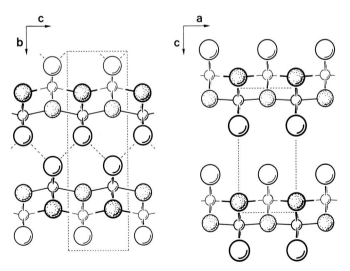

Fig. 130. The orthorhombic structures of boehmite γ-AlO(OH) (left) and FeOCl (right) viewed parallel to the layers. The asymmetric hydrogen bonds between the (OH) ions in the boehmite structure are indicated by broken lines.
All oxygen atoms are represented by dotted spheres: the smallest spheres are the cations.

TABLE 141
FeOCl structure, orthorhombic, D_{2h}^{13}—Pmmn (No. 59), $Z = 2$.
Fe in 2(b): $0, \frac{1}{2}, z; \frac{1}{2}, 0, z$
O and Cl in 2(a): $0, 0, z; \frac{1}{2}, \frac{1}{2}, \bar{z}$.

Compound	a(Å)	b(Å)	c(Å)	z_M	z_X	z_{Hal}	Ref.
ScOCl							
ScOBr							
ScOI							
TiOCl	3.79	3.38	8.03				[4]
TiOBr	3.787	3.487	8.529	0.110 8	0.945 0	0.328 0	[717]
VOCl	3.77	3.30	7.91	0.111	0.960	0.335	[718]
	3.780	3.300	7.91	0.114 8	0.954 2	0.327 9	[859]
VOBr	3.775	3.380	8.425	0.104	0.963	0.328	[718]
CrOCl	3.863	3.182	7.694	0.105	0.944	0.328	[719]
CrOBr	3.863	3.232	8.360	0.102	0.965	0.320	[720]
CrSBr(?)	4.78	3.52	7.98				[728]
MnOCl							
FeOCl	3.75	3.3	7.95	0.097	0.917	0.305	[4]
	3.780	3.302	7.917	0.115 7	0.951 7	0.330 0	[721]
	3.780	3.307	7.917				[773]
4.2 K	3.775	3.303	7.864				[773]
FeOBr							
FeO(NH$_2$)?							
FeO(OCH$_3$)	3.83	3.99	9.97				[1040]
InOCl	4.065	3.523	8.080	0.121	0.97	0.345	[722]
InOBr	4.049	3.611	8.649	0.113	0.98	0.336	[722]
InOI	4.05	3.71	9.49				[727]
TiNCl	3.937	3.258	7.803	0.100	0.950	0.330	[723]
TiNBr	3.927	3.349	8.332	0.080	0.950	0.330	[723]
TiNI	3.941	3.515	8.955	0.080	0.970	0.315	[723]
α-ZrNCl	4.08	3.52	8.57				[723]
α-ZrNBr	4.116	3.581	8.701	0.095	0.965	0.350	[723]
ZrNI	4.114	3.724	9.431	0.092	0.977	0.335	[723]
ZrN(NH$_2$)	4.21	3.43	7.52	0.119	0.901	0.361	[724]
HfNCl							
YSCl							[1050]
DySCl							[1050]
HoSCl							[1050]
ErSCl	5.26	3.974	7.44	0.154 4	0.90	0.387	[725]
TmSCl							[1050]
YbSCl							[1050]
LuSCl							[1050]
YSBr	5.35	4.06	8.09				[702]
β-DySBr	5.35	4.02	8.08				[702]
β-HoSBr	5.34	4.03	8.09				[702]
ErSBr	5.32	4.01	8.08				[702]
TmSBr	5.29	3.99	8.08				
YbSBr	5.27	3.97	8.07				[702]
LuSBr	5.274	3.955	8.085	0.136	0.918	0.378	[726]

layers are shifted in such a way as to reduce the cation coordination to an irregular octahedron [FeO$_{4/4}$Cl$_{2/2}$]: Fe—2O at 1.96, 2O at 2.10, 2Cl at 2.37 Å. One layer unit is thus formed by octahedra which share half their edges. The interlayer Cl—Cl distance is 3.68 Å which closely approximates the Van der Waals distance. The Cl—Cl distances within the same layer are given by the axes a and b. The short distance 3.30 Å along b is obviously the consequence of the M—Cl bonding. Similarly to the a-axis of the PbFCl-type compounds, the values of the a and b axes remain virtually constant if in a series of compounds along the halogen atom is varied. Interesting in view of the antiferromagnetic properties is the existence of nearly linear chains

$$-\text{Fe}-\text{O}-\text{Fe}-\text{O}- \quad \text{along [100]}.$$

As follows from Table 141, the FeOCl structure is met with cations that prefer octahedral coordination. Evidently Table 141 is very incomplete. The Hf analogs of the listed Zr^{4+} compounds will certainly have the same structure. Perhaps the anions S or Se will be electronegative enough to ensure a non-metallic phase NbNS or TaNSe? Needless to say that all the listed FeOCl representatives are non-metallic.

As in all true Van der Waals layer-type compounds intercalation phases with organic molecules are expected to exist. Thus, pyridine was intercalated in FeOCl at a ratio of 1:3 in the saturated phase [799]. The black complex 3FeOCl(C$_5$H$_5$N) exhibited a strongly elongated c-axis. The expansion corresponds to the amount expected when pyridine rings are perpendicularly placed between the layers. The orthorhombic unit cell had dimensions:

$$a = 3.78 \text{ Å}, \ b = 3.30 \text{ Å}, \ c = 13.45 \text{ Å} \quad [799]. \quad \text{See also [1040]}$$

In the lower part of Table 141, we have listed compounds of the type LnSCl and LnSBr. The most striking feature of these phases is the completely different axial ratio. Whereas the values for the a- and b-axes, which are determined by the (LnS) sheet only, are distinctly increased over the MOCl values, the c-axis, which is determined by the halogen atom, has shrunk. This shrinking, however, is a consequence of the larger Ln—S separation and the interlayer Cl—Cl distance in ErSCl is 3.68 Å exactly as in FeOCl.

It is noteworthy that the amides ZrN(NH$_2$) and possibly FeO(NH$_2$) also crystallize in the FeOCl structure whereas substitution of the halogen ion by the hydroxyl ion converts the structure to the boehmite type. The structure of orthorhombic γ-AlO(OH), boehmite, is made up of the same kind of puckered sheets as FeOCl but these sheets are stacked in a different manner. In the boehmite structure planar zig-zag chains link the sheets by means of H bonds. OH—OH = 2.70 Å. In addition to the true boehmite representatives we have listed in Table 142 also the data for Cu(OH)$_2$ which crystallizes in a closely related structure type. The structure of Cu(OH)$_2$, which contains the same

TABLE 142
Boehmite structure, orthorhombic, D_{2h}^{17}—Cmcm (No. 63)[a], $Z = 4$.
$(0, 0, 0; \tfrac{1}{2}, \tfrac{1}{2}, 0) +$
Al and O in 4(c): $\pm(0, y, \tfrac{1}{4})$

Compound	$a(Å)$	$b(Å)$	$c(Å)$	y_M	y_O	y_{OH}	Ref.
γ-AlO(OH)	2.866	12.227	3.700				[4]
(boehmite)	2.859	12.24	3.690	0.834	0.213	0.433	[1]
	2.86	12.24	3.69	0.822	0.209	0.420	[1]
				0.815[b]	0.213[b]	0.435[b]	
γ-ScO(OH)	3.24	13.01	4.01	0.818	0.218	0.429	[4]
	3.245	13.042	4.015				[1]
γ-FeO(OH)	3.06	12.4	3.87	0.822	0.21	0.425	[4]
(lepidocrocite)	3.06	12.50	3.86				[1]
related:							
Cu(OH)$_2$	2.949	10.59	5.256	0.817 5	0.198	0.435	[800]
25°C	2.936	10.54	5.238				[801]
100°C	2.936	10.52	5.290				[801]

[a] neglecting the H positions (the asymmetric O—H $\cdots$ O bonds require space group C_{2v}^{12}—Cmc2$_1$ (No. 36) [789])
[b] calculated according to Pauling's rules for the coordination of ionic crystals [789]

stepped layers composed of octahedra [CuO$_{2/2}$O$_{4/4}$], undergoes a minor reversible change at about 60°C [801].

e. THE AlOCl STRUCTURE

While a six-coordinate arrangement is adopted by medium-size cations ($r_{M^{3+}} \approx 0.64 \cdots 0.81$ Å or $r_{M^{3+}}/r_{O^{2-}} \approx 0.46 \cdots 0.58$ Å) a four-coordinate arrangement becomes stable for smaller cations. Thus, the sequence of the MXY layer structures

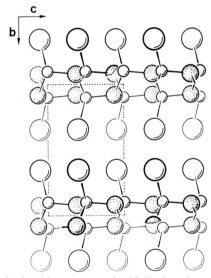

Fig. 131. The orthorhombic structure of AlOCl viewed parallel to the layers. The smallest spheres are cations, O atoms are dotted.

TABLE 143
AlOCl structure, orthorhombic, C_{2v}^5—$Pca2_1$ (No. 29), $Z=4$.
All atoms in 4(a): x, y, z; $\bar{x}, \bar{y}, \frac{1}{2}+z$; $\frac{1}{2}-x, y, \frac{1}{2}+z$; $\frac{1}{2}+x, \bar{y}, z$.

Compound	a(Å)	b(Å)	c(Å)	Ref.
AlOCl	5.50	8.23	4.92	[730]
AlOBr	6.73	10.02	4.92	[730]
AlOI	6.86	11.03	4.92	[730]
AlO(NH$_2$)?				
GaOCl*	5.653	8.328	5.081	[729]
GaOBr				
SiNCl?				
GeNCl?				

* GaOCl:

	x	y	z
Ga	0.094	0.093	0.000
O	0.087	0.085	0.376
Cl	0.995	0.655	0.395

Possibly related: AlSCl, AlSBr, AlSI, orthorhombic? [703]

AlSeCl type, monoclinic, C_{2h}^2—$P2_1/m$ (No. 11), $Z=40$.

Compound	a(Å)	b(Å)	c(Å)	$\beta(°)$	Ref.
AlSeCl	16.33	12.87	18.172	94.00	[731]
AlSeBr	16.97	12.87	18.732	94.00	[731]
AlSeI	18.11	12.87	19.973	94.00	[731]

PbFCl—SmSI—FeOCl— is continued by the AlOCl type with tetrahedral cation coordination. This sequence may end with the BSCl structure with boron in planar three-coordination.

In the AlOCl structure [729] the cation has 3 O neighbors in a sheet on one side and 1 Cl nearest neighbor on the other side. Again in a series with different halogen ions, the value of the c-axis remains constant, as it is solely determined by the M—O arrangement.

We speculate that hypothetical SiNCl, SiNBr, ... GeNCl might be further candidates for this structure type, possibly even SiOSe or ZnFBr, ZnFI, BeFBr.

In Table 143 we have also included the data for the monoclinic AlSeCl-type phases [731]. The constancy of the b-axes suggests that the yet unknown structure of these compounds is built up in a similar manner as the foregoing layer structure.

f. THE InTeCl STRUCTURE

In the monoclinic InTeCl structure [732] each cation is tetrahedrally coordinated. The [InTe$_{3/3}$Cl] tetrahedra form wavy layers parallel to the (b, c) plane. The structure may be derived from the $\frac{2}{3}$ double-tetrahedron structure (Figure 24d) by loosening the contacts at the doubly shared tetrahedron corners.

TABLE 144
InTeCl structure, monoclinic, C_{2h}^5—$P2_1/c$ (No. 14), $Z = 8$.
All atoms in 4(e): $\pm(x, y, z; x, \frac{1}{2}-y, \frac{1}{2}+z)$.

InTeCl: $a = 7.42$ Å, $b = 14.06$ Å, $c = 7.07$ Å, $\beta = 92.1°$ [732]

	x	y	z
In_I	0.794 7	0.023 9	0.128 2
In_{II}	0.818 0	0.312 0	0.126 6
Te_I	0.156 9	0.001 1	0.260 5
Te_{II}	0.605 0	0.172 5	0.275 0
Cl_I	0.068 1	0.234 7	0.988 6
Cl_{II}	0.391 3	0.396 9	0.271 7

The tetrahedra containing the In_{II} atoms form zig-zag chains along the c-axis with the Cl atoms all pointing in the same direction. Two such symmetry-related chains are connected by double tetrahedra with the In_I atoms. The Cl atoms in these double tetrahedra are in *trans* positions so that all Cl atoms point towards the channels parallel to the c-axis. Each Te atom has 3 In neighbors in trigonal-pyramid coordination. The nearest interlayer contacts are between Te and Cl at 3.68 Å corresponding to a Van der Waals type of bonding. The layer character of InTeCl is not very pronounced as it crystallizes in the form of reddish-brown columns. Platelets, as expected for a layer structure, on the other hand, are reported for InTeBr and InTeI. A monoclinic cell similar to that of InTeCl but with half its value for the b-axis has been derived for these two obviously isostructural compounds [733].

InSCl and InSeCl both crystallize in the $CdCl_2$ structure with hexacoordinated cations and equivalent anions. As Roos *et al.* [732] pointed out, the higher Lewis basicity of Te^{2-} compared with Cl^- is the reason for the higher coordination number of Te in InTeCl. The In^{3+} ion acts as a Lewis acid midway between hard and soft cationic acids.

15. Layer-Type Dihalides

As is evident from Table 145 most dihalides crystallize in layer structures as soon as the cation-to-anion-radius ratio becomes sufficiently small. Exceptions are found among the beryllium halides where the SiS_2 chain structure is stable. Further exceptions are the Mo, W, Pd and Pt halides and SnI_2. The Mo and W halides are exceptional anyway as they contain metal clusters similar to those met in the Nb and Ta halides. $PdCl_2$ and $PdBr_2$ each possess one modification which is a layered chain structure and, finally, the $PbCl_2$ and SnI_2 structures show at least a tendency towards layering. Among the layer-type phases the CdI_2 and the $CdCl_2$ structures predominate, together with their ordered mixtures, the so-called polytypes. No layer structures with a cation coordination number >6 exist, but 6 is by far most frequent. A trigonal-prismatic coordination around the cation is met in ThI_2 and $ZrCl_2$. The Jahn-Teller ions Cr^{2+} and Cu^{2+} are found in a

TABLE 145

Occurrence of layer structures among dihalides. Layer structures are printed in italics; borderline cases are underlined with broken lines. Difluorides are omitted since no layer-type difluoride is known. Where the structure is yet unknown the chemical formula only is given.

Cation	M(OH)$_2$	MCl$_2$	MBr$_2$	MI$_2$
Be	α-Be(OH)$_2$	α(r): SiS$_2$	α: SiS$_2$	α: SiS$_2$
	β: orthorh. Zn(OH)$_2$	β(h): orthorh.	β:	β'(290–370°): orthorh.
				β(>370°C): *tetragonal*
Mg	*brucite*	*CdCl$_2$*	*CdI$_2$*	*CdI$_2$*
	orthorhombic (?)			
Ca	*Mg(OH)$_2$*	orthorh. CaCl$_2$	CaCl$_2$	*CdI$_2$*
Sr	orthorh. Sr(OH)$_2$	r: CaF$_2$	tetrag. SrBr$_2$	CeSI
		p: PbCl$_2$		
Ba	r: Ba(OH)$_2$	r: CaF$_2$	r: PbCl$_2$	PbCl$_2$
	h: orthorh. Ba(OH)$_2$	p: PbCl$_2$	h: monoclinic	
Ra		PbCl$_2$	PbCl$_2$	PbCl$_2$
La				*MoSi$_2$*
Ce				*MoSi$_2$*
Pr				(t): *MoSi$_2$*, (r): *CdCl$_2$*
Nd		PbCl$_2$	PbCl$_2$	SrBr$_2$, (p): *MoSi$_2$*
Sm		PbCl$_2$	SrBr$_2$	monocl. EuI$_2$
			PbCl$_2$	
Eu		PbCl$_2$	SrBr$_2$	monocl. EuI$_2$
				SrI$_2$
Gd				*CdCl$_2$*?
Dy		SrI$_2$	SrI$_2$	*CdCl$_2$*
Ho		HoCl$_2$		
Tm		SrI$_2$	SrI$_2$	*CdI$_2$*
Yb		SrI$_2$	r: CaCl$_2$	*CdI$_2$*
			h: SrI$_2$	
Th				r: α-*ThI$_2$*
				h: β-*ThI$_2$*
Am		PbCl$_2$	SrBr$_2$	monocl. EuI$_2$
Cf			SrBr$_2$	monocl. EuI$_2$?
Ti		*CdI$_2$*	*CdI$_2$*	*CdI$_2$*
Zr		*3R–MoS$_2$*	*MoS$_2$*(?)	*ZrI$_2$*
Hf		*3R–MoS$_2$*?	*MoS$_2$*(?)	
V		*CdI$_2$*	*CdI$_2$*	*CdI$_2$*
				CdBr$_2$?
Nb		(Nb$_6$Cl$_{14}$)		(Nb$_6$I$_{11}$)
Ta		(Ta$_6$Cl$_{14}$)	(Ta$_6$Br$_{14}$)	(Ta$_6$I$_{14}$)
Cr		distorted CaCl$_2$	*CrBr$_2$*	*monocl. CrI$_2$*
				orthorh. CrI$_2$
Mo		(Mo$_6$Cl$_8$)Cl$_4$	*MoCl$_2$*	*MoCl$_2$*
W		*MoCl$_2$*	*MoCl$_2$*	*MoCl$_2$*
Mn	*Mg(OH)$_2$*	*CdCl$_2$*	*CdI$_2$*	*CdI$_2$*
			CdCl$_2$	
Re				ReI$_2$
Fe	*Mg(OH)$_2$*	r: *CdCl$_2$*	r: *CdI$_2$*	*CdI$_2$*
		p: *CdI$_2$*	h: *CdI$_2$*	
Ru		RuCl$_2$		
Co	*Mg(OH)$_2$*	*CdCl$_2$*	r: *CdI$_2$*	*CdI$_2$*
			h: *CdCl$_2$*	
Ni	*Mg(OH)$_2$*	*CdCl$_2$*	*CdCl$_2$*	*CdCl$_2$*
Pd		orthorh. PdCl$_2$	monocl. PdBr$_2$	α: orthorh. PdI$_2$
		h$_1$: Pt$_6$Cl$_{12}$	h$_1$:	β: monocl. PdI$_2$
		h$_2$:	h$_2$:	γ:
Pt		Pt$_6$Cl$_{12}$	Pt$_6$Cl$_{12}$	PtI$_2$
		PdCl$_2$		
		h$_2$:		

Table 145 (Continued)

Cation	M(OH)$_2$	MCl$_2$	MBr$_2$	MI$_2$
Cu	γ-AlO(OH) brucite (?)	CuBr$_2$	monocl. CuBr$_2$	
Zn	α: Mg(OH)$_2$ α': hexag. polytype β: orthorhombic γ: orthorhombic δ: orthorhombic ε: orthorh. (C31)	α: tetrag. (ccp, τ$_1$) β: monoclinic γ: tetrag. HgI$_2$ δ: C19τ	CdCl$_2$ tetragonal ZnBr$_2$ tetrag. HgI$_2$	CdI$_2$ CdCl$_2$ tetrag. HgI$_2$ tetrag. ZnBr$_2$
Cd	α: Mg(OH)$_2$ β: α'-Zn(OH)$_2$ γ: monoclinic	rhomboh. CdCl$_2$	CdCl$_2$ Cd(OH)Cl polytypes	Cd(OH)Cl CdI$_2$ polytypes
Hg		orthorh. HgCl$_2$	orthorh. HgBr$_2$	tetrag. HgI$_2$ tetrag. polytypes HgBr$_2$
Ge				CdI$_2$ polytypes
Sn		PbCl$_2$	orthorh. SnBr$_2$	monocl. SnI$_2$
Pb	hcp orthorhombic	orthorh. PbCl$_2$	PbCl$_2$	CdI$_2$ CdCl$_2$ polytypes

distorted octahedral coordination similar but less pronounced than in the corresponding Pd^{2+} compounds. A coordination number 4 with tetrahedral neighborhood is adopted by Zn, Cd and Hg in certain modifications of their halides. In the orthorhombic HgCl$_2$ and HgBr$_2$, finally, the idealized coordination octahedron is deformed in such a way that structures with linear molecules results similar to krennerite AuTe$_2$.

a. HALIDES WITH TRIGONAL-PRISMATIC CATION COORDINATION

NbSCl and ZrCl$_2$ are isoelectronic analogs of MoS$_2$. Therefore, these phases might well be insulators which contain the cations in trigonal-prismatic coordination. Indeed, the 3R—MoS$_2$ structure has been reported for ZrCl$_2$ that we have already listed in Table 122. HfCl$_2$ is probably isostructural and ZrBr$_2$, ZrI$_2$ [786], HfBr$_2$ and HfI$_2$ may crystallize in related structures.

ThI$_2$ might adopt the same structure. This compound undergoes a structural transition at 600–700°C and decomposes at 864°C [790]. Only the high-temperature modification, β-ThI$_2$, is structurally characterized. The low-temperature modification, α-ThI$_2$, is described as black. Therefore, it still might be a semiconductor and crystallize in a MoS$_2$ polytype structure. Its powder pattern, at least, is similar to that of β-ThI$_2$.

In its high-temperature form ThI$_2$ is a golden diamagnetic metal. Its structure is closely related to the 4H$_b$, 4H$_{d_1}$ and 4H$_{d_2}$ types met in the Nb and Ta chalcogenides. The position of the octahedral layers is identical with that in the 4H$_b$ type but the trigonal-prism layers are shifted. This structure offers no chance for semiconductivity as half the d^2 cations are octahedrally coordinated which

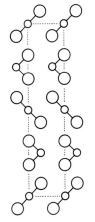

Fig. 132. (110) section of the hexagonal β-ThI$_2$ structure.

TABLE 145a

Pr$_2$I$_5$ structure, monoclinic, C$_{2h}^2$—P2$_1$/m (No. 11), Z = 2.
All atoms in 2(e); ±(x, ¼, z)

Pr$_2$I$_5$: a = 8.655 Å, b = 4.317 Å, c = 16.798 Å, β = 120.48°.

	x	z	
Pr$_I$	0.5755	0.1581	
Pr$_{II}$	0.5640	0.6555	
I$_I$	0.6846	0.9906	[1018]
I$_{II}$	0.2312	0.1820	
I$_{III}$	0.7384	0.3792	
I$_{IV}$	0.2793	0.4277	
I$_V$	0.1197	0.7766	

gives rise to an incompletely filled d band. We notice that not only the [ThI$_6$] prisms but to a much higher degree the octahedra are elongated. However, the deformation to a trigonal antiprism is not sufficient to split the original t_{2g} band.

The Th—I distances in β-ThI$_2$ are 3.20 Å in the trigonal-prismatic layers and 3.22 Å in the trigonal-antiprismatic layers, so that we cannot detect a difference due to localized d^2 electrons in the former and d^0 cations with delocalized electrons in the latter. These Th—I distances are equivalent to those found in ThI$_4$ where they range from 3.13 to 3.29 Å. The shortest interlayer I—I contact distance is 4.13 Å.

Purple to golden-colored metallic diiodides occur also with rare-earth cations [791]. The structure of PrI$_2$(r) is of the CdCl$_2$ type [784]. Pr in PrI$_2$ is trivalent in contrast to tetravalent Th in ThI$_2$. Thus, in order to obtain a similar metallic d-band as in β-ThI$_2$, PrI$_2$ has to contain all cations in octahedral coordination. The metallic LaI$_2$, CeI$_2$, PrI$_2$(t) and NdI$_2$(p) crystallize in the elongated MoSi$_2$ structure (Table 132).

It is interesting that a Pr phase with half as much excess valence electrons, the congruently melting Pr$_2$I$_5$ [792], crystallizes in a layer structure [784] built up of monocapped trigonal prisms, [PrI$_{7/3}$]∞ prisms along [010] and [PrI$_{2/2}$I$_{5/3}$] rippled prism rows; layer plane (100). Possibly, the reported LaI$_{2.4}$ and CeI$_{2.4}$ [792] are isotypic Ln$_2$I$_5$ phases.

b. CdCl$_2$- AND CdI$_2$-TYPE HALIDES

The CdCl$_2$ structure is based on a ccp anion packing and thus is a subtractive NaCl derivative. As a consequence NaCl-type halides form complete solid solutions with CdCl$_2$-type halides such as LiCl—MgCl$_2$, LiBr—MnBr$_2$ [796], etc. Increasing polarizability of the anions favors the CdI$_2$ structure which is based on a hcp anion packing. A considerable covalent part of the bonding in the CdI$_2$

TABLE 146

β-ThI_2 structure, hexagonal, D_{6h}^4—$P6_3/mmc$ (No. 194), $Z=4$.
I_{II} in 4(f): $\pm(\frac{1}{3}, \frac{2}{3}, z; \frac{2}{3}, \frac{1}{3}, \frac{1}{2}+z)$
I_I in 4(e): $\pm(0, 0, z; 0, 0, \frac{1}{2}+z)$
Th_{II} in 2(c): $\pm(\frac{1}{3}, \frac{2}{3}, \frac{1}{4})$
Th_I in 2(a): $0, 0, 0; 0, 0, \frac{1}{2}$.

$ThI_2(h)$: $a = 3.97$ Å, $c = 31.75$ Å, $c/4a = 2.00$; $z(I_I) = 0.1796$, $z(I_{II}) = 0.0713$ [650]

---

structure is also indicated by the fact that compounds with much too low cation-to-anion-radius ratios such as MgI_2 (0.33 compared with the theoretical limit 0.41) still are stable with the cations in octahedral coordination.

The transition-metal dihalides are most attractive for their anisotropic magnetic properties. All are antiferromagnetic or metamagnetic at low temperatures. The dihalides are non-metallic ranging from insulators to medium-gap semiconductors. For NiI_2, for example, an energy gap of 0.6 eV has been derived from resistivity measurements [797]. The lattice constants in series such as TCl_2 or TI_2 with $T =$ Ti $\cdots$ Zn nicely reflect the gradual filling of the cation d shell superposed to the general valence shell contraction with increasing atomic number. The existence of metallic $CdCl_2$-type rare-earth iodides, PrI_2 and DyI_2 is remarkable.

The values for the axial ratio c/a (and $c/3a$, resp.) vary in the CdI_2 structure from 1.53 (PbI_2) to 1.79 ($TiBr_2$) and in the $CdCl_2$ structure from 1.51 (DyI_2) to 1.68 (NiI_2). The reported z-values scatter around 0.255 regardless of whether $c/a \gtrless 1.63$. For symmetric cations we expect a compressed cation coordination octahedron above all for a large cation/anion-radius ratio. In d^2, d^6 and d^7 cations distortions will in part be due to the reduction of the d-electron energies. In d^4 and d^9 halides the distortions are such as to change the crystal symmetry.

It is interesting that the ordered anti-types of the CdI_2 and $CdCl_2$ structures are met also among carbides, nitrides, oxides and fluorides. While Cs_2O and Ag_2O appear to be normal-valence semiconductors all the remaining phases are expected to be metallic. We have already pointed to the puzzle of the high resistivity found in Sr_2N and Ba_2N [35, 795]. The carbides, on the other hand, are metallic. Ag_2F is even a superconductor [802].

The CdI_2 and the $CdCl_2$ structures both originate from the same type of XMX sandwiches, differing only in the way of stacking. The Madelung constants of both structures are nearly the same, depending on the axial ratio c/a. It is therefore at a first glance surprising that other stacking arrangements, so-called polytypes, have been reported only for a rather restricted number of halides. By far the most polytypes, nearly 200, have been reported for CdI_2 [805–809, 993]. The CdI_2-type modification, the 2H polytype itself, is only the next common type after 4H, the Cd(OH)Cl-type modification. All other polytypes of CdI_2 are of rather rare occurrence. The most common modification of $CdBr_2$ is 6R, the $CdCl_2$ type. Crystals grown by sublimation were all of pure 6R type while crystals grown from solution exhibited polytypism [765]. About 40 polytypes of PbI_2 have been

TABLE 147
CdI$_2$-type halides

Compound	a(Å)	c(Å)	c/a	z_X	Ref.
MgBr$_2$	3.81	6.26	1.64	0.25	[4]
	3.839	6.296	1.640		[1]
	3.815	6.256	1.640		[1]
	3.841	6.294	1.639		[767]
MgI$_2$	4.14	6.88	1.66	0.25	[4]
	4.157	6.877	1.654		[767]
CaI$_2$	4.48	6.96	1.55	0.25	[4]
	4.502	6.976	1.550		[767]
TmI$_2$	4.520	6.967	1.541	~0.25	[4, 739]
YbI$_2$	4.503	6.972	1.548	~0.25	[4, 739]
TiCl$_2$	3.561	5.875	1.650	0.25	[4]
TiCl$_2$(h)	3.43	6.10	1.78		[1]
TiBr$_2$	3.629	6.492	1.789	~0.25	[4]
TiI$_2$	4.110	6.820	1.659		[4]
VCl$_2$	3.601	5.835	1.620	~0.25	[4]
VBr$_2$	3.768	6.180	1.640		[4]
VI$_2$	4.000	6.670	1.668		[4]
	4.058	6.753	1.664		[762]
MnBr$_2$	3.820	6.188	1.620	0.25	[1, 4]
	3.868	6.272	1.622	0.25	[2]
	3.869	6.271	1.621		[796]
MnI$_2$	4.16	6.82	1.64	0.25	[4]
25°C	4.146	6.829	1.647		[2]
FeCl$_2$					
6 kbar	3.585	5.735	1.560	0.239 0	[744]
FeBr$_2$(r)	3.740	6.171	1.650	0.25	[4]
(<360°C)	3.78	6.20	1.64		[1]
	3.772	6.223	1.650		[749]
FeI$_2$	4.04	6.75	1.67	0.25	[4]
CoBr$_2$(r)	3.685	6.120	1.661	0.25	[1, 4]
(<370°C)	3.728	6.169	1.655		[749]
CoI$_2$	3.96	6.65	1.68	0.25	[4]
ZnI$_2$	4.25	6.54	1.54	0.25	[4]
CdBr$_2$					
CdI$_2$	4.24	6.84	1.61		[4]
	4.240	6.835	1.612		[2]
	4.244	6.859	1.616	0.249 2	[783]
(Cd, Pb)I$_2$	Végard's law				[827]
GeI$_2$	4.13	6.79	1.64	0.25	[1, 4]
	4.245	6.837	1.611		[804]
	4.249	6.833	1.608		[826]
PbI$_2$ 25°C	4.555	6.977	1.532	0.265	[4]
BiTeBr	4.23	6.47	1.53		[1]
BiTeI	4.31	6.83	1.58		[1]
anti-type					
Ag$_2$F	2.989	5.710	1.910	~0.3	[4]
	2.996	5.691	1.900	0.305	[766]

TABLE 148
$CdCl_2$ structure, trigonal, D_{3d}^5—$R\bar{3}m$ (No. 166). $Z = 1(3)$.
rhombohedral cell: Cl in 2(c): $\pm(x, x, x)$
Cd in 1(a): 0, 0, 0.
hexagonal cell: $(0, 0, 0; \frac{1}{3}, \frac{2}{3}, \frac{2}{3}; \frac{2}{3}, \frac{1}{3}, \frac{1}{3})+$
Cl in 6(c): $\pm(0, 0, z)$
Cd in 3(a): 0, 0, 0.

Compound	a_{rh}(Å)	α(°)	a_h(Å)	c_h(Å)	$c/3a$	z	Ref.
Mg(OH)Cl (?)	6.078	32.1	3.36	17.3	1.73		[4]
$MgCl_2$	6.220	33.61	3.596	17.589	1.630		[4]
	6.254	33.85	3.641	17.67	1.62		[767]
	6.291	33.55	3.632	17.795	1.633		[771]
PrI_2(r)	7.87	31.3	4.25	22.43	1.76	0.2436	[784]
DyI_2	7.445	36.1	4.61	20.86	1.51		[737]
$MnCl_2$	6.243	34.58	3.711	17.59	1.580		[771]
	6.200	34.59	3.686	17.470	1.580		[4]
$MnBr_2$(Li)	6.669	33.75	3.872	18.850	1.623		[796]
$FeCl_2$	6.200	33.55	3.579	17.536	1.633		[4]
	6.204	33.76	3.603	17.536	1.622	0.2543	[749]
1.04 kbar	6.186	33.81	3.598	17.481	1.620	0.2550	[744]
3.15 kbar	6.155	33.93	3.592	17.384	1.613	0.2546	[744]
5.70 kbar	6.155	34.13	3.589	17.260	1.603	0.2534	[744]
4.2 K	6.165	33.82	3.586	17.42	1.62		[750]
$FeBr_2$(h) 450°C	6.71	33.3	3.85	19.0	1.65		[2]
$Fe_{0.65}Cr_{0.35}I_2$	7.17	32.6	4.057	20.33	1.67		[996]
$CoCl_2$	6.16	33.43	3.544	17.430	1.639	0.25	[4]
	6.139	33.64	3.553	17.359	1.629	0.2558	[749]
$CoBr_2$(h) 433°C	6.61	33.05	3.76	18.73	1.66		[1]
Ni(OH)Cl(?)	5.963	31.78	3.265	16.99	1.73		[4]
$NiCl_2$	6.130	33.60	3.543	17.335	1.631		[4]
	6.139	32.96	3.483	17.40	1.665	0.2551	[771]
$NiBr_2$	6.465	33.33	3.708	18.300	1.645		[4]
	6.466	33.39	3.715	18.30	1.65	0.255	[2]
NiI_2	6.920	32.67	3.892	19.634	1.682		[4]
	6.919	32.70	3.895	19.63	1.68	0.250	[2]
	6.988	32.64	3.927	19.83	1.683	0.233	[783]
$ZnBr_2$	6.64	34.33	3.92	18.73	1.59		[4]
ZnI_2	7.57	32.6	4.25	21.5	1.69		[4]
$CdCl_2$	6.230	36.04	3.854	17.457	1.510	0.25	[4]
	6.235	35.92	3.8457	17.48	1.515		[771]
$CdCl_{0.75}(OH)_{1.25}$	5.910	35.33	3.587	16.605	1.543		[4]
$CdBr_2$	6.63	34.7	3.95	18.67	1.58		[4]
	6.688	34.66	3.985	18.841	1.576		[1, 765]
$CdBr_{0.6}(OH)_{1.4}$	6.198	33.53	3.58	17.55	1.63		[4]
CdI_2							[2]
PbI_2	7.374	35.87	4.54	20.7	1.52	0.26	[4]
	7.458	35.57	4.557	20.937	1.531	0.245	[4]
InSCl	6.31	34.4	3.728	17.78	1.59		[733]
InSBr	6.58	33.8	3.820	18.59	1.62		[733]
InSeCl	6.58	34.1	3.860	18.58	1.60		[733]
InSeBr	6.80	33.8	3.935	19.13	1.62		[733]
$CuTiCl_4$			2×3.61	17.64	1.63		[768]
$CuVCl_4$			2×3.545	17.53	1.65		[768]
$CuFeCl_4$(?)							[769]

Table 148 (Continued)

Compound	a_{rh}(Å)	α(°)	a_h(Å)	c_h(Å)	$c/3a$	z	Ref.
Ni(ClO$_4$)$_2$	7.80	35.53	4.76	21.90	1.53		[798]
anti-type							
Y$_2$C	6.339	33.15	3.617	17.96	1.655	0.258 5	[774]
Tb$_2$C 300 K	6.41	32.58	3.595	18.19	1.687	0.259 3	[775]
200 K	6.394	32.56	3.585	18.15	1.688		[775]
5 K	6.425	32.35	3.579	18.25	1.700		[775]
Ho$_2$C	6.247	33.07	3.556	17.70	1.66	0.256	[776]
Ca$_2$N[a]	6.603	31.98	3.638	18.78	1.721	0.268 0	[794]
Sr$_2$N[a]	7.246	30.83	3.852	20.69	17.90	0.267 8	[795]
Ba$_2$N[a]							
Cs$_2$O	6.79	36.53	4.256	18.99	1.487	0.256	[3]
	6.74	36.93	4.269	18.819	1.469		[1]

[a] true composition M$_2$NH?

reported up to now [810]. For this compound the most common type is 2H, the CdI$_2$ type. The CdCl$_2$-type modification results from growth by sublimation.

The origin of the polytype structures is attributed to spiral growth around suitable screw dislocations in the basic structure. In order to account for all observed polytypes it is necessary to assume that the parent matrix contains stacking faults near its surface at the time of generation of the screw dislocation ledge [806]. The resulting polytypes represent structures with low stacking-fault energy and this is the reason why the Zhdanov numbers in the dihalides are limited to 1, 2 and 3 (∞ for the CdCl$_2$ type). (The Zhdanov number counts the equivalent steps in stacking the anion layers: $1 \equiv h$, $2 \equiv hc$, $3 \equiv hcc$, etc. Thus, $2H = hh = (11)$, $4H = hchc = (22)$, $6H_a = hchchh = (2211)$, $6H_b = hcc\,hcc = (33)$, $12R_a = (hhcc)_3 = (13)_3$, $12R_b = (hcch)_3 = (31)_3$, $24R = (hchchcc)_3 = (2213)_3$, etc.). An ordered version of the 4H polytype is realized in the Cd(OH)Cl structure (Table 155 and Figure 135).

c. Cr and Cu dihalides

CrCl$_2$ is the only 3d dihalide that does not crystallize in a layer structure, although it does show a Jahn-Teller distortion similar to the other d^4 and d^9 dihalides. The energy gain due to the splitting of the threefold t_{2g} level of the cation is attained by elongating one diagonal of the coordination octahedron. Thus in CrCl$_2$ the deformation of the rutile structure does not lead to the CaCl$_2$ structure but to a specific type which approximates a chain structure.

The structure of CrBr$_2$ is derived from the CdI$_2$ type. Weakening of the bonds along one diagonal of the coordination octahedron would lead to an orthorhombic cell but the tilting of the remaining square induces the anion layers to repel each other which results in a slight relative displacement of the sandwiches.

The monoclinic CrI$_2$ structure is isopuntal with CrBr$_2$ but the monoclinic angle β corresponds to a much larger displacement of adjacent ICrI sandwiches. As can

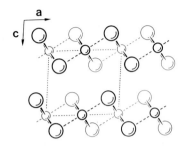

Fig. 133. Projection onto the (010) plane of the monoclinic structure of CrBr$_2$. The loosened bonds are indicated by broken lines. Cr: small spheres.

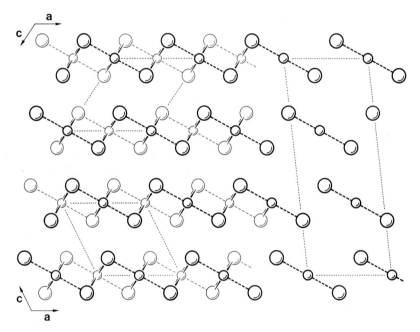

Fig. 134. The monoclinic structure of CrI$_2$ viewed parallel to the layers. The two weakened octahedral bonds are indicated by broken lines. The unit cell in the lower left belongs to CrI$_2$ while the upper cell corresponds to CuBr$_2$. The right part shows the distorted (110) section of the hexagonal cell of the underlying CdCl$_2$ structure. Large spheres: anions.

be seen in Figure 134 this structure is in fact derived from the CdCl$_2$ type. The CuBr$_2$ structure is of the same type, merely described in another cell.

A second modification, orthorhombic CrI$_2$ [781], is based on a mixed anion stacking of the type *hchc*. This stacking is characteristic of the C27 CdI$_2$ polytype and the Cd(OH)Cl structure (Figure 135, Table 155). CrBr$_2$, CrI$_2$, CuCl$_2$ and CuBr$_2$ crystallize as reddish-brown fibrous flakes. Crystals of CuCl$_2$ are strongly pleochroic. For the systems CrI$_2$—MI$_2$ (M = Ti, V, Mn, Fe, Co, Ni, Zn) see [1032].

TABLE 149

$CrBr_2$ structure, monoclinic, C_{2h}^3—$C2/m$ (No. 12), $Z = 2$.
$(0, 0, 0; \frac{1}{2}, \frac{1}{2}, 0)$ + Br in 4(i): $\pm(x, 0, z)$
Cr in 2(a): 0, 0, 0.

The CrI_2(I) type is isopuntal. It corresponds to the $CuBr_2$ type described with a different cell:

$CuBr_2$ structure Br in 4(i): as above
Cu in 2(b): 0, $\frac{1}{2}$, 0.

Compound	a(Å)	b(Å)	c(Å)	β(°)	x	z	Ref.
$CrBr_2$	7.114	3.649	6.217	93.88	0.648 9	0.240 9	[751]
CrI_2	7.545	3.929	7.505	115.52	0.733 6	0.242 0	[752]
$Mn_{0.26}Cr_{0.74}I_2$	7.441	3.978	7.448	113.66	0.737 4	0.241 7	[994]
$Fe_{0.3}Cr_{0.7}I_2$	7.388	3.965	7.410	114	0.735	0.239	[996]
$CuCl_2$	6.85	3.30	6.70	121	~0	~$\frac{1}{4}$	[1, 4]
$CuBr_2$	7.18	3.46	7.14	121.25	0.015	0.240	[1, 4]

d. THE BRUCITE-TYPE HYDROXIDES

The $Mg(OH)_2$ brucite structure is a CdI_2 type with OH dipoles instead of single anions. The OH dipoles are oriented parallel to the hexagonal axis with the H atoms on the outer surface of the layers. Thus, each proton is located in the tetrahedral hole formed by the oxygen atom, to which it is covalently bonded, and three O atoms of the neighboring layer. The values for the O—H distance

TABLE 150

CrI_2(II) structure, orthorhombic, C_{2v}^{12}—$Cmc2_1$ (No. 36), $Z = 4$.
$(0, 0, 0; \frac{1}{2}, \frac{1}{2}, 0)$ +
all atoms in 4(a): 0, y, z; 0, $\bar{y}$, $\frac{1}{2}+z$.

CrI_2: $a = 3.915$ Å, $b = 7.560$ Å, $c = 13.553$ Å;

	y	z	
Cr	0.164	0.380	
I_I	0.207	0	[781]
I_{II}	0.530	0.259	

TABLE 150a

Interatomic distances in Cu and Cr dihalides

	$CuCl_2$	$CuBr_2$	$CrBr_2$	monocl. CrI_2	orthorh. CrI_2
M—X(Å)	2.35 (4)	2.40 (4)	2.55 (4)	2.74 (4)	2.72 (2), 2.75 (2)
	2.94 (2)	3.17 (2)	3.00 (2)	3.24 (2)	3.22, 3.24
Shortest intralayer X—X(Å)	3.30	3.32	3.55	3.82	3.82
interlayer X—X(Å)	3.35	3.83	3.89 (2)	4.22	4.23 (2)
	3.72 (2)	3.89 (2)	3.97	4.25 (2)	4.28

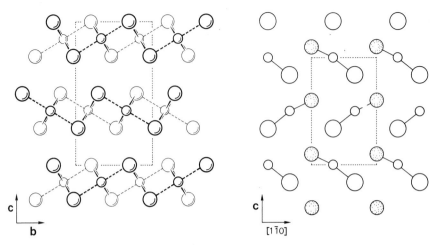

Fig. 135. The orthorhombic structure of CrI_2 (left) as compared with the hexagonal Cd(OH)Cl structure (right). The weakened bonds in CrI_2 are shown by broken lines. OH: stippled

calculated from reported atomic positions range from 0.8 to 1.05 Å. Interlayer H—H distances of 2.0–2.2 Å result which are comparable to twice the Van der Waals radius of hydrogen (2×1.2 Å). The oxygen atom has 3M+1H neighbors and thus uses all four valence orbitals in $p,^3 s \cdots sp^3$ bonds which accounts for the absence of interlayer H-bridges [11]. The Ca—O distance in $Ca(OH)_2$ fairly well corresponds to the sum of the ionic radii. Thus, it looks as if the surprisingly low axial ratios were the consequence of the pear form of the OH dipoles, but actually they help to increase the Madelung constant and obviously correspond to the optimum in Coulomb and polarization energy [783].

A mixture of halide layers drastically increases the c/a value. The situation is less clear in the case of oxygen. MnO(OH) can practically not be distinguished from $Mn(OH)_2$ whereas on oxidizing $Ni(OH)_2$ to NiO(OH) its a-axis strongly contracts while the lattice expands perpendicularly to the layers. On dehydration brucite retains its structure $Mg(OH)_{2-2x}O_x\square_x$ up to $x \approx 1$ [1030a].

It is interesting in view of the theory of structural polytypism that no polytypes have been detected among the crystals of $Mg(OH)_2$, $Ca(OH)_2$ and $Cd(OH)_2$ [741].

Ordered substitutions lead to various superstructures. The superstructures $M_n X_{2n-x} X'_x$ derived from the CdI_2 structure by anion substitution have been discussed theoretically by Aebi [759]. Aebi has given possible coordinations and space groups within the limits $1 \leq x \leq n$, $n \leq 4$, for all structures with single-layer unit cells and for two-layer structures with screw axes normal to the layer plane.

Ordered cation substitution is possible as well and the simplest superstructure of the brucite type is found in $Li_2Pt(OH)_6$. The three cations require a $\sqrt{3}a_0$ cell with the original c-axis. A rhombohedral cation ordering is realized in the $Na_2Sn(OH)_6$ structure. The anions, however, are markedly shifted towards the

TABLE 151

$Mg(OH)_2$ (brucite) structure, trigonal D_{3d}^3—$P\bar{3}m1$ (No. 164), $Z=1$.
O and H in 2(d): $\pm(\frac{1}{3}, \frac{1}{3}, z)$
Mg in 1(a): 0, 0, 0.

Compound	a(Å)	c(Å)	c/a	z_O	z_H	Ref.
$Mg(OH)_2$	3.147	4.768	1.515	0.217		[780]
26 °C	3.148	4.768	1.515			[2, 4]
	3.142	4.766	1.517	0.221 6	0.430 3	[779]
					0.426	[787]
$Ca(OH)_2$	3.593	4.909	1.366			[741]
	3.584 4	4.896 2	1.366	0.233 0		[4]
26 °C	3.592 9	4.908 2	1.366			[2]
	3.585 3	4.895	1.365	0.233 0	0.395	[788]
20 °C	3.591 8	4.906 3	1.366	0.234 1	0.424 8	[740]
133 K	3.586 2	4.880 1	1.361	0.234 6	0.428 0	[740]
77 K					0.437	[787]
$Ca_{0.5}Cd_{0.5}(OH)_2$	3.50	4.83	1.38			[1]
$Mn(OH)_2$	3.34	4.68	1.40			[4]
	3.322	4.734	1.425	0.226	0.446	[742]
$Fe(OH)_2$	3.258	4.605	1.413			[4]
$Co(OH)_2$	3.173	4.640	1.462	0.22		[1, 4]
$Co_{0.75}Zn_{0.25}(OH)_2$	3.15	4.66	1.48			[1]
$Ni(OH)_2$	3.117	4.595	1.474			[4]
	3.13	4.63	1.48	0.24	0.47	[745]
	3.126	4.605	1.473	0.23	0.415 2	[746]
$Ni(OD)_2$					0.439 2	[746]
$Ni_{0.8}Zn_{0.2}(OH)_2$	3.125	4.605	1.474			[1]
$Cu(OH)_2$	3.30	(4.85)	1.47			[2]
$Zn(OH)_2$	3.03	4.85	1.60			[1]
	3.19	4.645	1.46			[2]
$Cd(OH)_2$	3.48	4.67	1.34			[4]
	3.500	4.710	1.346			[741]
	3.496	4.702	1.345	0.27		[788]
	3.499	4.701	1.344	0.241		[783]
$Mg(OH)_2$ – CdI_2-type mixtures						
α-$Mg(OH)_{3/2}Cl_{1/2}$	3.247	5.733	1.766			[747]
α-$Mn(OH)_{3/2}Cl_{1/2}$	3.37	5.55	1.65			[747]
$MnO(OH)$	3.32	4.71	1.42			[748]
α-$Fe(OH)_{3/2}Cl_{1/2}$	3.32	5.52	1.66			[747]
α-$Co(OH)_{3/2}Cl_{1/2}$	3.22	5.50	1.71			[4]
	3.22	5.56	1.73			[747]
$Co(OH)_{3/2}Br_{1/2}$	3.23	5.91	1.83			[4]
$Co(OH)_{3/2}(NO_3)_{1/2}$	3.17	6.95	2.19			[757]
β-$NiO(OH)$	2.82	4.85	1.72			[2]
$Ni(OH)_{3/2}Cl_{1/2}$	3.15	5.36	1.70			[4]
	3.19	5.51	1.73			[747]
$Ni(OH)_{3/2}Br_{1/2}$	3.18	5.80	1.82			[757]
$Ni(OH)N_3$	3.12	7.2	2.3			[757]
$Cu_{0.7}Mg_{0.3}(OH)_{3/2}Cl_{1/2}$	3.14	5.80	1.85			[757]
$Cu_{0.75}Co_{0.25}(OH)_{3/2}Cl_{1/2}$	3.18	5.77	1.81			[757]
$Cu_{0.75}Zn_{0.25}(OH)_{3/2}Cl_{1/2}$	3.21	5.78	1.80			[757]
$Zn(OH)_{3/2}F_{1/2}$	3.19	4.65	1.46			[757]
$Zn(OH)N_3$	3.20	7.4	2.3			[757]
$Cd(OH)_{1.75}Cl_{0.25}$	3.53	5.03	1.42			[4]
$Cd(OH)_{1.33}Br_{0.67}$	3.61	6.01	1.67			[757]
$Cd(OH)_{3/2}I_{1/2}$	3.64	6.60	1.81			[757]

TABLE 152
Brucite superstructures based on cation ordering.
$Li_2Pt(OH)_6$ type, hexagonal, D_{3d}^1—P$\bar{3}$1m (No. 162), $Z=1$.
O in 6(k): $\pm(x, 0, z; 0, x, z; x, x, \bar{z})$
Li in 2(c): $\pm(\frac{1}{3}, \frac{2}{3}, 0)$
Pt in 1(a): 0, 0, 0.
(Positions of H not determined)

Compound	a(Å)	c(Å)	$\sqrt{3}c/a$	x_O	z_O	Ref.
$Li_2Pt(OH)_6$	5.362	4.647	1.501	0.325	0.254	[777]
$Na_2Pt(OH)_6$	5.831	4.755	1.412			[777]

$Na_2Sn(OH)_6$ type, trigonal, C_{3i}^2—R$\bar{3}$ (No. 148), $Z=3$.
hexagonal axes: $(0, 0, 0; \frac{1}{3}, \frac{2}{3}, \frac{2}{3}; \frac{2}{3}, \frac{1}{3}, \frac{1}{3})+$
(OH) in 18(f): $\pm(x, y, z; \bar{y}, x-y, z; y-x, \bar{x}, z)$
Na in 6(c): $\pm(0, 0, z)$
Sn in 3(a): 0, 0, 0.

Compound	a_{rh}(Å)	α(°)	a_h(Å)	c_h(Å)	$c/\sqrt{3}a$	z_{M^+}	x_{OH}	y_{OH}	z_{OH}	Ref.
$Na_2Sn(OH)_6$	5.84	61.2	5.96	14.20	1.38	0.320	0.404	0.317	0.417	[4]
$K_2Sn(OH)_6$	5.68	70.1	6.54	12.78	1.13	0.315	(0.404)	(0.317)	(0.417)	[4]
	5.697	70.02	6.537	12.803	1.13					[992]
$K_2Ir(OH)_6$	5.632	69.37	6.410	12.736	1.147					[992]
$K_2Pt(OH)_6$	5.65	69.0	6.41	12.84	1.16					[4]
	5.676	68.92	6.42	12.892	1.16					[992]
anion hcp						$\frac{1}{3}$	$\frac{1}{3}$	$\frac{1}{3}$	$\frac{1}{4}$	

smaller cations reflecting some tendency towards complex-anion formation (Table 152.

It is worth noting that on replacing the dipole OH$^-$ by the amide ion NH$_2^-$ the layer character of the structure gets lost. Ca(NH$_2$)$_2$, Sr(NH$_2$)$_2$ and Eu(NH$_2$)$_2$ crystallize with the tetragonal anatase structure which is a three-dimensional NaCl derivative. In the tetragonal structure of Mg(NH$_2$)$_2$, on the other hand, the cations

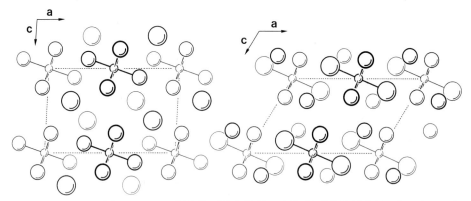

Fig. 136. The monoclinic structures of MgCl$_2$·6H$_2$O (left) and CoCl$_2$·6H$_2$O (right) projected onto the (010) plane. The spheres of medium size represent the O atoms of the H$_2$O molecules; the H atoms are omitted. Smallest spheres: cations, largest spheres: Cl.

TABLE 153

$MgCl_2 \cdot 6H_2O$ structure, monoclinic, C_{2h}^3—C2/m (No. 12), $Z = 2$.

$(0, 0, 0; \frac{1}{2}, \frac{1}{2}, 0) +$
H_{III}, H_{IV} and O_{II} in 8(j): $\pm(x, y, z; x, \bar{y}, z)$
Cl, H_I, H_{II}, and O_I in 4(i): $\pm(x, 0, z)$
Mg in 2(a): 0, 0, 0.

	a(Å)	b(Å)	c(Å)	β(°)	x(Cl, Br)	z(Cl, Br)	$x(O_I)$	$z(O_I)$	$x(O_{II})$	$y(O_{II})$	$z(O_{II})$	References
$MgCl_2 \cdot 6H_2O$ type												
$MgCl_2 \cdot 6H_2O$ 23°C	9.90	7.15	6.10	94.0	0.318	0.615	0.20	0.11	0.96	0.20	0.23	[4]
	9.8607	7.1071	6.0737	93.758	0.3176	0.6122	0.2018	0.1095	0.9571	0.2066	0.2233[a]	[852]
$MgClBr \cdot 6H_2O$	10.053	7.243	6.160	93.85	0.3185	0.5211	0.2025	0.1027	0.9572	0.2047	0.2135	[853]
$MgBr_2 \cdot 6H_2O$	10.25	7.40	6.30	93.5	0.320	0.615	0.20	0.11	0.96	0.20	0.23	[4]
$CoCl_2 \cdot 6H_2O$ type												
$CoCl_2 \cdot 6H_2O$[b,c]	10.34	7.06	6.67	122.33	0.274	0.171	0.288	0.702	0.0312	0.208	0.251	[4, 854]
$CoBr_2 \cdot 6H_2O$[c]	11.00	7.16	6.90	124								[855, 856, 1058]
$NiCl_2 \cdot 6H_2O$[d]	10.23	7.05	6.57	122.17	0.273	0.170	0.285	0.697	0.030	0.208	0.253	[857]
$NiBr_2 \cdot 6H_2O$	11.63	7.185	6.83	128.83								[1]

[a] $MgCl_2 \cdot 6H_2O$:
$x(H_I) = 0.2372$, $z(H_I) = 0.2583$; $x(H_{II}) = 0.2693$, $z(H_{II}) = 0.0083$;
$x(H_{III}) = 0.0209$, $y(H_{III}) = 0.2998$, $z(H_{III}) = 0.2784$; $x(H_{IV}) = 0.8840$, $y(H_{IV}) = 0.1984$, $z(H_{IV}) = 0.3151$ [852]

[b] $CoCl_2 \cdot 6H_2O$:
$x(H_I) = 0.190$, $z(H_I) = 0.548$; $x(H_{II}) = 0.271$, $z(H_{II}) = 0.834$;
$x(H_{III}) = 0.128$, $y(H_{III}) = 0.266$, $z(H_{III}) = 0.286$; $x(H_{IV}) = 0.454$, $y(H_{IV}) = 0.197$, $z(H_{IV}) = 0.222$ [855]
Thermal expansion around the Néel temperature $T_N = 2.28$ K: [858]

[c] For the crystal data of the monoclinic room-temperature and the triclinic low-temperature modifications of $CoCl_2 \cdot 6D_2O$ and $CoBr_2 \cdot 6D_2O$ their magnetic structures see [1058].

[d] For the hydrogen parameters at 300 K and 4.2 K (below the Néel point of 5.3 K).

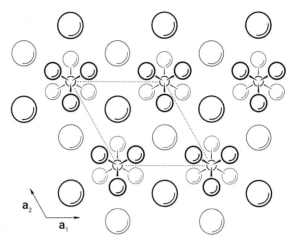

Fig. 137. Projection of the hexagonal structure of $NiI_2 \cdot 6H_2O$. Spheres of medium size represent oxygen atoms of the water molecules, the H atoms of which are omitted. Smallest spheres: cations, largest spheres: iodine.

occupy the tetrahedral holes of a ccp N stacking but again the bonding is three dimensional [785].

In $M_2M'X_6$ compounds with small M' cations $[M'X_6]$ complex formation is likely to take place. Layered structures then may form which are based on close-packed sandwiches M—$[M'X_6]$—M. Thus the $(NH_4)_2[SiF_6]$ structure, which is observed with large alkali ions and M' = Ti, Zr, Hf, Re, Ru, Pt, ..., may be interpreted as an anti-CdI_2 structure though the layer character is not well pronounced.

In Table 148 we have listed $Ni(ClO_4)_2$ as a $CdCl_2$-like representative. Instead of complex anions we may as well form complex cations, for example by the addition of $6H_2O$ or $6NH_3$. Thus, the monoclinic structure adopted by $MgCl_2 \cdot 6H_2O$ and $MgBr_2 \cdot 6H_2O$ can be interpreted as a distorted CdI_2 structure built up from $[Mg(H_2O)_6]^{2+}$ ions while the isopuntal structure of $CoCl_2 \cdot 6H_2O$ represents the $CdCl_2$ analog (Figure 136). The hexagonal structure of $NiI_2 \cdot 6H_2O$ can be described in the same terms, however, in this case the H_2O molecules almost fit into the iodine sublattice. Although it yields a puckered layer we may, in our description, integrate the H_2O molecules into the anion sublattice, $(3H_2O+I)_n$, and call the structure an ordered defective CdI_2 type,

TABLE 154

$NiI_2 \cdot 6H_2O$ structure, trigonal, D_{3d}^3—$P\bar{3}m1$ (No. 164), $Z = 1$.
O in 6(i): $\pm(x, \bar{x}, z; x, 2x, z; 2\bar{x}, \bar{x}, z)$
I in 2(d): $\pm(\frac{1}{3}, \frac{2}{3}, z)$
Ni in 1(a): 0, 0, 0.

$NiI_2 \cdot 6H_2O$: $a = 7.638$ Å, $c = 4.876$ Å, $2c/a = 1.277$;
$x(O) = 0.126\ 4$, $z(O) = 0.754\ 9$; $z(I) = 0.214\ 2$ [831]

TABLE 155

Cd(OH)Cl structure (C27 type), hexagonal. C_{6v}^4—P6$_3$mc (No. 186). $Z=2$
Cd and Cl in 2(b): $\frac{1}{3}, \frac{2}{3}, z; \frac{2}{3}, \frac{1}{3}, \frac{1}{2}+z$ ($z_M=0$)
(OH) in 2(a): $0, 0, z; 0, 0, \frac{1}{2}+z$.

Compound	a(Å)	c(Å)	$c/2a$	z_{OH}	z_{Hal}	Ref.
Ca(OH)Cl	3.86	9.90	1.28			[753]
Cd(OH)Cl	3.66	10.27	1.40	0.10	0.337	[4, 753]
	3.665	10.235	1.40			[754]
CdBr$_2$	3.985	12.561	1.576	$\frac{1}{8}$	$\frac{3}{8}$	[765]
CdI$_2$	4.24	13.67	1.61	$\frac{1}{8}$	$\frac{3}{8}$	[4]
GeI$_2$	4.17	13.33	1.60		$\frac{3}{8}$	[803]
PbI$_2$	4.557	13.958	1.53	0.133	0.367	[4]

Ni$_{1/4}$[(H$_2$O)$_{3/4}$I$_{1/4}$]$_2$, where ordering leads to a supercell, $a=2a_0$, in which only $\frac{1}{4}$ of the cation sites is occupied (Figure 137, Table 154).

e. MIXED HYDROXYHALIDES

Mixing of dihalides and dihydroxides leads to a number of distorted ordered structures. Simple layer structures occur for M(OH)X halides. Apart from Co and Cd phases only the chlorides appear to be investigated. For the smaller cations Mg, Mn, Co and Ni, a CdI$_2$/CdCl$_2$-related hexagonal 6-layer structure (β-Mn(OH)Cl type) has been reported without details [753]. Moreover, Co(OH)Cl and Mn(OH)Cl exist in an orthorhombic (pseudohexagonal) cell with the same c-value [753]. An ordered version of the 4H CdI$_2$ polytype structure is realized with Cd and Ca (Figure 135, Table 155). The alternation of pure OH and Cl layers, as met in the Cd(OH)Cl structure, is possible only with relatively large cations. A statistical arrangement of Cl and OH ions on both sides exerts less strain onto the cation layer [757], so that quite a number of mixed phases adopt a disordered CdI$_2$- or CdCl$_2$-type structure. It is noteworthy that the a-axis of the Cd(OH)Cl cell corresponds to the arithmetic mean of those of Cd(OH)$_2$ and CdCl$_2$. For a model based on close-packed anion layers one should expect a value close to that of CdCl$_2$.

The same anion stacking $hchc$ as in Cd(OH)Cl, but an ordered distribution of equal amounts of Cl and OH ions in both anion layers is found in orthorhombic β-Cu(OH)Cl [782] (Table 156). The anions are arranged in alternating zigzag rows. One tooth of this sawtooth distribution has the length $2a_0$. In terms of a CdI$_2$-like cell the unit cell of the orthorhombic β-Cu(OH)Cl is roughly

$$a = 3\sqrt{3}a_0, \quad b = 2a_0, \quad c \approx 2c_0.$$

As is to be expected for a Cu^{2+} ion its coordination octahedron is deformed in such a way that 2OH+2Cl form four short bonds while 1OH+1Cl form the

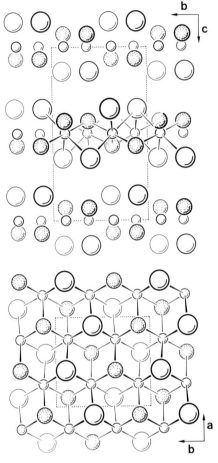

Fig. 138. Projection on (100) (above) and middle layer (below) of the orthorhombic β-Zn(OH)Cl structure. Largest spheres: Cl, smallest spheres: Cu, stippled spheres: OH.

elongated diagonal. Usually the weakened bonds are formed by the weaker anions only. The mean distances are

$$\text{Cu}-2\,\text{OH} \quad \text{at} \quad \sim 2.02\,\text{Å} \quad \text{and} \quad 1\,\text{OH} \quad \text{at} \quad \sim 2.46\,\text{Å}$$
$$2\,\text{Cl} \quad \text{at} \quad \sim 2.30\,\text{Å} \quad \text{and} \quad 1\,\text{Cl} \quad \text{at} \quad \sim 2.72\,\text{Å}$$

The OH—Cl distances between neighboring sandwiches are 3.18 and 3.22 Å. The small difference relative to the sum of the ionic radii (Table 161) is interpreted as being due to some O—H ··· Cl bonding [782].

A second kind of ordering of Cl and OH within the same layer is found in the orthorhombic β-Zn(OH)Cl structure (Table 157) reproduced in Figure 138. In this structure the anions are also ordered within zigzag rows but here the sawteeth

TABLE 156
β-Cu(OH)Cl structure, orthorhombic, C_{2v}^9—Pna2_1 (No. 33), $Z = 16$.
atoms in 4(a): x, y, z; $\bar{x}, \bar{y}, \frac{1}{2}+z$; $\frac{1}{2}+y, \frac{1}{2}+z$; $\frac{1}{2}+x, \frac{1}{2}-y, z$.

β-Cu(OH)Cl: $a = 11.72$ Å, $b = 6.45$ Å, $c = 11.75$ Å [782]

	x	y	z
Cu$_I$	0.052	0.950	0
Cu$_{II}$	0.052	0.450	0
Cu$_{III}$	0.302	0.250	0
Cu$_{IV}$	0.302	0.750	0
Cl$_I$	0.964	0.700	0.892
Cl$_{II}$	0.141	0.200	0.108
Cl$_{III}$	0.873	0.950	0.853
Cl$_{IV}$	0.232	0.450	0.147
Cl$_V$	0.962	0.200	0.057
Cl$_{VI}$	0.143	0.700	0.943
Cl$_{VII}$	0.228	0.950	0.108
Cl$_{VIII}$	0.873	0.450	0.892

have only a length of a_0. The resulting rippled layers are stacked along c in such a manner that the large anions fit onto the channels created by the smaller anions. The stacking can be described as

$$\cdots C\beta A' \; A\beta C' C\beta A' \; A\beta C' \cdots$$

Thus the closed-packed $(X_{1/2}X'_{1/2})$—M—$(X'_{1/2}X_{1/2})$ sandwiches themselves are not close packed. The rippling of the layers is evident from the difference in the Zn—O and Zn—Cl distances. Each cation is octahedrally surrounded by $2O + 1Cl$ on one side and $1O + 2Cl$ on the opposite side. For Zn(OH)Cl the distances are

$$\left.\begin{array}{l}\text{Zn—O} = 2.01 \text{ and } 2.17 \text{ Å} \\ \text{Cl} = 2.46 \text{ Å}\end{array}\right\} \text{on one side}$$

$$\left.\begin{array}{l}\text{Zn—O} = 2.08 \text{ Å} \\ \text{Cl} = 2.54 \text{ and } 2.54 \text{ Å}\end{array}\right\} \text{on the opposite side of the layer}$$

The separation between the layers is O—Cl = 3.24 Å compared with O—Cl = 3.08 Å, O—O = 2.56 Å and Cl—Cl = 3.48 Å within the layers. For Co(OH)Br the distances turn out to be

Co—3 O at 2.08 and 2.14 Å (2)
3 Br at 2.49, 2.56 and 2.65 Å
O—Br = 3.46 Å between the layers.

TABLE 157

β-Zn(OH)Cl structure, orthorhombic, D_{2h}^{15}—Pcab (No. 61), $Z=8$
all atoms in 8(c): $\pm(x, y, z; \frac{1}{2}-x, y+\frac{1}{2}, \bar{z}; x+\frac{1}{2}, \bar{y}, \frac{1}{2}-z; \bar{x}, \frac{1}{2}-y, z+\frac{1}{2})$

Compound	a(Å)	b(Å)	c(Å)	Ref.
α-Mn(OH)Cl	6.07	6.91	11.46	[4]
β-Fe(OH)Cl	5.928	6.661	11.330	[753]
Co(OH)Cl	5.75	6.60	11.38	[4]
Co(OH)Br†	5.903	6.700	11.86	[4]
(Cu(OH)Cl)^p	5.55	6.67	11.12	p
β-Zn(OH)Cl*	5.86	6.58	11.33	[4, 756]
β-Cd(OH)Br	6.440	7.325	12.096	[754]
β-Cd(OH)I	6.624	7.458	12.962	[754]

		x	y	z	
* β-Zn(OH)Cl:	Zn	0.250	0.131	0.000	
	O	0.594	0.126	0.071	[4, 756]
	Cl	0.583	0.107	0.357	

		x	y	z	
† Co(OH)Br:	Co	0.245	0.137	0.002	
	O	0.58	0.103	0.068	[4]
	Br	0.578	0.101	0.36	

		x	y	z	
^{p}: hypothetical					
β'-Cu(OH)Cl:	Cu	0.27	0.12	−0.015	taken from Figure 2b
	O	0.59	0.16	0.061	in [755]
	Cl	0.53	0.10	0.344	

Identical rippled layers and the same manner of packing is found in the monoclinic α-Cu(OH)Cl structure represented in Figure 139 (Table 158). The orientation of subsequent sandwiches, however, is different:

$$\cdots B'\gamma A\ A'\beta C\ C'\alpha B\ B'\gamma A\ A'\beta C \cdots$$

The difference in stacking between the α-Cu(OH)Cl and the β-Zn(OH)Cl type is analogous to the difference between the 4H type (C27) and the CdCl$_2$ type (C19). In the Cu compound the cation lies at the center of a distorted square formed by 3 OH + 1 Cl. Two more Cl at a considerably larger distance complete the coordination octahedron:

Cu—3 OH at 2.00, 2.01 and 2.03 Å
  1 Cl at 2.30 Å
  2 Cl at 2.70 and 2.73 Å

The pronounced deformation of the sandwiches is a peculiarity due to the d^9 Jahn-Teller ion Cu^{2+}. As can be seen in Table 158 α-Cu(OH)Cl is distinguished from the other members of this structure type by a shortened a-axis and a lengthened c-axis. To our knowledge no detailed structural analysis has been

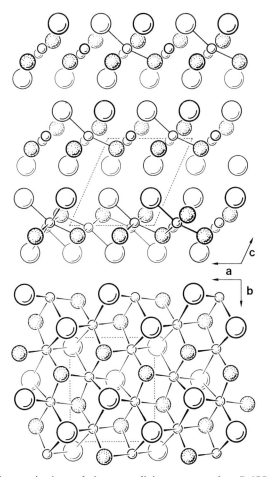

Fig. 139. Two projections of the monoclinic structure of α-Cu(OH)Cl.
Large spheres: Cl
smallest spheres: Cu
stippled spheres: OH

made for the other phases, but it is to be expected that their positional parameters will be somewhat different. This structure type has been predicted by Aebi [759].

Table 159 presents a survey of the structures occurring with compounds of composition $M_2(OH)_3X$. Although two kinds of 1:3 ordering are possible in a close-packed monolayer we are aware of only one ordered layer structure, the botallackite type shown in Figure 140, which is derived from the CdI_2 structure. It is rather non-trivial that for the chloride salts of Mg, Mn, Fe, Co and Ni both ordered structures, the atacamite and the β-$Co_2(OH)_3Cl$ type, are three-dimensional and based on a ccp anion packing —X—($\frac{1}{4}$M)—X—($\frac{3}{4}$M)—X— whereas in the disordered state the hcp layer-type stacking is stable.

The ordered substitution in the CdI_2 structure of one quarter of the anions offers the opportunity for the formation of the two long bonds adequate to the

TABLE 158

α-Cu(OH)Cl structure, monoclinic, C_{2h}^5—P2$_1$/a (No. 14), Z = 4.
all atoms in 4(e): ±(x, y, z; $x+\frac{1}{2},\frac{1}{2}-y, z$)

Compound	a(Å)	b(Å)	c(Å)	β(°)	Ref.
α-Fe(OH)Cl	5.929	6.662	6.006	109.18	[753]
α-Cu(OH)Cl*	5.555	6.671	6.127	114.88	[755]
α-Zn(OH)Cl	5.864	6.570	6.005	108.95	[753]
Cd(OH)Br	6.434	7.327	6.502	111.80	[754]
Cd(OH)I	6.620	7.473	6.918	110.48	[754]

* α-Cu(OH)Cl:		x	y	z	
	Cu	0.254 1	0.117 3	0.968 1	
	Cl	0.322 0	0.410 1	0.688 0	[755]
	O	0.154 9	0.349 3	0.122 7	

Cu^{2+} ion. The two kinds of cations in the botallackite structure have a different coordination, one has five OH neighbors while the second has only four. Again we have found positional parameters only for the Cu salts for which we have compiled the interatomic distances in Table 161. The shortening of the observed interlayer OH—X distances relative to the sum of the ionic radii is believed to be due to hydrogen bonds [758, 829]. These possible hydrogen bonds are indicated in Figure 140 by broken lines. All copper compounds crystallize in the form of greenish flat needles with b as the needle axis.

TABLE 159

Structure types observed on 3:1 hydroxide halides. Layer structures are printed in italics

Cation	M(OH)$_{3/2}$Cl$_{1/2}$	M(OH)$_{3/2}$Br$_{1/2}$	M(OH)$_{3/2}$I$_{1/2}$
Mg	α: CdI$_2$ β: atacamite		
Mn	α: CdI$_2$ β: atacamite	atacamite	*botallackite*
Fe	α: CdI$_2$ β: rh. β-Co(OH)$_{3/2}$Cl$_{1/2}$	*botallackite*	*botallackite*
Co	α: CdI$_2$ β: rhombohedral (~paratacamite)	α: *botallackite* β: rh. β-Co(OH)$_{3/2}$Cl$_{1/2}$	*botallackite*
Ni	α: CdI$_2$ β: atacamite	*botallackite*	*botallackite*
Cu	α: *mon. botallackite* β: γ: rh. paratacamite [829] δ: orth. atacamite	*botallackite*	*botallackite*
Zn			
Cd	α: orthorhombic β: atacamite	atacamite	atacamite
Sn	orthorhombic		
Pb	hexag. penfieldite		

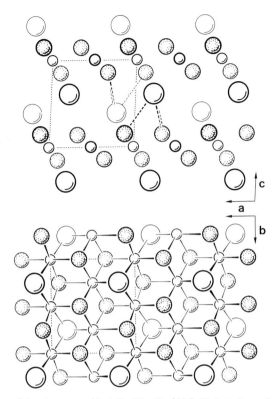

Fig. 140. Projections of the structure of botallackite $Cu_2(OH)_3Cl$. Interlayer O—H·····Cl bonds are indicated by broken lines
Largest spheres: Cl; smallest spheres: Cu; stippled spheres: OH

f. COMPOSITE LAYER STRUCTURES DERIVED FROM BRUCITE

Quite a number of hydroxides and hydroxyhalides possess structures which are reminiscent of the intercalation compounds. Thus the structure of green cobalt hydroxybromide $4Co(OH)_2 \cdot Co(OH)Br$ consists of an ordered $CdCl_2$-type $Co(OH)_2$ skeleton. Between each (OH)Co(OH) sandwich a monolayer of Co(OH)Br is intercalated in a disordered manner. The interlayer ions are very mobile, both anions and cations are easily exchangeable. Phases like $[4Co(OH)_2]$-CoOCl, $[4Co(OH)_2]CoOBr$, $[4Co(OH)_2]CoONO_3$, $[4Fe(OH)_2]FeOCl$, $[4Mg(OH)_2]AlOCl$, $[4Ni(OH)_2]NiO(OH)$, $[2Mn(OH)_2 \cdot 2AlO(OH)]AlOCl$ and $[Mn_3Al(OH)_7O]AlO(OH) \cdots [Mn_2Al_2(OH)_6O_2]AlOCl$ have been prepared [757] or are found as minerals, e.g. the bluish-green, soapy iowate $[4Mg(OH)_2]FeOCl \cdot x$ H_2O [832] and the yellow-green $[4Mg(OH)_2](Ni,Fe)O(OH)$ [833]. Moreover, it is possible to intercalate hydroxides of trivalent cations. Thus $[4M^{2+}(OH)_2]M^{3+}(OH)_3$ phases have been reported for $M^{2+}=$ Mg, Mn and $M^{3+}=$ Al, Mn, Fe, as well as for Co^{2+} and·Al, Fe^{3+}, Co^{3+}, for Ni^{2+} and Al and finally for Cd and Cr^{3+}. On precipitating these phases from aqueous solutions

TABLE 160

$Cu_2(OH)_3Cl$ (*botallackite*) structure, monoclinic, C_{2h}^2—$P2_1/m$ (No. 11), $Z=2$.
$(OH)_{II}$: in 4(f): $\pm(x, y, z; x, \frac{1}{2}-y, z)$
Cu_{II}, Cl and $(OH)_I$ in 2(e): $\pm(x, \frac{1}{4}, z)$
Cu_I in 2(a): $0, 0, 0; \frac{1}{2}, 0$.

Compound	a(Å)	b(Å)	c(Å)	β(°)	Ref.
$Mn_2(OH)_3I$	5.893	6.759	6.499	93.03	[747]
$Fe_2(OH)_3Br$	5.747	6.566	5.881	91.35	[747]
$Fe_2(OH)_3I$	5.773	6.585	6.434	94.93	[747]
α-$Co_2(OH)_3Br$	5.612	6.452	5.944	91.50	[747]
$Co_2(OH)_3I$	5.624	6.450	6.465	94.83	[747]
$Ni_2(OH)_3Br$	5.502	6.352	5.839	90.78	[747]
$Ni_2(OH)_3I$	5.547	6.386	6.417	94.53	[747]
$Cu_2(OH)_3Cl$	5.640	6.138	5.726	93.25	[747]
†	5.63	6.12	5.73	93.75	[760]
$Cu_2(OH)_3Br$*	5.640	6.139	6.056	93.50	[748]
$Cu_2(OH)_3I$**	5.653	6.157	6.560	95.17	[748]

† $Cu_2(OH)_3Cl$:

	x	y	z
Cu_{II}	0.50		0
Cl	0.210		0.392
$(OH)_I$	0.883		0.85
$(OH)_{II}$	0.324	0	0.857

* $Cu_2(OH)_3Br$:

	x	y	z
Cu_{II}	0.504 3		0.003 1
Br	0.210 4		0.376 7
$(OH)_I$	0.883 0		0.837 5
$(OH)_{II}$	0.300 0	0.004 3	0.862 9

** $Cu_2(OH)_3I$:

	x	y	z
Cu_{II}	0.509 0		0.003 0
I	0.207 7		0.380 6
$(OH)_I$	0.881 0		0.838 0
$(OH)_{II}$	0.311 0	0.995 0	0.861 0

additional H_2O molecules are incorporated. These form mixed octahedral groups or single $[M(OH)_3(H_2O)_3]$ octahedra which connect the parent sandwiches by H-bridges. Since each oxygen atom of the intercalated complexes possesses at least one free electron pair it is well disposed to bind an (OH) group of the parent $M(OH)_2$ lattice. The arrangement of the intercalated octahedra is therefore not close-packed but similar to that in the $NaHF_2$ structure.

TABLE 160a

Coalingite structure, trigonal, D_{3d}^5—$R\bar{3}m$ (No. 166), $Z=\frac{1}{2}$.
hexagonal cell: $(0, 0, 0; \frac{1}{3}, \frac{2}{3}, \frac{2}{3}; \frac{2}{3}, \frac{1}{3}, \frac{1}{3})+$
O_{III} in 18(h): $\pm(x, \bar{x}, z; x, 2x, z; 2\bar{x}, \bar{x}, z)$
M, O_I and O_{II} in 6(c): $\pm(0, 0, z)$
$Mg_{10}Fe_2(OH)_{24}CO_3 \cdot 2H_2O$: $a=3.12$ Å, $c=37.4$ Å [874a]

Atoms	x	y	z	
0.87 Mg + 0.17 Fe			0.438 0	
O_I			0.076	
O_{II}			0.799	[874a]
0.14 O_{III}	0.105	0.210	0	

TABLE 161
Interatomic distances in botallackite-type Cu salts. All distances in Å

	$Cu_2(OH)_3Cl$	$Cu_2(OH)_3Br$	$Cu_2(OH)_3I$
Cu_I—2(OH)$_I$	1.86	1.92	1.96
2(OH)$_{II}$	2.04	1.93	2.05
2X	2.91	2.94	3.08
Cu_{II}—4(OH)$_{II}$	1.98	2.01, 2.05(3)	1.99, 2.03, 2.10(2)
1(OH)$_I$	2.37	2.41	2.45
1X	2.85	2.89	3.13
intralayer OH—X	3.40	3.36	3.41
interlayer OH—X	3.11(2), 3.29	3.32(2), 3.44	3.52(2), 3.66
sum of ionic radii $r(OH^-) + r(X^-)$ [758]	3.28	3.42	3.63

A CrO(OH)-like stacking of the brucite-type sandwiches caused by interlayer H-bonding is met also in the pyroaurite group. The excess charge due to a partial replacement of the divalent by trivalent cations is compensated by intercalation of the appropriate amount of hydrated $(CO_3)^{2-}$ ions which are well suited for a planar arrangement. Examples are [874]:

	3R polytype	2H polytype
$[Mg_{3/4}Fe_{1/4}(OH)_2](CO_3)_{1/8}(H_2O)_{1/2}$	pyroaurite	sjögrenite
$[Mg_{3/4}Al_{1/4}(OH)_2](CO_3)_{1/8}(H_2O)_{1/2}$	hydrotalcite	manasseite
$[Mg_{3/4}Cr_{1/4}(OH)_2](CO_3)_{1/8}(H_2O)_{1/2}$	stichtite	barbertonite
$[Ni_{3/4}Fe_{1/4}(OH)_2](CO_3)_{1/8}(H_2O)_{1/2}$	reevesite	

In certain double layer structures the intercalated layers are also ordered. Thus, a superstructure of the $[4Co(OH)_2]Co(OH)Br$ type is realized in $[4Zn(OH)_2]ZnCl_2$ with $a = 2a_0$. In $[6Zn(OH)_2]ZnCl_2$ the main layers are arranged as in the brucite type [757].

An interesting combination of the brucite and sjögrenite structures is realized in the minerals coalingite and coalingite-K [874a]. The structure of coalingite $Mg_{10}Fe_2(OH)_{24}CO_3 \cdot 2H_2O$ is made up of two brucite layers $[Mg_5Fe(OH)_{12}]^+$ displaced relative to each other, and one intercalated disordered $[CO_3 \cdot 2H_2O]^{2-}$ layer. This sandwich unit is rhombohedrally stacked to yield brucite-type layer pairs. Hydrogen bonding is effective only to the intercalated layers (O—O = 2.89 Å) but not between the brucite layers (O—O = 3.15 Å) so that coalingite represents a true layer-type phase.

Incomplete oxidation and carbonation or a lower iron concentration may lead to coalingite-K in which three brucite-type layers occur between each carbonate/water layer. The idealized formula for the fully oxidized mineral thus is

$Mg_{16}Fe_2(OH)_{36}CO_3 \cdot 2H_2O$. It is to be expected that other members of this mineral group exist or can be synthesized.

g. WHITE LEAD $Pb(OH)_2 \cdot 2PbCO_3$

Basic lead carbonate (white lead) has a slightly disordered structure insofar as the composite sandwiches are nearly but not exactly stacked in a rhombohedral sequence. The unit cell for one composite layer is trigonal (Table 161a). This composite layer is bounded by planes of triangular CO_3^{2-} ions. The Pb atoms which connect the CO_3 and OH layers have surroundings similar to those in the $PbCO_3$ structure: six O atoms of three CO_3 groups (Pb—O ≈ 2.75 Å) on one side and four OH ions (instead of three equidistant O atoms) on the other side, one directly below the Pb atom at a distance of 2.75 Å and the other three at ~3.1 Å. The central layer is of a strongly distorted brucite type. The lead atoms are displaced by about 0.8 Å from the ideal positions. As a consequence each Pb atom has only three near OH neighbors at 2.4 and 2.6 Å(2). This obviously reflects the influence of the lone electron pair.

The translations between the layer units differ from the ideal rhombohedral $\pm(\frac{2}{3}, \frac{1}{3})$ by 10%. Whereas the disordered crystal has an average unit cell of trigonal symmetry, a completely ordered structure will be of lower symmetry, possibly triclinic [909].

TABLE 161a
Subcell of the approximately rhombohedral structure of white lead.
$Pb(OH)_2 \cdot 2PbCO_3$ structure, trigonal, C_{3v}^2—P31m (No. 157), $Z = 3$.
O_{III} and O_{IV} in 6(d): $x, y, z; \bar{y}, x-y, z; y-x, \bar{x}, z;$
$y, x, z; \bar{x}, y-x, z; x-y, \bar{y}, z$.
$Pb_{III}, Pb_{IV}, C_{III}, O_I, O_{II}$ and $(OH)_{III}$ in 3(c): $x, 0, z; 0, x, z; \bar{x}, \bar{x}, z$.
Pb_{II}, C_{II} and $(OH)_{II}$ in 2(b): $\frac{1}{3}, \frac{2}{3}, z; \frac{2}{3}, \frac{1}{3}, z$.
Pb_I, C_I and $(OH)_I$ in 1(a): 0, 0, z.
$Pb(OH)_2 \cdot 2PbCO_3$: $a' = 9.06$ Å, $c' = 8.27$ Å [909]

Atoms	x	y	z (estimated)
Pb_I			0.77
Pb_{II}			0.77
Pb_{III}	0.255		0.50
Pb_{IV}	0.660		0.23
C_I			0.14
C_{II}			0.14
C_{III}	0.66		0.86
O_I	0.16		0.14
O_{II}	0.51		0.86
O_{III}	0.53	0.16	0.14
O_{IV}	0.34	0.18	0.86
$(OH)_I$			0.42
$(OH)_{II}$			0.42
$(OH)_{III}$	0.70		0.58

TABLE 162
Koenenite structure [837]

Brucite-type hydroxide sublattice $[Mg_7Al_4(OH)_{22}]^{4+}$	$CdCl_2$-type Chloride sublattice $[Na_4(Ca_{2/3}Mg_{1/3})_2Cl_{12}]^{4-}$
P$\bar{3}$m1 (No. 164)	R$\bar{3}$m (No. 166)
M in 1(a): 0, 0, 0	M in 3(a): 0, 0, 0, rh.
OH in 2(d): $\pm(\frac{1}{3}, \frac{2}{3}, z)$	Cl in 6(c): $\pm(0, 0, z)$, rh.
$a = 3.052$ Å, $c = 10.88$ Å	$a = 4.072$ Å, $c = 3 \times 10.88$ Å
$z = 0.091\ 4$	$z = 0.379\ 7$

h. COMPOSITE LAYER STRUCTURES WITH INCOMMENSURATE SUBLATTICES

An intergrowth of two true layer structures is found in the scaly mineral koenenite. Brucite-like sandwiches $[Mg_7Al_4(OH)_{22}]^{4+}$ alternate with $CdCl_2$-type sandwiches $[Na_4(Ca, Mg)_2Cl_{12}]^{4-}$. Both sublattices are ordered but since the subcells are incommensurate a supercell cannot exist. The characteristic data for each sublattice are given in Table 162. Elsewhere the composition of koenenite is given as 1.78 $Mg_{0.64}Al_{0.36}(OH)_2 \cdot Na_{0.65}Mg_{0.35}Cl_2$ [834]. Hydrogen bridges O—H $\cdots$ Cl are assumed to be effective between the sandwiches in spite of the irregular relative orientation of the sublattices [836].

A very similar structure is adopted by valleriite which is a soft flaky mineral with a dull, bronzy lustre. In valleriite positively charged brucite-type $Mg_{2/3}Al_{1/3}(OH)_2$ layers alternate with sulfide sandwiches $Cu_{0.93}Fe_{1.07}S_2$ (Figure 141). In these two-dimensional chalcopyrite layers all the tetrahedral sites are

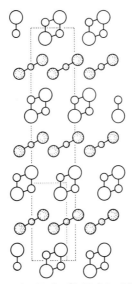

Fig. 141. (110) section of the structure of valleriite [1.53 (Mg, Al) (OH)$_2$] [CuFeS$_2$]. The atoms of the brucite-type layers are stippled.

Large empty circles: S
small empty circles: (Cu, Fe)

TABLE 163
Crystallographic data for the subcells of valleriite [834]

Hydroxide subcell (Mg(OH)$_2$ type)	Sulfide subcell
P$\bar{3}$m1 (No. 164)	R$\bar{3}$m (No. 166)
M in 1(b): 0, 0, $\frac{1}{2}$	M and S in 6(c): $\pm$(0, 0, z), rh.
OH in 2(d): $\pm(\frac{1}{3}, \frac{2}{3}, z)$	
a' = 3.070 Å, c' = 11.37 Å	a'' = 3.792 Å, c'' = 34.10 Å
z(OH) = 0.411 2	z(M) = 0.308 3, z(S) = 0.379 1
Super cell	
R$\bar{3}$m, R3m or R32	
a = 64.46 Å = 17a'' = 21a'	
c = 34.10 Å = c'' = 3c'	

occupied statistically by Cu and Fe atoms as in β-NiTe. We are thus happy to have found another example for Figure 24b. The stacking of the CuFeS$_2$ sandwiches is rhombohedral but not ccp as in β-NiTe. The CuFeS$_2$ sublattice can be described by

$$A\beta\alpha B \quad B\gamma\beta C \quad C\alpha\gamma A.$$

The S positions correspond to the anion positions in the NaHF$_2$ or γ-CrO(OH) (grimaldiite) structure. The definition of the independent sublattices is given in Table 163. In this case a supercell appears to exist. Since the a-axes of the two subcells are in the ratio 21:17 (though 26:21 is also fairly close to the experimental value) the composition of the mineral is [1.526 Mg$_{0.68}$Al$_{0.32}$(OH)$_2$][Cu$_{0.93}$Fe$_{1.07}$S$_2$]. The brucite layer thus carries a charge of 0.5 per formula. In order to compensate this charge the sulfide layer should contain 0.64 Fe^{2+} if we assume the remaining Fe to be trivalent and Cu monovalent. For steric reasons these Fe^{2+} ions cannot go into octahedral holes since the Fe—3Fe distance in the normal sites is already as short as 2.77 Å. Based on the accessible information we have to expect metallic properties. We wonder whether this deviation from charge balance is necessary for the stability of this hybrid structure or could a stoichiometric (non-metallic?) phase 1.5 Mg(OH)$_2$·CuFeS$_2$ exist as well? Quite a number of valleriite-type minerals have been analyzed ([1.56 M$_{0.83}$Fe$_{0.17}$(OH)$_2$]-Cu$_{0.81}$Fe$_{1.19}$S$_2$, [1.44 Mg$_{0.79}$Fe$_{0.21}$(OH)$_2$]Cu$_{0.87}$Fe$_{1.13}$S$_2$ both from Sweden, [1.57 Mg$_{0.73}$Al$_{0.27}$(OH)$_2$]Cu$_{0.95}$Fe$_{1.05}$S$_2$ and [1.67Mg$_{0.75}$Fe$_{0.16}$Al$_{0.09}$(OH)$_2$]Cu$_{1.04}$Fe$_{0.96}$S$_2$ from South Africa, [1.64 Mg$_{0.71}$Al$_{0.23}$Fe$_{0.06}$(OH)$_2$]Cu$_{1.19}$Fe$_{0.81}$S$_2$ from Canada, [1.49 Mg$_{0.68}$Fe$_{0.18}$Al$_{0.16}$(OH)$_2$]Cu$_{0.65}$Fe$_{1.35}$S$_2$ from California, and Cu-free species such as [1.58 Mg$_{0.53}$Fe$_{0.47}$(OH)$_2$] 2FeS from Cyprus, [1.64 Mg$_{0.23}$Fe$_{0.77}$(OH)$_2$] 2 FeS from California and [1.85 Mg$_{0.58}$Fe$_{0.26}$Al$_{0.16}$(OH)$_2$] 2 FeS) but only one, [1.47 Fe(OH)$_2$]Cu$_{1.00}$Fe$_{1.00}$S$_2$ from Western Siberia [835], appears to contain neutral layers, provided that all Fe atoms of the hydroxide sheet are divalent and those of the sulfide sheet trivalent. It is striking that according to these analyses the cation sublattices do not contain vacancies which might restore the charge balance. We might speculate that the valleriite family offers the

TABLE 164

Lithiophorite structure, monoclinic, C_{2h}^3—$C2/m$ (No. 12), $Z = 2$.

$(0, 0, 0; \frac{1}{2}, \frac{1}{2}, 0) +$
O and (OH) in 4(i): $\pm(x, 0, z)$
(Al, Li) in 2(d): $0, \frac{1}{2}, \frac{1}{2}$
Mn in 2(a): $0, 0, 0$

$Li_{0.32}Al_{0.68}(OH)_2 \cdot MnO_2$: $a = 5.06$ Å, $b = 2.91$ Å, $c = 9.55$ Å, $\beta = 100.5°$
$x(O) = 0.696$, $z(O) = 0.103$;
$x(OH) = 0.794$, $z(OH) = 0.397$ [838]

---

opportunity to create unique two-dimensional metals built up from insulating $M^{2+}(OH)_2$ layers alternating with metallic (Cu,Fe)S layers.

i. THE COMPOSITE LAYER STRUCTURE OF LITHIOPHORITE

The monoclinic structure of the black mineral lithiophorite $Li_{0.32}Al_{0.68}(OH)_2 \cdot Mn_{0.18}^{2+}Mn_{0.82}^{4+}O_2$ is built up from brucite-type sandwiches (Li, Al)(OH)$_2$ and MnO$_2$ alternating along the c-axis (Figure 142). It is however not simply a deformed brucite superstructure but the particular sandwiches are displaced so much that the anions of one sheet are exactly below the adjacent anions of the upper sheet. If the C-centering is ignored the symmetry of the mineral is indeed close to hexagonal. The lithiophorite structure thus is an ordered superstructure of the NaHF$_2$-type grimaldiite γ-CrOOH $(Cr_2^{III}O_2(OH)_2 \rightarrow M^{II}(OH)_2 \cdot M^{IV}O_2)$ with $c' = 2c_0$. The distance between adjacent O and OH ions is 2.76 Å which corresponds exactly to a hydroxyl bond. Therefore lithiophorite is in fact not a true layer compound according to our definition but is only layered. However, these hydrogen bonds are rather weak and

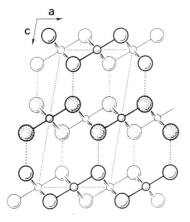

Fig. 142. The monoclinic structure of lithiophorite (Li, Al)(OH)$_2 \cdot$MnO$_2$. The OH ions of the brucite-type layer (Li, Al) (OH)$_2$ are stippled.

may be readily disrupted to cause the prominent basal cleavage. In addition to the H-bonds the electrostatic charge of the sandwiches may also be essential for the occurrence of this structure. Otherwise one would rather expect a composition LiAl(OH)$_4$ or isomorphous Mg(OH)$_2$ for the hydroxide layers. The same amount of excess charge could be achieved with, say, Mg$_{0.64}$Al$_{0.36}$(OH)$_2$. In natural minerals up to 3% of Li and Al are replaced by Cu, Co, Ni and Fe^{3+}.

The cation coordination octahedra in lithiophorite are similarly compressed as in Mg(OH)$_2$. The distances are

$$(Li, Al)—6(OH) \text{ at } 1.93(2) \text{ and } 1.95 \text{ Å } (4)$$
$$(OH)—(OH) = 2.55(4) \text{ and } 1.59 \text{ Å } (2)$$
$$Mn—6 \text{ O at } 1.93 (4) \text{ and } 1.97 \text{ Å } (2)$$
$$O—O = 2.53 (2) \text{ and } 2.58 \text{ Å } (4)$$
$$O—(OH) = 2.76 \text{ Å}$$

One is tempted to try to replace Mn (or part of Mn) by Cr, V or Ti, to form for example Li$_{1/3}$Al$_{2/3}$(OH)$_2$·Mn$^{2+}_{1/6}$Ti$^{4+}_{5/6}$O$_2$ or Li$_{2/3}$Al$_{2/3}$(OH)$_2$·Cr$^{3+}_{1/3}$Mn$^{4+}_{2/3}$O$_2$.

16. MX$_3$ Halides

Table 165 offers a survey of the structures occurring with trihalides and hydroxides. A graphical representation showing the repartition of the structures with respect to the relative size of cations and anions and their electronegativity difference might be more instructive. Nevertheless, we see that no layer structures form among the fluorides although some of the fluoride structures show a clear tendency towards layering. Among the chlorides, bromides and iodides of the rare-earth ions the structures change with decreasing cation-to-anion radius ratio according to the following sequence:

$$UCl_3 \rightarrow PuBr_3 \rightarrow AlCl_3 \rightarrow BiI_3.$$

In the hexagonal UCl$_3$ structure the cation coordination number is 9 and the bonding is threedimensional. A different stacking of the trigonal prisms and loosening of one bond leads to the improper layer structure of the PuBr$_3$ type. In the AlCl$_3$ and BiI$_3$ structures, finally, the bonding is truly two-dimensional. These latter structure types are based on anion close-packings with the cations in octahedral coordination. This kind of coordination is adequate for many transition-element cations. Distortions or deviations are due to the particular d-electron configurations. Thus, excess d-electrons may be bonded in pairs in d^1, d^3 and d^5 compounds. This is possible either in a layer structure (e.g. in the α-MoCl$_3$ type) or in a chain structure (e.g. in the RuBr$_3$ type, which is the orthorhombic variant, or in the MoBr$_3$ type, which is the hexagonal variant of the TiI$_3$ chain type). Chains are appropriate for d^2 compounds such as NbI$_3$ whereas

TABLE 165

Survey of the structures occurring in trihalides. Layer structures are underlined; dashed underlining indicates intermediate cases.

M	MF$_3$	M(OH)$_3$	MCl$_3$	MBr$_3$	MI$_3$
Sc	def. ReO$_3$	Sc(OH)$_3$	BiI$_3$	BiI$_3$	BiI$_3$
Y	r: orth. YF$_3$ h: α-UO$_3$	UCl$_3$	AlCl$_3$	BiI$_3$	BiI$_3$
La	t: r: tysonite	UCl$_3$	UCl$_3$	UCl$_3$	PuBr$_3$
Ce	t: r: LaF$_3$	UCl$_3$	UCl$_3$	UCl$_3$	PuBr$_3$
Pr	LaF$_3$	UCl$_3$	UCl$_3$	UCl$_3$	PuBr$_3$
Nd	LaF$_3$	UCl$_3$	UCl$_3$	PuBr$_3$	r: PuBr$_3$ h: [884]
Sm	r: YF$_3$ h: LaF$_3$	UCl$_3$	UCl$_3$	PuBr$_3$	BiI$_3$ h: [884]
Eu	r: YF$_3$ h: LaF$_3$	UCl$_3$	UCl$_3$	PuBr$_3$	
Gd	r: YF$_3$ h: LaF$_3$	UCl$_3$	r: PuBr$_3$ h: UCl$_3$	r: BiI$_3$ h: AlCl$_3$	r: BiI$_3$ h: AlCl$_3$(?)
Tb	r: YF$_3$ h: LaF$_3$	UCl$_3$	r: PuBr$_3$ h: AlCl$_3$	BiI$_3$	r: BiI$_3$ h: AlCl$_3$(?)
Dy	r: YF$_3$ h: LaF$_3$	UCl$_3$	PuBr$_3$ AlCl$_3$	BiI$_3$	BiI$_3$ h: [884]
Ho	r: YF$_3$ h: LaF$_3$	UCl$_3$	AlCl$_3$	BiI$_3$	BiI$_3$
Er	r: YF$_3$ h: α-UO$_3$	UCl$_3$	AlCl$_3$	BiI$_3$	BiI$_3$
Tm	r: YF$_3$ h: α-UO$_3$	UCl$_3$	AlCl$_3$	BiI$_3$	BiI$_3$
Yb	r: YF$_3$ h: α-UO$_3$	UCl$_3$	AlCl$_3$	BiI$_3$	BiI$_3$
Lu	r: YF$_3$ h: α-UO$_3$	UCl$_3$	AlCl$_3$	BiI$_3$	BiI$_3$
Ac	LaF$_3$		UCl$_3$	UCl$_3$	
Th					ThI$_3$
Pa					PuBr$_3$
U	LaF$_3$		UCl$_3$	UCl$_3$	PuBr$_3$
Np	LaF$_3$		UCl$_3$	α: UCl$_3$ β: PuBr$_3$	PuBr$_3$
Pu	LaF$_3$		UCl$_3$	orth. PuBr$_3$	PuBr$_3$
Am	LaF$_3$	UCl$_3$	UCl$_3$	PuBr$_3$	BiI$_3$ PuBr$_3$
Cm	LaF$_3$		UCl$_3$	PuBr$_3$	BiI$_3$
Bk	r: YF$_3$ h: LaF$_3$		UCl$_3$	r: PuBr$_3$ h$_1$: BiI$_3$(?) h$_2$: AlCl$_3$	BiI$_3$
Cf	r: YF$_3$ h: LaF$_3$		UCl$_3$ h(?): PuBr$_3$	r: BiI$_3$ h: AlCl$_3$	BiI$_3$
Es			UCl$_3$ PuBr$_3$	r: AlCl$_3$	BiI$_3$?
Ti	VF$_3$		α: BiI$_3$ β: TiI$_3$ γ': CrCl$_3$? γ: AlCl$_3$ t: α-MoCl$_3$	α: BiI$_3$ β: TiI$_3$	hexag. TiI$_3$
Zr	ReO$_3$		α: BiI$_3$ β: TiI$_3$	r: TiI$_3$ h: BiI$_3$	TiI$_3$ h: BiI$_3$?
Hf			TiI$_3$?	TiI$_3$?	TiI$_3$ h: BiI$_3$?

Table 165 (Continued)

M	MF$_3$	M(OH)$_3$	MCl$_3$	MBr$_3$	MI$_3$
V	rhomboh. VF$_3$		BiI$_3$	BiI$_3$	BiI$_3$
Nb	ReO$_3$		α-Nb$_{3-x}$Cl$_8$	α-Nb$_{3-x}$Cl$_8$	α-Nb$_{3-x}$Cl$_8$
					TiI$_3$
Ta	ReO$_3$		α-Nb$_{3-x}$Cl$_8$	α-Nb$_{3-x}$Cl$_8$	
Cr	VF$_3$		t: BiI$_3$	r: BiI$_3$	BiI$_3$
			r: AlCl$_3$	h: AlCl$_3$	
			CrCl$_3$(?)		
Mo	VF$_3$		α: def. AlCl$_3$	TiI$_3$	TiI$_3$
			β: monocl. (hcp)		
W			[W$_6$Cl$_{12}$]Cl$_6$	[W$_6$Cl$_{12}$]Cl$_6$	
Mn	monocl. dist. VF$_3$				
Tc					
Re			monocl. Re$_3$Cl$_9$	Re$_3$I$_9$	monocl. Re$_3$I$_9$
			rhomboh. Re$_3$Cl$_9$	rhomboh. Re$_3$Cl$_9$	
Fe	VF$_3$	monocl. (?)	r: BiI$_3$	BiI$_3$	
		Sc(OH)$_3$(?)	t: ccp		
Ru	VF$_3$		α: AlCl$_3$	orthorh. RuBr$_3$	RuBr$_3$
			β: RuBr$_3$		
Os			AlCl$_3$?		RuBr$_3$
Co	VF$_3$	Co(OH)$_3$			
Rh	VF$_3$		α: AlCl$_3$	AlCl$_3$	AlCl$_3$
			β: polytype		
Ir	Vf$_3$		α: AlCl$_3$	AlCl$_3$	AlCl$_3$
			β: orthorh. IrCl$_3$		
Pd	VF$_3$				
Pt				PtBr$_3$	rhomboh. PtBr$_3$
Au	hexag. AuF$_3$	Au(OH)$_3$	monocl. Au$_2$Cl$_6$	monocl. Au$_2$Br$_6$	Au$_2$I$_6$
B		triclinic B(OH)$_3$	hexag. BCl$_3$	BCl$_3$	BCl$_3$
Al	rhomboh. AlF$_3$	α: bayerite	r: monocl. AlCl$_3$	monocl. Al$_2$Br$_6$	Al$_2$I$_6$ (hexag.?)
	(VF$_3$?)	β: nordstrandite	h: disordered?		
		γ: hydrargillite			
		γ': gibbsite			
		p: ~Sc(OH)$_3$			
Ga	VF$_3$	Sc(OH)$_3$	triclinic Ga$_2$Cl$_6$		InI$_3$?
In	VF$_3$	Sc(OH)$_3$	AlCl$_3$	r: BiI$_3$	red: BiI$_3$ or AlCl$_3$
				h: AlCl$_3$	yellow: monocl. In$_2$I$_6$
Tl	YF$_3$		AlCl$_3$		orthorh. Tl$^+$I$_3$
P				r: dimers	PI$_3$
				h: sim. SbCl$_3$	
As			AsCl$_3$	orthorh. AsBr$_3$	red: BiI$_3$
					yellow:
Sb	orthorh. SbF$_3$		α: orthorh. SbCl$_3$	α: orthorh. AsBr$_3$	BiI$_3$
	(monomeric)		β:	β: orthorh. SbCl$_3$	
Bi	cubic BiF$_3$		α: cubic	r: cubic BiCl$_3$	r: BiI$_3$
	YF$_3$		β: orthorh. SbCl$_3$	h:	h:
			p: [936]	p: [936]	

triangles are found in d^4 Re halides. We wonder why no defective MoS$_2$-type structures occur. The existence of ZrCl$_2$ with the MoS$_2$ structure indicates that this absence of defect MoS$_2$ structures is not due to the electronegativity difference. For compounds like NbCl$_3$, NbBr$_3$, ..., TaI$_3$ a $\sqrt{3}a_0$ supercell would be appropriate. The trigonal-prismatic coordination of the 4d^2 and 5d^2 cations should lead to diamagnetic insulators.

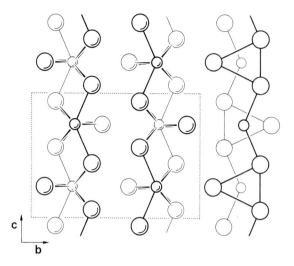

Fig. 143. Projection on (100) of the orthorhombic PuBr$_3$ structure. The 6+2 trigonal-prismatic coordination of the cations is emphasized on the right. Large spheres: Br.

With smaller cations the coordination number drops to 4 and 3. Molecular structures are formed which are at best layered.

a. THE PuBr$_3$ STRUCTURE

We meet here a similar situation as in the case of the PbFCl-type compounds. The PuBr$_3$ structure is not *a priori* a layer structure. The layer character depends on the atomic-site parameters (Table 166). Each cation is located at the center of a trigonal prism formed by the anions. These prisms are stacked on each other to infinity. Neighboring prism columns are shifted by c/2 so that each cation obtains another (2+1) neighbors in the equatorial plane. As can be checked in Table 167 one of the equatorial neighbors is roughly 1 Å farther away and this gives rise to the layer character of this structure. The PuBr$_3$ structure is related to that of ZrSe$_3$ but the mode of linking the prismatic columns is different. Very similar pairs of prism columns are found in the hexagonal UCl$_3$ structure but the hexagonal arrangement of the prism pairs in the UCl$_3$ structure does not allow the formation of true layers although one can detect some kind of layering of [MX$_2$]$_n$ and [X]$_n$ also in that structure [878].

It is worth noting that in UI$_3$ the three excess electrons are well localized: UI$_3$ is an insulator [877] in contrast to metallic ThI$_2$. ThI$_3$ is described as black, faintly translucent, showing a strong blue-green dichroism [790]. Obviously the one excess valence electron is localized, i.e. Th is present in the form of Th^{3+}. The structure of this peritectic iodide is yet unknown. The reported X-ray diagram [790] corresponds neither to the expected PuBr$_3$ structure nor to the hexagonal UCl$_3$ structure.

TABLE 166
$PuBr_3$ structure, orthorhombic, D_{2h}^{17}—Cmcm (No. 63), $Z=4$.

$(0, 0, 0; \frac{1}{2}, \frac{1}{2}, 0)+$

Br_{II} in 8(f): $\pm(0, y, z; 0, y, \frac{1}{2}-z)$

Pu and Br_I in 4(c): $\pm(0, y, \frac{1}{4})$

Compound	$a(Å)$	$b(Å)$	$c(Å)$	$y(M)$	$y(X_I)$	$y(X_{II})$	$z(X_{II})$	Ref.
LaI_3	4.37	14.01	10.04					[842]
CeI_3	4.341	14.00	10.015					[842]
PrI_3	4.309	13.98	9.958					[842]
$NdBr_3$	4.115	12.659	9.158					[841]
NdI_3	4.284	13.979	9.948					[842]
$PmBr_3$	4.08	12.65	9.12					[2]
$SmBr_3$	4.04	12.65	9.08					[2]
$EuBr_3$	4.013	12.66	9.12					[876]
	4.019	12.712	9.128					[2]
$GdCl_3(r)$	3.88	11.73	8.52					[875]
$TbCl_3(r)$	3.86	11.71	8.48	0.244	0.583	0.145	0.569	[872]
$DyCl_3$								[872]
PaI_3	4.33	14.00	10.02					[863]
UI_3	4.328	14.011	10.005	0.243 8	0.578 9	0.144 3	0.566 1	[870]
	4.32	14.01	10.01					[2]
β-$NpBr_3$	4.12	12.66	9.16					[4]
NpI_3	4.30	14.03	9.95					[2]
$PuBr_3$	4.10	12.64	9.14	0.25	0.57	0.14	0.55	[4]
PuI_3	4.33	12.95	9.96					[842]
	4.30	14.03	9.92					[2]
$AmBr_3$	4.064	12.661	9.144					[843]
AmI_3	4.31	14.0	9.9					[4]
$CmBr_3$	4.048	12.66	9.124					[843]
	4.041	12.70	9.135	0.243 1	0.583 1	0.146 1	0.565 8	[850]
$BkBr_3(r)$	4.1	12.6	9.1					[844]
	4.03	12.71	9.12					[850]
$CfCl_3$	3.869	11.748	8.561	0.243 4	0.584 3	0.146 6	0.566 7	[871]
$EsCl_3$								
$La(OH)_2$ NO_3(h) (substructure)	4.076	13.070	7.208					[1014]

b. Rare-earth Hydroxychlorides

While all the hydroxides as well as the chlorides of the larger rare-earth ions crystallize in the hexagonal UCl_3 structure, layer structures form in their mixtures. In the $Ln(OH)_2Cl$ structures each cation is surrounded by 8 anions, 6 of them (4 OH + 2 Cl) define the apices of a distorted trigonal prism and two are located in the equatorial plane. The third equatorial neighbor, a Cl atom of the adjacent layer, is more remote. The geometrical arrangement is thus similar to that in the $PuBr_3$ structure.

The $Ln(OH)_2Cl$ structures may be considered as consisting of two kinds of layers, $[Ln(OH)_2]$ and Cl. The high symmetry of the Cl layers gives rise to some kind of polytypism [917]. Thus, a monoclinic and orthorhombic modification are known with the medium-size rare-earth ions. The relation between the two

TABLE 167
Cation—anion distances in $PuBr_3$-type compounds

Compound	Trigonal-prism neighbors		Equatorial neighbors	
	M—$2X_1$(Å)	M—$4X_2$(Å)	M—$2X_2$(Å)	M—$1X_1$(Å)
$TbCl_3$	2.70	2.79	2.95	3.97
UI_3	3.165	3.244	3.456	4.696
$PuBr_3$	3.06	3.08	3.07	4.04
$CmBr_3$	2.866	2.982	3.137	4.318
$CfCl_3$	2.689	2.805	2.940	4.005

polytypes was discussed by Dornberger-Schiff and Klevtsova [917]. In the monoclinic form successive [Ln(OH)$_2$] layers L_{2n} and L_{2n+2} are connected by the symmetry center and the screw axis parallel to **b** which transform the Cl layer L_{2n+1} lying between them into itself. The symmetry operations thus convert the pair L_{2n}, L_{2n+1} into the pair L_{2n+2}, L_{2n+1}. In the orthorhombic modification, the cation sublattice has half the repeat distance along the c-axis, therefore it is similar to that in the monoclinic version. The hydroxide layers L_{2n} and L_{2n+2} are connected by the n-glide plane and the screw axis parallel to **a**, which transform L_{2n+1} into itself. Thus again the pair L_{2n}, L_{2n+1} is transformed by the symmetry operations into the pair L_{2n+2}, L_{2n+1}. Successive Cl layers L_{2n-1} and L_{2n+1} are related by the symmetry center and the screw axis parallel to **b** which converts the hydroxide layer L_{2n} into itself [917]. Analogous polytypic forms are known for Bi(OH)CrO$_4$.

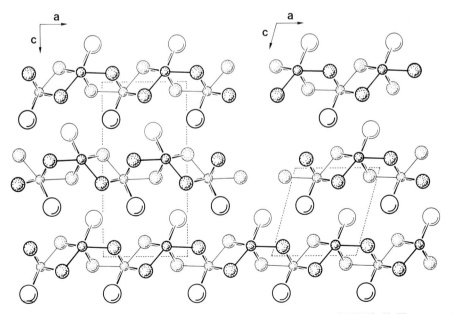

Fig. 144. The orthorhombic 2s (left) and monoclinic 1s polytype structure of Y(OH)$_2$Cl. The atoms of the central [Y(OH)$_2$]$_n$ skeleton are stippled. Largest spheres: Cl, smallest spheres: Y.

TABLE 168a
$1s$—$Ln(OH)_2Cl$ structure, monoclinic, C_{2h}^2—$P2_1/m$ (No. 11), $Z = 2$[a]

all atoms in 2(e): $\pm(x, \frac{1}{4}, z)$

Cation	a(Å)	b(Å)	c(Å)	β(°)	x(Ln)	z(Ln)	x(Cl)	z(Cl)	x(O$_I$)	z(O$_I$)	x(O$_{II}$)	z(O$_{II}$)	Ref.
Y	6.14	3.62	6.60	107	0.286	0.112	0.766	0.573	0.896	0.090	0.450	0.838	[917]
La	6.34	3.98	6.83	114	0.312	0.133	0.753	0.558	0.887	0.100	0.442	0.833	[918]
	6.320	3.98	6.932	113.92									[919]
	6.324 6	3.992 2	6.934 0	113.900									[920]
	6.293	3.966	6.901	113.83									[921]
Ce	6.278	3.945	6.867	113.50									[921]
Pr	6.21	3.89	6.82	113.42	0.311 3	0.134 0	0.755	0.560	0.898	0.082	0.425	0.831	[922]
	6.243	3.91	6.846	113.38									[919]
	6.211	3.894	6.820	113.42									[921]
Nd	6.189	3.877	6.805	113.25									[921]
	6.213	3.876	6.821	113.18									[919]
	6.214 4	3.876 5	6.823 7	113.187									[920]
Sm	6.16	3.80	6.74	112.18	0.306 4	0.130 0	0.754	0.561	0.904	0.081	0.427	0.832	[922]
	6.186	3.81	6.753	112.25									[919]
	6.183 8	3.812 1	6.755 9	112.301									[920]
	6.161	3.802	6.743	112.17									[921]
Eu	6.16	3.79	6.72	111.33									[923]
	6.166	3.79	6.734	112.87									[919]
Gd	6.15	3.74	6.70	111	0.302 4	0.126 0	0.755	0.563	0.901	0.083	0.435	0.843	[922]
	6.150	3.75	6.704	110.68									[919]
	6.147	3.739	6.698	110.00									[921]
	6.164 7	3.749 0	6.715 3	111.271									[920]
Tb	6.15	3.68	6.65	109.67									[923]
	6.180	3.71	6.661	109.65									[919]
Dy	6.15	3.64	6.63	108.58									[923]
	6.197	3.65	6.626	107.88									[919]
Ho	6.14	3.62	6.62	107.08									[923]
	6.198	3.61	6.603	107.07									[919]
Er	6.14	3.60	6.61	106.42									[923]

[a] possibly C_2^2—$P2_1$ (No. 4) like the closely related $Pr(OH)_2NO_3$[1054]

TABLE 168b
$2s$—$Ln(OH)_2Cl$ structure, orthorhombic, D_{2h}^{16}—$Pcmn$ (No. 62), $Z = 4$.

all atoms in 4(c): $\pm(x, \frac{1}{4}, z; \frac{1}{2}-x, \frac{1}{4}, \frac{1}{2}+z)$

Compound	a(Å)	b(Å)	c(Å)	Ref.
Y(OH)$_2$Cl	6.21	3.62	12.56	[917, 924]
	6.244	3.605	12.645	[620]
Dy(OH)$_2$Cl	6.24	3.65	12.62	[923]
Ho(OH)$_2$Cl	6.22	3.63	12.59	[923]
Er(OH)$_2$Cl	6.20	3.61	12.55	[923]

	Atom	x	z	
Y(OH)$_2$Cl:	Y	0.25	0.055	
	Cl	0.892	0.785	[917, 924]
	O$_I$	0.875	0.050	
	O$_{II}$	0.492	0.917	

TABLE 168c

$Y_3O(OH)_5Cl_2$ structure, orthorhombic D_{2h}^{13}—Pmmn (No. 59), Z = 4.

Cl, Y_{II}, $(OH)_{II}$ and $(OH)_{III}$ in 4(f): $\pm(x, \frac{1}{4}, z; \frac{1}{2}-x, \frac{1}{4}, z)$
Y_I in 2(b): $\pm(\frac{1}{4}, \frac{3}{4}, z)$
O and $(OH)_I$ in 2(a): $\pm(\frac{1}{4}, \frac{1}{4}, z)$

Cation	a(Å)	b(Å)	c(Å)	Ref.
Y	13.23	3.73	8.24	[925]
Ho	13.34	3.76	8.35	[926]
Er	13.25	3.75	8.28	[926]
Tm				
Yb	13.09	3.72	8.21	[926]
	13.108 0	3.609 8	8.240 7	[920]
Lu	13.08	3.71	8.19	[926]

$Y_3Cl_2O(OH)_5$:	Atom	x	z	
	Y_I		0.398 4	
	Y_{II}	0.112 8	0.083 5	
	Cl	0.128	0.574	[925]
	O		0.248	
	OH_I		0.924	
	OH_{II}	0.937	0.083	
	OH_{III}	0.908	0.743	

In orthorhombic (monoclinic) $Y(OH)_2Cl$ the two equatorial OH neighbors are at a distance Y—O_I = 2.34 (2.36) Å and Y—O_{II} = 2.31 (2.31) Å, while the trigonal prism is defined by 2 O_I at 2.37 (2.33) Å, 2 O_{II} at 2.44 (2.39) Å and 2 Cl at 2.85 (2.84) Å. The third equatorial neighbor is at Y—Cl = 3.67 (3.56) Å and this distance is decisive for the layer character of these compounds. It is claimed that the layers are held together by OH $\cdots$ Cl hydrogen bonds, as confirmed by infra-red measurements [921]. The interlayer Cl—OH contacts are however comparable with the intralayer contacts as may be seen from the following compilation:

Cl—1 O_I at 3.35 (3.27) Å $\Big\}$ within the same prism
    1 O_{II} at 3.00 (2.97) Å

2 O_I at 3.12 (3.14) Å   in neighboring prism on the left side

1 O_I at 3.41 (3.51) Å $\Big\}$ in the adjacent layer
2 O_{II} at 3.19 (3.21) Å

O_I—1 O_{II} at 2.92 (2.76) Å $\Big\}$ within the same prism
    1 Cl at 3.35 (3.27) Å

2 O_I at 2.70 (2.68) Å $\Big\}$ in the neighboring prism on the left side
2 Cl at 3.12 (3.14) Å

2 O_{II} at 2.95 (2.93) Å   in neighboring prism on the right side
1 Cl at 3.41 (3.51) Å   in the adjacent layer

O_{II}—1 O_I at 2.92 (2.76) Å
　　 1 Cl at 3.00 (2.97) Å } within the same prism

2 O_I at 2.95 (2.93) Å
2 O_{II} at 2.77 (2.73) Å } in neighboring prism on the right side

2 Cl at 3.19 (3.21) Å   in the adjacent layer

Ln(OH)$_2$Cl crystals grow under hydrothermal conditions at temperatures of 350–600 °C as elongated plates, sometimes of needle form [919–921, 926]. Up to now only the hydroxychlorides of Y, Dy, Ho and Er have been obtained in both modifications. The needle form of the orthorhombic crystals is accounted for by the glide planes parallel to the (001) layers with [010] as glide direction which is the direction of maximal growth rate due to preferred emergence of screw dislocations along [010]. These dislocations arise from lattice defects produced by incorporation of impurities or sudden temperature changes and act as growth centers [924].

Hydrothermal preparations using Y$_2$O$_3$, Ho$_2$O$_3$ and Er$_2$O$_3$ yielded in addition to the thin clear Ln(OH)$_2$Cl platelets other crystals in the form of thin blades or needles which had the composition Ln$_3$O(OH)$_5$Cl$_2$. With Yb and Lu only crystals with this stoichiometry and LnO(OH) grew under the given conditions but no Ln(OH)$_2$Cl phases [920, 925, 926]. The Ln$_3$O(OH)$_5$Cl$_2$ structure represented in Figure 145 consists of puckered layers which are weakly linked through Cl and OH ions by hydrogen bonds. There are two kinds of cation sites. Ln$_I$ has a coordination similar to that of the cations in Ln(OH)$_2$Cl except that the prism is now made up of 4 Cl + 2 O and the remote ninth neighbor is missing. Each Cl has

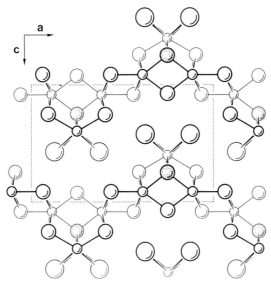

Fig. 145. The orthorhombic Y$_3$O(OH)$_5$Cl$_2$ structure projected onto the (010) plane. Smallest spheres: cations, largest spheres: Cl; O atoms are stippled.

now 3 OH neighbors in the adjacent layer available for hydrogen bonding. In $Y_3O(OH)_5Cl_2$ the distances are:

$$\left.\begin{array}{l} \text{Cl—1 O at 3.13 Å} \\ \text{1 Cl at 3.23 Å} \end{array}\right\} \text{in the same prism}$$

$2 OH_{III}$ at 3.24 Å in the $[Y_{II}(OH)_{5/3}(OH)_{1/2}O_{1/4}]$ unit

$$\left.\begin{array}{l} 1\,OH_I \text{ at 3.30 Å} \\ 2\,OH_{II} \text{ at 3.49 Å} \\ 1\,OH_{III} \text{ at 3.23 Å} \end{array}\right\} \text{in the adjacent layer}$$

The cation distances are similar to those in $Y(OH)_2Cl$:

Y_I—2 OH_{III} at 2.39 Å Y_{II}—1 O at 2.27 Å
  2 O at 2.24 Å   1 OH_I at 2.24 Å
  4 Cl at 2.86 Å   3 OH_{II} at 2.33 and 2.41 Å(2)
                                   2 OH_{III} at 2.37 Å

Probably it is not possible to replace the O atom by another OH and Ln_I by Yb^{2+} or Ca as the distance O—OH_I is 2.67 Å only (hydrogen bond?) compared with O—4 OH_{III} = 2.80 Å. Nothing is known about analogous Br compounds.

c. $CdCl_2$-/CdI_2-TYPE DERIVATIVES ($TiCl_3$, BiI_3, $AlCl_3$, $CrCl_3$)

Starting with a close-packed sandwich structure $M_\Omega \Box X_2$ we may create superstructures either by adding atoms into the formerly empty octahedral sites or we may, what is complementary, add holes in the formerly completely filled layer (compare Figure 7). Thus we expect to find similar structures for complementary compositions:

$$M(M_{1/2}\Box_{1/2})X_2 \leftrightarrow \Box(M_{1/2}\Box_{1/2})X_2$$
$$Cr_3S_4 \qquad\qquad NbCl_4, MoCl_4, NbI_4$$

$$M(M_{2/3}\Box_{1/3})X_2 \leftrightarrow M(M_{1/3}\Box_{2/3})X_2 \leftrightarrow \Box(M_{2/3}\Box_{1/3})X_2 \leftrightarrow \Box(M_{1/3}\Box_{2/3})X_2$$
$$Cr_5S_6 \qquad\qquad Cr_2S_3 \qquad \alpha\text{-}TiCl_3, BiI_3, CrCl_3, AlCl_3 \qquad WCl_6$$

$$M(M_{1/4}\Box_{3/4})X_2 \leftrightarrow M(M_{3/4}\Box_{1/4})X_2 \leftrightarrow \Box(M_{1/4}\Box_{3/4})X_2 \leftrightarrow \Box(M_{3/4}\Box_{1/4})X_2$$
$$V_5S_8 \qquad\qquad Cr_7S_8, Fe_7Se_8 \qquad MX_8 \text{ nonexisting} \qquad Nb_3Cl_8$$

Indeed, the α-$TiCl_3$ structure (Table 169) is complementary to the trigonal Cr_2S_3 structure. The occupied positions in the cation layers of the α-$TiCl_3$ structure (or the holes in the trigonal Cr_2S_3 structure) are described by the sequence:

—0—(A+B)—0—(A+C)—0—

TABLE 169

α-TiCl$_3$ structure, trigonal, D_{3d}^2—$P\bar{3}1c$ (No. 163), $Z = 4$

Cl in 12(i): $\pm(x, y, z; \bar{y}, x-y, z; y-x, \bar{x}, z; y, x; \frac{1}{2}+z;$
$\bar{x}, y-x, \frac{1}{2}+z; x-y, \bar{y}, \frac{1}{2}+z)$

Ti$_{II}$ in 2(d): $\pm(\frac{2}{3}, \frac{1}{3}, \frac{1}{4})$

Ti$_I$ in 2(a): $\pm(0, 0, \frac{1}{4})$

Compound	a(Å)	c(Å)	x	y	z	Ref.
TiCl$_3$	6.121	11.67	$(\sim\frac{2}{3})$	(0)	$(\sim\frac{1}{8})$	[888]
Mo$_{3/4}$Cl$_3$ (MoCl$_4$)	6.058	11.674	0.673	0.006	0.131	[888]

On the other hand the trigonal Cr$_5$S$_6$ structure is a filled-up α-TiCl$_3$ structure with the formerly empty cation layers now completely occupied. In the MoCl$_4$(= Mo$_{3/4}$□$_{1/4}$Cl$_3$) structure $\frac{3}{4}$ of the cation sites of the TiCl$_3$ structure are occupied at random. The columnar habit of the MoCl$_4$ crystals, however, points to the possibility that Mo atoms are also incorporated within the empty cation layers of the CdI$_2$ structure. It thus looks more like a NiAs-type derivative.

The rhombohedral BiI$_3$ structure (Table 170) is the complementary analog of the rhombohedral Cr$_2$S$_3$ structure both being derived from the CdI$_2$ structure as well. Moreover it corresponds to the Fe$_2$Se$_6$ sublattice of the Fe$_2$(P$_2$)Se$_6$ structure (Figure 112). The rhombohedral symmetry of the BiI$_3$ structure is due to the trigonal arrangement of the holes. No MX$_3$ example is known where the holes are arranged in line along the c-axis.

The free positional parameters of the BiI$_3$ structure allow a gradual transition from regular octahedral coordination as realized in BiI$_3$ to a molecular arrangement as in the case of AsI$_3$. In AsI$_3$ each As atom has 3I neighbors at 2.56 Å, the angle I—As—I being as large as 102.0°. The structure of SbI$_3$ represents an intermediate case. In SbI$_3$ too the cations are displaced from the centers of the iodine octahedra so that each Sb atom obtains three closer neighbors at 2.87 Å with an I—Sb—I angle of 95.8° while the next three iodine neighbors are at 3.32 Å.

It is noteworthy that the first ferromagnetic insulator, CrBr$_3$($T_C \approx$ 36 K), belongs to the BiI$_3$-type family. Isostructural CrCl$_3$, on the other hand, is antiferromagnetic. No low-temperature X-ray studies have been reported for the ferromagnetic CrI$_3$ ($T_C = 68$ K).

There appears to be much confusion about the structures of some of these layer-type compounds. Two ordered structures can be derived from the CdCl$_2$ structure. With a trigonal hole sublattice one of the two enantiomorphous forms of the CrCl$_3$ structure (Table 171) is created. On the other hand, stacking the three cation layers all with the same orientation leads to the monoclinic AlCl$_3$ structure (Table 172). Most of the compounds originally assigned to the CrCl$_3$ type belong in fact to the AlCl$_3$ type [848]. The following relation is used to

TABLE 170

BiI_3 structure, trigonal, C_{3i}^2—$R\bar{3}$ (No. 148), $Z = 2(6)$

X in 6(f): $\pm(x, y, z; z, x, y; y, z, x)$
M in 2(c): $\pm(x, x, x)$
hexagonal axes: $(0, 0, 0; \frac{1}{3}, \frac{2}{3}, \frac{2}{3}; \frac{2}{3}, \frac{1}{3}, \frac{1}{3}) +$
X in 18(f): $\pm(x, y, z; \bar{y}, x-y, z; y-x, \bar{x}, z)$
M in 6(c): $\pm(0, 0, z)$

Compound	a_h (Å)	c_h (Å)	c/a	a_{rh} (Å)	α (°)	z_M	x_X	y_X	z_X	Ref.
$ScCl_3$	6.384	17.78	2.785	6.979	54.43					[1]
$ScBr_3$	6.656	18.803	2.825	7.352	53.83					[840]
ScI_3	7.149	20.401	2.854	7.955	53.42					[840]
YBr_3	7.072	19.150	2.708	7.575	55.67					[841]
YI_3	7.505	20.88	2.782	8.198	54.48					[842]
SmI_3	7.503	20.81	2.774	8.178	54.61					[2]
$GdBr_3$	7.490	20.80	2.777	8.172	54.44					[842]
$GdI_3(r)$	7.261	19.189	2.643	7.633	56.40					[841]
$TbBr_3$	7.539	20.83	2.763	8.196	54.77					[842]
$TbI_3(r)$	7.159	19.163	2.677	7.608	56.13					[841]
$DyBr_3$	7.526	20.838	2.769	8.193	54.68					[842]
DyI_3	7.107	19.161	2.696	7.592	55.83					[841]
	7.488	20.833	2.782	8.179	54.48					[842]
$HoBr_3$	7.506	20.843	2.777	8.188	54.56					[784]
HoI_3	7.072	19.150	2.708	7.576	55.67					[841]
$ErBr_3$	7.474	20.817	2.785	8.171	54.43					[842]
ErI_3	7.045	19.148	2.718	7.568	55.47					[841]
$TmBr_3$	7.451	20.78	2.789	8.155	54.37					[842]
TmI_3	7.002	19.111	2.729	7.544	55.33					[841]
$YbBr_3$	7.415	20.78	2.802	8.141	54.18					[842]
YbI_3	6.981	19.115	2.738	7.540	55.17					[841]
$LuBr_3$	7.434	20.72	2.787	8.132	54.40					[842]
LuI_3	6.950	19.109	2.749	7.527	55.00					[841]
	7.395	20.71	2.800	8.117	54.20					[842]

(continued overleaf)

TABLE 170 (Continued)

Compound	a_h (Å)	c_h (Å)	c/a	a_{rh} (Å)	α (°)	z_M	x_X	y_X	z_X	Ref.
AmI$_3$	7.42	20.55	2.77	8.08	54.67					[842, 843]
CmI$_3$	7.44	20.4	2.74	8.04	55.1					[843]
BkBr$_3(h_1)$	7.26	19.23	2.88	7.66	56.6					[850]
BkI$_3$	7.5	20.4	2.7	8.06	55.4					[844]
CfBr$_3$	7.14	19.08	2.67	7.58	56.2					[850]
CfI$_3$	7.55	20.8	2.75	8.19	54.9					[845]
α-TiCl$_3$	6.12	17.50	2.86	6.82	53.3	$\frac{1}{3}(h)$	$\frac{1}{3}(h)$	$0(h)$	$0.079(h)$	[1]
TiBr$_3$	6.459	18.698	2.895	7.263	52.80					[1]
	6.47	18.65	2.88	7.25	53.0					[846]
ZrCl$_3$(?)										[4]
ZrBr$_3(h)$										[847]
VCl$_3$	6.012	17.34	2.884	6.735	53.02					[847]
	6.045	17.45	2.887	6.783	52.92					[847]
VCl$_2$Br	6.186	17.90	2.894	6.954	52.82					[847]
VBr$_3$	6.400	18.53	2.895	7.198	52.80					[762]
VBr$_2$I	6.589	19.30	2.929	7.474	52.31					[849]
VI$_3$	6.925	19.91	2.875	7.748	53.09					[848]
	6.919	19.91	2.877	7.746	53.05					
CrCl$_3(t)$ 225 K	5.942	17.333	2.917	6.719	52.48	$0.332\,3(h)$	$0.650\,7(h)$	$-0.007\,5(h)$	$0.075\,7(h)$	[883]
CrCl$_{0.66}$Br$_{2.34}$	6.225	18.19	2.922	7.048	52.41					[883]
CrBr$_3(r)$	6.26	18.20	2.91	7.06	52.6					[4, 848]
	6.308	18.35	2.909	7.119	52.60					[883]

Compound	a	c							Ref
FeCl$_3$	6.065	17.44	2.876	6.786	53.08	$\frac{1}{3}(h)$	0.653(h)	0(h)	0.077(h)	[890]
4K									0.076(h)	[890]
	6.06	17.38	2.87	6.768	53.2	$\frac{1}{3}(rh)$	0.744(rh)	0.410(rh)	0.077(rh)	[4]
						$\frac{2}{3}(h)$		0(h)	0.077(h)	[890]
FeCl$_2$Br	5.94	17.20	2.90	6.68	52.8					[889]
FeBr$_3$	6.12	17.88	2.92	6.93	52.4					[2]
InBr$_3$(r)	6.42	18.40	2.87	7.159	53.3					[4]
InI$_3$ red(?)										[868]
AsI$_3$	7.208	21.436	2.974	8.269	51.68	0.1985(h)	0.3485(h)	0.3333(h)	0.0822(h)	[869]
	7.187	21.39	2.976	8.25	51.33	$\sim\frac{1}{3}(rh)$	0.75(rh)	0.422(rh)	0.078(rh)	[851]
SbI$_3$	7.48	20.90	2.79	8.20	54.3	0.1820(h)	0.3415(h)	0.3395(h)	0.0805(h)	[4]
	7.466	20.89	2.798	8.18	54.23					[851]
(Sb, Bi)I$_3$ Végard's law										[4]
BiI$_3$	7.516	20.718	2.757	8.156	54.87	$\sim 0.1667(h)$	(0.3415(h))	(0.3395(h))	(0.08005(h))	[881]
	7.498	20.68	2.758	8.13	54.83	$\frac{1}{3}(rh)$	0.755(rh)	0.421(rh)	0.088(rh)	[851]
ideal hcp of the anions			$2\sqrt{2} =$ 2.8284			$\frac{1}{3}(h)$	$\frac{1}{3}(h)$	$\frac{1}{12}(h)$	$\frac{1}{12}(h)$	[4]
						$\frac{1}{3}(rh)$	$\frac{3}{4}(rh)$	$\frac{5}{12}(rh)$	$\frac{1}{12}(rh)$	

calculate the monoclinic cell from the pseudohexagonal cell:

$$\mathbf{a}_{mon.} = \mathbf{a}_1 \qquad\qquad a_{mon.} = a_{hex.}$$

$$\mathbf{b}_{mon.} = \mathbf{a}_1 + 2\mathbf{a}_2 \quad \text{or} \quad b_{mon.} = \sqrt{3}\, a_{hex.}$$

$$3\mathbf{c}_{mon.} = \mathbf{c}_h - \mathbf{a}_1 \qquad c_{mon.} = \tfrac{1}{3}\sqrt{a_{hex.}^2 + c_{hex.}^2}$$

$$\text{tg}\,\beta = -c_{hex.}/a_{hex.}$$

The Fe_2S_6 sublattice of the $Fe_2(P_2)S_6$ structure (Fig. 122) is identical with the $AlCl_3$ structure. In $Fe_2(P_2)S_6$ the cation holes are filled-up with (P_2) molecules.

In Table 172 we have included $TiCl_3$ and some chromium halides which had been reported with the $CrCl_3$ structure. In the case of CrI_3 intensity calculations decided in favor of the $AlCl_3$ structure and for $TiCl_3$ a similar conclusion was reached from the reported continuous solid solutions with $AlCl_3$. For $CrCl_3$ and $CrBr_3$ the $AlCl_3$-type modification is stable at higher temperatures and transforms to the BiI_3 type at lower temperatures. By analogy one might expect a similar transition in CrI_3 although the analogy is not complete as the structure changes near 240 K in $CrCl_3$, but above room temperature ($\sim 150\,°C$) in $CrBr_3$ [848].

While the diamagnetic d^6 configuration of the rhodium and iridium halides as well as the high-spin d^3 and d^5 configurations of the chromium and iron halides, respectively, are adequate for the octahedral coordination in these structures, the $AlCl_3$ structure of $TiCl_3(d^1)$, α-$RuCl_3$ and possibly $OsCl_3$ (low-spin d^5) may pertain to metastable modifications. In the case of $RuCl_3$ this appears to contradict the observed stability of α-$RuCl_3$. The brown diamagnetic β-$RuBr_3$ transforms above 500°C irreversibly into the α-modification [907]. The black lustrous crystals of α-$RuCl_3$ are semiconducting [892] and order antiferromagnetically at $T_N = 13\,K$ [891]. The magnetic susceptibility of $OsCl_3$ [873] above 77 K is

TABLE 171

$CrCl_3$ structure, hexagonal, D_3^3–$P3_112$ (No. 151), $Z=6$,
(or D_3^5–$P3_212$ (No. 153), $Z=6$)
Cl in 6(c): $x, y, z;\ x, x-y;\ \bar{z};\ \bar{y}, x-y, \tfrac{1}{3}+z;\ y-x, y, \tfrac{1}{3}-z;$
$y-x, \bar{x}, \tfrac{2}{3}+z;\ \bar{y}, \bar{x}, \tfrac{2}{3}-z.$
Cr in 3(a): $x, \bar{x}, \tfrac{1}{3};\ x, 2x, \tfrac{2}{3};\ 2\bar{x}, \bar{x}, 0.$

for ideal anion ccp:		x	y	z
	M_I	$\tfrac{2}{9}$		
	M_{II}	$\tfrac{5}{9}$		
	(hole)	$(\tfrac{8}{9})$		
	X_I	$\tfrac{2}{9}$	$\tfrac{1}{9}$	$\tfrac{1}{4}$
	X_{II}	$\tfrac{5}{9}$	$\tfrac{7}{9}$	$\tfrac{1}{4}$
	X_{III}	$\tfrac{8}{9}$	$\tfrac{4}{9}$	$\tfrac{1}{4}$

Compound	$a(\text{Å})$	$c(\text{Å})$	$c/\sqrt{3}a$	z_X	Ref.
$TiCl_3$	6.14	17.40	1.64	0.254	[880]
$CrCl_3$	6.00	17.3	1.66	0.26	[4]
CrI_3	6.85	19.89	1.676		[850]

TABLE 172

AlCl$_3$ structure, monoclinic, C_{2h}^3—C2/m (No. 12), Z = 4.

$(0, 0, 0; \frac{1}{2}, \frac{1}{2}, 0)+$
Cl$_{II}$ in 8(j): ±(x, y, z; x, $\bar{y}$, z)
Cl$_I$ in 4(i): ±(x, 0, z)
Al in 4(g): ±(0, y, 0)

Compound	a(Å)	b(Å)	c(Å)	β(°)	y(M)	x(X$_I$)	z(X$_I$)	x(X$_{II}$)	y(X$_{II}$)	z(X$_{II}$)	Ref.
YCl$_3$	6.92	11.94	6.44	111.0	0.166	0.211	0.247	0.229	0.179	0.760	[862]
GdBr$_3$(h)	7.224	12.512	6.84	110.6	0.167	0.210	0.210	0.250	0.167	0.750	[861]
GdI$_3$(h)											
DyCl$_3$	6.91	11.97	6.40	111.2							[862]
HoCl$_3$	6.85	11.85	6.39	110.8							[862]
ErCl$_3$	6.80	11.79	6.39	110.7							[862]
TmCl$_3$	6.75	11.73	6.39	110.6							[862]
YbCl$_3$	6.73	11.65	6.38	110.4							[862]
LuCl$_3$	6.72	11.60	6.39	110.4							[862]
BkBr$_3$(h$_2$)	7.23	12.53	6.83	110.6							[850]
CfBr$_3$(h)	7.214	12.423	6.825	110.70	0.166 6	0.228 0	0.255 6	0.243 6	0.177 0	0.756 0	[850]
EsBr$_3$(r)	7.27	12.59	6.81	110.8							[987]
γ-TiCl$_3$	6.168	10.68[a]	6.257[a]	109.18[a]							[879]
(Ti, Al)Cl$_3$											[2, 879]
CrCl$_3$(r)	5.953	10.311[a]	6.143[a]	108.85[a]							[883]
298 K	5.959	10.321	6.114	108.49	0.166 7	0.220 5	0.231 6	0.248 85	0.176 9	0.771 1	[848]
CrCl$_2$Br[a]	6.058	10.493[a]	6.274[a]	108.78[a]							[883]
CrCl$_2$I[a]	6.235	10.799[a]	6.571[a]	108.44[a]							[883]
CrBr$_3$(h)											[848]
CrBr$_2$I[a]	6.523	11.298[a]	6.740[a]	108.82[a]							[883]
CrI$_3$[a]	6.859	11.880[a]	7.010[a]	109.04[a]							[883]
(Cr, Ru)Cl$_3$											[2]
FeCl$_3$(t)[b]											[889]
α-RuCl$_3$											[866]
(Ru, Ir)Cl$_3$											[866]
OsCl$_3$											

(continued overleaf)

Table 172 (Continued)

Compound	$a(\text{Å})$	$b(\text{Å})$	$c(\text{Å})$	$\beta(°)$	$y(M)$	$x(X_I)$	$z(X_I)$	$x(X_{II})$	$y(X_{II})$	$z(X_{II})$	Ref.
α-RhCl$_3$	5.95	10.30	6.03	109.2	0.167	0.226	0.219	0.250	0.175	0.781	[864]
RhBr$_3$	6.27	10.85	6.35	109.0	0.167	0.229	0.224	0.248	0.175	0.778	[865]
RhI$_3$	6.77	11.72	6.83	109.3	0.167	0.229	0.224	0.248	0.178	0.778	[865]
α-IrCl$_3$	5.99	10.37	5.99	109.4	0.167	0.227	0.248	0.249	0.176	0.779	[866]
IrBr$_3$ (disordered)	6.30	10.98	6.34	108.7		0.233	0.226	0.247	0.172	0.772	[867]
IrI$_3$ (disordered)	6.74	11.75	6.80	108							[867]
AlCl$_3$	5.90	10.21	6.17	107.97	0.167	0.226	0.219	0.250	0.175	0.781	[882]
	5.93	10.24	6.17	108							[5]
	5.921	10.25	6.183	108.6							[879]
AlCl$_2$Br	6.06	10.50	6.34	108.5							[2]
InCl$_3$	6.41	11.10	6.31	109.8							[862]
InBr$_3(h)$											[868]
InI$_3$ red(?)											[869]
TlCl$_3$	6.54	11.33	6.32	110.2							[862]

[a] values calculated from data given for a hexagonal cell
[b] a fcc modification with $a = 9.46$ Å has been reported to be metastable between 160 and 260 K [889]

similar to that of α-RuCl₃ from which we conclude that OsCl₃ has the same crystal structure. In the orthorhombic RuBr₃ structure there would be no localized magnetic moment.

As in the case of the CdI₂-type phases disorder phenomena are to be expected. Stacking disorder has indeed been observed in RhCl₃ [864], IrCl₃, IrBr₃ and IrI₃ [867].

Based on the similar crystal structure one might expect extended solid solutions between CdI₂-type dihalides and BiI₃-type trihalides, as well as between CdCl₂-type and AlCl₃-type halides. This, however, appears not to be the case. The systems MgCl₂—CrCl₃, FeCl₂—CrCl₃ and MnCl₂—CrCl₃ were found to be similar to eutectic-type systems [906]. Moreover, the different anion stacking is sufficient to prevent mixed-crystal formation between CrCl₃ and FeCl₃ [906].

An interesting property of AlCl₃ is the increase in the electrical conductivity between 140° and the melting point of 190°C [882] which is reminiscent of the behavior of AgBr. Obviously the mobility of the Al ions between octahedral and tetrahedral sites is greatly enhanced before melting. On fusion the molar volume of AlCl₃ increases sharply by a factor of almost two and the electrical conductivity drops to an extremely low value. Liquid AlCl₃ consists of dimers like solid AlBr₃.

d. MoCl₃

An excess d-electron might give rise to deformations of the AlCl₃ and BiI₃ structures such as are observed in α-MoCl₃ (Table 173) where Mo(d³) obviously is in a low-spin state [888]. Mo—Mo pair formation leads to monoclinic symmetry also in the case of the BiI₃-type-related β-MoCl₃. The Mo—Mo distance is reduced to 2.76 Å compared with the distance of the octahedron centers of 3.24 Å. Mo—Cl distances are 2.40(2), 2.45(2) and 2.54 Å(2). α-MoCl₃ is diamagnetic [888]. The second modification, β-MoCl₃ is paramagnetic due to a slight cation deficiency. Its stable composition appears to be MoCl$_{3.08}$ (= Mo$_{\sim 0.97}$Cl₃). Thus, 5–6% of the cation pairs are destroyed and the remaining single Mo atoms carry now a magnetic moment. The reported paramagnetism [888] can be explained only by assuming a high-spin moment on each of the single molybdenum atoms while the Mo pairs are evidently in a low-spin state. Disregarding the monoclinic distortions due to the cation-pair formation the β-MoCl₃

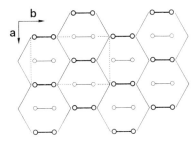

Fig. 146. The cation sublattice of monoclinic β-MoCl₃ showing the Mo—Mo pair formation.

TABLE 173
α-$MoCl_3$ structure, monoclinic, C_{2h}^3—C2/m (No. 12), $Z = 4$.
  (0, 0, 0; $\frac{1}{2}$, $\frac{1}{2}$, 0)+
  Cl_{II} in 8(j): ±(x, y, z; x, ȳ, z)
  Cl_I in 4(i): ±(x, 0, z)
  Mo in 4(g): ±(0, y, 0)

α-$MoCl_3$: $a = 6.092$ Å, $b = 9.745$ Å, $c = 7.275$ Å, $\beta = 124.6°$

Atom	x	y	z	
Mo		0.141 7		
Cl_I	0.008		0.274	[888]
Cl_{II}	0.502	0.170	0.223	

$a_{MoCl_3} = [\bar{1}00]_{AlCl_3}$, $b_{MoCl_3} = [0\bar{1}0]_{AlCl_3}$, $c_{MoCl_3} = [101]_{AlCl_3}$.

structure is complementary to the trigonal Cr_2S_3 structure:

$$\Box(\Box_{1/3}Mo_{2/3})Cl_2 \rightleftharpoons Cr(Cr_{1/3}\Box_{2/3})S_2.$$

We expect the $MoCl_3$ structures to exist also in low-temperature modifications of d^1 and low-spin d^5 layer compounds. This was indeed postulated [893] for the layer-type modifications of $TiCl_3$ which undergo a transition to a non-magnetic state near 217 K [894]. In semiconducting α-$TiCl_3$ discontinuities at the structural transition have been recorded for the lattice parameters [894], the electrical resistivity [895] and the optical absorption [893, 896], whereas the antiferromagnetic transition at $T_N = 265$ K has practically no influence on these properties.

e. THE NbS_2Cl_2 STRUCTURE

A unique substitutional variant of the β-$MoCl_3$ or $AlCl_3$ layer type is realized in the monoclinic NbS_2Cl_2 structure. One of the three Cl atoms of the simple halides is replaced by the polyanions $(S_2)^{2-}$ (Figure 147). The single d-electron of Nb which is not engaged in cation—anion bonding gives rise to the formation of the

TABLE 174
β-$MoCl_3$ structure, monoclinic, C_{2h}^6—C/2c (No. 15), $Z = 8$.
  (0, 0, 0; $\frac{1}{2}$, $\frac{1}{2}$, 0)+
  Cl in 8(f): ±(x, y, z; z, ȳ, $\frac{1}{2}$+z)
  Mo in 4(e): ±(0, y, $\frac{1}{4}$)

$MoCl_{3.08}$: $a = 6.115$ Å, $b = 9.814$ Å, $c = 11.906$ Å, $\beta = 91.0°$

Atom	y	Atom	x	y	z	
Mo_I	0.025	Cl_I	0.815	0.167	0.387	[888]
Mo_{II}	0.309	Cl_{II}	0.850	0.497	0.363	
		Cl_{III}	0.850	0.837	0.363	

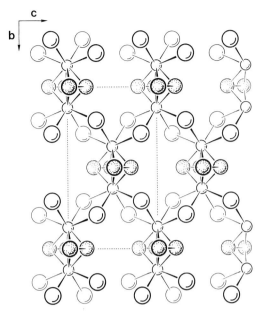

Fig. 147. Basal layer of the monocline NbS$_2$Cl$_2$ structure. Small spheres: Nb; S atoms are stippled.

cation pairs in the unit [Nb$_2$(S$_2$)$_2$Cl$_{8/2}$]:

Nb—1 Nb at 2.90 Å S—S = 2.03 Å
4 Cl at 2.60(2) and 2.62 Å(2)
4 S at 2.51 Å

The formation of a polysulfide is not trivial. It would be conceivable if only S$_2$Cl$_2$ were used in the reaction with niobium. However, all these phases can be prepared by starting from the elements. Crystals grow in a temperature gradient at the lower temperatures [886, 887]. The sulfohalides are dark red transparent flakes whereas the selenohalides are black [886]. NbSe$_2$I$_2$ was obtained in the form of gray metal-like platelets. These had a resistivity of ~10 MΩ·cm at room temperature with an extrinsic activation energy of 0.26 eV and a photo-conductivity maximum near 2 eV [887]. The Nb compounds listed in Table 175 crystallize either with the same or a closely related structure. Polytypism may occur also. It is rather surprising that the corresponding Ta compounds did not form [886].

Substitution of Cl by a single S atom instead of the polyanion would in fact lead to the same electron configuration of Nb and would at the same time require less distortion of the structure. Such compounds as NbSCl$_2$, NbSeBr$_2$, are unknown, but the structures of NbOCl$_2$, NbOBr$_2$, NbOI$_2$ and MoOCl$_2$ are indeed layer structures (Table 198, Appendix) corresponding to Fig. 14a and related to the chain structure of NbOCl$_3$ [901, 954]. M—M pairs, adequate in the Nb compounds, are present also in MoOCl$_2$ [1022].

TABLE 175

NbS_2Cl_2 structure, monoclinic, C_{2h}^3—C2/m (No. 12), $Z = 4$,
description given in C_{2h}^3—F2/m, $Z = 8$:
$(0, 0, 0; \frac{1}{2}, \frac{1}{2}, 0; \frac{1}{2}, 0, \frac{1}{2}; 0, \frac{1}{2}, \frac{1}{2}) +$ Cl in 16(j): $\pm(x, y, z; x, \bar{y}, z)$
S_I and S_{II} in 8(i): $\pm(x, 0, z)$
Nb in 8(g): $\pm(0, y, 0)$

NbS_2Cl_2: $a = 12.603$ Å, $b = 11.113$ Å, $c = 6.301$ Å, $\beta = 96.9°$ [885]
$a = 12.56$ Å, $b = 11.09$ Å, $c = 6.27$ Å, $\beta = 96.70°$ [1017]

$y(Nb) = 0.130$; $x(S_I) = 0.073$, $z(S_I) = 0.230(0.730?$ [1017]); $x(S_{II}) = 0.163$, $z(S_{II}) = 0.019$;
$x(Cl) = 0.117$; $y(Cl) = 0.208$, $z(Cl) = 0.340$ [885]

C2/m:

NbS_2Cl_2: $a = 6.301$ Å, $b = 11.113$ Å, $c = 6.698$ Å, $\beta = 110.94°$ [885]
$a = 6.27$ Å, $b = 11.09$ Å, $c = 6.68$ Å, $\beta = 111.07°$ [1017]
NbS_2Br_2: $a = 6.54$ Å, $b = 11.32$ Å, $c = 6.91$ Å, $\beta = 110.54°$ [1017]
$NbSe_2Cl_2$: $a = 6.65$ Å, $b = 11.44$ Å, $c = 6.96$ Å, $\beta = 109.02°$ [1017]
$NbSe_2Br_2$: $a = 6.76$ Å, $b = 11.53$ Å, $c = 7.20$ Å, $\beta = 113.90°$ [1017]
$NbSe_2I_2(h)$: $a = 6.89$ Å, $b = 12.34 - 12.46$ Å, $c = 7.51$ Å, $\beta = 112.26°$ [1017]

NbS_2I_2 structure, triclinic, layer plane (010) [1017]

NbS_2Br_2: $a = 6.589$ Å, $b = 7.254$ Å, $c = 6.528$ Å, $\alpha = 112.63$ Å, $\beta = 120.05°$, $\gamma = 67.72°$ [1017]
NbS_2I_2: $a = 6.80$ Å, $b = 7.23$ Å, $c = 6.77$ Å, $\alpha = 102.3°$, $\beta = 117.4°$, $\gamma = 73.8°$ [1017]
$NbSe_2Cl_2$: $a = 6.538$ Å, $b = 7.261$ Å, $c = 6.350$ Å, $\alpha = 111.35°$, $\beta = 119.01°$, $\gamma = 66.93°$ [1017]
$NbSe_2Br_2$: [1017]
$NbSe_2I_2(r)$: $a = 7.207$ Å, $b = 7.757$ Å, $c = 7.060$ Å, $\alpha = 113.20°$, $\beta = 121.10°$, $\gamma = 67.59°$ [1017]

No cation pairs are needed in $TiSCl_2$, $MnSCl_2$, $PtSeBr_2$ and their analogs which therefore could crystallize in a structure derived directly from $AlCl_3$ or BiI_3. Of course, the corresponding polychalcogenides $Zr(S_2)Cl_2$, etc., might exist too.

f. Nb_3Cl_8, Nb_3Br_8, $NbCl_4$

For $4d^2$ and $5d^2$ trihalides one should expect diamagnetic modifications with cation chains or triangles. Cation chains are not observed in layer compounds but triangles are indeed found in niobium halides. However, triangles do not fit into the 1:3 stoichiometry. Instead, the composition of these Nb halides is $Nb_3X_8(= Nb_{3/4}\square_{1/4}X_2)$. Two stacking variants have been detected. The hexagonal Nb_3Cl_8 type (Table 176) is derived from the CdI_2 structure. The one quarter empty cation sites require a supercell with $a = 2a_0$. Doubling along the trigonal axis ($c = 2c_0$) is due to a symmetric arrangement of the shortened cation triangles. The β-Nb_3Br_8 structure (Table 177) is based on a $(cchh)_3$ stacking instead of the pure h stacking of the α-form. In both modifications the coordination of the trinuclear unit M_3X_8 may be described by $Nb_3X_4X_{6/2}X_{3/3}$. In Nb_3Cl_8, β-Nb_3Br_8 and β-Nb_3I_8 the Nb—Nb distances are 2.81, 2.88 and 3.00 Å, respectively. This is to be compared with the corresponding distances of the octahedron centers, 3.37, 3.54 and 3.80 Å. In β-Nb_3Br_8 the Br—Nb distances are as follows: 2.56 Å for the three Br

TABLE 176

α-Nb_3Cl_8 structure, trigonal, D_{3d}^3—$P\bar{3}m1$ (No. 164), $Z = 2$.
Nb, Cl_{III} and Cl_{IV} in 6(i): $\pm(x, \bar{x}, z; x, 2x, z; 2\bar{x}, \bar{x}, z)$
Cl_I and Cl_{II} in 2(d): $\pm(\frac{1}{3}, \frac{2}{3}, z)$
cation holes in 2(c): $\pm(0, 0, z_{Nb})$

Nb_3Cl_8 [901]:

Compound	a(Å)	c(Å)	c/a	Ref.		x	z
Nb_3Cl_8	6.744	12.268	1.819	[899]	Nb	0.5278	0.247
α-Nb_3Br_8	7.227	12.93	1.789	[902]	Cl_I		0.904
$NbBr_{3.04}$	7.258	12.94	1.783	[902]	Cl_{II}		0.354
α-Nb_3I_8					Cl_{III}	0.165	0.137
$TaCl_{2.9-3.1}$				[901]	Cl_{IV}	0.833	0.379
$TaBr_{2.9-3.1}$				[901]			

$NbCl_4$ structure, monoclinic, C_{2h}^3—$C2/m$ (No. 12), $Z = 4$.

Nb in 4(g)   ($y = 0.224$)

Cl_I and Cl_{II} in 4(i) ($x_I = 0.217, z_I = 0.238; x_{II} = 0.693, z_{II} = 0.212$)

Cl_{III} in 8(j)   ($x = 0.995, y = 0.238, z = 0.274$)

$a = 12.32$ Å, $b = 6.82$ Å, $c = 8.21$ Å, $\beta = 134°$ [901]

(see Table 173 for atomic positions)

atoms of each unit which make two intranuclear bonds, 2.70 Å for the 6 Br atoms which link two different units and, finally, 2.80 Å for the 3 Br atoms which are in contact with three M_3X_8 units [900]. If the stoichiometry were Nb_3X_9, these compounds would be diamagnetic. Actually each Nb_3X_8 unit carries one excess d-electron and the magnetic susceptibility of α-Nb_3Cl_8 indeed obeys a Curie-Weiss law with an effective moment of 1.86 Bohr magnetons [900]. The magnetic behavior of the β-modification is less clear. Both β-Nb_3Br_8 and β-Nb_3I_8 exhibit a low paramagnetism, practically constant between 90 and 350 K and slightly

TABLE 177

β-Nb_3Br_8 structure, trigonal, D_{3d}^5—$R\bar{3}m$ (No. 166), $Z = 2(6)$
hexagonal axes: $(0, 0, 0; \frac{1}{3}, \frac{2}{3}, \frac{2}{3}; \frac{2}{3}, \frac{1}{3}, \frac{1}{3})+$
M, X_I and X_{II} in 18(h): $\pm(x, \bar{x}, z; x, 2x, z; 2\bar{x}, \bar{x}, z)$
X_{III} and X_{IV} in 6(c): $\pm(0, 0, z)$

Compound	a_h(Å)	c_h(Å)	$c/3a$	a_{rh}(Å)	$\alpha(°)$	Ref.
β-$Nb_{3-x}Cl_8$						[908]
β-Nb_3Br_8	7.080	38.975	1.835	13.620	30.13	[900]
β-Nb_3I_8	7.600	41.715	1.830	14.580	30.22	[900]
β-$Ta_{3-x}Cl_8$						[908]

	β-Nb_3Br_8		β-Nb_3I_8	
Atom	x	z	x	z
Nb	0.469	0.0826	0.465	0.0826
X_I	0.167	0.1269	0.165	0.1276
X_{II}	0.837	0.0455	0.837	0.0456
X_{III}		0.2152		0.2162
X_{IV}		0.3648		0.3646

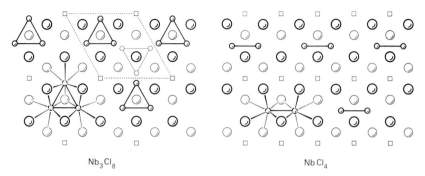

Fig. 148. The defective CdI$_2$-type derivatives Nb$_3$Cl$_8$ and NbCl$_4$. Squares indicate empty cation sites. The distortion of the anion sublattice is neglected and the cation-pair formation is emphasized.

increasing up to 600 K [900]. In contrast to the α-modifications, the lustrous black leaflets of β-Nb$_3$Br$_8$ and β-Nb$_3$I$_8$ are stoichiometric. The excess d-electron appears to be delocalized within the Nb$_3$ triangle.

It is not clear why in a layer structure the stacking mode can have such a decisive influence on composition. The CdI$_2$-type derivatives all have homogeneity ranges: NbCl$_{2.67\cdots3.13}$($=$Nb$_{3-x}$Cl$_8$ with $x \leqslant 0.44$), NbBr$_{2.67\cdots3.03}$($x \leqslant 0.36$), NbI$_{2.89\cdots3.05}$($x = 0.23 \cdots 0.38$), TaCl$_{2.9\cdots3.1}$ and TaBr$_{2.9\cdots3.1}$($x = 0.24 \cdots 0.42$) [900]. This can be understood as a mixed-crystal formation with NbCl$_4$($=$Nb$_2\square_2$Cl$_8$) where all triangles are reduced to cation pairs (Figure 148). Miscibility is, however, not complete since below about Nb$_{2.5}$Cl$_8$ a heterogeneous domain is reached with the limiting phases NbCl$_{3.13}$ and NbCl$_4$. In NbCl$_4$ these cation pairs are appropriate to bind all d-electrons, in agreement with the observed diamagnetism [901]. The Nb—Nb distance within the pairs is 3.06 Å compared with the distance between the octahedron centers of 3.41 Å. The other Nb and Ta tetrahalides are analogous to NbCl$_4$, all are diamagnetic [901]. The activation energy for electronic conduction increases from 0.21 eV in Nb$_3$Cl$_8$ to 0.44 eV in NbCl$_4$ [903]. This change is of considerable interest if it corresponds to the increase of the intrinsic gap. NbCl$_4$ structural data are added in Table 176.

It might be possible to create a diamagnetic M$_3$X$_8$-type compound by replacing either the central halogen atom of each M$_3$X$_8$ unit or the three halogen atoms which are common to three units by S, Se or Te. In an analogous Mo compound we would have to replace either all inner or all outer halogen atoms by a chalcogen atom.

g. Al(OH)$_3$

Aluminium hydroxide occurs in a number of modifications which differ by the mode of superposing the brucite-like layers. Bayerite, α-Al(OH)$_3$, is the stable modification of pure Al(OH)$_3$. Hydrargillite, γ-Al(OH)$_3$, always contains some alkali ions. It occurs with a monoclinic structure and a triclinic superstructure (gibbsite).

TABLE 178
Bayerite structure, monoclinic, C_{2h}^5—$P2_1/a$ (No. 14), $Z = 4$
all atoms in 4(e): $\pm(x, y, z; \frac{1}{2}+x, \frac{1}{2}-y, z)$

α-Al(OH)$_3$: $a = 5.062$ Å, $b = 8.671$ Å, $c = 4.713$ Å, $\beta = 90.27$ Å [910]

Atom	x	y	z	
Al	0.527	0.167	−0.015	
O$_I$	0.365	−0.011	0.215	
O$_{II}$	0.204	0.176	0.777	
O$_{III}$	0.344	0.308	0.229	
H$_I$	0.32	0.52	0.77	[910]
H$_{II}$	0.79	0.40	0.73	
H$_{III}$	0.32	0.31	0.39	

Bayerite is built up of defective Mg(OH)$_2$-type layers similar to those in BiI$_3$. The cation holes, however, are not arranged in a rhombohedral sequence as in BiI$_3$ but superposed in c-direction. Distortions of the [AlO$_6$] octahedra possibly due to H-bonds require a monoclinic unit cell with $a \approx \sqrt{3} a_0$, $b \approx 3 a_0$, $c \approx c_0$, instead of the ($\sqrt{3} a_0, c_0$) supercell of the idealized structure. Hexagonal, orthorhombic and triclinic variants have been reported as well [2]. The distortion of the [AlO$_6$] octahedron in the monoclinic structure is apparent from the Al—O distances which are 1.74, 1.90(2), 1.92, 1.98 and 2.06 Å [910]. As in brucite there are no hydrogen bonds which connect the O—Al—O sandwiches. Part of the H atoms are arranged as in brucite but other H atoms are claimed to occupy positions within the empty octahedra [910] (compare [1026]).

In the monoclinic structure of hydrargillite the O packing is no longer close-packed. Instead, the oxygen ions of adjacent sheets are exactly superposed which is suggestive for the existence of directed bonds between the OH groups. The mode of connecting the sandwiches is the same as in NaHF$_2$ or γ-CrO(OH), however the orientation of the octahedron layers is different. The anion stacking is ABBA in γ-Al(OH)$_3$ but ABBCCA in γ-CrO(OH).

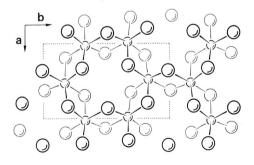

Fig. 149. The structure of bayerite α-Al(OH)$_3$. Small spheres: Al.

Nordstrandite, the rare triclinic modification of Al(OH)$_3$, represents a case intermediate between bayerite and hydrargillite. This is evident already from the interlayer spacings, which are 4.72, 4.79 and 4.85 Å for bayerite, nordstrandite and hydrargillite/gibbsite, respectively. The brucite-like layers are displaced with respect to each other. The hydroxyl groups of one layer therefore are no longer directly above the interstices of the adjacent layer as in bayerite, but they are not exactly aligned with the OH groups of adjacent layers as in hydrargillite either. Nevertheless, close approach suggests hydrogen bonding as in the latter case. In contrast to nordstrandite the orientation of neighboring octahedron layers is not reversed but the stacking is similar to that in BiI$_3$.

h. Rhenium trihalides

ReCl$_3$, ReBr$_3$ and ReI$_3$ form dark red or black crystals either columnar or lamellar. The common structural feature of all rhenium trihalides and a host of compounds derived therefrom is the unique Re$_3$X$_9$ cluster based on a Re$_3$ triangle with Re—Re distances of 2.44–2.49 Å. In the monoclinic structure of Re$_3$I$_9$ and Re$_3$Br$_9$ only two of the three Re atoms are connected with a neighboring unit which gives rise to a chain structure [912, 913]. In the rhombohedral modification of ReCl$_3$, however, all Re atoms are linked via a Cl atom to a neighboring Re$_3$Cl$_9$ cluster which leads to six-membered rings containing the Re$_3$ triangles at two different heights (Figure 150). The coordination of the trinuclear unit is as follows:

Re—2 Re at 2.49 Å
 1 Cl$_I$ at 2.29 Å   (terminal, out of Re$_3$ plane)
 2 Cl$_{II}$ at 2.46 Å (intermolecular bridging Cl atom)
 1 Cl$_{III}$ at 2.40 Å (out-of-plane Cl atom involved in intercluster bridging)
 1 Cl$_{III'}$ at 2.66 Å (in plane, belongs to neighboring cluster)

Four d-electrons are left on the cation for cation-cation bonds and these bonds are indeed interpreted as having bond-order 2 [915]. The intercluster Re—Cl

TABLE 179
Re$_3$Cl$_9$ structure, trigonal, D$_{3d}^5$—R$\bar{3}$m (No. 166), Z = 6.
hexagonal axes: $(0, 0, 0; \frac{1}{3}, \frac{2}{3}, \frac{2}{3}; \frac{2}{3}, \frac{1}{3}, \frac{1}{3})+$
all atoms in 18(h): $\pm(x, \bar{x}, z; x, 2x, z; 2\bar{x}, \bar{x}, z)$

ReCl$_3$: a = 10.326 Å, c = 20.364 Å, c/a = 1.971 [2, 912]

Atom	x	z	
Re	0.253 0	0.392 9	
Cl$_I$	0.237 6	0.504 5	[912]
Cl$_{II}$	0.492^a	0.388 0	
Cl$_{III}$	0.229 2	0.276 9	

ReBr$_3$ [914]

^{a} calculated from given distances

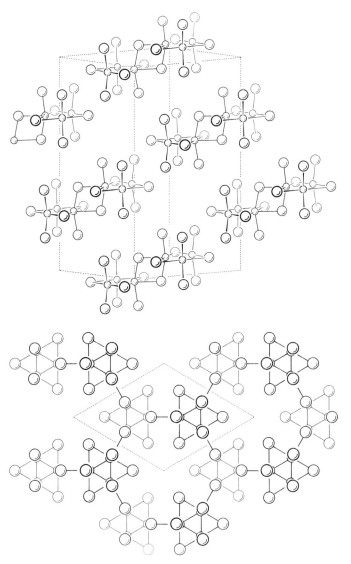

Fig. 150. Two projections of the hexagonal cell of rhombohedral ReCl$_3$. The Re—Re bonds are indicated only in the projection on (001). Large spheres: Cl.

bonds are abnormally long which may be due to steric strain. Although no crystallographic data were given [914] we deduce from the reported lamellar habit of the black-brown ReBr$_3$ crystals that this modification may have the same rhombohedral structure.

i. THE AsBr$_3$ STRUCTURE

The volatile crystals of AsBr$_3$ are colorless plates elongated along c [932] whereas α-SbBr$_3$ grown from carbon disulfide solution has the form of needles with the

TABLE 180
AsBr$_3$ structure, orthorhombic, D_2^4—$P2_12_12_1$ (No. 19), $Z = 4$
all atoms in 4(a): x, y, z; $\frac{1}{2}-x, \bar{y}, \frac{1}{2}+z$; $\frac{1}{2}+x, \frac{1}{2}-y, \bar{z}$; $\bar{x}, \frac{1}{2}+y, \frac{1}{2}-z$.

AsBr$_3$: $a = 10.24$ Å, $b = 12.20$ Å, $c = 4.33$ Å

	x	y	z	
As	0.303 0	0.288 2	−0.000 4	
Br$_I$	0.301 0	0.120 6	0.282 4	[932]
Br$_{II}$	0.482 5	0.369 3	0.250 4	
Br$_{III}$	0.138 0	0.380 2	0.275 2	

α-SbBr$_3$: $a = 10.12$ Å, $b = 12.30$ Å, $c = 4.42$ Å

	x	y	z	
Sb	0.300 8	0.285 9	−0.016 0	
Br$_I$	0.296 0	0.114 9	0.272 1	[933]
Br$_{II}$	0.488 8	0.367 4	0.270 7	
Br$_{III}$	0.126 7	0.386 5	0.289 4	

needle axis parallel to the c-axis [933]. Their structure is of the layered molecular type with pseudo-layer formation parallel to (010), as is evident from Figure 151. The pyramidal molecules are adequate for these lone-pair cations. Bond lengths in AsBr$_3$ are 2.35(2) and 2.38 Å and the bond angles 97–98° lie well between the p^3 and sp^3 values. The coordination in SbBr$_3$ is less regular: Sb—Br = 2.46, 2.50 and 2.54 Å, bond angles are 93.6, 95.5 and 97.4°. The bond distances do not

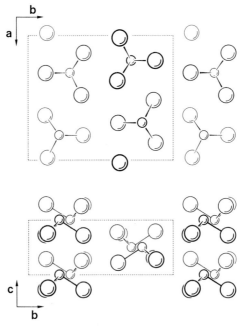

Fig. 151. The orthorhombic AsBr$_3$ structure projected along two axes to show the layering parallel to (010). Large spheres: Br.

differ significantly from the values determined by electron diffraction of the vapor (2.33 Å in AsBr$_3$ and 2.51 Å in SbBr$_3$) but the bond angles in the crystals are distinctly lower than in the free molecules (100–101° in AsBr$_3$ and 97° in SbBr$_3$).

If in this structure we consider the full neighborhood then the cation coordination is roughly trigonal prismatic (SbBr$_3$: three more neighbors at 3.75, 3.79 and 3.82 Å) with three additional 'equatorial' neighbors (at 3.85, 4.10 and 4.19 Å in SbBr$_3$). Whereas the largest two distances are close to the Van der Waals distance the closer neighbors are interpreted as bridging [934] as deduced from optical measurements on BiCl$_3$ where the bridging-mode vibration for the chlorine atoms has clearly been detected.

j. Group III halides (AlBr$_3$, GaCl$_3$, InI$_3$)

The common feature of these three structure types is the occurrence of dimeric molecules M$_2$X$_6$ with the adequate tetrahedral coordination of the cations.

The anion sublattice of the AlBr$_3$ structure is a distorted hcp array with the layers parallel to (10$\bar{1}$). Since AlBr$_3$ crystallizes in flakes [2] one might expect a layered structure with the cations between every second layer according to Figure 21. This is however not the case but the cations are equally distributed between all anion layers as is evident from Figure 152. The main layering occurs parallel to (001) though (100) is also a cleavage plane but a less prominent one. The distances Al—Br are 2.22 and 2.34 Å to the terminal anions and 2.33 Å and 2.42 Å to the bridging bromine atoms. The distance between the two bridging anions is 3.58 Å, the remaining innermolecular Br—Br distances range from 3.77 to 3.86. Br—Br distances to neighboring dimeric units are 3.93—4.05 Å. AlBr$_3$ appears to be the only representative of this structure type. A related structure with dimeric Al$_2$Br$_6$ molecules is found in the adduct Al$_2$Br$_6$·C$_6$H$_6$. In this triclinic structure Al$_2$Br$_6$ molecules are packed in infinite bands with benzene molecules arranged between them.

The hexagonal structure of AlI$_3$ [882] is claimed to consist of a close-packed anion array with I—I = 4.29 Å and similar Al$_2$I$_6$ molecules. In AlCl$_3$ these tetrahedral molecules form only in the liquid state. AlCl$_3$, like PCl$_5$, becomes

TABLE 181
AlBr$_3$ structure, monoclinic, C_{2h}^5—P2$_1$/c (No. 14), Z = 4
all atoms in 4(e): $\pm(x, y, z; x, \frac{1}{2}-y, \frac{1}{2}+z)$

AlBr$_3$: a = 7.41 Å, b = 7.03 Å, c = 10.17 Å, β = 96°50' [882]
a = 7.48 Å, b = 7.09 Å, c = 10.20 Å, β = 96° [4]

	x	y	z	
Al	0.183	0.095	0.050	
Br$_\text{I}$	0.917	0.075	0.150	[4]
Br$_\text{II}$	0.411	0.922	0.169	
Br$_\text{III}$	0.252	0.392	0.008	

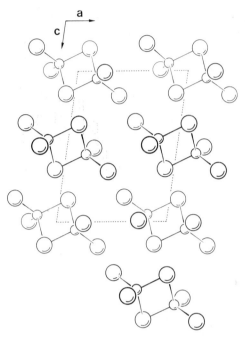

Fig. 152. The monoclinic structure of AlBr$_3$ projected onto the (010) plane. Large spheres: Br.

more covalent on melting. These low-melting, highly volatile aluminium halides contain the same dimers in the vapor state.

GaCl$_3$ crystallizes in a unique triclinic structure illustrated in Figure 153. The anion sublattice can be interpreted as a distorted hcp array parallel to (001) from which one quarter of the atoms has been removed. The cations are inserted in tetrahedral holes such as to give rise to the formation of dimers Ga$_2$Cl$_6$. The structure thus represents a combination of the elements of Fig. 18d and Figure 24. The Ga—Cl distance to terminal Cl atoms is 2.06 Å, that to the bridging anions is 2.29 Å. The Cl—Ga—Cl angle is 123° towards peripheral and 94° towards

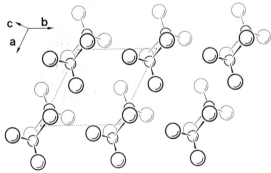

Fig. 153. The triclinic structure of GaCl$_3$ projected onto the (001) plane. Large spheres: Cl atoms at $z = \frac{3}{4}$ and $\frac{1}{4}$.

TABLE 182
GaCl₃ structure, triclinic, C_i^1—$P\bar{1}$ (No. 2), $Z = 2$
all atoms in 2(i): $\pm(x, y, z)$

GaCl₃: $a = 6.94$ Å, $b = 6.84$ Å, $c = 6.82$ Å, $\alpha = 119.5°$, $\beta = 90.8°$, $\gamma = 118.6°$

	x	y	z	
Ga	0.218	0.071	0.159	
Cl$_I$	−0.032	0.233	0.234	[940]
Cl$_{II}$	0.554	0.360	0.190	
Cl$_{III}$	0.184	0.871	0.315	

bridging chlorine atoms. The four peripheral chlorine atoms of each dimer make only eight instead of the twelve contacts with neighboring Cl atoms. It was therefore argued [940] that the possession of a partial negative charge by the chlorine atoms is the reason for not forming a complete close-packed structure especially since only four of the eight contacts are with other terminal chlorine atoms. The large free space in this structure may account for the relatively high conductivity of GaCl₃ near the melting point of 78 °C which is higher than in the liquid.

The monoclinic structure of yellow InI₃ contains again tetrahedral dimers In₂I₆. It is based on a nearly undistorted cubic close-packing of the iodine atoms. The (010) plane of the monoclinic cell represented in Figure 154 corresponds to the (001) plane of the usual fcc description of the anion sublattice. Every third layer of tetrahedral cation sites is completely empty in InI₃ and of the remaining sites one quarter is occupied by indium atoms. Pairs $(\tau_1 + \tau_2)$ are arranged perpendicular to the layer plane (100). In—I distances are 2.64 Å to the terminal and 2.84 Å to the bridging iodine atoms. The corresponding angles I—In—I are 125° and 94°,

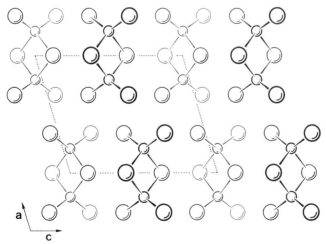

Fig. 154. The monoclinic structure of InI₃ projected on (010). Large spheres: iodine atoms at $z = \frac{3}{4}, \frac{1}{4}$ and $-\frac{1}{4}$; small spheres: indium atoms at $z = \frac{1}{2}$ and 0.

TABLE 183
InI_3 structure, monoclinic, C_{2h}^5—$P2_1/c$ (No. 14), $Z = 4$
all atoms in 4(e): $\pm(x, y, z; x, \frac{1}{2}-y, \frac{1}{2}+z)$

InI_3: $a = 9.837$ Å, $b = 6.102$ Å, $c = 12.195$ Å, $\beta = 107.69°$

	x	y	z	
In	0.207 2	−0.000 5	0.051 0	
I_I	0.000 1	0.236 0	0.122 0	[941]
I_{II}	0.336 1	0.729 2	0.219 8	
I_{III}	0.661 7	0.730 6	0.053 1	

respectively. The angles between In and bridging and terminal iodine atoms are 107° and 109°, close to the ideal tetrahedral angle.

The red low-temperature and high-pressure modification of InI_3 contains the cation in octahedral coordination [869, 944]. The transition red → yellow occurs at 69 °C.

17. Layer-Type Halides of Composition MX_4, MX_5 and MX_6

The number of possible layer structures is drastically reduced on going to compositions with such low cation-to-anion ratios. Only large cations with coordination number 8 are still able to form true layer compounds. The high oxidation number limits the possible cations to Group IV–VII 4d and 5d transition elements and the actinides mainly. The actinides thus offer the best chance for the occurrence of true layer structures. Among the MX_4 actinide halides two true layer structures are indeed found, namely the monoclinic structures of the UBr_4 type and ThI_4 types. All tetrafluorides with Ce, Pr, Tb, Th—Cf crystallize in the three-dimensional tetragonal ZrF_4 structure. The structure of the known actinide tetrachlorides is also three-dimensional with a dodecahedral cation coordination $ThBr_4$(h) and $PaBr_4$ also adopt the bc tetragonal UCl_4 structure.

Among the other MX_4 halides we meet the layer-like structure of the SnF_4 type, the layered chain structure of $ZrCl_4$, the layered triclinic structure of tetrameric $MoNCl_3(=[MoN_{2/2}Cl_3]_4)$ [966] and other chain or dimeric structures such as the γ-NbI_4 structure with cation pairs, occurring also in $NbCl_4$, $TaCl_4$, α-$MoCl_4$ [967], WCl_4, α-$ReCl_4$ and α-PtI_4, though without cation pairs in the latter. A nicely layered molecular structure is found in trans-$PtCl_2 \cdot 2NH_3$ [4].

No true layer structures form with compositions MX_5 and MX_6. Pentahalides (mainly fluorides) form only with V, Nb, Ta, Pa (d^0, f^0 cations), Cr, Mo, W, U (d^1 or f^1 cations), Mn, Tc, Re, Np (d^2 or f^2 cations), Ru, Os, Pu (d^3 or f^3 cations), Rh, Ir, Am (d^4 or f^4 cations), Pt, (d^5 cations) and phosphorus group cations. Hexavalent elements may form chalcohalides such as $WSBr_4$, etc. Although with this composition monomeric molecular structures are possible they seem to form only exceptionally, as in the structure of SbCl$_5$ which contains trigonal-bipyramidal molecules. Whether a subdivision into tetrahedral and octahedral coordination, as in PCl$_5$ = [PCl$_4$]$^+$[PCl$_6$]$^-$ and SbCl$_2$F$_3$, is possible or not depends upon the polarity of these complex molecules. The most common coordination is still a more or less distorted octahedral one. The chemical composition then requires sharing of two anions [MX$_4$X$_{2/2}$] located either at adjacent (cis) or opposite (trans) vertices or at a common edge. In the first case cyclic tetramers M$_4$X$_{20}$ (as in WOF$_4$ = [WF$_4$O$_{2/2}$]$_4$, in monoclinic hcp RuF$_5$ and isostructural OsF$_5$, RhF$_5$, IrF$_5$ and PtF$_5$, in monoclinic SbF$_5$, in orthorhombic NbF$_5$, TaF$_5$, MoF$_5$ and WF$_5$) or zigzag chains (as in orthorhombic VF$_5$, CrF$_5$, TcF$_5$ and ReF$_5$ or in MoOF$_4$, TcOF$_4$ and ReOF$_4$, or in the NbI$_5$ structure) are the structure-building elements whereas linear chains (as in tetragonal α-UF$_5$ or in tetragonal WOCl$_4$ and isostructural WOBr$_4$) and M$_2$X$_{10}$ dimers (as in NbCl$_5$, TaCl$_5$, ferromagnetic MoCl$_5$, WCl$_5$, in the ReCl$_5$ structure of ReBr$_5$, in the UCl$_5$ structure, in orthorhombic NbBr$_5$ and TaBr$_5$, or in the layered monoclinic structures of WSCl$_4$ and WSBr$_4$ [952, 969]) will occur in the second and third case, respectively. A layered structure composed of dimers and chains is realized in monoclinic ReOCl$_4$ [953]. Of this variety of structures we will discuss only a few examples which show the highest degree of layering.

The limiting composition for compounds whose structure is based on a close-packed anion array is MX$_6$. Hexavalent halides are rather seldom, most of them are fluorides which occur in a cubic high-temperature and an orthorhombic low-temperature modification (M = Cr, Mo, W, Mn, Tc, Re, Ru, Os, Rh, Ir, Pt, U, Np, Pu). The only layered type is realized in the α-WCl$_6$ structure. The molecular structure of cis-PtCl$_4\cdot$2NH$_3$ is also layered. This monoclinic structure is based on close-packed Cl/NH$_3$ layers in which the Pt atoms occupy the distorted octahedral holes [4].

a. THE UBr$_4$ STRUCTURE

Since in most actinide compounds the cation coordination number is higher than six their structures cannot be based on an anion close-packing. In the monoclinic UBr$_4$ structure the configuration around the cation is a pentagonal bipyramid. The U—Br distances for the Br atoms in the pentagonal ring are 2.85(2), 2.93(2) and 2.95 Å. One of the two apical Br atoms is terminal with U—Br = 2.61 Å. The second apical Br atom is at the same time an equatorial atom of the adjoining bipyramid; its distance is therefore larger, 2.78 Å. These bipyramids are linked into infinite chains parallel to **b** by sharing edges. The chains in turn are further crosslinked into sheets parallel to (001) by edge-sharing involving the bridging apical bromine atom. The stacking of these sheets is illustrated in Figure 155. Interlayer Br—Br approaches are 3.76, 3.79 and 3.89 Å. A second example for this structure type is found in NpBr$_4$ (Table 184).

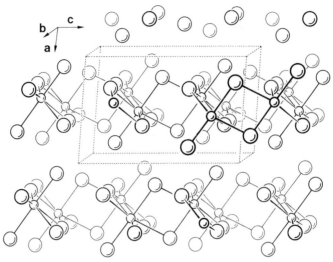

Fig. 155. The monoclinic structure UBr$_4$ viewed parallel to the layers. Large spheres: Br.

b. The ThI$_4$ structure

In the monoclinic structure of the yellow crystals of ThI$_4$ the cation is located at the center of a strongly distorted square antiprism. Each iodine atom therefore is bridging. The I—Th distances for the four sets of anions are 3.19 and 3.28, 3.16 and 3.21, 3.13 and 3.23, and 3.13 and 3.29 Å. The ThI$_4$ cell is described in space group P2$_1$/n. For comparison we have indicated in Figure 156 the cell for a description in P2$_1$/c which would be more appropriate for this kind of stacking.

The structures of dark green PaI$_4$ and black UI$_4$ appear to be yet unknown. These two compounds might be further candidates for the ThI$_4$ structure.

TABLE 184

UBr$_4$ structure, monoclinic, C_{2h}^3—C2/m (No. 12), Z = 4.
(0, 0, 0; $\frac{1}{2}$, $\frac{1}{2}$, 0)+
Br$_{III}$ in 8(j): ±(x, y, z; x, ȳ, z)
remaining atoms in 4(i): ±(x, 0, z)

Compound	a(Å)	b(Å)	c(Å)	β(°)	Ref.
UBr$_4$*	10.92	8.69	7.05	93.9	[1]
	11.04	8.76	7.04	94.05	[945]
NpBr$_4$	10.89	8.74	7.05	94.19	[946]
PuBr$_4$?					

		x	y	z	
* UBr$_4$:	U	0.192		0.432	
	Br$_I$	0.942		0.247	[945]
	Br$_{II}$	0.328		0.142	
	Br$_{III}$	0.863	0.311	0.670	

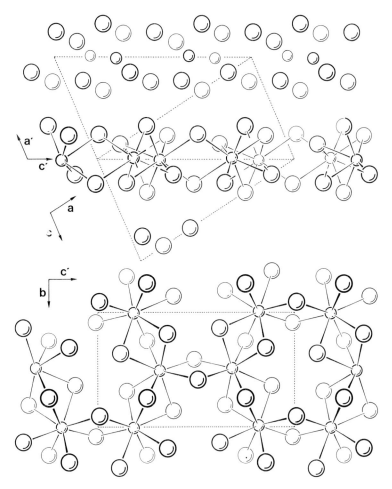

Fig. 156. Two projections of the monoclinic structure of ThI$_4$. The axes a,b,c refer to a cell of the structure described in P2$_1$/n whereas a',b',c' refer to a description in P2$_1$/c. Small spheres: Th.

TABLE 185

ThI$_4$ structure, monoclinic, C_{2h}^5—P2$_1$/n (No. 14'), $Z=4$.
all atoms in 4(e): $\pm(x, y, z; \frac{1}{2}-x, \frac{1}{2}+y, \frac{1}{2}-z)$

ThI$_4$: $a = 13.216$ Å, $b = 8.068$ Å, $c = 7.766$ Å, $\beta = 98.68°$ [947]

Atom	x	y	z	
Th	0.183 5	0.014 9	0.176 9	
I$_I$	0.058 7	0.909 8	0.809 4	[947]
I$_{II}$	0.180 1	0.253 5	0.498 4	
I$_{III}$	0.097 2	0.691 7	0.325 1	
I$_{IV}$	0.151 7	0.363 8	0.001 4	

c. $SrCl_2 \cdot 2H_2O$ AND $BaCl_2 \cdot 2H_2O$

Chemically, these compounds do not belong to the tetrahalides but if we disregard the H atoms their composition is also MX_4. The reason why we add them after ThI_4 is that they contain the same square-antiprismatic cation coordination polyhedra, though less distorted. In the platy crystals of $SrCl_2 \cdot 2H_2O$, Sr—4Cl = 2.71 Å and Sr—$4H_2O$ = 2.99 Å. The closest approach of atoms in adjacent sheets is Cl—H_2O = 3.34 Å. As Figure 157 demonstrates the distribution of the water molecules is rather unilateral. It is more regular in the structure of $BaCl_2 \cdot 2H_2O$. In the Ba salt the Ba—Cl bond distances lie between 3.11 and 3.27 Å, the Ba—OH_2 separations between 2.78 and 2.82 Å. The closest approaches between Cl ions and H_2O molecules of adjacent sheets are 3.19 and 3.22 Å. There is however a Ba—Cl separation across the sheets of only 3.38 Å which does not

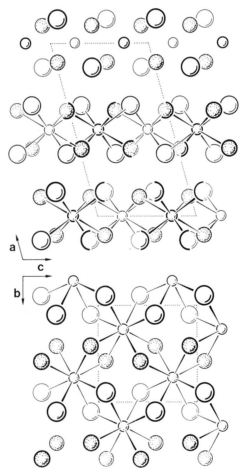

Fig. 157. Projection parallel and perpendicular to the layers of the monoclinic structure of $SrCl_2 \cdot 2H_2O$. Small spheres: Sr, stippled spheres: H_2O, large open spheres: Cl.

TABLE 186

$SrCl_2 \cdot 2H_2O$ structure, monoclinic, C_{2h}^6—$C2/c$ (No. 15), $Z = 4$.
(0, 0, 0; $\frac{1}{2}$, $\frac{1}{2}$, 0)+
Cl and H_2O in 8(f): $\pm(x, y, z; x, \bar{y}, \frac{1}{2}+z)$
Sr in 4(e): $\pm(0, y, \frac{1}{4})$

$SrCl_2 \cdot 2H_2O$: $a = 11.71$ Å, $b = 6.39$ Å, $c = 6.67$ Å, $\beta = 105.7°$ [4]
$a = 11.69$ Å, $b = 6.38$ Å, $c = 6.66$ Å, $\beta = 105.9°$ [1] (kX?)

	x	y	z	
Sr		0.25		
Cl	0.14	0.10	0.64	[4]
H_2O	0.11	0.60	0.48	

greatly exceed the normal bond distance. The situation is reminiscent of the PbFCl-type compounds with their gradual transition from layer-type to three-dimensional phases.

d. THE $ZrCl_4$ STRUCTURE

This monoclinic structure is based on a cubic anion close-packing where the cations occupy half the octahedral holes of every second Ω layer. It may be considered as a kind of defective reorganized $AlCl_3$ structure. Only the zigzag chains of the honeycomb cation sublattice are left but they are rearranged in $ZrCl_4$. The structure which is illustrated in Figure 158 accounts well for the needle form of the crystals. Since Zr^{4+} is a d^0 cation metal–metal bonds cannot exist and the shortest Zr—Zr distance is indeed as long as 3.96 Å. The $[ZrCl_2Cl_{4/2}]$ coordination gives rise to considerable distortion of the octahedron. The Zr—Cl distance to the terminal Cl atoms is 2.31 Å which is comparable with the 2.32 Å found in the tetrahedral gas molecule. The Zr—Cl bridging bonds have two rather different lengths: 2.50 and 2.66 Å.

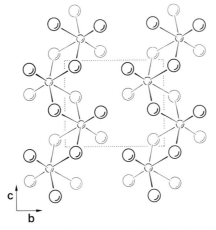

Fig. 158. The monoclinic structure of $ZrCl_4$ projected on (100).

TABLE 187

ZrCl$_4$ structure, monoclinic, C$_{2h}^4$–P2/c (No. 13), Z = 2
Cl in 4(g): ±(x, y, z; x, ȳ, $\frac{1}{2}$+z)
Zr in 2(e): ±(0, y, $\frac{1}{4}$)

Compound	a(Å)	b(Å)	c(Å)	β(°)	Ref.
ZrCl$_4$	6.361	7.407	6.256	109.30	[948]
HfCl$_4$					[948]
ZrBr$_4$					[948]
HfBr$_4$(?)					[948]

ZrCl$_4$:		x	y	z	
	Zr		0.164 1		
	Cl$_I$	0.226 3	0.107 6	−0.002 2	[948]
	Cl$_{II}$	0.255 2	0.637 1	−0.020 5	

The structure of ZrCl$_4$ represents the joint between the AlCl$_3$ structure of YCl$_3$ and the dimeric structure of NbCl$_5$, the octahedral cation coordination being maintained in the whole series. The analogy in the 5d series extends even farther:

$$\text{LuCl}_3 \rightarrow \text{HfCl}_4 \rightarrow \text{TaCl}_5 \rightarrow \alpha\text{-WCl}_6.$$

e. THE SnF$_4$ STRUCTURE

The body-centered tetragonal cell of SnF$_4$ shown in Figure 159 can be built up of two NaCl cells from which $\frac{3}{4}$ of the cations are removed. [SnF$_2$F$_{4/2}$] octahedra are thus corner-linked to form sheets parallel to (001). Since the upper free corners of the octahedra of one layer lie in the same plane as the lower free corners of the octahedra of the adjacent layer the puckered sheets mesh together (Figure 10b).

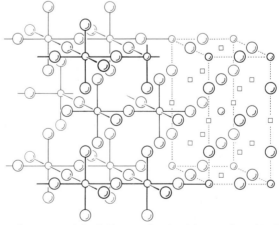

Fig. 159. The tetragonal structure of the SnF$_4$ type. On the right part the unit cell and its relation to a cubic close-packed array of the NaCl type are indicated: squares represent empty octahedral cation sites.

TABLE 188

SnF_4 structure, tetragonal, D_{4h}^{17}—$I4/mmm$ (No. 139), $Z = 2$.
$(0, 0, 0; \frac{1}{2}, \frac{1}{2}, \frac{1}{2})+$
F_{II} in 4(e): $\pm(0, 0, z)$
F_I in 4(c): $0, \frac{1}{2}, 0; \frac{1}{2}, 0, 0$
Sn in 2(a): $0, 0, 0$

Compound	a(Å)	c(Å)	c/a	z	Ref.
SnF_4	4.048	7.930	1.959	0.245	[955]
PbF_4	4.247	8.030	1.891		[955]
NbF_4	4.081	8.162	2.000	$\frac{1}{4}$	[956]
	4.083	8.161	1.999		[957]

It is interesting that the black hygroscopic NbF_4 also adopts this structure, which might be more adequate to TiF_4, $ZrF_4(d^0)$ or $MnF_4(d^3)$. As in this structure no metal-metal bonds can develop the temperature-independent paramagnetism [901] of NbF_4 may point to a metallic character. TaF_4 which might crystallize in the same structure is yet unknown.

f. $UO_2(OH)_2$ AND UO_2F_2

The orthorhombic structure originally reported for β-$UO_2(OH)_2$ (prepared by thermal hydrolysis of uranyl acetate solutions) is very similar to the SnF_4 structure. The linking of the cation coordination octahedra takes place via hydroxyl ions: $[UO_2(OH)_{4/2}]$. As the free z-parameter (Table 189) is definitely lower than $\frac{1}{4}$ the free octahedron corners of adjacent sheets now lie no longer exactly in the same plane. The layer character is thus more pronounced than in SnF_4. The symmetry is so high that no single H bonds can develop in this modification which we termed β'. Either the structure is incorrect or belongs to a high-temperature modification. A more recent structure determination [959] led to a similar but more distorted structure. The UO_2 units are no longer exactly parallel to the c-axis. The layers are H-bonded together [961], the H bond extending from an OH ion in one layer to the uranyl oxygen atom in the adjacent layer. Distances are

U—O (uranyl) = 1.7–1.8 Å
O (hydroxyl) = 2.3–2.4(2) and 2.3 Å(2)
O—OH = 2.7–2.8 Å (interlayer) [959–961]

TABLE 189

β'—$UO_2(OH)_2$ structure, orthorhombic, D_{2h}^{23}—$Fmmm$ (No. 69), $Z = 4$
$(0, 0, 0; 0, \frac{1}{2}, \frac{1}{2}; \frac{1}{2}, 0, \frac{1}{2}; \frac{1}{2}, \frac{1}{2}, 0)+$
O in 8(i): $\pm(0, 0, z)$
OH in 8(e): $\frac{1}{4}, \frac{1}{4}, 0; \frac{1}{4}, \frac{1}{4}, \frac{1}{2}$
U in 4(a): $0, 0, 0$.

$UO_2(OH)_2$: $a = 6.295$ Å, $b = 5.636$ Å, $c = 9.929$ Å; $z = 0.20$ [958]

TABLE 190
β-$UO_2(OH)_2$ structure, orthorhombic, D_{2h}^{15}—Pbca (No. 61), $Z=4$.
H, O_I and O_{II}(OH) in 8(c): $\pm(x, y, z; \frac{1}{2}-x, \frac{1}{2}+y, z; x, \frac{1}{2}-y, \frac{1}{2}+z; \frac{1}{2}-x, \bar{y}, \frac{1}{2}+z)$
U in 4(a): $0, 0, 0; \frac{1}{2}, \frac{1}{2}, 0; 0, \frac{1}{2}, \frac{1}{2}; \frac{1}{2}, 0, \frac{1}{2}$

β-$UO_2(OH)_2$	a(Å)	b(Å)	c(Å)	Ref.
r.t.	5.635	6.285	9.919	[959]
21°C	5.643 8	6.286 7	9.937 2	[960]
170°C	5.621	6.337	9.950	[960]
280°C	5.545	6.155	9.950	[960]

T	$x(O_I)$	$y(O_I)$	$z(O_I)$	$x(O_{II})$	$y(O_{II})$	$z(O_{II})$	x(H)	y(H)	z(H)	Ref.
r.t.	0.137 0	0.467 2	0.334 8	0.184 8	0.279 1	0.092 5				[959]
21°C	0.154	0.464	0.342	0.196	0.288	0.082				[960]
>260°C	0.180	0.490	$\frac{1}{2}-1.511/c$	0.247	0.247	0.905/c				[960]
r.t.	0.136	0.472	0.350	0.203	0.304	0.091	0.206	0.344	0.172	[961]

Thermal expansion of this modification is strongly anisotropic [960] up to 260°C, but almost isotropic at higher temperatures. The anisotropy is caused by a rotation of the octahedra towards the orientation in the more symmetric structure reported by Bergström and Lundgren [958]. A closely related modification, monoclinic γ-$UO_2(OH)_2$ is stable at 0–125°C [962]. In this structure the layers are somewhat shifted relative to each other. Each uranyl oxygen is still hydrogen-bonded to one hydroxyl oxygen of the neighboring layer. Both β- and γ-$UO_2(OH)_2$ transform below 0°C or on application of pressure to the orthorhombic α-modification. In α-$UO_2(OH)_2$ the coordination number of the cation is 8: [$UO_2(OH)_{6/3}$]. Each UO_2^{2+} group is surrounded at a right angle by a puckered hexagon of hydroxyl oxygen atoms. The linear uranyl group (U—O = 1.71 Å) is inclined by 70° to the layer plane (010). The distances from the cation to the equatorial hydroxyl oxygen atoms are 2.51(2) and 2.46 Å(4). Again each uranyl oxygen atom is hydrogen-bonded to a hydroxyl oxygen atom of the adjacent layer: O—O = 2.76 [961] or 2.88 Å [963].

The trigonal structure of UO_2F_2 is closely related to that of α-$UO_2(OH)_2$. The absence of the hydrogen bonds allows the arrangement to be more symmetric.

TABLE 191
γ-$UO_2(OH)_2$ structure, monoclinic, C_{2h}^5—$P2_1/c$ (No. 14), $Z=2$.
O and OH in 4(e): $\pm(x, y, z; x, \frac{1}{2}+z)$
U in 2(a): $0, 0, 0; 0, \frac{1}{2}, \frac{1}{2}$

γ-$UO_2(OH)_2$: $a = 5.560$ Å, $b = 5.522$ Å, $c = 6.416$ Å, $\beta = 112.71°$ [962]

	x	y	z	
O	0.302 6	0.169 6	0.107 2	[962]
OH	0.171 6	0.732 0	0.295 5	

β-$(UO_2(OH)_2$ in this description [960, 962]:
21°C: $a = 5.879$ Å, $b = 5.644$ Å, $c = 6.287$ Å, $\beta = 122.3°$
280°C: $a = 5.920$ Å, $b = 5.545$ Å, $c = 6.419$ Å, $\beta = 122.8°$

TABLE 192

α-$UO_2(OH)_2$ structure, orthorhombic, D_{2h}^{18}–Cmca (No. 64) or
C_{2v}^{17}–C2bc (No. 41), Z = 4

$(0, 0, 0; \frac{1}{2}, \frac{1}{2}, 0)+$
O_I, O_{II} and H in 8(f): $\pm(0, y, z; \frac{1}{2}, \bar{y}, \frac{1}{2}+z)$
U in 4(a): $0, 0, 0; 0, \frac{1}{2}, \frac{1}{2}$.

α-$UO_2(OH)_2$: $a = 4.242$ Å, $b = 10.302$ Å, $c = 6.868$ Å [963]

$y(O_I)$	$z(O_I)$	$y(O_{II})$	$z(O_{II})$	$y(H)$	$z(H)$	Ref.
0.155	0.089	0.069	0.649			[963]
0.158	0.109	0.076	0.654	0.351	0.138	[961]

The uranyl groups here are perpendicular and the puckered F hexagons parallel to the layers. Puckering of the F hexagons was more pronounced in the original structure proposal by Zachariasen [4], which was closer to a strongly distorted (hcch)$_3$ packing of OFFO layers. This formal relation may be evident from Figure 160 although the additional two bonds U—O brought in by the distortion are the strongest of all. The double-bond distance U—O is 1.74 Å whereas U—6 F = 2.43 Å. F—3 F = 2.46 Å, O—3 F = 3.11 Å within the same layer and O—F = 2.86 Å to the next layer. A characteristic feature of the structure is the occurrence of stacking faults. The arrangement shown in Figure 160 as well as the structural data given in Table 193 refer to an idealization.

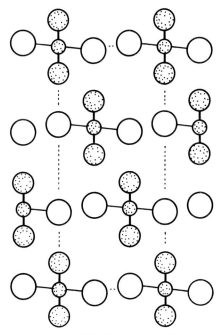

Fig. 160. (110) section of the rhombohedral UO_2F_2 structure. The double bond of the uranyl group is emphasized and its atoms are stippled. The large open circles represent F atoms. These pairs repeat thrice around the threefold axis defined by the O—U—O group.

TABLE 193

UO_2F_2 structure, trigonal, D_{3d}^5—$R\bar{3}m$ (No. 166), $Z = 3$.
Hexagonal axes: $(0, 0, 0; \frac{1}{3}, \frac{2}{3}, \frac{2}{3}; \frac{2}{3}, \frac{1}{3}, \frac{1}{3})+$
O and F in 6(c): $\pm(0, 0, z)$
U in 3(a): 0, 0, 0.

Compound	a_{rh}(Å)	α(°)	a_h(Å)	c_h(Å)	z(O)	z(F)	Ref.
UO_2F_2	5.766	42.78	4.206	15.692	0.122	0.294	[4]
	5.754	42.73	4.192	15.660	0.111	0.320	[964]
NpO_2F_2	5.793	42.27	4.178	15.80			[4]
PuO_2F_2	5.797	42.0	4.154	15.84			[4]

g. WO_2Cl_2 AND WO_2I_2

The colorless oxychloride WO_2Cl_2 crystallizes in a true layer structure which is a distorted version of that represented in Figure 10a. The cations are shifted from the centers of the coordination octahedra $[WCl_2O_{4/2}]$ as we met already in the MoO_3-type and related structures. Distances are: W—2 Cl = 2.31 Å, W—O = 1.63, 1.70, 2.22 and 2.34 Å. In $WOCl_4$ the W—Cl distance is 2.29 Å and the W—O distances range from 1.8 to 2.2 Å, in WO_3 the W—O distances are between 1.72 and 2.16 Å, for comparison. The WO_2Cl_2 crystals are highly disordered and Figure 161 and Table 194 represent an ideal case.

A related layer structure is expected for WO_2I_2. This oxyiodide crystallizes in a temperature gradient as thin dark platelets with a greenish metallic lustre [951]. The stacking of the layers is strongly disordered as in WO_2Cl_2. The arrangement of the $[MoO_2Cl_2]$ 'tetrahedra' is different in MoO_2Cl_2 [1024].

TABLE 194

WO_2Cl_2 structure, monoclinic, C_s^4—Cc (No. 9), $Z = 4$
$(0, 0, 0; \frac{1}{2}, \frac{1}{2}, 0)+$
all atoms in 4(a): $x, y, z; x, \bar{y}, \frac{1}{2}+z$

WO_2Cl_2: $a = 14.42$ Å, $b = 3.89$ Å, $c = 7.68$ Å, $\beta = 105.4°$

	x	y	z	
W	0	−0.071 2	0	
Cl_I	0.162	0.009	0.128	[949]
Cl_{II}	−0.162	0.009	−0.034	
O_I	0.0	0.083	0.279	
O_{II}	0.0	0.478	0.049	

related(?):
WO_2I_2, monoclinic, C2/c or Cc, $Z = 4$;
$a = 17.095$ Å, $b = 3.899$ Å, $c = 7.492$ Å, $\beta = 102.66$ Å [950, 951]

h. THE NbI_5 STRUCTURE

Niobium pentaiodide crystallizes in elongated thin fragile platelets of brassy color. The {100} faces are well developed and show a metallic lustre. As Nb is a

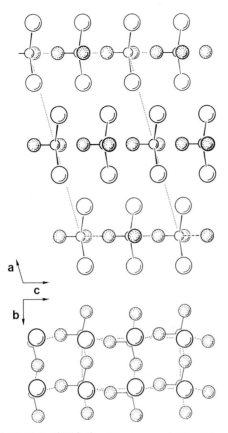

Fig. 161. The monoclinic structure of WO_2Cl_2. The two weak W—O bonds are dashed or omitted. Large open spheres Cl; stippled spheres: O; small spheres: W

d^0-cation no cation pairs are to be expected. The formation of dimers is required by the octahedral coordination of the cation only. At the first glance we expected a normal layered structure based on a hexagonal close-packed anion array with every second Ω-layer completely empty and $\frac{2}{3}$ occupied in the others as represented in Figure 8. The actual structure [965] has the expected projection on (010), however, there are no dimers but zigzag chains parallel to [010]. The coordination octahedra share two cis-corners with two different neighbors instead of the expected common edge. Layering thus occurs parallel to the (100) planes at a right angle to the plane of anion stacking (Figure 162). As only approximate atomic parameters (Table 195) were derived we abstain from listing interatomic distances.

i. Triclinic UCl_5

The layer structure expected for NbI_5 is indeed realized in β-UCl_5, the triclinic modification that can be obtained at room temperature by slow reduction of UCl_6

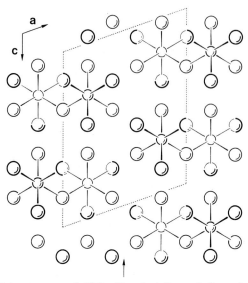

Fig. 162. The monoclinic structure of NbI$_5$. The deviations of the anion sublattice from ideal hexagonal close-packing are neglected. The arrow indicates the (100) plane of layering.

with H$_2$CCl$_2$ [970]. The structure of β-UCl$_5$ contains the same dimeric molecules as the monoclinic α-UCl$_5$ that forms on cooling a hot saturated solution in CCl$_4$. The cations which have a f^1 configuration are slightly displaced from the octahedron centers but in contrast to 4d^1 and 5d^1 compounds the cation-cation distance is increased: U—U = 4.16 Å compared with Cl—Cl = 3.89 Å of the corresponding apical anions. The U—Cl distances are 2.43(2) and 2.45 Å(2) for the terminal and 2.70 Å for the two bridging anions.

TABLE 195
NbI$_5$ structure, monoclinic, C_{2h}^5—P2$_1$/c (No. 14), Z = 4.
all atoms in 4(e): ±(x, y, z; x, $\frac{1}{2}$−y, $\frac{1}{2}$+z)

NbI$_5$: a = 10.58 Å, b = 6.58 Å, c = 13.88 Å, β = 109.14° [965]

	x	y	z
Nb	0.200	$\frac{1}{2}$	0.300
I$_I$	0.000	$\frac{1}{4}$	0.333
I$_{II}$	0.200	$\frac{1}{4}$	0.133
I$_{III}$	0.400	$\frac{1}{4}$	0.433
I$_{IV}$	0.600	$\frac{1}{4}$	0.233
I$_V$	0.800	$\frac{1}{4}$	0.033

j. THE α-WCl$_6$ STRUCTURE

The rhombohedral structure of α-WCl$_6$ is a CdI$_2$-type derivative complementary to the BiI$_3$ type:

$$(Bi_{2/3}\square_{1/3})I_2 \leftrightarrows (\square_{2/3}W_{1/3})Cl_2$$

TABLE 196
β-UCl$_5$ structure, triclinic, C_i^1—P$\bar{1}$ (No. 2), Z = 2.
all atoms in 2(i): ±(x, y, z)

β-UCl$_5$: a = 7.07 Å, b = 9.65 Å, c = 6.35 Å
α = 89.1°, β = 117.35°, γ = 108.54° [970]

	x	y	z
U	0.220 1	0.219 3	0.112 6
Cl$_I$	0.815	0.045	0.066
Cl$_{II}$	0.397	0.152	0.513
Cl$_{III}$	0.013	0.258	0.699
Cl$_{IV}$	0.588	0.334	0.128
Cl$_V$	0.190	0.436	0.266

The attraction of the chlorine atoms by the tungsten atoms gives rise to a distortion of the hcp anion array, leading to a W—Cl distance of 2.24 Å. The empty octahedral holes are correspondingly expanded. Cl—Cl contacts within an octahedron are much shorter (3.11 and 3.21 Å) than between the different octahedra (3.42 and 3.72 Å).

Above 200°C α-WCl$_6$ transforms into β-WCl$_6$ with the hexagonal UCl$_6$ structure. The high-temperature modification is based on the same hexagonal anion close-packing but the cations are now distributed between all layers. Thus the β-WCl$_6$ structure is in fact a NiAs-type derivative.

TABLE 197
α-WCl$_6$ structure, trigonal, C_{3i}^2—R$\bar{3}$ (No. 148), Z = 1(3)
rhombohedral cell: Cl in 6(f): ±(x, y, z; z, x, y; y, z, x)
W in 1(a): 0, 0, 0
hexagonal cell: $(0, 0, 0; \frac{1}{3}, \frac{2}{3}, \frac{2}{3}; \frac{2}{3}, \frac{1}{3}, \frac{1}{3})+$
Cl in 18(f): ±(x, y, z; ȳ, x−y, z; y−x, x̄, z)
W in 3(a): 0, 0, 0

WCl$_6$(r): a_{rh} = 6.58 Å, α = 55.0°; x = 0.37, y = 0.29, z = 0.21
a_h = 6.088 Å, c = 16.68 Å, c/√3a = 1.582; x = 0.295, z = 0.080 [4]
WBr$_6$(?) [971]

As Taylor and Wilson [972] pointed out the octahedral coordination is observed in both modifications of WCl$_6$ although its radius ratio 0.32 is in the range for tetrahedral coordination. We expect that even WBr$_6$ adopts the α-WCl$_6$ structure. The hexabromide crystals are described as gray lamellar with a metallic lustre [971]. It dissociates above 200° and in the gas phase WBr$_5$ is the highest bromide whereas WCl$_6$ does exist in the vapor state.

k. Cr(OH)$_3$·3H$_2$O

The structure of hydrated chromium hydroxide is complementary to the structure of bayerite α-Al(OH)$_3$:

$$(Cr\square_2)(OH)_3(OH_2)_3 \leftrightharpoons (\square Al_2)(OH)_6$$

The isolated $[Cr(OH)_3(OH_2)_3]$ octahedra are linked by hydrogen bonds [973]. The same kind of layers may also be found in composite brucite phases of the type $[Mg_3(OH)_6] Cr(OH)_3 \cdot 3H_2O$.

18. Layered Silicates Derived From Brucite and Bayerite

A fascinating group of composite sandwich structures can be generated by combining brucite- or bayerite-type layers with silicate layers of the type represented in Figure 18c. In this way $\frac{2}{3}$ of the OH ions are replaced by $[OSiO_{3/2}]^-$ tetrahedra. Since the apical oxygen atoms of the perforated tetrahedron sheets have to fit to the octahedron corners of the close-packed OH layers such combinations are possible only with properly selected cations. The Si—Si distance in the tetrahedron layer is equal to the O—O distance of the hydroxide layer.

The wurtzite-related hexagonal SiO_2 modification high-tridymite contains the same tetrahedron layers though connected into a three-dimensional array. From its lattice constant $a = 5.04$ Å we obtain the Si—Si distance. Checking of Table 151 for appropriate cations is at first glance slightly frustrating as no divalent cation has exactly the correct size (Mn: $a = 3.34$ Å, Fe 3.26, Co 3.17, Ni 3.12, Mg 3.14 Å). Al^{3+}, on the other hand, exactly fits into the frame as is evident from Table 178. However, Al is in fact slightly too small since we have neglected the contraction of the hydroxide layer due to the OH → O substitution. Obviously, the lattice can bear some strain since both Al and Mg minerals are rather frequent. Nature has its own tricks to avoid unpleasant situations. Crystals of one-sided silicates are usually curved in order to reduce the strain and, moreover, they contain only a restricted number of layers. Compounds with silicate sheets on both sides of the hydroxide layers crystallize only in minute and badly ordered crystallites (clays!). It is also possible to replace a certain amount of Si atoms by larger trivalent cations such as Al^{3+} or even Fe^{3+}. A simultaneous substitution in the hydroxide layer has of course to restore charge neutrality. Geometrically, substitution of Si by Ge should easily be possible since GeO_2 is only a few % larger than SiO_2, not even enough to provide an exact fit to $Mg(OH)_2$.

Schematically we may represent the generation of composite silicates as follows:

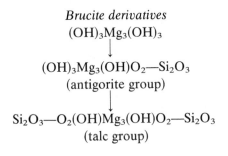

bayerite (hydrargillite) derivatives
(OH)$_3$(☐Al$_2$)(OH)$_3$
↓
(OH)$_3$(☐Al$_2$)(OH)O$_2$—Si$_2$O$_3$
(kaolinite group)
↓
Si$_2$O$_3$—O$_2$(OH)(☐Al$_2$)(OH)O$_2$—Si$_2$O$_3$
(pyrophyllite)

Figure 163 offers an illustration of the structure of one polytype of the kaolinite group. Al$_2$(OH)$_4$Si$_2$O$_5$ occurs in various modifications which differ in stacking and distribution of the empty octahedral sites. Kaolin exists in a triclinic form with one layer (1 T) and a monoclinic form with 3 layers. One layer is present also in metahalloysite. For dickite (Figure 163) and nacrite the most common polytype is the monoclinic 2 M type; for nacrite also a 6 R type. As pointed out by Pauling in 1930 and by Newnham and Brindley [974] there is a considerable misfit between the octahedral Al layer as found in gibbsite, where the parameter corresponding to b is 8.64 Å (dickite: $a = 5.15$ Å, $b = 8.95$ Å, $c = 14.42$ Å, $\beta = 96.8°$, Cc, $Z = 4$), and an ideal hexagonal net of tetrahedra with Si—O = 1.62 Å, which has a corresponding b value of 9.16 Å. Therefore, both the Si—O tetrahedral layer and the Al—O,OH octahedral layer are considerably distorted. The layers in these

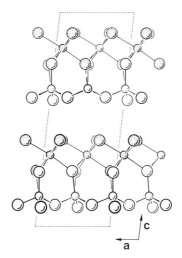

Fig. 163. The monoclinic structure of dickite Al$_2$(OH)$_4$Si$_2$O$_5$.
Large open spheres: O
large stippled spheres: OH
small stippled spheres: Al
small empty spheres: Si

one-sided silicates are held together by hydroxyl-hydrogen bonds. The chemical composition in the kaolin group is rather constant. Chromium kaolinite is a green variety with 1–2% Al replaced by Cr^{3+}. Donbassite represents a combination of the kaolin and antigorite type since substitution of $\frac{1}{4}$ of the Si atoms by Al atoms is compensated by an additional insertion of Na, Ca and Mg into the hydrargillite-type octahedral layer. A gradual transition to the antigorite compositions $LiAl_2(OH)_4Si_{3/2}Al_{1/2}O_5$ and $MgAl_2(OH)_4SiAlO_5$ should be possible.

The antigorite group with completely filled octahedral holes offers a greater chemical variation. Since the dimensions of the brucite part of the composite layers do not exactly match those of the Si_2O_5 sheets, the layers curl up, the larger brucite part being on the outside. In chrysotile (asbestos) fibres built of curled ribbons from cylinders several thousand Å long and some ten layers thick. In antigorite the OH and Si_2O_5 layers alternate so as to form a corrugated-iron-like arrangement [977]. The different behavior of the layers may be due to certain substitutions. Antigorite is said to contain some Fe and Al in the brucite layers. Chrysotile contains the same kind of foreign atoms but to a lesser amount and, moreover, Al substitutes only for Si. Lizardite is similar with a higher impurity level. Antigorite-type compounds exist also with other cations, either grown synthetically in a pure state (with Co and Ni [975, 976]) or as minerals more or less pure and transitional to the kaolinite composition:

Cariopilite $Mn_3(OH)_4Si_2O_5$
Greenalite $(Fe^{2+}, Fe^{3+})_{3-\delta}(OH)_4Si_2O_5$
Synthetic $Co_3(OH)_4Si_2O_5$
Garnierite ⎫
Nepouite ⎬ $Ni_3(OH)_4Si_2O_5$
Berthierite $(Fe, Mg, Al)_3(OH)_4(Al, Si)_2O_5$
Amesite $Mg_{1.6}Fe_{0.4}Al(OH)_4AlSiO_5$
Grovesite $(Mn, Mg, Al)_3(OH)_4(Si, Al)_2O_5$
Cronstedtite $Fe^{2+}_2Fe^{3+}(OH)_4SiFeO_5$

The mineral cronstedtite is found in a large number of polytypes: 1M, 1H, 2M, 2H, 3H, 6H, 9R. It is interesting that on heating in air pure ferro-berthierite transforms at 400°C reversibly into ferri-berthierite by oxydation of the outer OH layers while the inner OH ions resist up to 450–500°C [5]:

$$(OH)_3Fe^{2+}_3(OH)Si_2O_5 + \tfrac{3}{4}O_2 \xrightarrow{400°} O_3Fe^{3+}_3(OH)Si_2O_5 + \tfrac{3}{2}H_2O.$$

Oxydation removes the hydrogen bonds and thus generates a true layer compound.

True two-dimensional structures only are possible in the pyrophyllite (Figure 164) and talc group. Pyrophyllite $Al_2(OH)_2(Si_2O_5)_2$ usually contains some Mg, Fe^{2+} and Fe^{3+}. In talc $Mg_3(OH)_2(Si_2O_5)_2$ Mg may partly or completely be

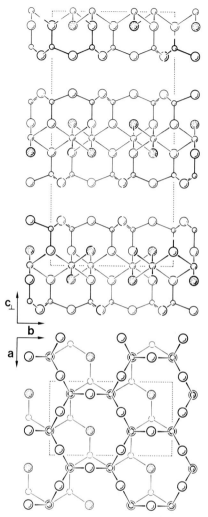

Fig. 164. The monoclinic structure of pyrophyllite $Al_2(OH)_2Si_4O_{10}$. *Above:* Projection of the structure along the a-axis. *Below:* Top view of the bottom layer. Only the upper part is reproduced which represents at the same time a dickite layer.

large empty spheres: O
large stippled spheres: OH
smallest spheres: Si

substituted by Fe, Co and Ni [975]:

Minnesotaite	$Fe_3(OH)_2(Si_2O_5)_2$
Synthetic	$Co_3(OH)_2(Si_2O_5)_2$
Willemseite	$Ni_3(OH)_2(Si_2O_5)_2$

The charge balance due to substitution of elements in the Si_2O_5 layers may be restored internally or by intercalation of foreign ions either in a strictly stoichiometric ratio as in muscovite $KAl_2(OH)_2Si_3AlO_{10}$, margarite $CaAl_2(OH)_2Si_2Al_2O_{10}$, biotite $KMg_3(OH)_2Si_3AlO_{10}$ and xanthophyllite $CaMg_3(OH)_2Si_2Al_2O_{10}$, or non-stoichiometrically with additional water as in the group of montmorillonite $[(Mg_{1/3}Al_{5/3})(OH)_2(Si_2O_5)_2][Na_{1/3}(H_2O)_4]$. Hydrogen-linked hybrid layer structures exist in which silicate layers alternate with pure H_2O layers as in monoclinic halloysite $[Al_2(OH)_4Si_2O_5](H_2O)_2$. In the structures of the members of the chlorite group pure neutral or positively charged hydroxide layers are inserted between the silicate layers:

neutral: Manandonite $[LiAl_2(OH)_2AlBSi_2O_{10}][Al_2(OH)_6]$
Orthochlorite $[(Mg, Fe^{2+})_3(OH)_2Si_3AlO_{10}][(Mg, Fe^{2+})_3(OH)_6]$
Talc chlorite $[Mg_3(OH)_2(Si_2O_5)_2][Mg_3(OH)_6]$
Clinochlore $[Mg_2Al(OH)_2Si_3AlO_{10}][Mg_3(OH)_6]$
Corundophilite $[(Mg, Al, Fe)(OH)_2(Al, Si)_4O_{10}][Mg_3(OH)_6]$

charged: Sudoite $[Al_2(OH)_2Si_3AlO_{10}]^-[Al_{7/3}(OH)_6]^+$
Cookeite $[Al_2(OH)_2Si_3AlO_{10}]^-[LiAl_2(OH)_6]^+$

Obviously all combinations are possible. Many more representatives should be synthesizable. The missing link between sudoite and cookeite for example would contain an $[AlMg_2(OH)_6]$ brucite layer. We wonder whether it might be possible to replace the inner OH ions in these phases by F ions. Hydrothermal preparations should lead to many new compositions with other transition-element cations. Should it be possible to exchange the brucite or hydrargillite layers in the chlorite family by layers such as MCl_2 or MCl_3 which are not able to form hydrogen bonds? Intercalation of organic radicals like imidazole and methylimidazole [978], pyridine-N-oxide and picolin-N-oxides [979] into the kaolinite minerals has recently been reported. This intercalation expands the layers considerably. The first group produces an increase of the layer separation of 4.2–4.3 and 5.0–5.1 Å, the latter an increase of 5.4 Å to ~12.6 Å. Obviously the organic radicals link the silicate layers via H bridges. H-bridging is effective also in the silicic acid $H_2Si_{14}O_{29} \cdot 5H_2O$ which is able to form intercalation compounds with dimethylsulfoxide (eg. $2(CH_3)_2SO \cdot H_2Si_{14}O_{29}$), acid amides, urea and its derivatives and bases such as hydrazine and pyridine [980]. Recently the structure of a formamide intercalate of dickite, $Al_2(OH)_4Si_2O_5 \cdot HCONH_2$, was reported [1027].

A huge number of new improper layer-type phases should be obtainable from members of the muscovite, margarite, biotite and xanthophyllite group as well as from the montmorillonite and saponite group by exchanging the positively charged alkali or alkaline-earth cations (including the water of hydration in the second group) by organic radicals. Thus Wyoming montmorillonite reacts with 4,4'-diaminostilbene dihydrochloride to form the intercalate

$$[Mg_{0.25}Al_{1.75}(OH)_2(Si_2O_5)_2][\tfrac{1}{8}H_3NC_6H_4—CH=CH—C_6H_4NH_3]$$

which on heating produces aniline [981].

Many of the layered silicates retain their structures on dehydration. Thus, pyrophyllite $Al_2(OH)_2(Si_2O_5)_2$ forms a well-ordered relict structure $Al_2(O\square)$-$(Si_2O_5)_2$ [1030b].

Improper layer structures quite analogous to the mica derivatives can be generated by replacing the alkali cations in the tetragonal K_2NiF_4-type representatives by an organic radical such as $C_nH_{2n+1}NH_3$. These phases exist with divalent cations such as Mn, Fe, Co, Ni, Cu, Cd, ... and with Cl and Br as anions. They have attracted a great deal of interest for their magnetic properties. The phases containing transition-element cations exhibit magnetic order (antiferromagnetic with Mn, ferromagnetic with Cu) and approximate two-dimensional Heisenberg systems [982, 982a].

APPENDIX

TABLE 198

$NbOCl_2$ structure, monoclinic, C_2^3—C2 (No. 5), $Z = 4$.

$(0, 0, 0; \frac{1}{2}, \frac{1}{2}, 0)+$

All atoms in 4(c): $\pm(x, y, z; \bar{x}, y, \bar{z})$

$NbOCl_2$: $a = 12.79$ Å, $b = 3.933$ Å, $c = 6.704$ Å, $\beta = 105.2°$ [1022]
$NbOBr_2$: $a = 14.03$ Å, $b = 3.93$ Å, $c = 7.102$ Å, $\beta = 105°$ [1022]
$NbOI_2$: $a = 15.08$ Å, $b = 3.923$ Å, $c = 7.484$ Å, $\beta = 104.6°$ [1022]

		x	y	z	
$NbOCl_2$:	Nb	0.000	0.047	0.266	
	O	0.500	0.012	0.250	[1022]
	Cl_I	0.147	0.000	0.573	
	Cl_{II}	0.143	0.000	0.072	

$MoOCl_2$ structure, monoclinic, C_{2h}^3—C2/m (No. 12), $Z = 4$.

$(0, 0, 0; \frac{1}{2}, \frac{1}{2}, 0)+$

All atoms in 4(i): $\pm(x, 0, z)$

$MoOCl_2$: $a = 12.77$ Å, $b = 3.569$ Å, $c = 6.540$ Å, $\beta = 104.8°$ [1022]

	x	z		x	z	
Mo	0.000	0.276	Cl_I	0.150	0.579	[1022]
O	0.500	0.250	Cl_{II}	0.135	0.069	

3i. $CuTe_2Cl$

The layer structure of the diamagnetic semiconductor $CuTe_2Cl$ ($= Te_2 + CuCl$) represents an interesting example of a lone-pair tetrahedral structure. Tellurium spiral chains (Te—Te = 2.737 and 2.792 Å, bond angles of 99.9 and 102.70°) as occur in elemental tellurium are linked to form layers parallel (001) by

$$\text{>Cu<}\genfrac{}{}{0pt}{}{Cl}{Cl}\text{>Cu<}$$

bridges, where Cu is tetrahedrally coordinated [$CuTe_2Cl_{2/2}$].

A related layer structure (but with every third Se atom bonded to 2Se only) is to be expected for orthorhombic $CuBrSe_3$ which crystallizes in red flakes [1025].

TABLE 199

$CuTe_2Cl$ structure, monoclinic, C_{2h}^5—$P2_1/c$ (No. 14), $Z=4$.
All atoms in 4(e): $\pm(x, y, z; x, \frac{1}{2}-y, \frac{1}{2}+z)$

$CuTe_2Cl$: $a=8.207$ Å, $b=4.9347$ Å, $c=15.279$ Å, $\beta=134.92$ [1023]

	x	y	z
Cu	0.3283	0.6403	0.2489
Te_I	−0.0020	0.4980	0.3652
Te_{II}	0.2958	0.2795	0.3587
Cl	0.3296	0.3912	0.1185

$CuTe_2Br$: $a=8.358$ Å, $b=4.951$ Å, $c=15.704$ Å, $\beta=135.1°$ [1025]
$CuTe_2I$: $a=8.672$ Å, $b=4.881$ Å, $c=16.493$ Å, $\beta=135.0°$ [1025]

TABLE 200

$Th(NH)_2$ structure, trigonal, D_{3d}^5—$R\bar{3}m$ (No. 166), $Z=6$.

$(0,0,0; \frac{1}{3},\frac{2}{3},\frac{2}{3}; \frac{2}{3},\frac{1}{3},\frac{1}{3})+$
All atoms in 6(c): $\pm(0, 0, z)$

$Th(NH)_2$: $a=3.95$ Å, $c=27.58$ Å;
$z(Th)=0.220$; $z(NH_I)\approx0.132$; $z(NH_{II})\approx0.377$ [1029]
(z-values for NH were taken from α-Th_3N_4)

7e. AuSe

AuSe exists in two modifications which both are monoclinic and contain mono- and trivalent gold [1031]. $Au^I Au^{III} Se_2$ thus represents a normal valence compound. Two parallel $PdCl_2$-type $Au^{III}Se_2$ units are linked via Au^I to form infinite square rods in α-AuSe. In β-AuSe the Au^+ cations which connect the one-dimensional $[Au^{III}Se_2]_\infty$ units are located on different sides thus giving rise to the formation of corrugated sheets parallel to (100). This explains the plate-like growth of the β-AuSe crystals parallel to (100) and the observed bending of the crystal face. The layer structure of β-AuSe may be interpreted as a strongly distorted NaCl structure in which the $[Au^+Se_{6/6}]$ octahedra are compressed and the $[Au^{III}Se_{6/6}]$ octahedra are stretched. Bonding distances are Au^I—Se = 2.43 Å and Au^{III}—Se = 2.50 Å, while the elongated distances within the deformed octahedra range from 3.40 to 3.52 Å.

TABLE 201

β-AuSe structure, monoclinic, C_{2h}^3—C2/m (No. 12), Z = 4.

$(0, 0, 0; \frac{1}{2}, \frac{1}{2}, 0) +$

Se in 4(i): $\pm(x, 0, z)$
Au$_{II}$ in 2(d): $0, \frac{1}{2}, \frac{1}{2}$
Au$_I$ in 2(a): $0, 0, 0$

β = AuSe: a = 8.355 Å, b = 3.663 Å, c = 6.262 Å, β = 106.03° [1031]

TABLE 202

Si_2Te_3 structure, triclinic, C_i^1—P1 (No. 1), Z = 4.

atoms in 1(a); x, y, z

Si_2Te_3: a = 7.428 Å, b = 7.428 Å, c = 13.488 Å, α = 90.0°, β = 90.0°, γ = 120.0° [1034]

	x	y	z		x	y	z
Te$_I$	0.0	0.3333	0.1191	Si$_I$	0.996	0.003	0.165
Te$_{II}$	0.3207	0.9926	0.1186	Si$_{II}$	0.993	0.006	0.338
Te$_{III}$	0.6621	0.6759	0.1176	Si$_{III}$	0.996	0.012	0.668
Te$_{IV}$	0.6581	0.9907	0.3816	Si$_{IV}$	0.004	0.008	0.835
Te$_V$	0.9996	0.6723	0.3831	Si$_V^{(1)}$	0.316	0.511	0.219
Te$_{VI}$	0.3179	0.3305	0.3825	Si$_V^{(2)}$	0.190	0.683	0.219
Te$_{VII}$	0.9972	0.3380	0.6185	Si$_V^{(3)}$	0.490	0.804	0.220
Te$_{VIII}$	0.3329	0.0007	0.6179	Si$_{VI}^{(1)}$	0.319	0.809	0.274
Te$_{IX}$	0.6607	0.6629	0.6170	Si$_{VI}^{(2)}$	0.486	0.683	0.279
Te$_X$	0.6620	0.0027	0.8813	Si$_{VI}^{(3)}$	0.193	0.512	0.276
Te$_{XI}$	0.9980	0.6649	0.8823	Si$_{VII}^{(1)}$	0.805	0.484	0.719
Te$_{XII}$	0.3375	0.3384	0.8814	Si$_{VII}^{(2)}$	0.656	0.181	0.709
				Si$_{VII}^{(3)}$	0.519	0.348	0.726
				Si$_{VIII}^{(1)}$	0.510	0.194	0.771
				Si$_{VIII}^{(2)}$	0.649	0.477	0.771
				Si$_{VIII}^{(3)}$	0.818	0.353	0.780

(According to [1045] the Si$_2$Te$_3$ structure is trigonal, P$\bar{3}$1c, a = 7.430 Å, c = 13.482 Å; atomic positions are given.)

Orthorhombic GeAsSe obviously crystallizes also in a layer structure [1052]. This Mooser-Pearson phase may contain As—As pairs in order to saturate the bonds.

A new type of intercalation compounds is based on the phosphates and arsenates $M^{4+}(HXO_4)_2 \cdot nH_2O$, with M = Zr, Si, Sn, Ce, and X = P, As, of which Zr(HPO$_4$)$_2 \cdot n$H$_2$O (n = 0, 1, 2) is the best known example [1057]. In the monoclinic structure of Zr(HPO$_4$)$_2 \cdot H_2$O, e.g., the Zr(HPO$_4$)$_2$ layers are held together by H-bonds only and the intercalated hydrogen-bonded H$_2$O molecules can be replaced by a variety of organic molecules such as dimethyl sulfoxide, piperidine, etc. The dihydrate reacts with even a larger number of guest molecules than the monohydrate because the phosphate layers are separated by bimolecular water layers [1057].

The structure of the shiny black Mooser-Pearson phase In$_2$Te$_5$ is built up of corrugated layers (In$_4$Te$_4$)$_n^{2+}$(Te$_3$)$_{2n}^{2-}$ with In in tetrahedral coordination. [In$_2$Te$_2$]$_n$ chains running parallel to the a-axis are interlinked by Te$_3$ bridges to form two sheets parallel to the (a, b) plane of the monoclinic cell [1060].

REFERENCES

1. Donnay, J. D. H. and Ondik, H. M. (ed.): *Crystal Data Determinative Tables, Inorganic Compounds*, third ed., JCPDS 1973.
2. Landolt-Börnstein: *Numerical Data and Functional Relationships in Science and Technology*, New Series, Group III, Springer, Berlin, Vol. 6 (1971) and Vol. 7 (1973).
3. Pearson, W. B.: *A Handbook of Lattice Spacings and Structures of Alloys*, Pergamon, Oxford, Vol. 1, 1964 and Vol. 2, 1967.
4. Wyckoff, R. W. G.: *Crystal Structures*, Wiley-Interscience, New York, Vols. 1, 2 and 3, second ed. 1965.
5. Strunz, H.: *Mineralogische Tabellen*, Akad. Verlagsges. Geest und Portig, Leipzig, 5. Auflage 1973.
6. Povarennikh, A. S.: *Crystal-Chemical Classification of Minerals*, Vols. 1 and 2, Plenum, New York, 1972.
7. Schubert, K.: *Kristallstrukturen zweikomponentiger Phasen*, Springer, Berlin, 1964.
8. Donohue, J.: *The Structures of the Elements*, Wiley-Interscience, New York, 1974.
9. Pearson, W. B.: *The Crystal Chemistry and Physics of Metals and Alloys*, Wiley-Interscience, New York, 1972.
10. Wells, A. F.: *Structural Inorganic Chemistry*, Clarendon Press, Oxford, fourth ed. 1975.
11. Krebs, H.: *Grundzüge der anorganischen Kristallchemie*, Enke, Stuttgart, 1968.
12. Canterford, J. H. and Colton, R.: *Halides of the First-Row Transition Metals*, Wiley-Interscience, New York, 1969.
13. Canterford, J. H. and Colton, R.: *Halides of the Second- and Third-Row Transition Metals*, Wiley-Interscience, New York, 1968.
14. Brown, D.: *Halides of the Lanthanides and Actinides*, Wiley-Interscience, New York, 1968.
15. Brown, D.: 'The Actinide Halides and Their Complexes', in *MTP Int. Rev. Sci., Inorg. Chem.*, Series One, Vol. 7, p. 87.
16. Winfield, J. M.: 'Fluorides', in *MTP Int. Rev. Sci., Inorg. Chem.*, Series One, Vol. 5, p. 271.
17. Colton, R.: 'Chlorides, Bromides and Iodides', in *MTP Int. Rev. Sci., Inorg. Chem.*, Series One, Vol. 5, p. 299.
18. Jellinek, F.: 'Sulphides, Selenides and Tellurides of the Transition Elements', in *MTP Int. Rev. Sci., Inorg. Chem.*, Series One, Vol. 5, p. 339.
19. Dell, R. M. and Bridger, N.: 'Actinide Chalcogenides and Pnictides', in *MTP Int. Rev. Sci., Inorg. Chem.*, Series One, Vol. 5, p. 211.
20. Jellinek, F.: 'Sulphides', in *Inorganic Sulphur Chemistry*, Elsevier, New York, 1969.
21. Wilson, J. A. and Yoffe, A. D.: *Adv. Phys.* **18** (1969), 193.
22. Yoffe, A. D.: *Ann. Rev. Materials Sci.*, **3** (1973), 147; *Festkörperprobleme* **13** (1973), 1.
23. Hulliger, F.: *Structure and Bonding* **4** (1968), 83.
24. Troitskaya, N. V. and Pinsker, Z. G.: *Sov. Phys. – Cryst.* **6** (1961), 34.
25. Khitrova, V. I. and Pinsker, Z. G.: *Sov. Phys. – Cryst.* **6** (1962), 712.
26. Takéuchi, Y. and Nowacki, W.: *Schweiz. Min. Petrogr. Mitt.* **44** (1964), 105.
27. Brown, B. E. and Beerntsen, D. J.: *Acta Cryst.* **18** (1965), 31.
28. Zvyagin, B. B. and Soboleva, S. V.: *Sov. Phys. – Cryst.* **12** (1967), 46.
29. Wickman, F. E. and Smith, D. K.: *Amer. Mineral.*, **55** (1970), 1843.
30. Jellinek, F.: *J. Less-Common Met.* **4** (1962), 9.
31. Pinsker, Z. G.: *Sov. Phys.-Cryst.* **5** (1960), 600.
32. Khitrova, V. I. and Pinsker, Z. G.: *Sov. Phys. – Cryst.* **5** (1960), 679.
33. Poulin, M., Poulin, M., and Lucas, J.: *J. Solid State Chem.* **8** (1973), 132.
34. Takéuchi, Y. and Donnay, G.: *Acta Cryst.* **12** (1959), 465.
35. Keve, E. T. and Skapski, A. C.: *Inorg. Chem.* **7** (1968), 1757.
36. Weiss, A. and Weiss, A.: *Z. Naturforsch.* **11b** (1956), 604.

37. Jagodzinski, H.: *Z. Krist.* **112** (1959), 80.
38. Strähle, J. and Lörcher, K. P.: *Z. Naturforsch.* **29b** (1974), 266.
39. Janssen, E. M. W., Folmer, J. C. W., and Wiegers, G. A.: *J. Less-Common Met.* **38** (1974), 71.
40. Faltens, M. O. and Shirley, D. A.: *J. Chem. Phys.* **53** (1970), 4249.
41. Zachariasen, W. H.: *Acta Cryst.* **7** (1954), 305.
42. Craven, B. M. and Sabine, T. M.: *Acta Cryst.* **20** (1966), 214.
43. Peters, C. H. and Milberg, M. E.: *Acta Cryst.* **17** (1964), 229.
44. Coulson, C. A.: *Acta Cryst.* **17** (1964), 1086.
44a. Krebs, B., and Diercks, H.: *Acta Cryst.* **A31** (1975), S66, and private communication.
44b. Chen, H. Y., Conard, B. R., and Gilles, P. W.: *Inorg. Chem.* **9** (1970), 1776.
44c. Hillel, R.: Thesis, Univ. Lyon 1971.
45. El Goresy, A. and Donnay, G.: *Sci.* **161** (1969), 363.
46. Whittaker, A. G. and Wolten, G. M.: *Sci.* **178** (1972), 54.
47. Dutta, A. K.: *Phys. Rev.* **90** (1953), 187.
48. Wallace, P. R.: *Phys. Rev.* **71** (1947), 622; Coulson, C. A. and Taylor, R.: *Proc. Phys. Soc. (London)* **65A** (1952), 815; Lomer, W. M.: *ibid.* **227A** (1955), 330.
49. Slonczewski, J. C. and Weiss, P. R.: *Phys. Rev.* **109** (1958), 272.
50. McClure, J. W. and Smith, L. B.: *Proc. 5th Conf. on Carbon*, Pergamon, New York, 1963, Vol. II, p. 3.
51. Hennig, G. R.: *Progr. Inorg. Chem.* **1** (1959), 125.
52. Rüdorff, W.: *Advan. Inorg. Chem. Radiochem.* **1** (1959), 223.
53. Ubbelohde, A. R. and Lewis, F. A.: *Graphite and its Crystal Compounds*, Oxford Univ. Press, London, 1960.
54. Haering, R. R. and Mrozowski, S.: *Progr. Semicond.* **5** (1960), 273.
55. Painter, G. S. and Ellis, D. E.: *Phys. Rev.* **B1** (1970), 4747.
56. McFeely, F. R., Kowalczyk, S. P., Ley, L., Cavell, R. G., Pollak, R. A., and Shirley, D. A.: *Phys. Rev.* **B9** (1974), 5268.
56a. Rasor, N. S. and McClelland, J. D.: *J. Phys. Chem. Solids* **15** (1960), 17.
57. Dulin, I. N., Al'tshuler, L. V., Vashchenko, V. Ya., and Zubarev, V. N.: *Sov. Phys.-Solid State* **11** (1969), 1016.
57a. Hérold, A., Marzluf, B., and Pério, P.: *Comptes rendus (Paris)* **246** (1958), 1866.
58. Khusidman, M. B.: *Sov. Phys.-Solid State* **14** (1973), 2791.
59. Hall, H. T. and Compton, L. A.: *Inorg. Chem.* **4** (1965), 1213.
60. Lipscomb, W. N. and Britton, D.: *J. Chem. Phys.* **33** (1960), 275.
61. Naslain, R., Guette, A., and Barret, M.: *J. Solid State Chem.* **8** (1973), 68.
62. Goodman, C. H. L.: in *Chemical Bonds in Solids*, Plenum, N.Y. (1972), p. 62.
63. Rüdorff, W.: *Chimia* **19** (1965), 489.
64. Daumas, N. and Hérold, A.: *Comptes rendus (Paris)* **268C** (1969), 373.
65. Nixon, D. E. and Parry G. S.: *J. Phys.* **D1** (1968), 291.
66. Juza, R. and Wehle, V.: *Naturwiss.* **52** (1965), 560.
67. Stein, C., Poulenard, J., Bonnetain, L., and Golé, J.: *Comptes rendus (Paris)* **260** (1965), 4503.
68. Guérard, D. and Hérold, A.: *Comptes rendus (Paris)* **275C** (1972), 571.
69. Lagrange, P., Métrot, A., and Hérold, A.: *Comptes rendus (Paris)* **278C** (1974), 701.
69a. Billaud, D. and Hérold, A.: *Bull. Soc. Chim. France* (1974), 2715.
70. Billaud, D., Carton, B., Métrot, A., and Hérold, A.: *Bull. Soc. Chim. France* (1973), 2259.
71. Nixon, D. E. and Parry, G. S.: *J. Phys.* **C2** (1969), 1732.
72. Aronson, S., Salzano, F. J., and Bellafiore, D.: *J. Chem. Phys.* **49** (1968), 434.
73. Carton, B. and Hérold, A.: *Bull. Soc. Chim. France* (1972), 1337.
74. Rüdorff, W. and Schulze, E.: *Z. anorg. allg. Chem.* **277** (1954), 156.
75. Billaud, D. and Hérold, A.: *Bull. Soc. Chim. France* (1972), 103.
76. Carver, G. P.: *Phys. Rev.* **B2** (1970), 2284.
77. Hérold, A.: *Bull. Soc. chim. France* (1955), 999.
78. Furdin, G., Carton, B., Billaud, D., Zeller, C., and Hérold, A.: *Comptes rendus (Paris)* **278B** (1974), 1025.
79. Hennig, G. R.: *J. Chem. Phys.* **43** (1965), 1201.

80. Hannay, N. B., Geballe, T. H., Matthias, B. T., Andres, K., Schmidt, P., and Mac Nair, D.: *Phys. Rev. Lett.* **14** (1965), 225.
81. Parry, G. S.: *Conf. Ind. Carbons Graphite* (3rd) (ed. by J. G. Gregory), Soc. Chem. Ind., London, 1970, p. 58.
81a. Boersma, M. A. M.: *Cat. Rev. - Sci. Eng.* **10** (1974), 243.
82. Ubbelohde, A. R.: *Proc. Roy. Soc.* **A309** (1969), 297.
83. Ubbelohde, A. R., Parry, G. S., and Nixon, D.: *Nature* **206** (1965), 1352.
84. Dzurus, M. and Hennig, G. R.: *J. Amer. Chem. Soc.* **79** (1957), 1051.
85. Fuzellier, H. and Hérold, A.: *Comptes rendus (Paris)* **267C** (1968), 607.
85a. Chun-hsu, Lin, Selig, H., Rabinovitz, M., Agranat, I., and Sarig, S.: *Inorg. Nucl. Chem. Lett.* **11** (1975), 601.
86. Lalancette, J. M. and Lafontaine, J.: *J. Chem. Soc., Chem. Commun.* (1973), 815.
87. Opalovskii, A. A., Nazarov, A. S., Uminskii, A. A., and Chichagov, Yu. V.: *Russ. J. Inorg. Chem.* **17** (1972), 1227.
88. Selig, H. and Gani, O.: *Inorg. Nucl. Chem. Lett.* **11** (1975), 75.
89. Opalovskii, A. A., Kuznetsova, Z. M., Chichagov, Yu. V., Nazarov, A. S., and Uminskii, A. A.: *Russ. J. Inorg. Chem.* **19** (1974), 1134.
90. Mizutani, Y.: *Kyoto Daigaku Kogaku Kenkyusho Iho* **41** (1972), 61; *Chem. Abstracts* **78** (1973), 8952.
91. Turnbull, J. A. and Eeles, W. T.: Papers 2nd. Conf. Ind. Carbon Graphite, London, 1965, p. 173.
92. Hohlwein, D., Grigutsch, F. D., and Knappwost, A.: *Angew. Chem.* **18** (1969), 333.
93. Sasa, T., Takahashi, Y., and Mukaibo, T.: *Bull. Chem. Soc. Japan* **45** (1972), 937, 2250, 2267.
94. Balestreri, C., Vangelisti, R., Melin, J., and Hérold, A.: *Comptes rendus (Paris)* **279C** (1974), 279.
95. Rüdorff, W. Stumpp, E., Spriessler, W., and Siecke, F. W.: *Angew. Chem.* **75** (1963), 130.
96. Cowley, J. M. and Ibers, J. A.: *Acta Cryst.* **9** (1956), 421.
97. Liengme, B. V., Bartlett, M. W., Hooley, J. G., and Sams, J. R.: *Phys. Lett.* **25A** (1967), 127.
98. Novikov, Yu. N., Vol'pin, M. E., Prusakov, V. E., Stukan, R. A., Gol'danskii, V. I., Semion, V. A., and Struchkov, Yu. T.: *Russ. J. Struct. Chem.* **11** (1970), 970.
99. Karimov, Yu. S., Zvarykina, A. V., and Novikov, Yu. N.: *Sov. Phys. - Solid State* **13** (1972), 2388.
100. Hohlwein, D. and Metz, W.: *Z. Krist.* **139** (1974), 279.
101. Tominaga, T., Sakai, T., and Kimura, T.: *Chem. Lett.* (1974), 853.
102. Ohhashi, K. and Tsujikawa, I.: *J. Phys. Soc. Japan* **36** (1974), 422, 980.
102a. Metz, W. and Hohlwein, D.: *Carbon* **13** (1975), 84.
103. Freeman, A. G.: *J. Chem. Soc., Chem. Commun.* (1974), 746.
104. Johnson, A. W. S.: *Acta Cryst.* **23** (1967), 770.
105. Zvarykina, A. V., Karimov, Yu. S., Vol'pin, M. E., and Novikov, Yu. N.: *Sov. Phys. - Solid State* **13** (1971), 21.
106. Novikov, Yu. N., Semion, V. A., Struchkov, Yu. T., and Vol'pin, M. E.: *Russ. J. Struct. Chem.* **11** (1970), 814.
107. Ohhashi, K. and Tsujikawa, I.: *J. Phys. Soc. Japan* **37** (1974), 63.
108. Hohlwein, D., Readman, P. W., Chamberod, A., and Coey, J. M. D.: *Phys. Stat. Sol.* **64B** (1974), 305.
109. Stukan, R. A., Prusakov, V. A., Novikov, Yu. N., Vol'pin, M. E., and Gol'danskii, V. I.: *Russ. J. Struct. Chem.* **12** (1971), 567.
110. Knappwost, A. and Metz, W.: *Naturwiss.* **56** (1969), 85; *Z. phys. Chem.* **64** (1969), 178.
111. Stukan, R. A., Novikov, Yu. N., Povitskii, V. A., and Salugin, A. N.: *Sov. Phys. - Solid State* **14** (1973), 2914.
112. Takahashi, Y., Yamagata, H., and Mukaibo, T.: *Carbon* **11** (1973), 19.
113. Bach, B. and Ubbelohde, A. R.: *J. Chem. Soc. A* (1971), 3669.
114. Boeck, A. and Rüdorff, W.: *Z. anorg. allg. Chem.* **392** (1972), 236.
115. Vangelisti, R. and Hérold, A.: *Comptes rendus (Paris)* **276C** (1973), 1109.
116. Vangelisti, R., Furdin, G., Carton, B., and Hérold, A.: *Comptes rendus (Paris)* **278C** (1974), 869.

117. Boissonneau, J. F. and Collin, G.: *Carbon* **11** (1973), 567.
118. Boeck, A. and Rüdorff, W.: *Z. anorg. allg. Chem.* **397** (1973), 179.
119. Freeman, A. G. and Larkindale, J. P.: *J. Chem. Soc. A* (1969), 1307.
120. Kirkinskii, V. A.: *Eksp. Issled. Mineral.* (1968), 9; Kirkinskii, V. A. and Yakushev, V. G.: *Chem. Abstr.* **77** (1972), 10273.
120a. Anderson, T. L. and Krause, H. B.: *Acta Cryst.* **B30** (1974), 1307.
121. Bercha, D. M., Pankevich, Z. V., Savitskii, A. V., and Tovstyuk, K. D.: *Sov. Phys. – Solid State* **7** (1966), 1968.
122. Zav'yalova, A. A. and Imamov, R. M.: *Russ. J. Struct. Chem.* **13** (1972), 811.
122a. Atabaeva, É. Ya., Itskevich, E. S., Mashkov, S. A., Popova, S. V., and Vereshchagin, L. F.: *Sov. Phys. – Solid State* **10** (1968), 43.
123. Atabaeva, É. Ya., Mashkov, S. A., and Popova, S. V.: *Sov. Phys. – Cryst.* **18** (1973), 104.
123a. Atabaeva, É. Ya., Bendeliani, N. A., and Popova, S. V.: *Sov. Phys. – Solid State* **15** (1974), 2346.
123b. Yakushev, V. G. and Kirkinskii, V. A.: *Inorg. Mater.* **10** (1974), 1025.
124. Il'ina, M. A. and Itskevich, E. S.: *Sov. Phys. – Solid State* **13** (1972), 2098.
124a. Vereshchagin, L. F., Atabaeva, É. Ya., and Bendeliani, N. A.: *Sov. Phys. – Solid State* **13** (1972), 2051.
125. Becker, K. A., Plieth, K., and Stranski, I. N.: *Progr. Inorg. Chem.* **4** (1962), 1.
125a. Pertlik, F.: *Monatsh. Chem.* **106** (1975), 755.
126. Malmros, G.: *Acta. Chem. Scand.* **24** (1970), 384.
127. White, W. B., Dachille, F., and Roy, R.: *Z. Krist.* **125** (1967), 450.
128. Mullen, D. J. E. and Nowacki, W.: *Z. Krist.* **136** (1972), 48.
128a. Carron, G. A.: *Acta Cryst.* **16** (1963), 338.
129. Kirkinskii, V. A. and Yakushev, V. G.: *Dokl. Akad. Nauk SSSR* **182** (1968), 1083.
130. Renninger, A. L. and Averbach, B. L.: *Acta Cryst.* **B29** (1973), 1583.
130a. Zhukov, É. G., Dzhaparidze, O. I., Dembovskii, S. A., and Popova, N. P.: *Inorg. Mater.* **10** (1974), 1619.
131. Kupčik, V.: *Naturwiss.* **54** (1967), 114, and private communication.
132. Weissberg, B. G.: *Amer. Min.* **50** (1965), 1817.
133. Guillermo, T. R. and Wuensch, B. J.: *Acta Cryst.* **B29** (1973), 2536.
134. Soklakov, A. I. and Illarionov, V. V.: *Russ. J. Struct. Chem.* **5** (1964), 218.
135. Krebs, H. and Gruber, H. U.: *Z. Naturforsch.* **22A** (1967), 96.
136. Vincent, H.: *Bull. Soc. chim. France* (1972), 4517.
137. Thurn, H. and Krebs, H.: *Acta Cryst.* **B25** (1969), 125.
138. Brown, A. and Rundqvist, S.: *Acta Cryst.* **19** (1965), 684.
139. Jamieson, J. C.: *Sci.* **139** (1963), 1291.
140. Wittig, J. and Matthias, B. T.: *Sci.* **160** (1968), 994.
141. Berman, I. V. and Brandt, N. B.: *JETP Lett.* **7** (1968), 323.
142. Duggin, M. J.: *J. Phys. Chem. Solids* **33** (1972), 1267.
143. Berman, I. V. and Brandt, N. B.: *JETP Lett.* **10** (1969), 55.
144. McDonald, T. R. R., Gregory, E., Barberich, C. S., McWhan, D. B., Geballe, T. H., and Hull, G. W.: *Phys. Lett.* **14A** (1965), 16.
145. Kabalkina, S. S., Kolobyanina, T. N., and Vereshchagin, L. F.: *Sov. Phys. – JETP* **31** (1970), 259.
146. Jaggi, R.: *Helv. Phys. Acta* **37** (1964), 618.
147. Jayaraman, A. and Cohen, L. H.: *Phase Diagrams*, Vol. 1 (1970), Academic Press New York.
148. Schaufelberger, P., Merx, H., and Contré, M.: *High Temp. – High Press.* **4** (1972), 111.
149. Il'ina, M. A. and Itskevich, E. S.: *JETP Lett.* **11** (1970), 218.
150. Falicov, L. M. and Golin, S.: *Phys. Rev.* **137** (1965), A871.
151. McWhan, D. B.: *Sci.* **176** (1972), 751.
152. Voronov, F. F. and Stal'gorova, O. V.: *Phys. Met. Metallogr.* **34** (1972), 45.
153. Averkin, A. A., Vorov, Yu. G., Ivanov, G. A., and Regel', R. R.: *Sov. Phys. – Solid State* **13** (1971), 309.
154. Khvostantsev, L. G., Vereshchagin, L. F., and Uliyanitskaya, N. M.: *High Temp. – High Press.* **5** (1973), 261.

155. Broïde, E. L.: *Sov. Phys. – JETP* **36** (1973), 307.
156. Schirber, J. E. and O'Sullivan, W. J.: *Solid State Commun.* **7** (1969) 709.
157. Wehrli, L.: *Phys. kondens. Materie* **8** (1968), 87.
158. Brandt, N. B., Dittmann, KH., and Ponomarev, Ya. G.: *Sov. Phys. – Solid State* **13** (1972), 2408.
159. Tichovolsky, E. J. and Mavroides, J. G.: *Solid State Commun.* **7** (1969), 927.
160. Beneslavskii, S. D., Brandt, N. B., Golyamina, E. M., Chudnov, S. M., and Yakovlev, G. D.: *JETP Lett.* **19** (1974), 154.
161. Ohyama, M.: *J. Phys. Soc. Japan* **21** (1966), 1126.
162. Jain, A. L.: *Phys. Rev.* **114** (1959), 1518, Oelgart, G., Schneider, G., Kraak, W., and Herrmann, R.: *Phys. Stat. Sol.* **74B** (1976) K 75.
163. Jaggi, R.: *Naturwiss.* **51** (1964), 459.
164. Souers, P. C. and Jura, G.: *Sci.* **143** (1964), 467.
165. Venttsel', V. A. and Rakhmanina, A. V.: *Sov. Phys. – Solid State* **14** (1973), 2325.
166. Skinner, B. J.: *Economic Geology* **60** (1965), 228.
167. Dismukes, J. P., Paff. R. J., Smith, P. T., and Ulmer, R.: *J. Chem. Eng. Data* **13** (1968), 317.
168. Cucka, P. and Barrett, C. S.: *Acta Cryst.* **15** (1962), 865.
169. Vecchi, M. P. and Dresselhaus, M. S.: *Phys. Rev.* **B10** (1974), 771.
170. Krebs, H., Holz, W., and Worms, K. H.: *Chem. Ber.* **90** (1957), 1031.
171. Schieferl, D. and Barrett, C. S.: *J. Appl. Cryst.* **2** (1969), 905.
172. Taylor, J. B., Bennett, S. L., and Heyding, R. D.: *J. Phys. Chem. Solids* **26** (1965), 69.
173. Barrett, C. S., Cucka, P., and Haefner, K.: *Acta Cryst.* **16** (1963), 451.
174. Kolobyanina, T. N., Kabalkina, S. S., Vereshchagin, L. F., and Fedina, L. V.: *Sov. Phys. – JETP* **28** (1969), 88.
175. Kabalkina, S. S., Vereshchagin, L. F., and Shulenin, B. M.: *Sov. Phys. – JETP* **18** (1964), 1422.
176. Krebs, H., Gruen, K., Kallen, D., and Lippert, W.: *Z. anorg. allg. Chem.* **308** (1961), 200.
177. Donohue, P. C. and Young, H. S.: *J. Solid State Chem.* **1** (1970), 143.
178. Pospelov, Yu. A.: *Sov. Phys. – Solid State* **12** (1971), 1695.
179. Brugger, R. M., Bennion, R. B., and Worlton, T. G.: *Phys. Lett.* **24A** (1967), 714.
180. Steigmeier, E. F. and Harbeke, G.: *Solid State Commun.* **8** (1970), 1275.
181. Brillson, L. J., Burstein, E., and Muldawer, L.: *Phys. Rev.* **B9** (1974), 1547.
182. Zitter, R. N. and Watson, P. C.: *Phys. Rev.* **B10** (1974), 607.
183. Schubert, K. and Fricke, H.: *Z. Metallk.* **44** (1953), 457.
184. Zhukova, T. B. and Zaslavskii, A. I.: *Sov. Phys. – Cryst.* **12** (1967), 28.
185. Novikova, S. I., Shelimova, L. E., Abrikosov, N. Kh., and Evseev, B. A.: *Sov. Phys. – Solid State* **13** (1972), 2310.
185a. Novikova, S. I., Dzhabua, Z. U., Shelimova, L. E., and Abrikosov, N. Kh.: *Sov. Phys. – Solid State* **16** (1975), 2346.
186. Kabalkina, S. S., Vereshchagin, L. F., and Serebryanaya, N. R.: *Sov. Phys. – JETP* **24** (1967), 917.
186a. Iizumi, M., Hamaguchi, Y., Komatsubara, K. F., and Kato, Y.: *J. Phys. Soc. Japan* **38** (1975) 443.
187. Novikova, S. I. and Shelimova, L. E.: *Sov. Phys. – Solid State* **9** (1967), 1046; **7** (1966), 2052.
188. Muldawer, L.: *J. Nonmetals* **1** (1973), 177.
188a. Valassiades, O. and Economou, N. A.: *Phys. Stat. Sol.* **30a** (1975), 187.
189. Kolomoets, N. V., Lev, E. Ya. and Sysoeva, L. M.: *Sov. Phys. – Solid State* **5** (1964), 2101.
190. Muir, J. A. and Cashman, R. J.: *J. Phys. Chem. Solids* **28** (1967), 1009.
191. Lewis, J. E.: *Phys. Stat. Sol.* **42**, K 97 (1970); **59B** (1973), 367; **38** (1970), 131; **35** (1969), 737.
192. Hein, R. A., Gibson, J. W., Mazelsky, R., Miller, R. C., and Hulm, J. K.: *Phys. Rev. Lett.* **12** (1964), 320.
193. Hockings, E. F. and White, J. G.: *Acta Cryst.* **14** (1961), 328.
194. Man, L. I. and Semiletov, S. A.: *Sov. Phys. – Cryst.* **7** (1962), 686.
195. Semiletov, S. A. and Man, L. I.: *Sov. Phys. – Cryst.* **4** (1959), 385.
196. Dembovskii, S. A., Lisovskii, L. G., and Bunin, V. M.: *Inorg. Mat.* **4** (1968), 115.
197. Brandt, N. B., Gitsu, D. V., Popovich, N. S., Sidorov, V. I., and Chudinov, S. M.: *Sov. Phys. Semicond.* **8** (1974), 390.

198. Chumak, G. H., Dyntu, M. P., Kantser, Ch. T., Donika, F. G., Stratan, G. I., and Gitsu, D. V.: *Inorg. Mat.* **7** (1971), 276.
199. Knowles, C. R.: *Amer. Min.* **51** (1966), 264.
200. Fleet, M. E.: *Z. Krist.* **138** (1973), 147.
201. Pinsker, Z. G. and Khitrova, V. I.: *Sov. Phys. – Cryst.* **1** (1956), 231.
202. Pinsker, Z. G., Semiletov, S. A., and Belova, E. N.: *Dokl. Akad. Nauk SSSR* **106** (1956), 1003.
203. Goldak, J. and Barrett, C. S.: *J. Chem. Phys.* **44** (1966), 3323.
204. Muir, J. A. and Beato, V.: *J. Less-Common Met.* **33** (1973), 333.
205. Krebs, H. and Langner, D.: *Z. anorg. allg. Chem.* **334** (1964), 37.
206. Bierly, J. N., Muldawer, L., and Beckman, O.: *Acta Met.* **11** (1963), 447.
207. Krebs, H., Gruen, K., and Kallen, D.: *Z. anorg. allg. Chem.* **312** (1961), 307.
208. Hohnke, D., Holloway, H., and Kaiser, S.: *J. Phys. Chem. Solids* **33** (1972), 2053.
209. Antcliffe, G. A., Bate, R. T., and Buss, D. D.: *Solid State Commun.* **13** (1973), 1003.
210. Johnson, W. D. and Sestrich, D. E.: *J. Inorg. Nucl. Chem.* **19** (1961), 229.
211. Lewis, J. E. and Rodot, M.: *J. Phys. (Paris)* **29** (1968), 352.
212. Glazov, V. M., Nagiev, V. A., and Zargarova, M. I.: *Inorg. Mat.* **6** (1970), 503.
213. Mazelsky, R. and Lubell, M. S.: *Adv. Chem.* **39** (1963), 210.
214. Leonard, B. F., Mead, C. W., and Finney, J. J.: *Amer. Min.* **56** (1971), 1127.
215. Gullman, J. and Olofsson, O.: *J. Solid State Chem.* **5** (1972), 441.
216. Johan, Z., Laforêt, C., Picot, P., and Feraut, J.: *Bull. Soc. franç. Min. Crist.* **96** (1973), 131.
217. Mariano, A. N. and Chopra, K. L.: *Appl. Phys. Lett.* **10** (1967), 282.
218. Karbanov, S. G., Zlomanov, V. P., and Ukrainskii, Yu. M.: *Inorg. Mater.* **6** (1970), 104.
219. Wakabayashi, I., Kobayashi, H., Nagasaki, H., and Minomura, S.: *J. Phys. Soc. Japan* **25** (1968), 227.
220. Wiedemeier, H. and Siemers, P. A.: *Z. anorg. allg. Chem.* **411** (1975), 90.
221. Karbanov, S. G., Zlomanov, V. P., and Ukrainskii, Yu. M.: *Inorg. Mater.* **5** (1969), 997.
222. Blachnik, R. and Kasper, F. W.: *Z. Naturforsch.* **29B** (1974), 159.
223. Donaldson, J. D., Moser, W., and Simpson, W. B.: *Acta Cryst.* **16** (1963), A22.
224. Kabalkina, S. S., Serebryanaya, N. R., and Vereshchagin, L. F.: *Sov. Phys. – Solid State* **10** (1968), 574.
225. Samara, G. A. and Drickamer, H. G.: *J. Chem. Phys.* **37** (1962), 1159.
226. Kafalas, J. A. and Mariano, A. N.: *Sci.* **143** (1964), 952.
227. Okazaki, A.: *J. Phys. Soc. Japan* **13** (1958), 1151.
228. Schiferl, D.: *Phys. Rev.* **B10** (1974), 3316.
229. Smith, P. M., Leadbetter, A. J., and Apling, A. J.: *Phil. Mag.* **31** (1975), 57.
230. Johan, Z.: *Chem. Erde* **20** (1959), 71.
231. Keyes, R. W.: *Phys. Rev.* **92** (1953), 580.
232. Bridgman, P. W.: *Proc. Amer. Acad. Sci.* **70** (1935), 71.
233. Kannewurf, C. R., Kelly, A. and Cashman, R. J.: *Acta Cryst.* **13** (1960), 449.
234. Karbanov, S. G., Zlomanov, V. P., and Novoselova, A. V.: *Inorg. Mater.* **5** (1969), 997.
235. Dutta, S. N. and Jeffrey, G. A.: *Inorg. Chem.* **4** (1965), 1363.
236. Malyushitskaya, Z. V., Kabalkina, S. S., and Vereshchagin, L. F.: *Sov. Phys. – Solid State* **14** (1972), 1040.
237. Okazaki, O. and Ueda, I.: *J. Phys. Soc. Japan* **11** (1956), 470.
238. Strauss, A. J.: *Trans. Met. Soc. AIME* **242** (1968), 354.
239. Dickens, B.: *J. Inorg. Nucl. Chem.* **27** (1965), 1495.
240. Andersson, S. and Åström, A.: *NBS Spec. Publ.* **364,** *Solid State Chem.* (1972), 3.
241. Leciejewicz, J.: *Acta Cryst.* **14** (1961), 66.
242. Kay, M. I.: *Acta Cryst.* **14** (1961), 80.
243. Leciejewicz, J.: *Acta Cryst.* **14** (1961), 1304.
244. Bystrøm, A.: *Arkiv Kemi, Min. Geol.* **20A** (1945), 1.
245. Moore, W. J. and Pauling, L.: *J. Amer. Chem. Soc.* **63** (1941), 1392.
246. Straumanis, M. and Strenk, C.: *Z. anorg. allg. Chem.* **213** (1933), 301.
247. Vereshchagin, L. F., Kabalkina, S. S., and Lityagina, L. M.: *Sov. Phys. – Doklady* **10** (1966), 622.

248. Boivin, J. C., Thomas, D., and Tridot, G.: *Comptes rendus (Paris)* **268C** (1969), 1149.
249. Panek, P. and Hoppe, R.: *Z. anorg. allg. Chem.* **400** (1973), 219.
250. Scholder, R., Malle, K. G., Triebskorn, B., and Schwarz, H.: *Z. anorg. allg. Chem.* **364** (1969), 41.
251. Byström, A. and Evers, L.: *Acta Chem. Scand.* **4** (1950), 613.
252. Bouvaist, J. and Weigel, D.: *Acta Cryst.* **A26** (1970), 501.
253. Lawson, F.: *Nature* **215** (1967), 955.
254. Murken, G. and Trömel, M.: *Z. anorg. allg. Chem.* **397** (1973), 117.
255. Fayek, M. K. and Leciejewicz, J.: *Z. anorg. allg. Chem.* **336** (1965), 104.
256. Witteveen, H. T.: *Solid State Commun.* **9** (1971), 1313.
257. Boivin, J. C., Thomas, D., and Tridot, G.: *Comptes rendus (Paris)* **276C** (1973), 1105.
258. Hoppe, R. and Schwedes, B.: *Rev. Chim. minér.* **8** (1971), 583.
259. Schwedes, B. and Hoppe, R.: *Z. anorg. allg. Chem.* **391** (1972), 313.
260. Schwedes, B. and Hoppe, R.: *Z. anorg. allg. Chem.* **392** (1972), 97.
261. Menary, J. W.: *Acta Cryst.* **11** (1958), 742.
261a. Imamov, R. M., Pinsker, Z. G., and Ivchenko, A. I: *Sov. Phys.– Cryst.* **9** (1965), 721.
261b. Kupčik, V.: *Referate der 8. Diskussionstagung der Sektion für Kristallkunde der DMG*, Marburg (1965), 16.
261c. Nuffield, E. W.: *Amer. Min.* **37** (1952), 447.
261d. Niizeki, N. and Buerger, M. J.: *Z. Krist.* **109** (1957), 129.
261e. Burri, G., Graeser, S., Marumo, F., and Nowacki, W.: *Chimia* **19** (1965), 499.
262. Ganne, M. and Tournoux, M.: *Comptes rendus (Paris)* **271C** (1970), 828.
263. Sabrowsky, H.: *Naturwiss.* **56** (1969), 562; **57** (1970), 355.
264. Sabrowsky, H.: *Z. Naturforsch.* **27B** (1972), 1567.
265. Sabrowsky, H.: *Naturwiss.* **57** (1970), 244.
266. Wandji, R. and Kamsu Kom, J.: *Comptes rendus (Paris)* **275C** (1972), 813.
266a. Sahl, K.: *Z. Krist.* **132** (1970), 99.
266b. Bode, H. and Voss, E.: *Electrochim. Acta* **1** (1959), 318.
266c. Miyazawa, S. and Iwasaki, H.: *J. Crystal Growth* **8** (1971), 359.
266d. Sahl, K.: *Z. Krist.* **141** (1975) 145.
267. Süsse, P. and Buerger, M. J.: *Z. Krist.* **131** (1970), 161.
268. Schwarz, H.: *Naturwiss.* **52** (1965), 206.
269. Brixner, L. H., Bierstedt, P. E., Jaep, W. F., and Barkley, J. R.: *Mat. Res. Bull.* **8** (1973), 497.
269a. Tolédano, J. C., Pateau, L., Primot, J., Aubrée, J., and Morin, D.: *Mat. Res. Bull.* **10** (1975), 103; Torres, J., Aubrée, J., and Brandon, J.: *Optics Commun.* **12** (1974), 416.
269b. Ng, H. N. and Calvo, C.: *Can. J. Phys.* **53** (1975), 42.
270. Keppler, U.: *Z. Krist.* **132** (1970), 228.
271. Bachmann, H. G. and Kleber, W.: *Fortschr. Min.* **31** (1952), 9.
272. Bachmann, H. G.: *N. Jahrb. Min., Mh.* (1953), 209.
272a. Sinyakov, E. V., Dudnik, E. F., Gene, V. V., and Litvin, B. N.: *Kristallografiya* **20** (1975), 423.
272b. Gene, V. V., Dudnik, E. F., and Sinyakov, E. V.: *Sov. Phys.– Solid State* **16** (1975), 2299.
272c. Isupov, V. A., Krainik, N. N., and Kosenko, É. L.: *Inorg. Mater.* **9** (1973), 139.
273. Brixner, L. H. and Foris, C. M.: *J. Solid State Chem.* **7** (1973), 149.
274. Brixner, L. H., Bierstedt, P. E., and Foris, C. M.: *J. Solid State Chem.* **6** (1973), 430.
275. Brixner, L. H., Bierstedt, P. E., and Foris, C. M.: *Mat. Res. Bull.* **7** (1972), 883.
276. Kinberger, B.: *Acta Chem. Scand.* **24** (1970), 320.
277. Brown, I. D.: *J. Solid State Chem.* **11** (1974), 214.
278. Aurivillius, B. and Malmros, G.: *Trans. Roy. Inst. Technol. Stockholm*, Nr. 291 (1972), 545.
279. Dachet, J. P., D'Yvoire, F., and Guérin, H.: *Comptes rendus (Paris)* **268C** (1969), 1767.
280. D'Yvoire, F., Prades, F., and Guérin, H.: *Comptes rendus (Paris)* **268C** (1969), 1514.
281. Gavrilova, N. D., Karyakina, N. F., Koptsik, V. A., and Novik, V. K.: *Sov. Phys.– Doklady* **15** (1971), 1075.
282. Popolitov, V. I., Lobachev, A. N., Peskin, V. F., Syrkin, L. N., and Feoktistova, N. N.: *Sov. Phys.– Cryst.* **18** (1973), 258.
283. Aurivillius, B.: *Arkiv. Kemi* **3** (1951), 153.
284. Gründer, W., Pätzold, H. and Strunz, H.: *N. Jahrb. Min., Monatsh.* (1962), 93.

285. Rogers, D. and Skapski, A. C.: *Proc. Chem. Soc.* (1964), 400.
286. Keve, E. T. and Skapski, A. C.: *J. Solid State Chem.* **8** (1973), 159.
287. Jeitschko, W. and Sleight, A. W.: *Acta Cryst.* **B30** (1974), 2088.
288. Mooney-Slater, R. C. L.: *Z. Krist.* **117** (1962), 371.
289. Donaldson, J. D. and Puxley, D. C.: *Acta Cryst.* **B28** (1972), 864.
290. Qurashi, M. M. and Barnes, W. H.: *Amer. Min.* **38** (1953), 489.
291. Roth, R. S. and Waring, J. L.: *Amer. Min.* **48** (1963), 1348.
292. Berndt, A. F.: *Acta Cryst.* **B30** (1974), 529.
293. Berndt, A. F. and Lamberg, R.: *Acta Cryst.* **B27** (1971), 1092.
294. Meunier, G., Darriet, J., and Galy, J.: *J. Solid State Chem.* **5** (1972), 314; **6** (1973), 67.
295. Galy, J.: *NBS Spec. Publ.* **364** (Proc. 5th Materials Res. Symp. 1972) p. 29.
296. Lindqvist, O.: *Acta Chem. Scand.* **22** (1968), 977.
297. Malyutin, S. A., Samplavskaya, K. K., and Karapet'yants, M. Kh.: *Russ. J. Inorg. Chem.* **16** (1971), 781.
298. Beyer, H.: *Z. Krist.* **124** (1967), 228.
298a. Lindqvist, O., Mark, W., and Moret, J.: *Acta Cryst.* **B31** (1975), 1255.
299. Swink, L. N. and Carpenter, G. B.: *Acta Cryst.* **21** (1966), 578.
300. Lindqvist, O. and Moret, J.: *Acta Cryst.* **B29** (1973), 643; 956.
301. Folger, F.: *Z. anorg. allg. Chem.* **411** (1975), 103.
302. Furuseth, S., Selte, K., Hope, H., Kjekshus, A., and Klewe, B.: *Acta Chem. Scand.* **A28** (1974), 71.
303. Van den Berg, J. M.: *Acta Cryst.* **14** (1961), 1002.
304. Sabrowsky, H.: *Z. anorg. allg. Chem.* **381** (1971), 266.
305. Tournoux, M., Marchand, R., and Bouchama, M.: *Comptes rendus (Paris)* **270C** (1970), 1007.
306. Sabrowsky, H.: *Naturwiss.* **56** (1969), 414.
307. Touboul, M., Marchand, R., and Tournoux, M.: *Bull. Soc. chim. France* (1972), 570.
307a. Marchand, R. and Tournoux, M.: *Comptes rendus (Paris)* **277C** (1973), 863.
308. Strehlow, W. H. and Cook, E. L.: *J. Phys. Chem. Ref. Data* **2** (1973), 163.
309. Vasil'ev, V. P., Nikol'skaya, A. V., and Gerasimov, Ya. I.: *Inorg. Mater.* **9** (1973), 498.
310. Leclerc, B. and Bailly, M.: *Acta Cryst.* **B29** (1973), 2334.
311. Lévy, F. and Mooser, E.: *Helv. Phys. Acta* **45** (1972), 902.
312. Busing, W. R.: *J. Chem. Phys.* **23** (1955), 933.
313. Jones, R. E. and Templeton, D. H.: *Acta Cryst.* **8** (1955), 847.
314. Ungelenk, J.: *Naturwiss.* **49** (1962), 252.
315. Stehr, H.: *Z. Krist.* **125** (1967), 332.
316. Barlow, M. and Meredith, C. C.: *Z. Krist.* **130** (1969), 304.
317. Alcock, N. W. and Jenkins, H. D. B.: *J. Chem. Soc. Dalton Trans.* (1974), 1907.
318. Pistorius, C. W. F. T. and Clark, J. B.: *Phys. Rev.* **173** (1968), 692.
319. Schulz, L. G.: *Acta Cryst.* **4** (1951), 487.
320. Zahner, J. C. and Drickamer, H. G.: *J. Phys. Chem. Solids* **11** (1959), 92.
321. Samara, G. A., Walters, L. C., and Northrop, D. A.: *J. Phys. Chem. Solids* **28** (1967), 1875.
322. Samara, G. A.: *Phys. Rev.* **165** (1968), 959.
323. Bradley, R. S., Grace, J. D., and Munro, D. C.: *Z. Krist.* **120** (1964), 349.
324. Pistorius, C. W. F. T.: *Z. phys. Chem.* **65** (1969), 51.
325. Dachs, H.: *Z. Krist.* **112** (1959), 60.
326. Shomate, C. H. and Cohen, A. J.: *J. Amer. Chem. Soc.* **77** (1955), 285.
326a. Itkina, L. S., Rannev, N. V., Portnova, S. M., and Demidova, T. A.: *Russ. J. Inorg. Chem.* **19** (1974), 774.
327. Moore, M. J. and Kasper, J. S.: *J. Chem. Phys.* **48** (1968), 2446.
328. Binnie, W. P.: *Acta Cryst.* **9** (1956), 686.
329. Takano, K. and Sato, T.: *Phys. Lett.* **44A** (1973), 309.
330. Berner, R.: *Sci.* **137** (1962), 669.
331. Bertaut, E. F., Burlet, P., and Chappert, J.: *Solid State Commun.* **3** (1965), 335.
332. Kjekshus, A., Nicholson, D. G., and Mukherjee, A. D.: *Acta Chem. Scand.* **26** (1972), 1105.
333. Grønvold, F., Haraldsen, H., and Vihovde, J.: *Acta Chem. Scand.* **8** (1954), 1927.
334. Ward, J. C.: *Rev. Pure Appl. Chem.* **20** (1970), 175.

335. Takeno, S., Zôka, H., and Niihara, T.: *Amer. Min.* **55** (1970), 1639.
336. De Médicis, R.: *Rev. Chim. minér.* **7** (1970), 723.
337. Taylor, L. A. and Finger, L. W.: *C.I.W. Year Book* **69** (1971), 319.
338. Baranova, R. V. and Pinsker, Z. G.: *Sov. Phys. – Cryst.* **9** (1964), 83.
339. Stevels, A. L. N. and Wiegers, G. A.: *Rec. Trav. chim. Pays-Bas* **90** (1971), 352.
340. Avilov, A. S. and Baranova, R. V.: *Sov. Phys. – Cryst.* **17** (1972), 180.
341. De Médicis, R.: *Comptes rendus (Paris)* **272D** (1971), 513.
342. Arunsingh, Srivastava, O. N. and Dayal, B.: *Acta Cryst.* **B28** (1972), 635.
343. Avilov, A. S., Imamov, R. M., and Pinsker, Z. G.: *Sov. Phys. – Cryst* **17** (1972), 237.
344. Dvoryankina, G. G. and Pinsker, Z. G.: *Sov. Phys. – Cryst.* **8** (1964), 448.
345. Carpenter, C. D., Diehl, R., and Nitsche, R.: *Naturwiss.* **57** (1970), 393.
346. Brehler, B.: *Z. Krist.* **115** (1961), 373.
347. Prewitt, C. T. and Young, H. S.: *Sci.* **149** (1965), 535.
348. Jeffrey, G. A. and Vlasse, M.: *Inorg. Chem.* **6** (1967), 396.
349. Turyanitsa, I. D. and Khiminets, V. V.: *Sov. Phys. – Cryst.* **18** (1974), 688.
350. Tonkov, E. D. and Tikhomirova, N. A.: *Sov. Phys. – Cryst.* **15** (1971), 945.
351. Dworsky, R. and Komarek, K. L.: *Monatsh. Chem.* **101** (1970), 984.
352. Schwarzenbach, D.: *Z. Krist.* **128** (1969), 97.
352a. Buck, P. and Carpentier, C. D.: *Acta Cryst.* **B29** (1973), 1864.
353. Weiss, A. and Schäfer, H.: *Naturwiss.* **47** (1960), 495.
354. Johnson, R. E., Staritzky, E., and Douglass, R. M.: *J. Amer. Chem. Soc.* **79** (1957), 2037.
355. Weiss, A. and Schäfer, H.: *Z. Naturforsch.* **18b** (1963), 81.
356. Rubenstein, M. and Roland, G.: *Acta Cryst.* **B27** (1971), 505.
356a. Burgeat, J., Le Roux, G., and Brenac, A.: *J. Appl. Cryst.* **8** (1975), 325.
356b. Dittmar, G. and Schäfer, H.: *Acta Cryst.* **B31** (1975), 2060.
357. Ch'ün-Hua, L., Pashinkin, A. S., and Novoselova, A. V.: *Russ. J. Inorg. Chem.* **7** (1962), 1117.
358. Zakharov, V. P., Gerasimenko, V. S., and Sheremet, G. P.: *JETP Lett.* **17** (1973), 351.
359. Viane, W. and Moh, G.: *N. Jahrb. Min., Abh.* **119** (1973), 113.
360. Wang, N. and Horn, E.: *N. Jahrb. Min., Monatsh.* (1973), 413.
361. Cueilleron, J. and Hillel, R.: *Bull. Soc. chim. France* (1968), 3635.
362. Wiberg, E. and Sturm, W.: *Z. Naturforsch.* **8b** (1953), 529, 530.
362a. Schwarz, W., Hausen, H. D., Hess, H.: *Z. Naturforsch.* **29b** (1974), 596.
363. Nakashima, S., Mishima, H., and Mitsuishi, A.: *J. Raman Spectrosc.* **1** (1973), 325.
364. Jellinek, F. and Hahn, H.: *Z. Naturforsch.* **16b** (1961), 713.
365. Schubert, K., Dörre, E., and Kluge, M.: *Z. Metallk.* **46** (1955), 216.
366. Schubert, K., Dörre, E., and Günzel, E.: *Naturwiss.* **41** (1954), 448.
366a. Kuliev, A. A., Kagramanyan, Z. G., Suleimanov, D. M., and Prud'ko, V. V.: *Inorg. Mater.* **10** (1974), 1628.
367. Semiletov, S. A. and Vlasov, V. A.: *Sov. Phys. – Cryst.* **8** (1964), 704.
368. Vezzoli, G. C.: *Mat. Res. Bull.* **6** (1971), 1201.
369. Wadsten, T.: *Acta Chem. Scand.* **21** (1967), 593.
370. Wadsten, T.: *Acta Chem. Scand.* **19** (1965), 1232.
371. Bryden, J. H.: *Acta Cryst.* **15** (1962), 167.
371a. Rau, J. W. and Kannewurf, C. R.: *Phys. Rev.* **B3** (1971), 2581.
372. Weis, J., Schäfer, H., Eisenmann, B., and Schön, G.: *Z. Naturforsch.* **29b**, (1974), 585.
373. Donohue, P. C. and Young, H. S.: *J. Solid State Chem.* **1** (1970), 143.
374. Donohue, P. C.: *Inorg. Chem.* **9** (1970), 335.
375. Katz, G., Kohn, J., and Broder, J.: *Acta Cryst.* **10** (1957), 607.
376. Losev, V. G., Kabalkina, S. S., and Vereshchagin, L. F.: *Sov. Phys. – Solid State* **16** (1974), 965.
377. Giessen, B. C. and Borromée-Gautier, C.: *J. Solid State Chem.* **4** (1972), 447.
378. Wadsten, T.: *Acta Chem. Scand.* **23** (1969), 331.
379. Wadsten, T.: *Chem. Commun. Univ. Stockholm*, No. 7 (1970).
380. Billy, M. and Goursat, P.: *Rev. Chim. minér.* **7** (1970), 193.
381. Semiletov, S. A.: *Sov. Phys. – Cryst.* **3** (1958), 292.
382. Hahn, H. and Frank, G.: *Z. anorg. allg. Chem.* **278** (1955), 340.
383. Terhell, J. C. J. M. and Lieth, R. M. A.: *Phys. Stat. Sol.* **5a** (1971), 719; **10a** (1972), 529.

REFERENCES

383a. Terhell, J. C. J. M., Lieth, R. M. A., and Van der Vleuten, W. C.: *Mat. Res. Bull.* **10** (1975), 577.
384. Suzuki, H. and Mori, R.: *Jap. J. Appl. Phys.* **13** (1974), 417.
385. Nasirov, Ya. N., Zargarova, M. I., Gamidov, R. S., and Akperov, M. M.: *Inorg. Mater.* **6** (1970), 865.
386. Khalilov, Kh. M., Nasirov, Ya. N., Kuliev, B. B., Akperov, M. M., and Gambarov, F. A.: *Phys. Stat. Sol.* **31** (1969), K 113.
387. Babaeva, B. K., Gamidov, R. S., and Rustamov, P. G.: *Inorg. Mater.* **3** (1966), 826.
388. Ugai, Ya. A., Miroshnichenko, S. N., Domashevskaya, É. P., and Vasil'evskaya, M. A.: *Inorg. Mater.* **9** (1973), 6.
389. Wadsten, T.: *Acta Chem. Scand.* **23** (1969), 2532.
390. Donohue, P. C., Siemons, W. J., and Gillson, J. L.: *J. Phys. Chem. Solids* **29** (1968), 807.
390a. Springthorpe, A. J.: *Mat. Res. Bull.* **4** (1969), 125.
391. Basinski, Z. S., Dove, D. B., and Mooser, E.: *Helv. Phys. Acta* **34** (1961), 373.
392. Aliev, N. G., Kerimov, I. G., Kurbanov, M. M., and Mamedov, T. A.: *Sov. Phys.–Solid State* **14** (1972), 1304.
393. Schlüter, M.: *Helv. Phys. Acta* **45** (1972), 874; *Nuovo Cim.* **13B** (1973), 313.
394. Mooser, E., Schlüter, I. Ch., and Schlüter, M.: *J. Phys. Chem. Solids* **35** (1974), 1269.
395. Müller, D., Eulenberger, G., and Hahn, H.: *Z. anorg. allg. Chem.* **398** (1973), 207.
396. Isaacs, T. J.: *J. Appl. Cryst.* **6** (1973), 413.
397. Likforman, A. and Guittard, M.: *Comptes rendus (Paris)* **279C** (1974), 33.
397a. Likforman, A., Carré, D., Etienne, J., and Bachet, B.: *Acta Cryst.* **B31** (1975), 1252.
398. Imai, K., Sato, M., and Abe, Y.: *J. Electrochem. Soc.* **121** (1974), 1674.
399. Ugai, Ya. A., Murav'eva, S. N., Goncharov, E. G., and Vagina, É. A.: *Inorg. Mater.* **8** (1972), 1461.
400. Bailey, L. G.: *J. Phys. Chem. Solids* **27** (1966), 1593.
401. Klein Haneveld, A. J.: Thesis, University of Groningen, 1970.
402. Chiragov, M. I. and Talybov, A. G.: *Sov. Phys.–Cryst.* **10** (1965), 331.
403. Nagpal, K. C. and Ali, S. Z.: *Acta Cryst.* **A31** (1975), S 67.
404. Lagow, R. J., Badachhape, R. B., Wood, J. L., and Margrave, J. L.: *J. Chem. Soc. Dalton Trans.* (1974), 1268.
405. Whittingham, M. S.: *J. Electrochem. Soc.* **122** (1975), 526.
406. Parry, D. E., Thomas, J. M., Bach, B., and Evans, E. L.: *Chem. Phys. Lett.* **29** (1974), 128.
407. Ebert, L., Huggins, R. A., and Brauman, J. T.: *J. Amer. Chem. Soc.* **96** (1974), 7841.
408. Hengge, E.: *Fortschr. chem. Forsch.* **9** (1967), 145.
409. Schmeisser, M. and Voss, P.: *Fortschr. chem. Forsch.* **9** (1967), 165.
410. Hengge, E. and Scheffler, G.: *Monatsh. Chem.* **95** (1964), 1450.
411. Hengge, E. and Olbrich, G.: *Monatsh. Chem.* **101** (1970), 1068.
412. Hengge, E. and Brychcy, U.: *Z. anorg. allg. Chem.* **339** (1965), 120.
413. Hengge, E. and Olbrich, G.: *Z. anorg. allg. Chem.* **365** (1969), 321.
414. Emons, H. H. and Theisen, L.: *Z. anorg. allg. Chem.* **361** (1968), 321.
415. Hengge, E.: *Z. anorg. allg. Chem.* **315** (1962), 298.
416. Hengge, E. and Pretzer, K.: *Chem. Ber.* **96** (1963), 470.
417. Hengge, E. and Grupe, H.: *Chem. Ber.* **97** (1964), 1783.
418. Hulliger, F.: *J. Phys. Chem. Solids* **26** (1965), 639.
419. Bither, T. A., Donohue, P. C., and Young, H. S.: *J. Solid State Chem.* **3** (1971), 300.
420. Kamsu Kom, J., Flahaut, J., and Domange, L.: *Comptes rendus (Paris)* **255** (1962), 701.
421. Kamsu Kom, J.: *Comptes rendus (Paris)* **265** (1967), 727.
422. Jeitschko, W.: *Acta Cryst.* **B30** (1974), 2565.
423. Haendler, H. M., Mootz, D., Rabenau, A., and Rosenstein, G.: *J. Solid State Chem.* **10** (1974), 175.
424. Rabenau, A., Rau, H., and Rosenstein, G.: *J. Less-Common Met.* **21** (1970), 395.
425. MacGillavry, C. H., De Decker, H. C. J., and Nijland, L. M.: *Nature* **164** (1949), 448.
426. Beyer, H., Glemser, O., Krebs, B., and Wagner, G.: *Z. anorg. allg. Chem.* **376** (1970), 87.
427. Böschen, I. and Krebs, B.: *Acta Cryst.* **B30** (1974), 1795.
428. Oswald, H. R., Günter, J. R., and Dubler, E.: *J. Solid State Chem.* **13** (1975), 330.

429. Krebs, B.: *Z. anorg. allg. Chem.* **380** (1971), 146.
430. Krebs, B., Müller, A., and Beyer, H. H.: *Inorg. Chem.* **8** (1969), 436.
431. Gloeikler, D. and Gleitzer, C.: *Comptes rendus (Paris)* **276C** (1973), 499.
432. Magnéli, A.: *J. Inorg. Nucl. Chem.* **2** (1956), 330.
433. Kihlborg, L.: *Arkiv Kemi* **21** (1963), 357.
434. Glemser, O. and Lutz, G.: *Z. anorg. allg. Chem.* **264** (1951), 17.
435. Wilhelmi, K. A.: *Acta Chem. Scand.* **23** (1969), 419.
436. Pierce, J. W. and Vlasse, M.: *Acta Cryst.* **B27** (1971), 158.
437. Glemser, O., Lutz, G., and Meyer, G.: *Z. anorg. allg. Chem.* **285** (1956), 173.
438. Krebs, B.: *Acta Cryst.* **B28** (1972), 2222.
439. Mitchell, R. S.: *Amer. Min.* **48** (1963), 935.
440. Günter, J. R.: *J. Solid State Chem.* **5** (1972), 354.
441. Kihlborg, L.: *Arkiv Kemi* **21** (1964), 443.
442. Evans, Jr., H. T. and Mrose, M. E.: *Acta Cryst.* **11** (1958) 56; *Amer. Min.* **45** (1960), 1144.
443. Bachmann, H. G., Ahmed, F. R., and Barnes, W. H.: *Z. Krist.* **115** (1961), 110.
444. Bursill, L. A. and Hyde, B. G.: *J. Solid State Chem.* **4** (1972), 430.
445. Magnéli, A.: *Acta Cryst.* **6** (1953), 495.
446. Semiletov, S. A.: *Sov. Phys. – Cryst.* **6** (1961), 158.
447. Hahn, H. and Frank, G.: *Naturwiss.* **44** (1957), 533; *Z. anorg. allg. Chem.* **278** (1958), 333.
448. Semiletov, S. A.: *Sov. Phys. – Cryst.* **6** (1961), 673; *Sov. Phys. – Solid State* **3** (1961), 544.
449. Chistov, S. F., Boryakova, V. A., and Grinberg, Ya. Kh.: *Inorg. Mater.* **10** (1974), 1319.
450. Miyazawa, H. and Sugaike, S.: *J. Phys. Soc. Japan* **12** (1957), 312.
451. Diehl, R. and Nitsche, R.: *J. Crystal Growth* **20** (1973), 38; **28** (1975), 306.
452. Osamura, K., Murakami, Y., and Tomiie, Y.: *J. Phys. Soc. Japan* **21** (1966), 1848.
453. Fitzgerald, A. G.: *Thin Solid Films* **13** (1972), S5.
454. Becker, W., Range, K. J., and Weiss, A.: *Z. Naturforsch.* **23b** (1968), 1545.
455. Barua, K. C. and Goswami, A.: *Indian J. Pure Appl. Phys.* **8** (1970), 258.
456. Popović S., Čelustka, B., and Bidjin, D.: *Phys. Stat. Sol.* **6a** (1971), 301.
457. Bidjin, D., Popović, S., and Čelustka, B.: *Phys. Stat. Sol.* **6a** (1971), 295.
458. Čelustka, B., Bidjin, D., and Popović, S.: *Phys. Stat. Sol.* **6a** (1971), 699.
459. Van Landuyt, J., Van Tendeloo, G., and Amelinckx, S.: *Phys. Stat. Sol.* **26a** (1974), K99 and K103; **30a** (1975), 299.
460. Muschinsky, V. P. and Ambross, V. P.: *Kristall Technik* **5** (1970), K5.
461. Krämer, V., Nitsche, R., and Ottemann, J.: *J. Crystal Growth* **7** (1970), 285.
462. Shand, W. A.: *Phys. Stat. Sol.* **3a** (1970), K77.
463. Donika, F. G., Radautsan, S. I., Semiletov, S. A., Donika, T. V., Mustya, I. G., and Zhitar', V. F.: *Sov. Phys. – Cryst.* **15** (1971), 695.
464. Gorter, E. W.: *J. Solid State Chem.* **1** (1970), 279.
465. Donika, F. G., Radautsan, S. I., Kiosse, G. A., Semiletov, S. A., Donika, T. V., and Mustya, I. G.: *Sov. Phys. – Cryst.* **16** (1971), 190.
466. Donika, F. G., Radautsan, S. I., Semiletov, S. A., Kiosse, G. A., and Mustya, I. G.: *Sov. Phys. – Cryst.* **17** (1972), 575.
467. Lappe, F., Niggli, A., Nitsche, R., and White, J. G.: *Z. Krist.* **117** (1962), 146.
468. Flahaut, J.: *Ann. Chim. (Paris)* **7** (1952), 632.
469. Tressler, R. E. and Stubican, V. S.: *NBS Spec. Publ.* **364** (1972), 695; Tressler, R. E., Hummel, F. A., and Stubican, V. S.: *J. Amer. Ceram. Soc.* **51** (1968), 648.
470. Yokota, M., Syono, Y., and Minomura, S.: *J. Solid State Chem.* **3** (1971), 520.
471. Donika, F. G., Radautsan, S. I., Semiletov, S. A., Donika, T. V., and Mustya, I. G.: *Sov. Phys. – Cryst.* **17** (1972), 578.
472. Donika, F. G., Radautsan, S. I., Kiosse, G. A., Semiletov, S. A., Donika, T. V., and Mustya, I. G.: *Sov. Phys. – Cryst.* **15** (1971), 698.
473. Donika, F. G., Kiosse, G. A., Radautsan, S. I., Semiletov, S. A., and Zhitar', V. F.: *Sov. Phys. – Cryst.* **12** (1968), 745.
474. Radautsan, S. I., Donika, F. G., Kiosse, G. A., and Mustya, I. G.: *Phys. Stat. Sol.* **37** (1970), K123.
475. Gnehm, C., Nitsche, R., and Wild, P.: *Naturwiss.* **56** (1969), 86.

476. Barnett, D. E., Boorman, R. S., and Sutherland, J. K.: *Phys. Stat. Sol.* **4a** (1971), K49.
477. Domingo, G., Itoga, R. S., and Kannewurf, C. R.: *Phys. Rev.* **143** (1966), 536.
478. Silverman, M. S.: *Inorg. Chem.* **5** (1966), 2067.
479. Horak, J. and Rodot, H.: *Comptes rendus (Paris)* **267B** (1968), 1427.
480. Borets, A. N., Puga, G. D., and Chepur, D. V.: *Sov. Phys. Solid State* **15** (1973), 1255.
481. Puga, G. D., Borets, A. N., and Chepur, D. V.: *Sov. Phys. Semicond.* **8** (1974), 748.
482. Guenter, J. R. and Oswald, H. R.: *Naturwiss.* **55** (1968), 177.
483. Mitchell, R. S., Fujiki, Y., and Ishizawa, Y.: *Nature* **247** (1974), 537.
484. Karakhanova, M. I., Pashinkin, A. S., and Novoselova, A. V.: *Inorg. Mater.* **2** (1966), 844.
485. Busch, G., Fröhlich, C., Hulliger, F., and Steigmeier, E.: *Helv. Phys. Acta* **34** (1961), 359.
485a. Rimmington, H. P. B. and Balchin, A. A.: *Phys. Stat. Sol.* **6a** (1971), K47.
486. Teramoto, I. and Takayanagi, S.: *J. Phys. Chem. Solids* **19** (1961), 124.
487. Krause, B. H.: *Acta Cryst.* **A31** (1975), S66.
488. Wilhelm, K. R. and Bale, H. D.: *J. Phys. Chem. Solids* **36** (1975), 624.
488a. Soonpa, H. H.: *J. Phys. Chem. Solids* **23** (1962), 407.
489. Nakajima, S.: *J. Phys. Chem. Solids* **24** (1963), 479.
489a. Schubert, K. and Fricke, H.: *Z. Metallk.* **44** (1953), 457.
490. Eckerlin, P. and Kischio, W.: *Z. anorg. allg. Chem.* **363** (1968), 1.
491. Birkholz, U.: *Z. Naturforsch.* **13a** (1958), 780.
492. Wiese, J. R. and Muldawer, L.: *J. Phys. Chem. Solids* **15** (1960), 13.
493. Kuznetsov, V. G. and Palkina, K. K.: *Russ. J. Inorg. Chem.* **8** (1963), 624.
494. Belotskii, D. P. and Legeta, L. V.: *Inorg. Mater.* **8** (1972), 1677.
495. Geller, S., Jayaraman, A., and Hull, Jr., G. W.: *J. Phys. Chem. Solids* **26** (1965), 353.
496. Beckmann, O., Boller, H., and Nowotny, H.: *Monatsh. Chem.* **101** (1970), 945.
497. Rudy, E.: *J. Less-Common Met.* **20** (1970), 49.
498. Rudy, E.: *Met. Trans.* **1** (1970), 1249.
499. Joffé, A.: *J. Phys. Chem. Solids* **8** (1959), 6.
500. Smith, M. J., Knight, R. J., and Spencer, C. W.: *J. Appl. Phys.* **33** (1962), 2186.
501. Abrikosov, N. Kh. and Poretskaya, L. V.: *Inorg. Mater.* **1** (1965), 462.
502. Bekebrede, W. R. and Guentert, O. J.: *J. Phys. Chem. Solids* **23** (1962), 1023.
503. Rosenberg, A. J. and Strauss, A. J.: *J. Phys. Chem. Solids* **19** (1961), 105.
504. Belotskii, D. P. and Demyanchuk, N. V.: *Inorg. Mater.* **5** (1969), 1289.
505. Abrikosov, N. Kh. and Makareeva, E. G.: *Inorg. Mater.* **10** (1974), 1029, 1032.
506. Belotskii, D. P. and Legeta, L. V.: *Inorg. Mater.* **10** (1974), 19.
507. Abrikosov, N. Kh., Danilova-Dobryakova, G. T., and Dobrynina, N. A.: *Inorg. Mater.* **9** (1973), 512.
508. Nitsche, R.: *Fortschr. Min.* **44** (1967), 231.
509. Fuschillo, N., Bierly, J. N., and Donahue, F. J.: *J. Phys. Chem. Solids* **8** (1959), 430.
510. Ioffe, A. V., Kuznetsov, V. G., and Palkina, K. K.: *Russ. J. Inorg. Chem.* **8** (1963), 1113.
511. Kuznetsov, V. G., Palkina, K. K., and Dmitriev, A. V.: *Russ. J. Inorg. Chem.* **8** (1963), 1116.
512. Bekdurdyev, Ch. D., Gol'tsman, B. M., and Kutasov, V. A.: *Sov. Phys. Solid State* **16** (1975), 1790.
513. Austin, I. G. and Sheard, A.: *J. Electronics* **3** (1957), 236.
514. Testardi, L. R. and Wiese, J. R.: *Trans. Met. Soc. AIME* **221** (1961), 647.
515. Drabble, J. R. and Goodman, C. H. L.: *J. Phys. Chem. Solids* **5** (1958), 142.
516. Black, J., Conwell, E. M., Seigle, L., and Spencer, C. W.: *J. Phys. Chem. Solids* **2** (1957), 240.
517. Rönnlund, B., Beckman, O., and Levy, H.: *J. Phys. Chem. Solids* **26** (1965), 1281.
517a. Il'ina, M. A. and Itskevich, E. S.: *Sov. Phys. Solid State* **17** (1975), 89.
518. Barnes, J. O.,. Rayne, J. A., and Ure, Jr., R. W.: *Phys. Lett.* **46A** (1974), 317.
518a. Pauling, L.: *Amer. Mineral.* **60** (1975), 994.
518b. Glatz, A. C.: *Amer. Mineral.* **52** (1967), 161.
519. Strunz, H.: *N. Jahrb. Mineral., Monatsh.* (1963), 154.
520. Gobrecht, H., Boeters, K. E., and Pantzer, G.: *Z. Phys.* **177** (1964), 68.
521. Stasova, M. M.: *Russ. J. Struct. Chem.* **5** (1964), 731.
522. Stasova, M. M. and Karpinskii, O. G.: *Russ. J. Struct. Chem.* **8** (1967), 69.
523. Stasova, M. M.: *Russ. J. Struct. Chem.* **8** (1967), 584.

524. Stasova, M. M.: *Inorg. Mater.* **4** (1968), 21.
525. Imamov, P. M. and Semiletov, S. A.: *Sov. Phys. – Cryst.* **15** (1971), 845.
526. Brebrick, R. F.: in *The Chemistry of Extended Defects in Non-Metallic Solids*, (LeRoy Eyring and O'Keeffe, M. (ed.), North-Holland, 1970), p. 183.
527. Abrikosov, N. Kh., Bankina, V. F., Kolomoets, L. A., Shubina, G. Yu., and Tskhadaya, R. A.: *Inorg. Mater.* **8** (1972), 1864.
528. Abrikosov, N. Kh., Skudnova, E. V., and Poretskaya, L. V.: *Inorg. Mater.* **8** (1972), 1401.
529. Lee, P. A. and Said, G.: *J. Phys.* **D1** (1968), 837.
530. Evans, B. L. and Hazelwood, R. A.: *J. Phys.* **D2** (1969), 1507.
531. Olofsson, O.: *Acta Chem. Scand.* **24** (1970), 1153.
532. Yvon, K. and Parthé, E.: *Acta Cryst.* **B26** (1970), 149.
533. Thevet, F., Nguyen Huy Dung, and Dagron, C.: *Comptes rendus (Paris)* **275C** (1972), 1279; **276C** (1973), 1787.
534. Blachnik, R. and Kasper, F. W.: *Z. Naturforsch.* **29b** (1974), 159.
535. Novoselova, A. V., Tordiya, M. K., Odin, I. N., and Popovkin, B. A.: *Inorg. Mater.* **7** (1971), 437.
536. Novoselova, A. V., Odin, I. N. and Popovkin, B. A.: *Russ. J. Inorg. Chem.* **14** (1969), 1402.
537. Krebs, B.: *Z. Naturforsch.* **25b** (1970), 223; *Z. anorg. allg. Chem.* **396** (1973), 137.
538. Hardt, H. D. and Scheepker, H.: *Naturwiss.* **57** (1970), 39.
539. Novoselova, A. V., Odin, I. N., Fedoseeva, I. N., and Popovkin, B. A.: *Inorg. Mater.* **6** (1970), 113.
540. Agaev, K. A. and Semiletov, S. A.: *Sov. Phys. – Cryst.* **13** (1968), 201.
541. Agaev, K. A. and Talybov, A. G.: *Sov. Phys. – Cryst.* **11** (1966), 400.
542. Agaev, K. A. and Semiletov, S. A.: *Sov. Phys. – Cryst.* **10** (1965), 86.
543. Zhukova, T. B. and Zaslavskii, A. I.: *Sov. Phys. – Cryst.* **16** (1972), 796.
544. Talybov, A. G.: *Sov. Phys. – Cryst.* **6** (1961), 40.
545. Geller, S. and Hull, Jr., G. W.: *Phys. Rev. Lett.* **13** (1964), 127.
546. Agaev, K. A., Talybov, A. G., and Semiletov, S. A.: *Sov. Phys. – Cryst.* **11** (1967), 630.
547. Petrov, I. I., Imamov, R. M., and Pinsker, Z. G.: *Sov. Phys. – Cryst.* **13** (1968), 339.
548. Petrov, I. I. and Imamov, R. M.: *Sov. Phys. – Cryst.* **14** (1970), 593.
549. Talybov, A. G. and Chiragov, M. I.: *Issled. Obl. Neorg. Fiz. Khim (Baku)* (1966) 101; *Chem. Abstr.* **67** (1967), 6486t.
550. Imamov, R. M., Semiletov, S. A., and Pinsker, Z. G.: *Sov. Phys. – Cryst.* **15** (1970), 239.
551. Petrov, I. I. and Imamov, R. M.: *Sov. Phys. – Cryst.* **15** (1970), 134.
552. Agaev, K. A., Talybov, A. G., and Semiletov, S. A.: *Sov. Phys. – Cryst.* **13** (1968), 44.
553. Zhukova, T. B. and Zaslavskii, A. I.: *Russ. J. Struct. Chem.* **11** (1970), 423.
554. Vainshtein, B. K., Imamov, R. M., and Talybov, A. G.: *Sov. Phys. – Cryst.* **14** (1970), 597.
555. Talybov, A. G. and Vainshtein, B. K.: *Sov. Phys. – Cryst.* **6** (1962), 432.
556. Talybov, A. G.: *Sov. Phys. – Cryst.* **9** (1964), 41.
557. Kutasov, V. A., Smirnov, I. A., Zhukova, T. B., and Zaslavskii, A. I.: *Inorg. Mater.* **7** (1971), 1657.
558. Erd, R. C., Evans, H. T., and Richter, D. H.: *Amer. Min.* **42** (1957), 309.
559. Taylor, L. A. and Williams, K. L.: *Amer. Min.* **57** (1972), 1571.
560. Yamaguchi, S. and Wada, H.: *Z. anorg. allg. Chem.* **392** (1972), 191.
561. Røst, E. and Åkesson, G.: *Acta Chem. Scand.* **26** (1972), 3662.
562. Åkesson, G. and Røst, E.: *Acta Chem. Scand.* **27** (1973), 79.
563. Bhan, S., Gödecke, T., and Schubert, K.: *J. Less-Common Met.* **19** (1969), 121.
564. Kjekshus, A.: *Acta Chem. Scand.* **27** (1973), 1452.
565. Troyanov, S. I.: *Vestnik Moskov. Gos. Univ., Ser. Khim* (1973), 370.
566. Troyanov, S. I. and Tsirel'nikov, V. I.: *Russ. J. Inorg. Chem.* **15** (1970), 1762.
567. Izmailovich, A. S., Troyanov, S. I., and Tsirel'nikov, V. I.: *Russ. J. Inorg. Chem.* **19** (1974), 1597.
568. Troyanov, S. I., Marek, G. S., and Tsirel'nikov, V. I.: *Russ. J. Inorg. Chem.* **18** (1973), 135.
569. Struss, A. W. and Corbett, J. D.: *Inorg. Chem.* **9** (1970), 1373.
570. Klingen, W., Eulenberger, G., and Hahn, H.: *Z. anorg. allg. Chem.* **401** (1973), 97.
571. Klingen, W., Ott, R., and Hahn, H.: *Z. anorg. allg. Chem.* **396** (1973), 271.

572. Taylor, B. E., Steger, J., and Wold, A.: *J. Solid State Chem.* **7** (1973), 461.
573. Golovei, M. I., Olekseyuk, I. D., and Voroshilov, Yu. V.: *Inorg. Mater.* **9** (1973), 1346.
574. Taylor, B., Steger, J., Wold, A., and Kostiner, E.: *Inorg. Chem.* **13** (1974), 2719.
575. Dittmar, G. and Schäfer, H.: *Z. Naturforsch.* **29b** (1974), 312.
576. Carpentier, C. D. and Nitsche, R.: *Mat. Res. Bull.* **9** (1974), 401.
577. Rimmington, H. P. B. and Balchin, A. A.: *J. Crystal Growth* **21** (1974), 171.
578. Lucovsky, G., White, R. M., Benda, J. A., and Revelli, J. F.: *Phys. Rev.* **B7** (1973), 3859.
579. Riekel, C. and Schöllhorn, R.: *Mat. Res. Bull.* **10** (1975), 629.
580. McTaggart, F. K. and Wadsley, A. D.: *Austral. J. Chem.* **11** (1958), 445.
581. Brattås, L. and Kjekshus, A.: *Acta Chem. Scand.* **27** (1973), 1290.
582. Whitehouse, C. R., Rimmington, H. P. B., and Balchin, A. A.: *Phys. Stat. Sol.* **18a** (1973), 623.
583. Gleizes, A. and Jeannin, Y.: *J. Less-Common Met.* **34** (1973), 165.
584. Rimmington, H. P. B. and Balchin, A. A.: *J. Mater. Sci.* **9** (1974), 343.
585. Jacquin, Y. and Huber, M.: *Comptes rendus* (Paris) **262C** (1966), 1059.
586. Smeggil, J. G. and Bartram, S.: *J. Solid State Chem.* **5** (1972), 391.
587. Kadijk, F. and Jellinek, F.: *J. Less-Common Met.* **23** (1971), 437.
588. Conroy, L. E. and Pisharody, K. R.: *J. Solid State Chem.* **4** (1972), 345.
589. Golubnichaya, A. A. and Kalikhman, V. L.: *Inorg. Mater.* **10** (1974), 1018.
590. Brouwer, R. and Jellinek, F.: *Acta Cryst.* **A31** (1975), S85; *Mat. Res. Bull.* **9** (1974), 827.
591. Al-Alamy, F. A. S. and Balchin, A. A.: *Mat. Res. Bull.* **8** (1973), 245.
592. Bjerkelund, E. and Kjekshus, A.: *Acta Chem. Scand.* **21** (1967), 513.
593. Huisman, R.: Thesis, Univ. Groningen, 1969.
594. Hulliger, F.: *Nature* **204** (1964), 644.
595. Coffin, P., Jacobson, A. J., and Fender, B. E. F.: *J. Phys.* **C7** (1974), 2781.
596. Witteman, W. G., Giorgi, A. L., and Vier, D. T.: *J. Phys. Chem.* **64** (1960), 434.
597. Kjekshus, A. and Grønvold, F.: *Acta Chem. Scand.* **13** (1959), 1767.
598. Muller, O. and Roy, R.: *J. Less-Common Met.* **16** (1968), 129.
599. Hoekstra, H. R., Siegel, S., and Gallagher, F. X.: *Adv. Chem. Series* **98** (1971), 39.
600. Grønvold, F., Haraldsen, H. and Kjekshus, A.: *Acta Chem. Scand.* **14** (1960), 1879.
600a. Furuseth, S., Selte, K., and Kjekshus, A.: *Acta Chem. Scand.* **19** (1965), 257.
601. Bither, T. A., Bouchard, R. J., Cloud, W. H., Donohue, P. C., and Siemons, W. J.: *Inorg. Chem.* **7** (1968), 2208.
601a. Bither, T. A.: private communication.
602. Gleizes, A. and Jeannin, Y.: *J. Solid State Chem.* **5** (1972), 42.
603. Myron, H. W.: *Solid State Commun.* **15** (1974), 395.
604. Greenaway, D. L. and Nitsche, R.: *J. Phys. Chem. Solids* **26** (1965), 1445.
605. Thompson, A. H., Pisharody, K. R., and Koehler, Jr., R. F.: *Phys. Rev. Lett.* **29** (1972), 163.
606. Benda, J. A.: *Phys. Rev.* **B10** (1974), 1409.
607. Wertheim, G. K., Di Salvo, F. J., and Buchanan, D. N. E.: *Solid. State Commun.* **13** (1973), 1225.
608. Beal, A. R., Knights, J. C., and Liang, W.: *J. Phys.* **C5** (1972), 3531.
609. Murray, R. B., Bromley, R. A., and Yoffe, A. D.: *J. Phys.* **C5** (1972), 746.
610. Geertsma, W., Haas, C., Huisman, R., and Jellinek, F.: *Solid State Commun.* **10** (1972), 75.
611. Van Landuyt, J., Van Tendeloo, G., and Amelinckx, S.: *Acta Cryst.* **A31** (1975), S85.
612. Wilson, J. A., Di Salvo, F. J., and Mahajan, S.: *Adv. Phys.* **24** (1975), 117.
613. Williams, P. M., Parry, G. S., and Scruby, C. B.: *Phil. Mag.* **29** (1974), 695.
613a. Thompson, A. H.: *Phys. Rev. Lett.* **34** (1975), 520.
614. Antonova, E. A., Kiseleva, K. V., and Medvedev, S. A.: *Sov. Phys. JETP* **32** (1971), 31.
615. Brown, B. E.: *Acta Cryst.* **20** (1966), 264.
616. Allakhverdiev, K. R., Antonova, E. A., and Kalyuzhnaya, G. A.: *Inorg. Mater.* **5** (1969), 1401.
617. Aslanov, L. A. and Novoselova, A. V.: *Russ. J. Inorg. Chem.* **12** (1967), 437.
618. Ukrainskii, Yu. M., Kovba, L. M., Simanov, Yu. P., and Novoselova, A. V.: *Russ. J. Inorg. Chem.* **4** (1959), 1305.
619. Van Maaren, M. H. and Schaeffer, G. M.: *Phys. Lett.* **24A** (1967), 645.
620. Tunell, G. and Pauling, L.: *Acta Cryst.* **5** (1952), 375.
621. Biswas, T. and Schubert, K.: *J. Less-Common Met.* **19** (1969), 223.

622. Brown, B. E.: *Acta Cryst.* **20** (1966), 268.
623. Brixner, L. H.: *J. Inorg. Nucl. Chem.* **24** (1962), 257.
624. Opalovskii, A. A., Fedorov, V. E., Lobkov, E. U., Érenburg, B. G., and Senchenko, L. N.: *Inorg. Mater.* **6** (1970), 495.
625. Wildervanck, J. C. and Jellinek, F.: *J. Less-Common Met.* **24** (1971), 73.
626. Alcock, N. W. and Kjekshus, A.: *Acta Chem. Scand.* **19** (1965), 79.
627. Jellinek, F., Brauer, G., and Müller, H.: *Nature* **185** (1960), 376.
628. Marezio, M., Dernier, P. D., Menth, A., and Hull, Jr., G. W.: *J. Solid State Chem.* **4** (1972), 425.
629. Kadijk, F., Huisman, R., and Jellinek, F.: *Rec. Trav. chim. Pays-Bas* **83** (1964), 768.
630. Revelli, Jr., J. F. and Phillips, W. A.: *J. Solid State Chem.* **9** (1974), 176.
631. Kalikhman, V. L., Lobova, T. A., and Pravoverova, L. L.: *Inorg. Mater.* **9** (1973), 826.
632. Kalikhman, V. L., Zelikman, A. N., Lobova, T. A., Pravoverova, L. L., Gladchenko, E. P., Nikolaeva, N. N., and Yashina, N. O.: *Inorg. Mater.* **7** (1971), 1032.
633. Kalikhman, V. L., Gladchenko, E. P., and Pravoverova, L. L.: *Inorg. Mater.* **8** (1972), 1020.
634. Franzen, H. F. and Graham, J.: *Z. Krist.* **123** (1966), 133.
635. Franzen, H. F., Smeggil, J., and Conard, B. R.: *Mat. Res. Bull.* **2** (1967), 1087.
636. Wildervanck, J. C. and Jellinek, F.: *Z. anorg. allg. Chem.* **328** (1964), 309.
637. Champion, J. A.: *Brit. J. Appl. Phys.* **16** (1965), 1035.
638. Brixner, L. H. and Teufer, G.: *Inorg. Chem.* **2** (1963), 992.
639. Brixner, L. H.: *J. Inorg. Nucl. Chem.* **24** (1962), 257.
640. Knop, O. and MacDonald, R. D.: *Canad. J. Chem.* **39** (1961), 897.
641. Puotinen, D. and Newnham, R. E.: *Acta Cryst.* **14** (1961), 691.
642. Al-Hilli, A. A. and Evans, B. L.: *J. Appl. Cryst.* **5** (1972), 221.
643. Semiletov, S. A.: *Sov. Phys.–Cryst.* **6** (1962), 428.
644. Towle, L. C., Oberbeck, V., Brown, B. E., and Stajdohar, R. E.: *Sci.* **154** (1966), 895.
645. Morosin, B.: *Acta Cryst.* **B30** (1974), 551.
646. Brixner, L. H.: *J. Electrochem. Soc.* **110** (1963), 289.
647. Troyanov, S. I. and Tsirel'nikov, V. I.: *Vestn. Mosk. Univ., Khim.*, **14** (1973), 67.
648. Di Salvo, F. J., Bagley, B. G., Voorhoeve, J. M., and Waszczak, J. V.: *J. Phys. Chem. Solids* **34** (1973), 1357.
648a. Murphy, D. W., Di Salvo, F. J., Hull, Jr., G. W., Waszczak, J. V., Mayer, S. F., Stewart, G. R., Early, S., Acrivos, J. V., and Geballe, T. H.: *J. Chem. Phys.* **62** (1975), 967.
648b. Meyer, S. F., Howard, R. E., Stewart, G. R., Acrivos, J. V., and Geballe, T. H.: *J. Chem. Phys.* **62** (1975), 4411.
649. Huisman, R. and Jellinek, F.: *J. Less-Common Met.* **17** (1969), 111.
650. Guggenberger, L. J. and Jacobson, R. A.: *Inorg. Chem.* **7** (1968), 2257.
651. Aslanov, L. A., Ukrainskii, Yu. M., and Simanov, Yu. P.: *Russ. J. Inorg. Chem.* **8** (1963), 937.
652. White, R. M. and Lucovsky, G.: *Solid State Commun.* **11** (1972), 1369.
653. Gamble, F. R.: *J. Solid State Chem.* **9** (1974), 358.
654. Madhukar, A.: *Solid State Commun.* **16** (1975), 383.
655. Moncton, D. E., Axe, J. D., and Di Salvo, F. J.: *Phys. Rev. Lett.* **34** (1975), 734.
656. Tidman, J. P., Singh, O., Curzon, A. E., and Frindt, R. F.: *Phil. Mag.* **30** (1974), 1191.
657. Yamaya, K. and Sambongi, T.: *Solid State Commun.* **11** (1972), 903.
658. Lee, H. N. S., Garcia, M., McKinzie, H., and Wold, A.: *J. Solid State Chem.* **1** (1970), 190.
659. Norling, B. K. and Steinfink, H.: *Inorg. Chem.* **5** (1966), 1488.
660. Pardo, M. P. and Flahaut, J.: *Bull. Soc. chim. France* (1967), 3658; Pardo, M. P., Gorochov, O., Flahaut, J., and Domange, L.: *Comptes rendus (Paris)* **260** (1965), 1666.
661. Chukalin, V. I., Yarembash, E. I., and Villenskii, A. I.: *Inorg. Mater.* **3** (1967), 1341.
662. Bucher, E., Andres, K., Maita, J. P., Cooper, A. S., and Longinotti, L. D.: *J. Phys. (Paris)* **32** Suppl. C1 (1971), 115.
663. Yarembash, E. I. and Vigileva, E. S.: *Inorg. Mater.* **7** (1971), 1388.
664. Yarembash, E. I., Vigileva, E. S., Eliseev, A. A., Zachatskaya, A. V., Aminov, T. G., and Chernitsyna, M. A.: *Inorg. Mater.* **10** (1974), 1212.
665. Cannon, J. F. and Hall, H. T.: *Inorg. Chem.* **9** (1970), 1639.
666. Damien, D.: *J. Inorg. Nucl. Chem.* **36** (1974), 307.

667. Damien, D.: *Inorg. Nucl. Chem. Lett.* **9** (1973), 453.
668. Damien, D.: *Inorg. Nucl. Chem. Lett.* **8** (1972), 501.
669. Haase, D. J., Steinfink, H., and Weiss, E. J.: *Inorg. Chem.* **4** (1965), 541.
670. Ramsey, T. H., Steinfink, H., and Weiss, E. J.: *Inorg. Chem.* **4** (1965), 1154.
671. Lin, W., Steinfink, H., and Weiss, E. J.: *Inorg. Chem.* **4** (1965), 877.
672. Yarembach, E. I.: *Les Eléments des Terres Rares*, Vol. I, p. 471 (Colloques Int. CNRS 1969, Grenoble).
673. Bucher, E., Andres, K., Di Salvo, F. J., Maita, J. P., Gossard, A. C., Cooper, A. S., and Hull, Jr., G. W.: *Phys. Rev.* **B11** (1975), 500.
674. Zhuravlev, N. N.: *Sov. Phys. JETP* **5** (1957), 1064.
675. Krönert, W. and Plieth, K.: *Z. anorg. allg. Chem.* **336** (1965), 207.
676. Haraldsen, H., Kjekshus, A., Røst, E., and Steffensen, A.: *Acta Chem. Scand.* **17** (1963), 1283.
676a. Grimmeiss, H. G., Rabenau, A., Hahn, H., and Ness, P.: *Z. Elektrochem.* **65** (1961), 776.
677. Brattås, L. and Kjekshus, A.: *Acta Chem. Scand.* **26** (1972), 3441.
678. Graham, J. and McTaggart, F. K.: *Austral. J. Chem.* **13** (1960), 67.
679. Suski, W. and Trzebiatowski, W.: *Bull. Acad. Polon. Sci., Sér. Sci. chim.* **12** (1964), 277.
680. Picon, M. and Flahaut, J.: *Bull. Soc. chim. France* (1958), 772.
681. Khodadad, P.: *Bull. Soc. chim. France* (1961), 133.
682. Grønvold, F., Haraldsen, H., Thurmann-Moe, T., and Tufte, T.: *J. Inorg. Nucl. Chem.* **30** (1968), 2117.
683. Ellert, G. V. and Slovyanskikh, V. K.: *Russ. J. Inorg. Chem.* **19** (1974), 756.
683a. Slovyanskikh, V. K., Yarembash, E. I., Ellert, G. V., and Eliseev, A. A.: *Inorg. Mater.* **4** (1968), 543.
684. Breeze, E. W., Brett, N. H., and White, J.: *J. Nucl. Mater.* **39** (1971), 157.
685. Marcon, J. P.: *Comptes rendus (Paris)* **265C** (1967), 235; Rapport CEA-R-3919 (1969).
686. Damien, D., Damien, N., Jove, J., and Charvillat, J. P.: *Inorg. Nucl. Chem. Lett.* **9** (1973), 649.
687. Furuseth, S., Brattås, L., and Kjekshus, A.: *Acta Chem. Scand.* **27** (1973), 2367.
688. Bjerkelund, A., Fermor, J. H., and Kjekshus, A.: *Acta Chem. Scand.* **20** (1966), 1836.
689. Meerschaut, A. and Rouxel, J.: *J. Less-Common Met.* **39** (1975), 197.
690. Karabekov, A. K., Yarembash, E. I., and Tyurin, E. G.: *Izv. Akad. Nauk Kirg. SSR* (1973), 68; *Chem. Abstr.* **80** (1974), 75371.
691. Flahaut, J.: *J. Solid State Chem.* **9** (1974), 124.
692. Tanguy, B., Pezat, M., Fontenit, C., and Portier, J.: *Comptes rendus (Paris)* **280C** (1975), 1019.
693. Sauvage, M.: *Acta Cryst.* **B30** (1974), 2786.
694. Lambrecht, Jr., V. G., Robbins, M., and Sherwood, R. C.: *J. Solid State Chem.* **10** (1974), 1.
695. Rulmont, A.: *Comptes rendus (Paris)* **276C** (1973), 775.
696. Batsanov, S. S., Filatkina, V. S., and Kustova, G. N.: *Bull. Acad. Sci. USSR, Div. Chem. Sci.* (1971), 1103.
697. Filatkina, V. S. and Kustova, G. N.: *Bull. Acad. Sci. USSR, Div. Chem. Sci.* (1972), 1213.
698. Peterson, J. R.: *J. Inorg. Nucl. Chem.* **34** (1972), 1603.
699. Copeland, J. C. and Cunningham, B. B.: *J. Inorg. Nucl. Chem.* **31** (1969), 733.
700. Fujita, D. K., Cunningham, B. B., and Parsons, T. C.: *Inorg. Nucl. Chem. Lett.* **5** (1969), 307.
701. Fujita, D. K.: USAEC 1970, UCRL-19507; *Chem. Abstr.* **74** (1971), 37835.
702. Dagron, C. and Thevet, F.: *Ann. Chim. (Paris)* **6** (1971), 67.
703. Schmid, R. and Hahn, H.: *Z. anorg..allg. Chem.* **373** (1970), 168; *Naturwiss.* **52** (1965), 475.
704. Dagron, C.: *Comptes rendus (Paris)* **273C** (1971), 352.
705. Filatkina, V. S., Kustova, G. N., and Batsanov, S. S.: *Bull. Acad. Sci. USSR, Div. Chem. Sci.* (1972), 2107.
706. Filatkina, V. S. and Batsanova, S. S.: *Russ. J. Inorg. Chem.* **14** (1969), 1204.
707. Juza, R. and Sievers, R.: *Z. anorg. allg. Chem.* **363** (1968), 258.
708. Juza, R. and Meyer, W.: *Z. anorg. allg. Chem.* **366** (1969), 43.
709. Yoshihara, K., Yamagami, S., Kanno, M., and Mukaibo, T.: *J. Inorg. Nucl. Chem.* **33** (1971), 3323.
710. Ballestracci, R., Bertaut, E. F., and Pauthenet, R.: *J. Phys. Chem. Solids* **24** (1963), 487.
711. Marcon, J. P.: *Comptes rendus (Paris)* **264C** (1967), 1475.
712. Elmaleh, D., Fruchart, D., and Joubert, J. C.: *J. Phys. (Paris)* **32** (1971), C1-741.

713. Mayer, L., Zolotov, S., and Kassierer, F.: *Inorg. Chem.* **4** (1965), 1637.
714. Étienne, J.: *Bull. Soc. franç. Min. Crist.* **92** (1969), 134.
715. Savigny, N., Adolphe, C., Zalkin, A., and Templeton, D. H.: *Acta Cryst.* **B29** (1973), 1532.
716. Brandt, G. and Diehl, R.: *Mat. Res. Bull.* **9** (1974), 411.
717. von Schnering, H. G., Collin, M., and Hassheider, M.: *Z. anorg. allg. Chem.* **387** (1972), 137.
718. Levayer, C. and Rouxel, J.: *Comptes rendus (Paris)* **268C** (1969), 167.
719. Nørlund Christensen, A., Johansson, T., and Quézel, S.: *Acta Chem. Scand.* **A28** (1975), 1171.
720. Danot, M. and Rouxel, J.: *Comptes rendus (Paris)* **262C** (1966), 1879.
721. Lind, M. D.: *Acta Cryst.* **B26** (1970), 1058.
722. Forsberg, H. E.: *Acta Chem. Scand.* **10** (1956), 1287.
723. Juza, R. and Heners, J.: *Z. anorg. allg. Chem.* **332** (1964), 159.
724. Blunck, H. and Juza, R.: *Z. anorg. allg. Chem.* **406** (1974), 145.
725. Sfez, G. and Adolphe, C.: *Bull. Soc. franç. Min. Crist.* **96** (1973), 37.
726. Collin, G., Dagron, C., and Thevet, F.: *Bull. Soc. chim. France* (1974), 418.
727. Goggin, P. L., McColm, I. J., and Shore, R.: *J. Chem. Soc.* **A** (1966), 1004.
728. Katscher, H. and Hahn, H.: *Naturwiss.* **53** (1966), 361.
729. Hardy, A. and Hardy, A. M.: *Comptes rendus (Paris)* **256** (1963), 3477.
730. Hagenmuller, P., Rouxel, J., David, J., Colin, A., and Le Neindre, B.: *Z. anorg. allg. Chem.* **323** (1963), 1.
731. Palvadeau, P. and Rouxel, J.: *Bull. Soc. chim. France* (1967), 2698.
732. Roos, G., Eulenberger, G. and Hahn, H.: *Z. anorg. allg. Chem.* **396** (1973), 284; *Naturwiss.* **59** (1972), 363.
733. Hahn, H. and Nickels, W.: *Z. anorg. allg. Chem.* **314** (1962), 307.
734. Baybarz, R. D., Asprey, L. B., Strouse, C. E., and Fukushima, E.: *J. Inorg. Nucl. Chem.* **34** (1972), 3427.
735. Kruse, F. H., Asprey, L. B., and Morosin, B.: *Acta Cryst.* **14** (1961), 541.
736. Bärnighausen, H.: *J. prakt. Chem.* **14** (1961), 313.
737. Bärnighausen, H. and Warkentin, E.: *Rev. Chim. minér.* **10** (1973), 141.
738. Rietschel, E. T. and Bärnighausen, H.: *Z. anorg. allg. Chem.* **368** (1969), 62.
739. Bärnighausen, H.: *Structure and Bonding* (in press).
740. Busing, W. R. and Levy, H. A.: *J. Chem. Phys.* **26** (1957), 563.
741. Mitchell, R. S.: *Z. Krist.* **123** (1966), 459; **125** (1967), 272.
742. Nørlund Christensen, A. and Ollivier, G.: *Solid State Commun.* **10** (1972), 609.
743. Takada, T., Bando, Y., Kiyama, M., and Miyamoto, H.: *J. Phys. Soc. Japan* **21** (1966), 2726 and 2745.
744. Vettier, C. and Yelon, W. B.: *J. Phys. Chem. Solids* **36** (1975), 401.
745. Szytuła, A., Murasik, A., and Bałanda, M.: *Phys. Stat. Sol.* **43B** (1971), 615.
746. Pieper, G. and Bartl, H.: *Acta Cryst.* **A31** (1975), S94.
747. Oswald, H. R. and Feitknecht, W.: *Helv. Chim. Acta* **47** (1964), 272.
748. Meldau, R., Newesely, H., and Strunz, H.: *Naturwiss.* **60** (1973), 387.
749. Wilkinson, M. K., Cable, J. W., Wollan, E. O., and Koehler, W. C.: *Phys. Rev.* **113** (1959), 497.
750. Dillon, Jr., J. F., Yi Chen, E., and Guggenheim, H. J.: *Solid State Commun.* **16** (1975), 371.
751. Tracy, J. W., Gregory, N. W., and Lingafelter, E. C.: *Acta Cryst.* **15** (1962), 672.
752. Tracy, J. W., Gregory, N. W., Stewart, J. M., and Lingafelter, E. C.: *Acta Cryst.* **15** (1962), 460.
753. Oswald, H. R. and Feitknecht, W.: *Helv. Chim. Acta* **44** (1961), 847.
754. Walter-Lévy, L. and Groult, D.: *Bull. Soc. chim. France* (1970), 3868.
755. Iitaka, Y., Locchi, S., and Oswald, H. R.: *Helv. Chim. Acta* **44** (1961), 2095.
756. Forsberg, H. E.: *Acta Cryst.* **13** (1960), 1024; Forsberg, H. E. and Nowacki, W.: *Acta Chem. Scand.* **13** (1959), 1049.
757. Feitknecht, W.: *Fortschr. chem. Forsch.* **2** (1953), 670.
758. Oswald, H. R., Iitaka, Y., Locchi, S., and Ludi, A.: *Helv. Chim. Acta* **44** (1961), 2103.
759. Aebi, F.: *Acta Cryst.* **7** (1954), 26.
760. Voronova, A. A. and Vainshtein, B. K.: *Sov. Phys. – Cryst.* **3** (1958), 445.
761. Berthold, H. J. and Knecht, H.: *Z. anorg. allg. Chem.* **348** (1966), 50.

762. Juza, D., Giegling, D., and Schäfer, H.: *Z. anorg. allg. Chem.* **366** (1969), 121.
763. Ashby, E. C., Kovar, R. A., and Kawakami, K.: *Inorg. Chem.* **9** (1970), 317.
764. Frevel, L. K., Rinn, H. W., and Anderson, H. C.: *Ind. Eng. Chem. Anal. Ed.* **18** (1946), 83.
765. Mitchell, R. S.: *Z. Krist.* **117** (1962), 309.
766. Argay, Gy. and Naray-Szabo, I.: *Acta Chim. Acad. Sci. Hung.* **49** (1966), 329; *Chem. Abstr.* **66** (1967), 32681b.
767. Hadenfeldt, C.: *Z. Naturforsch.* **30b** (1975), 165.
768. Lascelles, K. and Shelton, R. A. J.: *J. Less-Common Met.* **21** (1970), 181; **27** (1972), 423.
769. Jones, E. R., Hendricks, M. E., Finklea, S. L., Cathey, L., Auel, T., and Amma, E. L.: *J. Chem. Phys.* **52** (1970), 1922.
770. Bill, H.: University of Geneva (private communication).
771. Ferrari, A., Braibanti, A., and Bigliardi, G.: *Acta Cryst.* **16** (1963), 846.
772. Bärnighausen, H., Brauer, G., and Schultz, N.: *Z. anorg. allg. Chem.* **338** (1965), 250.
773. Adam, A. and Buisson, G.: *Phys. Stat. Sol.* **30a** (1975), 323.
774. Atoji, M.: *J. Chem. Phys.* **51** (1969), 3872.
775. Atoji, M. and Kikuchi, M.: *J. Chem. Phys.* **51** (1969), 3863.
776. Lallement, R.: Thesis Paris, 1966.
777. Trömel, M. and Lupprich, E.: *Naturwiss.* **60** (1973), 350.
778. Petch, H. E.: *Acta Cryst.* **14** (1961), 950.
779. Zigan, F. and Rothbauer, R.: *N. Jahrb. Min., Monatsh.* (1967), 137.
780. Isetti, G.: *Periodico Mineral. (Roma)* **34** (1965), 327.
781. Besrest, F. and Jaulmes, S.: *Acta Cryst.* **B29** (1973), 1560.
782. Voronova, A. A. and Vainshtein, B. K.: *Sov. Phys. – Cryst.* **18** (1973), 63.
783. De Haan, Y. M.: *NBS Spec. Publ.* **301** (1967), 233.
784. Bärnighausen, H.: University of Karlsruhe (private communication).
785. Jacobs, H.: *Z. anorg. allg. Chem.* **382** (1971), 97.
786. Sale, F. R. and Shelton, R. A. J.: *J. Less-Common Met.* **9** (1965), 60.
787. Henderson, D. M. and Gutowsky, H. S.: *Amer. Min.* **47** (1962), 1231.
788. Bertrand, G. and Dusausoy, Y.: *Comptes rendus (Paris)* **270C** (1970), 612.
789. Holm, C. H., Adams, C. R., and Ibers, J. A.: *J. Phys. Chem.* **62** (1958), 992.
790. Scaife, D. E. and Wylie, A. W.: *J. Chem. Soc.* (1964), 5450.
791. Corbett, J. D., Sallach, R. A., and Lokken, D. A.: *Adv. Chem. Ser.* **71** (1967), 56.
792. Corbett, J. D., Druding, L. F., Burkhard, W. J., and Lindahl, C. B.: *Discuss. Farad. Soc.* **32** (1961), 79.
793. Howie, R. A., Moser, W., and Trevena, I. C.: *Acta Cryst.* **B28** (1972), 2965.
794. Keve, E. T. and Skapski, A. C.: *Inorg. Chem.* **7** (1968), 1757.
795. Gaudé, J., L'Haridon, P., Laurent, Y., and Lang, J.: *Bull. Soc. franç. Min. Crist.* **95** (1972), 56.
796. Seifert, H. J. and Dau, E.: *Z. anorg. allg. Chem.* **391** (1972), 302.
797. McPherson, G. L., Wall, Jr., J. E., and Hermann, A. M.: *Inorg. Chem.* **13** (1974), 2230.
798. Le Borgne, G. and Weigel, D.: *Bull. Soc. chim. France* (1972), 3081.
799. Kanamaru, F., Shimada, M., Koizumi, M., Takano, M., and Takada, T.: *J. Solid State Chem.* **7** (1973), 297; Kanamaru, F. and Koizumi, M.: *Japan. J. Appl. Phys.* **13** (1974), 1319.
800. Jaggi, H. and Oswald, H. R.: *Acta Cryst.* **14** (1961), 1041.
801. Schönenberger, U. W., Günter, J. R., and Oswald, H. R.: *J. Solid State Chem.* **3** (1971), 190.
802. Andres, K., Kuebler, N. A., and Robin, M. B.: *J. Phys. Chem. Solids* **27** (1966), 1747.
803. Avilov, A. S. and Imamov, R. M.: *Sov. Phys. – Cryst.* **13** (1968), 52.
804. Novoselova, A. V., Todriya, M. K., Odin, I. N., and Popovkin, B. A.: *Inorg. Mater.* **7** (1971), 1125.
805. Gyaneshwar, Chadha, G. K., and Trigunayat, G. C.: *Z. Krist.* **141** (1975), 67.
806. Pandey, D. and Krishna, P.: *Phil. Mag.* **31** (1975), 1113.
807. Agrawal, V. K.: *Acta Cryst.* **A28** (1972), 93.
808. Srinivasan, R. and Parthasarathi, V.: *Z. Krist.* **137** (1973), 296.
809. Trigunayat, G. C. and Chadha, G. K.: *Phys. Stat. Sol.* **4a** (1971), 9.
810. Chand, M. and Trigunayat, G. C.: *Z. Krist.* **141** (1975), 59; *Acta Cryst.* **B31** (1975), 1222.
811. Srikrishnan, T. and Nowacki, W.: *Z. Krist.* **141** (1975), 174.
812. Myron, H. W. and Freeman, A. J.: *Phys. Rev.* **B11** (1975), 2735.

813. Goodenough, J. B.: *Mat. Res. Bull.* **3** (1968), 409.
814. Beal, A. R., Knights, J. C., and Liang, W. Y.: *J. Phys.* **C5** (1972), 3540.
815. Title, R. S. and Shafer, M. W.: *Phys. Rev. Lett.* **28** (1972), 808.
816. Harper, P. G. and Edmondson, D. R.: *Phys. Stat. Sol.* **44b** (1971), 59.
817. Kasowski, R. V.: *Phys. Rev. Lett.* **30** (1973), 1175.
818. Williams, R. H.: *J. Phys.* **C6** (1973), L32.
819. Wood, K. and Pendry, J. B.: *Phys. Rev. Lett.* **31** (1973), 1400.
820. Leveque, G., Robin-Kandare, S., and Martin, L.: *Phys. Stat. Sol.* **63b** (1974), 679.
821. Murray, R. B. and Williams, R. H.: *Phil. Mag.* **29** (1974), 473.
822. Hughes, H. P. and Liang, W. Y.: *J. Phys.* **C7** (1974), L162.
823. Smith, N. V., Traum, M. M., and Di Salvo, F. J.: *Solid State Commun.* **15** (1974), 211.
824. Shepherd, F. R. and Williams, P. M.: *J. Phys.* **C7** (1974), 4427.
825. Klose, W., Entel, P., and Nohl, H.: *Phys. Lett.* **50A** (1974), 186.
826. Jolly, W. L. and Latimer, W. M.: *J. Amer. Chem. Soc.* **74** (1952), 5752.
827. Belotskii, D. P., Sushkevich, T. N., and Lyuter, Ya. A.: *Inorg. Mater.* **10** (1974), 1632.
828. Damien, D., Wojakowski, A., and Müller, W.: *Inorg. Nucl. Chem. Lett.* **12** (1976), 441.
829. Fleet, M. E.: *Acta Cryst.* **B31** (1975), 183.
830. Juza, R. and Friedrichsen, H.: *Z. anorg. allg. Chem.* **332** (1964), 173.
831. Louër, M., Grandjean, D., and Weigel, D.: *J. Solid State Chem.* **7** (1973), 222.
832. Kohls, D. W. and Rodda, J. L.: *Amer. Min.* **52** (1967), 1261.
833. Lapham, D. M.: *Amer. Min.* **50** (1965), 1708.
834. Evans, Jr., H. T. and Allmann, R.: *Z. Krist.* **127** (1968), 73.
835. Harris, D. C. and Vaughan, D. J.: *Amer. Min.* **57** (1972), 1037.
836. Allmann, R.: *Fortschr. Miner.* **48** (1971), 24.
837. Allmann, R., Lohse, H. H., and Hellner, E.: *Z. Krist.* **126** (1968), 7.
838. Wadsley, A. D.: *Acta Cryst.* **5** (1952), 676.
839. Radke, A. S. and Dickson, F. W.: *Amer. Min.* **60** (1975), 559.
840. Men'kov, A. A. and Komissarova, L. N.: *Russ. J. Inorg. Chem.* **9** (1964), 425 and 952.
841. Brown, D., Fletcher, S., and Hola, D. G.: *J. Chem. Soc.* **A** (1968), 1889.
842. Asprey, L. B., Keenan, T. K., and Kruse, F. H.: *Inorg. Chem.* **3** (1964), 1137.
843. Asprey, L. B., Keenan, T. K., and Kruse, F. H.: *Inorg. Chem.* **4** (1965), 985.
844. Cohen, D., Fried, S., Siegel, S., and Tani, B.: *Inorg. Nucl. Chem. Lett.* **4** (1968), 257.
845. Fried, S., Cohen, D., Siegel, S., and Tani, B.: *Inorg. Nucl. Chem. Lett.* **4** (1968), 495.
846. Newland, B. G. and Shelton, R. A. J.: *J. Less-Common Met.* **22** (1970), 369.
847. McCarley, R. E., Roddy, J. W. and Berry, K. O.: *Inorg. Chem.* **3** (1964), 50.
848. Morosin, B. and Narath, A.: *J. Chem. Phys.* **40** (1964), 1958.
849. Berry, K. O., Smardzewski, R. R., and McCarley, R. E.: *Inorg. Chem.* **8** (1969), 1994.
850. Burns, J. H., Peterson, J. R., and Stevenson, J. N.: *J. Inorg. Nucl. Chem.* **37** (1975), 743.
851. Trotter, J. and Zobel, T.: *Z. Krist.* **123** (1966), 67.
852. Agron, P. A. and Busing, W. R.: *Acta Cryst.* **B25** (1969), S118.
853. Goodyear, J. and Ali, S. A. D.: *Acta Cryst.* **B25** (1969), 2664.
854. Mizuno, J.: *J. Phys. Soc. Japan* **15** (1960), 1412.
855. Saffar, Z. M.: *J. Phys. Soc. Japan* **21** (1966), 1844.
856. Mizuno, J., Ukei, K., and Sugawara, T.: *J. Phys. Soc. Japan* **14** (1959), 383.
857. Mizuno, J.: *J. Phys. Soc. Japan* **16** (1961), 1574.
858. Donaldson, R. H. and Lanchester, P. C.: *J. Phys.* **C1** (1968), 364.
859. Haase, A. and Brauer, G.: *Acta Cryst.* **B31** (1975), 2521.
860. Wadsten, T.: *Chem. Scripta* **8** (1975), 63.
861. Thomas, H. H. and Baker, Jr., W. A.: *Acta Cryst.* **B29** (1973), 1740.
862. Templeton, D. H. and Carter, G. F.: *J. Phys. Chem.* **58** (1954), 940.
863. Scherer, V., Weigel, F., and Van Ghemen, M.: *Inorg. Nucl. Chem. Lett.* **3** (1967), 589.
864. Bärnighausen, H. and Handa, B. K.: *J. Less-Common Met.* **6** (1964), 226.
865. Brodersen, K., Thiele, G., and Recke, I.: *J. Less-Common Met.* **14** (1968), 151.
866. Brodersen, K., Moers, F., and Schnering, H. G.: *Naturwiss.* **52** (1965), 205.
867. Brodersen, K., Thiele, G., Ohnsorge, H., Recke, I., and Moers, F.: *J. Less-Common Met.* **15** (1968), 347.

868. Okuda, T., Furukawa, Y., Shigemoto, H., and Negita, H.: *Bull. Chem. Soc. Japan* **46** (1973), 741.
869. Gynane, M. J. S. and Worrall, I. J.: *Inorg. Nucl. Chem. Lett.* **9** (1973), 903.
870. Levy, J. H., Taylor, J. C., and Wilson, P. W.: *Acta Cryst.* **B31** (1975), 880.
871. Burns, J. H., Peterson, J. R., and Baybarz, R. D.: *J. Inorg. Nucl. Chem.* **35** (1973), 1171.
872. Forrester, J. D., Zalkin, A., Templeton, D. H., and Wallmann, J. C.: *Inorg. Chem.* **3** (1964), 185.
873. Belova, V. I. and Semenov, I. N.: *Russ. J. Inorg. Chem.* **16** (1971), 1527.
874. Taylor, H. F. W.: *Miner. Mag.* **39** (1973), 377.
874a. Pastor-Rodriguez, J. and Taylor, H. F. W., *Mineral. Mag.* **38** (1971), 286.
875. Harris, A. L. and Veale, C. R.: *J. Inorg. Nucl. Chem.* **27** (1965), 1437.
876. Haschke, J. M. and Eick, H. A.: *J. Inorg. Nucl. Chem.* **32** (1970), 2153.
877. Corbett, J. D., Clark, R. J., and Munday, T. F.: *J. Inorg. Nucl. Chem.* **25** (1963), 1287.
878. Haschke, J. M.: *J. Solid State Chem.* **14** (1975), 238.
879. Tornqvist, E. G. M., Richardson, J. T., Wilchinsky, Z. W., and Looney, R. W.: *J. Catalysis* **8** (1967), 189.
880. Natta, G., Corradini, P., and Allegra, G.: *J. Polymer Sci.* **51** (1961), 399.
881. Viter, V. G., Belotskii, D. P., and Antipov, I. N.: *Russ. J. Inorg. Chem.* **13** (1968), 1614.
882. Semenenko, K. N. and Naumova, T. N.: *Russ. J. Inorg. Chem.* **9** (1964), 718.
883. Handy, L. L. and Gregory, N. W.: *J. Amer. Chem. Soc.* **74** (1952), 891.
884. Kutscher, J. and Schneider, A.: *Inorg. Nucl. Chem. Lett.* **7** (1971), 815.
885. v. Schnering, H. G. and Beckmann, W.: *Z. anorg. allg. Chem.* **347** (1966), 231.
886. Schäfer, H. and Beckmann, W.: *Z. anorg. allg. Chem.* **347** (1966), 225.
887. Weissenstein, J. and Horák, J.: *Czech. J. Phys.* **B24** (1974), 235.
888. Schäfer, H., v. Schnering, H. G., Tillack, J., Kuhnen, F., Wöhrle, H., and Baumann, H.: *Z. anorg. allg. Chem.* **353** (1967), 281.
889. Earls, D. E., Axtmann, R. C., and Hazoni, Y.: *J. Phys. Chem. Solids* **29** (1968), 1859.
890. Cable, J. W., Wilkinson, M. K., Wollan, E. O., and Koehler, W. C.: *Phys. Rev.* **127** (1962), 714.
891. Fletcher, J. M., Gardner, W. E., Fox, A. C., and Topping, G.: *J. Chem. Soc.* **A** (1967), 1038.
892. Binotto, L., Pollini, I., and Spinola, G.: *Phys. Stat. Sol.* **44b** (1971), 245.
893. Emeis, C. A., Reinders, F. J., and Drent, E.: *Solid State Commun.* **16** (1975), 239.
894. Ogawa, S.: *J. Phys. Soc. Japan* **15** (1960), 1901.
895. Cavallone, F., Pollini, I., and Spinola, G.: *Nuovo Cim.* **4** (1970), 764.
896. Cavallone, F., Pollini, I., and Spinola, G.: *Phys. Stat. Sol.* **45b** (1971), 405.
897. Kobayashi, K. L. I., Kato, Y., Katayama, Y., and Komatsubara, K. F.: *Solid State Commun.* **17** (1975), 875.
898. Bullett, D. W.: *Solid State Commun.* **17** (1975), 965.
899. v. Schnering, H. G., Wöhrle, H., and Schäfer, H.: *Naturwissensch.* **48** (1961), 159.
900. Simon, A. and v. Schnering, H. G.: *J. Less-Common Met.* **11** (1966), 31.
901. Schäfer, H. and v. Schnering, H. G.: *Angew. Chem.* **76** (1964), 833; and private communication.
902. Berdonosov, S. S. and Lapitskii, A. V.: *Russ. J. Inorg. Chem.* **10** (1965), 1525.
903. Kepert, D. L. and Marshall, R. E.: *J. Less-Common Met.* **34** (1974), 153.
904. Billaud, D. and Hérold, A.: *Comptes rendus* (Paris) **281C** (1975), 305.
905. Bishop, S. G. and Shevchik, N. J.: *Phys. Rev.* **B12** (1975), 1567.
906. Galitskii, N. V., Kudryavtsev, V. I., and Klyuchnikova, E. F.: *Chem. Abstr.* **79** (1973), 140164n.
907. Kolbin, N. I. and Ryabov, A. N.: *Vestn. Leningr. Univ., Ser. Fiz. Khim.*, **14** (1959), 121; *Chem. Abstr.* **54** (1960), 9581e.
908. Frère, P.: *Ann. Chim.* (Paris) **7** (1962), 85.
909. Cowley, J. M.: *Acta Cryst.* **9** (1956), 391.
910. Rothbauer, R., Zigan, F., and O'Daniel, H.: *Z. Krist.* **125** (1967), 317.
911. Bosmans, H. J.: *Acta Cryst.* **B26** (1970), 649.
912. Cotton, F. A. and Mague, J. T.: *Proc. Chem. Soc.* (1964), 233; *Inorg. Chem.* **3** (1964), 1402.
913. Bennett, M. J., Cotton, F. A., and Foxman, B. M.: *Inorg. Chem.* **7** (1968), 1563.
914. Gelinek, J. and Rüdorff, W.: *Naturwiss.* **51** (1964), 85.

915. Kolbin, N. I. and Ovchinnikov, K. V.: *Russ. J. Inorg. Chem.* **13** (1968), 1190.
916. Penfold, B. R.: in *Perspectives in Structural Chemistry* (ed. by Dunitz, J. D. and Ibers, J. A.). Wiley, New York, Vol. II (1968), 71.
917. Dornberger-Schiff, K. and Klevtsova, R. F.: *Acta Cryst.* **22** (1967), 435.
918. Tarkhova, T. N., Grishin, I. A., and Mironov, N. N.: *Russ. J. Inorg. Chem.* **15** (1970), 1340.
919. Dem'yanets, L. N., Bukin, V. I., Emel'yanova, E. N., and Ivanov, V. I.: *Sov. Phys. Cryst.* **18** (1974), 806.
920. Carter, F. L. and Levinson, S.: *Inorg. Chem.* **8** (1969), 2788.
921. Klevtsov, P. V., Bembel', V. M., and Grankina, Z. A.: *Russ. J. Struct. Chem.* **10** (1969), 543.
922. Klevtsova, R. F. and Glinskaya, L. A.: *Russ. J. Struct. Chem.* **10** (1969), 408.
923. Klevtsov, P. V., Kharchenko, L. Yu., Lysenina, T. G., and Grankina, Z. A.: *Russ. J. Inorg. Chem.* **17** (1972), 1512.
924. Klevtsova, R. F. and Klevtsov, P. V.: *Sov. Phys. – Doklady* **10** (1965), 487.
925. Klevtsova, R. F., Kozeeva, L. P., and Klevtsov, P. V.: *Inorg. Mater.* **3** (1967), 1247.
926. Klevtsov, P. V., Lysenina, T. G., and Kharchenko, L. Yu.: *Russ. J. Struct. Chem.* **14** (1973), 76.
927. Atoji, M. and Lipscomb, W. N.: *J. Chem. Phys.* **27** (1957), 195; **28** (1958), 355.
928. Ring, M. A., Donnay, J. D. H., and Koski, W. S.: *Inorg. Chem.* **1** (1962), 109.
929. Rollier, M. A. and Riva, A.: *Gazz. Chim. Ital.* **77** (1947), 361.
930. Binbrek, O. S., Krishnamurthy, N., and Anderson, A.: *J. Chem. Phys.* **60** (1974), 4400.
931. Clark, E. S. and Templeton, D.: *Acta Cryst.* **11** (1958), 284.
932. Trotter, J.: *Z. Krist.* **122** (1965), 230.
933. Cushen, D. W. and Hulme, R.: *J. Chem. Soc.* (1964), 4162.
934. Nyburg, S. C., Ozin, G. A., and Szymański, J. T.: *Acta Cryst.* **B27** (1971), 2298.
935. Cohen, S. T.: *Diss. Abstr.* **27** (1967), 4290-B.
936. Darnell, A. J. and McCollum, W. A.: *J. Phys. Chem.* **72** (1968), 1327.
937. Thompson, A. H.: *Solid State Commun.* **17** (1975), 1115.
938. Williams, P. M., Scruby, C. B., and Tatlock, G. J.: *Solid State Commun.* **17** (1975), 1197.
939. Corrigan, F. R. and Bundy, F. P.: *J. Chem. Phys.* **63** (1975), 3812.
940. Wallwork, S. C. and Worrall, I. J.: *J. Chem. Soc.* (1965), 1816.
941. Forrester, J. D., Zalkin, A., and Templeton, D. H.: *Inorg. Chem.* **3** (1964), 63.
942. Lörcher, K. P. and Strähle, J.: *Z. Naturforsch.* **30b** (1975), 662.
943. Furuseth, S., Brattås, L., and Kjekshus, A.: *Acta Chem. Scand.* **A29** (1975), 623.
944. Brooker, H. R. and Scott, T. A.: *J. Chem. Phys.* **41** (1964), 475.
945. Taylor, J. C. and Wilson, P. W.: *Acta Cryst.* **B30** (1974), 2664.
946. Brown, D., Hill, J., and Rickard, C. E. F.: *J. Chem. Soc.* **A** (1970), 476.
947. Zalkin, A., Forrester, J. D., and Templeton, D. H.: *Inorg. Chem.* **3** (1964), 639.
948. Krebs, B.: *Angew. Chem., Int. Ed.* **8** (1969), 146.
949. Jarchow, O., Schröder, F., and Schulz, H.: *Z. anorg. allg. Chem.* **363** (1968), 58.
950. Tillack, J., Eckerlin, P., and Dettingmeijer, J. H.: *Angew. Chem.* **78** (1966), 451.
951. Tillack, J.: *Z. anorg. allg. Chem.* **357** (1968), 11.
952. Drew, M. G. B. and Mandyczewsky, R.: *J. Chem. Soc.* **A** (1970), 2815.
953. Edwards, A. J.: *J. Chem. Soc. Dalton Trans.* (1972), 582.
954. v. Schnering, H. G. and Wöhrle, H.: *Angew. Chem.* **75** (1963), 684.
955. Hoppe, R. and Dähne, W.: *Naturwiss.* **49** (1962), 254.
956. Gortsema, F. P. and Didchenko, R.: *Inorg. Chem.* **4** (1965), 182.
957. Schäfer, H., v. Schnering, H. G., Niehues, K. J., and Nieder-Vahrenholz, H. G.: *J. Less-Common Met.* **9** (1965), 95.
958. Bergström, G. and Lundgren, G.: *Acta Chem. Scand.* **10** (1956), 673.
959. Roof, Jr., R. B., Cromer, D. T., and Larson, A. C.: *Acta Cryst.* **17** (1964), 701.
960. Bannister, M. J. and Taylor, J. C.: *Acta Cryst.* **B26** (1970), 1775.
961. Taylor, J. C. and Hurst, H. J.: *Acta Cryst.* **B27** (1971), 2018.
962. Siegel, S., Hoekstra, H. R., and Gebert, E.: *Acta Cryst.* **B28** (1972), 3469.
963. Taylor, J. C.: *Acta Cryst.* **B27** (1971), 1088.
964. Atoji, M. and McDermott, M. J.: *Acta Cryst.* **B26** (1970), 1540.
965. Littke, W. and Brauer, G.: *Z. anorg. allg. Chem.* **325** (1963), 122.
966. Strähle, J.: *Z. anorg. allg. Chem.* **375** (1970), 238; **380** (1971), 96.

967. Kepert, D. L., and Mandyczewsky, R.: *Inorg. Chem.* **7** (1968), 2091.
968. Seabaugh, P. W. and Corbett, J. D.: *Inorg. Chem.* **4** (1965), 176.
969. Britnell, D., Fowles, G. W. A., and Rice, D. A.: *J. Chem. Soc., Dalton Trans.* (1974), 2191.
970. Müller, U. and Kolitsch, W.: *Z. anorg. allg. Chem.* **410** (1974), 32.
971. Shchukarev, S. A. and Kokovin, G. A.: *Russ. J. Inorg. Chem.* **9** (1964), 715.
972. Taylor, J. C. and Wilson, P. W.: *Acta Cryst.* **B30** (1974), 1216.
973. Giovanoli, R., Stadelmann, W., and Feitknecht, W.: *Helv. Chim. Acta* **56** (1973), 839.
974. Newnham, R. E. and Brindley, G. W.: *Acta Cryst.* **9** (1956), 759.
975. Dalmon, J. A. and Martin, G. A.: *Comptes rendus (Paris)* **267C** (1968), 610.
976. Dalmon, J. A., Martin, G. A., and Imelik, B.: *J. chim. phys.* **70** (1973), 214.
977. Liebau, F.: *Acta Cryst.* **B24** (1968), 690.
978. Weiss, A., Ruthardt, R., and Orth, H.: *Z. Naturforsch.* **28b** (1973), 446.
979. Weiss, A. and Orth, H.: *Z. Naturforsch.* **28b** (1973), 252.
980. Lagaly, G., Beneke, K., and Weiss, A.: *Z. Naturforsch.* **28b** (1973), 234.
981. Tennakoon, D. T. B., Thomas, J. M., Tricker, M. J., and Graham, S. H.: *J. Chem. Soc., Chem. Commun.* (1974), 124.
982. De Jongh, L. J. and Miedema, A. R.: *Adv. Phys.* **23** (1974), 1.
982a. Miedema, A. R., Bloembèrgen, P., Colpa, J. H. P., Gorter, F. W., de Jongh, L. J., and Nordermeer, L.: *AIP Conf. Proc.* **18** (1974), 806.
983. Beck, H. P.: *J. Solid State Chem.* **17** (1976), 275, and private communication.
984. Trucano, P. and Chen, R.: *Nature* **258** (1975), 136.
985. Man, L. I.: *Sov. Phys.–Cryst.* **15** (1970), 399.
986. Batsanov, S. S., Sokolova, M. N., and Ruchkin, E. D.: *Bull. Acad. Sci. USSR, Div. Chem. Sci.* (1971), 1757.
987. Fellows, R. L., Peterson, J. R., Noé, M., Young, J. P., and Haire, R. G.: *Inorg. Nucl. Chem. Lett.* **11** (1975), 737.
988. Jellinek, F., Pollack, R. A., and Shafer, M. W.: *Mat. Res. Bull.* **9** (1974), 845.
989. Chianelli, R. R., Scanlon, J. C., and Thompson, A. H.: *Mat. Res. Bull.* **10** (1975), 1379.
990. Svensson, C.: *Acta Cryst.* **B30** (1974), 458.
991. Riekel, C.: *J. Solid State Chem.* **17** (1976), 389.
992. Chamberland, B. L.: *J. Less-Common Met.* **44** (1976), 239.
993. Jain, P. C. and Trigunayat, G. C.: *Z. Krist.* **142** (1975), 121.
994. Guen, L. and Nguyen-Huy-Dung.: *Acta Cryst.* **B32** (1976), 311.
995. Wang, R., Steinfink, H., and Raman, A.: *Inorg. Chem.* **6** (1967), 1298.
996. Guen, L., Nguyen-Huy-Dung, Eholie, R., and Flahaut, J.: *Ann. chim.* **10** (1975), 11.
997. Guérard, D. and Hérold, A.: *Comptes rendus (Paris)* **281C** (1975), 929.
998. Hooley, J. G. and Reimer, M.: *Carbon* **13** (1975), 401.
999. Kuhn, A., Chevalier, R., and Rimsky, A.: *Acta Cryst.* **B31** (1975), 2841; Kuhn, A., Chevy, A., and Chevalier, R.: *Phys. Stat. Sol.* **31a** (1975), 469.
1000. Portheine, J. C. and Nowacki, W.: *Z. Krist.* **141** (1975), 387.
1001. Beyreuther, C., Hierl, R., and Wiech, G.: *Ber. Bunsenges.* **79** (1975), 1081; Berg, U., Dräger, G., and Brümmer, O.: *Phys. Stat. Sol.* **74B** (1976), 341.
1002. Melin, J. and Hérold, A.: *Carbon* **13** (1975), 357.
1003. Guérard, D., Lelaurain, M., and Aubry, A.: *Bull. Soc. franç. Min. Crist.* **98** (1975), 43.
1004. Kuhn, A., Chevy, A., and Chevalier, R.: *Acta Cryst.* **B32** (1976), 983.
1005. Brandt, N. B., Kapustin, G. A., Karavaev, V. G., Kotosonov, A. S., and Svistova, E. A.: *Sov. Phys.–JETP* **40** (1975), 564.
1006. Kapustin, G. A., and Meilikhov, E. Z.: *Sov. Phys. Solid State* **17** (1975), 1979.
1007. Selig, H., Rabinovitz, M., Agranat, I., and Lin, Chun-Hsu: *J. Amer. Chem. Soc.* **98** (1976), 1601.
1008. Bragin, F. V., Novikov, Yu. N., and Ryabchenko, S. M.: *Sov. Phys.–JETP* **39** (1974), 172; **41** (1975), 681.
1009. Vol'pin, M. E. *et al.*: *J. Amer. Chem. Soc.* **97** (1975), 3366.
1010. Fourcroy, P. H., Rivet, J., and Flahaut, J.: *Comptes rendus (Paris)* **279C** (1974), 1035.
1011. Brandt, N. B., Gitsu, D. V., Popovich, N. S., Sidorov, V. I., and Chudinov, S. M.: *JETP Lett.* **22** (1975), 104.

1012. Göttlicher, S., and Kieselbach, B.: *Acta Cryst.* **A32** (1976), 185.
1013. Tsuji, T., Howe, A. T., and Greenwood, N. N.: *J. Solid State Chem.* **17** (1976), 157.
1014. Wyles, L. R., Deline, T. A., Haschke, J. M., and Peacor, D. R.: *Proc. 11th Rare Earth Conf., Traverse City, Mi.* (1974), 550.
1015. Soled, S. and Wold, A.: *Mat. Res. Bull.* **11** (1976), 657.
1016. Kanamura, F., Otani, S., and Koizumi, M.: *Proc. 5th Int. Conf. Solid Compds. Transition Elements* (Uppsala 1976), p. 36.
1017. Rijndorp, J.: *Proc. 5th Int. Conf. Solid Compds. Transition Elements* (Uppsala 1976), p. 45.
1018. Warkentin, E., and Bärnighausen, H.: *Third European Cryst. Meeting, Zürich,* 1976.
1019. Nguyen-Huy-Dung and Thévet, F.: *Acta Cryst.* **B32** (1976), 1108 and 1112.
1020. Diehl, R., Carpentier, C. D., and Nitsche, R.: *Acta Cryst.* **B32** (1976), 1257.
1021. Mercier, R., *Rev. Chim. minér.* **12** (1975), 508.
1022. v. Schnering, H. G.: Habilitationsschrift, University Münster (1963).
1023. Fenner, J.: *Acta Cryst.* **B32** (1976) 3084.
1024. Schroeder, F. A. and Scherle, J.: *Z. Naturforsch.* **28b** (1973), 216.
1025. Rabenau, A., Rau, H., and Rosenstein, G.: *Naturwiss.* **56** (1969), 137; *Z. anorg. allg. Chem.* **374** (1970), 43.
1026. Giese, Jr., R. F.: *Acta Cryst.* **B32** (1976), 1719.
1027. Adams, J. M. and Jefferson, D. A.: *Acta Cryst.* **B32** (1976), 1180.
1028. Thomas, D. and Tridot, G.: *Comptes rendus (Paris)* **258** (1964), 2587.
1029. Blunck, H. and Juza, R.: *Z. anorg. allg. Chem.* **410** (1974), 9.
1030. Kuhn, A., Chevalier, R., Desnoyers, C. and Terhell, J. C. J. M.: *Acta Cryst.* **B32** (1976), 1910.
1030a. Freund, F. and Sperling, V.: *Mat. Res. Bull.* **11** (1976), 621.
1030b. Wardle, R. and Brindley, G. W.: *Amer. Mineral.* **57** (1972), 732.
1031. Rabenau, A. and Schulz, H.: *J. Less-Common Met.* **48** (1976), 89.
1032. Guen, L., Nguyen Huy Dung, Eholie, R., and Flahaut, J.: *Ann. Chim. (Paris)* **1** (1976), 39.
1033. Corbett, J. D.: *Proc. 12th Rare Earth Res. Conf., Vail. (Colorado)*, 1976, p. 396.
1034. Dittmar, G.: Dissertation Techn. Hochschule Darmstadt 1976.
1034a. Stumpp, E. and Terlan, A.: *Carbon* **14** (1976), 89.
1035. Girifalco, L. A. and Montelbano, T. O.: *J. Mater. Sci.* **11** (1976), 1036.
1036. Fischer, J. E., Thompson, T. E., Foley, G. M. T., Guérard, D., Hoke, M., and Lederman, F. L.: *Phys. Rev. Lett.* **37** (1976), 769.
1037. Ubbelohde, A. R.: *Carbon* **14** (1976), 1.
1038. Kobayashi, K. L. I., Kato, Y., Katayama, Y., and Komatsubara, K. F.: *Phys. Rev. Lett.* **37** (1976), 772.
1039. Adolphson, D. G. and Corbett, J. D.: *Inorg. Chem.* **15** (1976), 1820.
1040. Kikkawa, S., Kanamaru, F., and Koizumi, M.: *Inorg. Chem.* **15** (1976), 2195.
1041. Bayard, M., Mentzen, B. F., and Sienko, M. J.: *Inorg. Chem.* **15** (1976), 1763.
1042. Mentzen, B. F. and Sienko, M. J.: *Inorg. Chem.* **15** (1976), 2198.
1043. Rea, M. and Domenicali, C. A.: *Solid State Commun.* **20** (1976), 325.
1044. Hoggins, J. T. and Steinfink, H.: *Inorg. Chem.* **15** (1976), 1682.
1045. Ploog, K., Stetter, W., Nowitzki, A., and Schönherr, E.: *Mat. Res. Bull.* **11** (1976), 1147.
1046. Fritz. I. J.: *Solid State Commun.* **20** (1976), 299.
1047. Ebert, L. B.: *Ann. Rev. Mat. Sci.* **6** (1976), 181.
1048. Buscarlet, E., Touzain, P., and Bonnetain, L.: *Carbon* **14** (1976), 75.
1049. Lance, E. T., Haschke, J. M., and Peacor, D. R.: *Inorg. Chem.* **15** (1976) 780.
1050. Flahaut, J.: *Ann. Chim (Paris)* **1** (1976), 27.
1051. Chaussy, J., Haen, P., Lasjaunias, J. C., Monceau, P., Waysand, G., Waintal, A., Meerschaut, A., Molinié, P., and Rouxel, J.: *Solid State Commun.* **20** (1976) 759.
1052. Pachali, K. E., Ruska, J., and Thurn, H.: *Inorg. Chem.* **15** (1976) 991.
1054. Lundberg, M. and Skarnulis, A. J.: *Acta Cryst.* **B32** (1976), 2944.
1055. Loub, J., Podlahová, J. and Novák, C.: *Acta Cryst.* **B32** (1976), 3115.
1056. Cachau-Herreillat, D. and Moret, J.: *Comptes rendus (Paris)* **282C** (1976), 511.
1057. Behrendt, D., Beneke, K. and Lagaly, G.: *Angew. Chem. Int. Ed.* **15** (1976), 544.
1058. Basten, J. A. J. and Bongaarts, A. L. M.: *Phys. Rev.* **B14** (1976), 2119.
1059. Kleinberg, R.: *J. Chem. Phys.* **50** (1969), 4690.
1060. Sutherland, H. H., Hogg, J. H. C. and Walton, P. D.: *Acta Cryst.* **B22** (1976), 2539.

MINERAL NAME INDEX

achavalite 133
amesite 344
anatase 38, 282
antigorite 342
apatite 109
arsenolite 63
asbestos 344
atacamite 38, 290

barbertonite 293
barite 111
bayerite 320, 343
berndtite (SnS_2) 196
berthierite 344
biotite 346
bismoclite (BiOCl) 161
black phosphorus 77, 87, 90
boehmite 257, 265, 267, 268
botallackite 290, 291, 292
brucite 271, 279, 280, 342

calaverite 38, 228
cariopilite 344
carlinite 125
chalcopyrite 295
chaoite 48
chlorite 346
chondrotite 39
chromium kaolinite 344
chrysotile 344
claudetite 66, 67
clinochlore 346
coalingite 292, 293
cookeite 346
corundophilite 346
covellite 135
cronstedtite 344
cyanite 39

dickite 345
doloresite 180
donbassite 344
duranusite 87
duttonite 178

emplectite 73
eulytite 109

feitknechtite (MnOOH) 281
fergusonite 111
ferro-berthierite 344

garnierite 344
getchellite 72
gibbsite 300, 320
graphite 48
greenalite 344
greigite 211
grimaldiite 296
grovesite 344
gypsum 119

häggite 178
halloysite 346
hedleyite ($Bi_{14}Te_6$) 210
herzenbergite 93
Hifforf's phosphorus 77, 94
humite 39
hydrargillite 300, 321
hydrotalcite 293

ikunolite 209
imhofite 77
iowaite 291

joseite 209

kaolin 343
kermesite 63, 70, 71
kitkaite 222
koenenite 295
kostovite 228
krennerite 228

laitakarite 209
lanarkite 105
lautite 159
lepidocrocite 268
litharge 97
lithiophorite 297
livingstonite 75
lizardite 344
lonsdalite 48
lorandite 86

mackinawite 133
manandonite 346
manasseite 293
manganostibite 39
margarite 346
massicot 95
matlockite 258
melonite ($NiTe_2$) 222

merenskyite 223
metaboric acid 43
metahalloysite 343
mica 347
minium 103
minnesotaite 345
molybdenite (MoS$_2$) 237, 238
monazite 111
moncheite 223
montmorillonite 346
muscovite 2, 346

nacrite 343
navajoite 182
nepouite 344
norbergite 39
nordstrandite 300

olivine 39
orpiment 63, 67, 68
orthochlorite 346

palmierite 108
paradocrasite 86, 87
paraguanajuatite Bi$_4$(Se, S)$_5$ 210
paratacamite 38, 290
penfieldite 290
protodoloresite 179, 180
pucherite 111
pyroaurite 293
pyrophyllite 343, 344, 347

quenselite 102

raguinite 105
raspite 111
reevesite 293
rickardite 134
rutile 38

saponite 346
sassolite 43
scheelite 111
schultenite 113
selenojoseite 209
senarmontite 63
servantite 110
sillimanite 39
sjögrenite 293
smythite 38, 210
spinel 39, 187
stannite 132
stibiotantalite 110
stibnite 63, 64
stichtite 293
stolzite 111
sudoite 346
sylvanite 228

talc 342, 344
talc chlorite 346
teallite 93
tellurite 114
tellurobismutite 186, 196, 200
tetradymite 63, 195, 200
thortveitite 168
topaz 39

valentinite 63
valleriite 295, 296
vulcanite 133, 134

white lead 294
willemseite 345
wolfsbergite 72
wulfenite 111

xanthophyllite 346

zavaritskite (BiOF) 98, 260
zircon 111

FORMULA INDEX

AcOBr 261
AcOCl 261
AcOI 262

AgAsS 159
AgAsSe 159
AgAuTe$_4$ 228
Ag$_2$F 28, 274, 275
Ag$_2$HgI$_4$ 187
AgI 132
AgIn$_5$Te$_8$ 38
AgMgAs 38
Ag$_2$O 28, 195, 196, 274
AgP$_2$ 159
AgPS 159
AgPSe 159
Ag$_2$PbO$_2$ 102
AgTe 134
Ag$_3$TlTe$_2$ 135

AlB$_2$ 4, 5, 52
AlBr$_3$ 24, 38, 315, *325*
Al$_2$Br$_6$·C$_6$H$_6$ 323
Al$_4$C$_3$ 28
AlCl$_3$ 13, 38, 298, *307*, 314, 315, 325
AlCl$_2$Br 314
AlCr$_2$ 247
AlF$_3$ 30, 300
AlI$_3$ 325
Al$_{3/4}$Mo$_2$S$_4$ 39
Al(NH$_3$)$_2$C$_n$ 53
Al$_2$O$_3$ 29, 38
AlOBr 269
AlOCl 19, 257, *268*
Al(OH)$_3$ 13, 300, *320*
Al$_2$(OH)$_2$(Si$_2$O$_5$)$_2$ 2, 344
Al$_2$(OH)$_4$Si$_2$O$_5$ 343
[Al$_2$(OH)$_2$Si$_3$AlO$_{10}$][LiAl$_2$(OH)$_6$] 346
Al$_2$(OH)$_4$Si$_2$O$_5$·2H$_2$O 346
Al$_2$(OH)$_4$Si$_2$O$_5$·HCONH$_2$ 346
AlOI 269
AlO(OH) 18, *265*, 268, 273
AlO(NH$_2$) 269
Al$_2$O(Si$_2$O$_5$)$_2$ 347
AlPS 145
AlPS$_4$ 38, 138, *141*
Al$_2$S$_3$ 183
AlSBr 269
AlSCl 257, 269
AlSI 269
Al$_2$Se$_3$ 183

AlSeBr 269
AlSeCl 257, 269
AlSeI 269
Al$_2$SiO$_5$ 39
Al$_2$SiO$_4$F$_2$ 39
AlSiS$_3$ 219
Al$_2$Te$_3$ 183
AlTeBr 257
AlTeCl 257
AlTeI 257

AmBr$_3$ 302
AmI$_3$ 302, 310
AmOBr 261
AmOCl 261
AmOI 262
AmTe$_3$ 244

As 26, 77, 88
AsBr$_3$ 323
As—GeSe 79
As—GeTe 79
AsI$_3$ 35, 308, 311
As$_2$NiO$_4$ 103
As$_2$O$_3$ 26, 28, 35, 62, 63, *66*
As$_2$O$_4$ 109
As$_2$O$_3$·SO$_3$ 67
As—P 79, 90
AsP 93
AsP$_3$ 87
AsPO$_4$ 109
As—P—Sb 79
As$_4$S 87
As$_2$S$_3$ 26, 35, 63, *67*
AsS 29
As—Sb 79, 80
AsSbS$_3$ 63, *72*
AsSb$_3$ 86, 87
As$_2$Se$_3$ 35, 63, 67, 71
As—SnTe 79
As$_2$Te$_3$ 35, 63, *69*

(Au, Ag)Te$_2$ 228
AuBr 40
AuBr$_3$ 165, 166
AuBrCl$_2$ 165
AuClI$_2$ 165
AuF$_3$ 30, 163, 165, 300
AuI 29, 40
AuIBr$_2$ 165
AuICl$_2$ 165

$Au_4In_3Sn_3$ 213
Au_2In_2Sn 213
AuP_2 159
AuPS 159
AuSb 214
AuSe 349
AuSeBr 163
AuSiS 159
$AuTe_2$ 38, *228*
$AuTe_2Br$ 164
$AuTe_2Cl$ 163
$AuTe_2I$ 164

BBr_3 48
BC_3 157
BCl_3 47
B_3F 51
BI_3 48
BN 26, *50*, 62
B_2O 51
$B(OH)_3$ *43*, 301
BPS_4 38, 138, 141, 142
B_2S_3 29, *44*
BSBr 143
BSCl 143, 257, 269
BSI 143, 257
BS(SH) 143
B_2Se_3 46
BSeBr 143, 257
BSeCl 143, 257
BSeI 143, 257

$BaBiO_2Cl$ 98
$BaBr_2$ 120
BaC_8 53
$BaCl_2$ 120
$BaCl_2 \cdot 2H_2O$ 332
BaFBr 260
BaFCl 260
BaFI 260
BaH_2 120
BaHBr 259, 260
BaHCl 260
BaHI 259, 260
BaI_2 120
BaMgSi 258
Ba_2N 28, 274, 277
Ba_2NH 277
$Ba(NH_3)_2C_n$ 53
$Ba_3(PO_4)_2$ 107
$BaPbO_2$ 101
$BaSO_4$ 111
BaSi 5, 28
$BaSnO_2$ 101
$Ba_5Ta_4O_{15}$ 1
$Ba_3(VO_4)_2$ 107

Be_2C 28
BeC_2 52, 55
$BeBr_2$ 141, 271
$BeCl_2$ 141, 271
BeFBr 269
BeI_2 141, 142, 271

Be_3N_2 28
$Be(NH_3)_2C_6$ 53
$Be(OH)_2$ 29, 136, 271
$BeSiN_2$ 38

Bi 77, 82, 90
$BiAsO_4$ 110, 111
$BiBr_3$ 300
$BiCl_3$ 300, 325
BiI_3 13, 35, 38, 298, 300, 307, *308*, 311, 340
BiI_3—SbI_3 311
Bi_2MnO_4 103
$BiNbO_4$ 15, 110, 111
Bi_2NiO_4 103
Bi_2O_3 63, 65, 109
Bi_2O_3—PbO 98
BiOBr 259, 262
BiOCl 259, 261
BiOF 98, 260
$Bi_2O(GeO_4)$ 107
$Bi(OH)CrO_4$ 303
BiOI 259, 262
Bi_2O_2Se 63
$BiPO_4$ 110
$BiPb_3(AsO_4)_3$ 109
$BiPb_3(AsO_4)(PO_4)_2$ 109
$BiPb_3(PO_4)_3$ 109
$BiPb_3(VO_4)_3$ 109
Bi_2S_3 63
Bi—Sb 80
$BiSbO_4$ 110
Bi_8Se_7 208
Bi_8Se_9 207, 209
Bi_4Se_3 207, 208, 209
Bi_2Se_3 63, 195, 197, 198
BiSe 207, 208
Bi_2Se_3—Bi_2S_3 197
BiSeCl 257
Bi_2Se_3—Bi_2Te_3 197
Bi_2Se_3—In_2Se_3 197
Bi_4Se_2S 209
Bi_2Se_3—Sb_2Se_3 197
Bi_2Se_3—Sb_2Te_3 197
Bi_2Se_2Te 197
$BiTaO_4$ 110, 111
$Bi_{2(m+n)}Te_{3n}$ 207, 210
Bi_4Te_3 209
Bi_2Te 208
Bi_2Te_3 38, 63, 186, *196*, 200
BiTe 207, 208
Bi_2Te_3—Bi_2S_3 197, 200
Bi_2Te_3—Bi_2Se_3 197, 199
BiTeBr 195, 196, 276
BiTeCl 257
BiTeI 195, 196, 275
Bi_2Te_3—In_2Se_3 197
$Bi_{14}Te_{13}S_8$ 200
$Bi_{14}Te_{15}S_6$ 201
$Bi_{14}Te_{20}S$ 201
$Bi_8Te_7S_5$ 197
Bi_4TeS_2 209
Bi_4Te_2S 209

Bi_2Te_2S 38, 63, *195*
Bi_2Te_2Se 197, 198
Bi_2TeSe_2 197
$BiVO_4$ 110, 111

$BkBr_3$ 299, 310, 313
BkI_3 299, 310
$BkOBr$ 262
$BkOCl$ 261
$BkOI$ 262

C 26, *48*
C_mAlBr_y 59
C_mAlCl_y 58, 59
C_mAuBr_y 60
C_mAuCl_y 59
C—Br 60
C—CF_3COOH 61
C_mCdCl_y 59
C_mCoCl_y 59
C_mCrCl_y 59
$C_{13}CrO_3$ 61
$C_{27}CrO_2Cl_2$ 61
C_mCuCl_y 59
CF 52, *153*
CF_2 155
C_mFeCl_y 59, 60, 61
C_mGaBr_y 59, 60
C_mGaCl_y 59
C—$HClO_4$ 61
$(C_nH_{2n+1}NH_3)_2CuCl_4$ 347
$(C_nH_{2n+1}NH_3)_2FeCl_4$ 347
$(C_nH_{2n+1}NH_3)_2MnCl_4$ 347
C—HNO_3 61
C—H_3PO_4 61
C—$H_4P_2O_7$ 61
C—H_2SO_4 61
$C_{24}HSO_4 \cdot 2H_2SO_4$ 57, 61
C_mHgCl_y 59
C—IBr 61
C_5ICl 61
C_mInCl_y 59
$C_{11}MgCl_2$ 59
C_mMoCl_5 59, 60
C_mMoOCl_4 59
$C_9N_2O_5$ 61
$C_{24}NO_3 \cdot 3HNO_3$ 57, 61
$C_{40}NbCl_{5.1}$ 59
C_mNiCl_2 62
$C_8O_2(OH)_2$ 155
$C_6O_2(OH)_2$ 155
C_mPtCl_y 59
C_mSbCl_5 59
C_mTaCl_y 59
C_mTiF_4 60
C_mUCl_5 59
$C_{70}WCl_6$ 59

$CaAl_2(OH)_2Si_2Al_2O_{10}$ 346
$CaAl_2Si_2O_8$ 22
$Ca_{1.25}Bi_{1.5}O_2Cl_3$ 99

$CaCl_2$ 38, 271, 277
$CaFBr$ 260
$CaFCl$ 260
$CaFI$ 260
CaH_2 120
$CaHAsO_4 \cdot 2H_2O$ 120
$CaHBr$ 260
$CaHCl$ 260
$CaHI$ 259, 260
CaI_2 275
$CaHPO_4 \cdot 2H_2O$ 120
$CaMg_3(OH)_2Si_2Al_2O_{10}$ 346
Ca_2N 28, 277
Ca_2NH 277
$Ca(NH_2)_2$ 282
$Ca(NH_3)_2C_n$ 53
$Ca(OH)_2$ 280
$Ca(OH)_2$—$Cd(OH)_2$ 281
$Ca(OH)Cl$ 285
$CaPb_2(PO_4)_2$ 108
$CaSO_4 \cdot 2H_2O$ 119
$CaSeO_4 \cdot 2H_2O$ 120
$CaSi_2$ 28, 156
$CaSnO_2$ 101
$CaWO_4$ 111

$CdBr_2$ 265, 272, 274-6, 285
$CdBr_{0.6}(OH)_{1.4}$ 276
$Cd(CN)_2$ 38
$CdCl_2$ 18, 19, 38, *273*, 276
$CdCl_{0.75}(OH)_{1.25}$ 276
$CdGaInS_4$ 188
$CdGa_2S_4$ 38
CdI_2 18, 38, *270*, 274-6, 280, 285
CdI_2—PbI_2 275
$CdIn_2Se_4$ 38
$Cd(OH)_2$ 272, 280, 281
$Cd(OH)Br$ 288, 290
$Cd(OH)Cl$ 38, 274, 277, 285
$Cd_4(OH)_8Cr(OH)_3$ 291
$Cd(OH)I$ 288, 290
$Cd(OH)_{3/2}I_{1/2}$ 281
$CdPS_3$ 217
$CdPSe_3$ 218

$Ce(HPO_4)_2 \cdot nH_2O$ 351
$CeHTe$ 260
Ce_2I_5 273
CeI_2 247, 273, 276
CeI_3 302
$CeOBr$ 261
$CeOCl$ 260
$CeOF$ 260
$Ce(OH)_2Cl$ 304
$CeOI$ 262
$CePO_4$ 111
$CeSF$ 262
$CeSI$ 255, 258, 264
$CeSeF$ 262
Ce_2Te_5 245
$CeTe_3$ 244

CfBr$_3$ 299, 310, 313
CfCl$_3$ 299, 302, 303
CfI$_3$ 310
CfOBr 262
CfOCl 261
CfOI 262

CmBr$_3$ 302, 303
CmI$_3$ 310
CmOBr 261
CmOCl 261
CmOI 262
CmTe$_3$ 244

CoAl$_2$Cl$_8$ 39
CoBr$_2$ 275, 276
CoBr$_2$·6H$_2$O 283
CoCl$_2$ 276
CoCl$_2$·6H$_2$O 282, 283
CoGe 29, 246
CoI$_2$ 275
CoMo$_2$S$_4$ 38
Co(OH)$_2$ 281
Co$_2$(OH)$_3$Br 281, 292
Co(OH)Br 287, 288
Co$_2$(OH)$_3$Cl 281
Co(OH)Cl 285, 288
Co$_4$(OH)$_8$Al(OH)$_3$ 291
Co$_4$(OH)$_8$CoOBr 291
Co$_4$(OH)$_8$CoOCl 291
Co$_4$(OH)$_8$Co(OH)Br 291, 293
Co$_4$(OH)$_8$CoONO$_3$ 291
Co$_2$(OH)$_3$I 292
Co$_2$(OH)$_3$NO$_3$ 281
Co$_3$(OH)$_2$(Si$_2$O$_5$)$_2$ 345
Co$_3$(OH)$_4$Si$_2$O$_5$ 344
CoPS 217
CoSeTe 222
CoTe 222
Co$_{3/4}$Zn$_{1/4}$(OH)$_2$ 281

CrB 5, 29, 127
CrBr$_2$ 271, 277
CrBr$_3$ 308, 310, 313
CrBr$_2$I 313
CrCl$_2$ 277
CrCl$_3$ 13, 38, 307, 308, 310, 312, 313
CrCl$_2$Br 313
CrCl$_3$—FeCl$_2$ 315
CrCl$_2$I 313
CrCl$_3$—MgCl$_2$ 315
CrCl$_3$—MnCl$_2$ 315
CrCl$_3$—RuCl$_3$ 313
CrI$_2$ 277, 278
CrI$_3$ 308, 312, 313
CrI$_2$—FeI$_2$ 278, 279
CrI$_2$—MnI$_2$ 278, 279
Cr$_2$NiB$_4$ 7
CrO$_3$ 30, 167
CrOBr 266
CrOCl 266
Cr(OH)$_3$·3H$_2$O 341

CrO(OH) 296, 297
CrPSe$_3$ 217
Cr$_7$S$_8$ 307
Cr$_5$S$_6$ 38, 307
Cr$_3$S$_4$ 38, 307
Cr$_2$S$_3$ 38, 307, 316
CrS 29, 38
CrSBr 266
CrSiSe$_3$ 219
CrTi$_2$O$_3$(OH)$_5$ 180
CrTi$_2$S$_4$ 38

Cs$_2$AgAuCl$_6$ 158
CsBiO$_2$ 104
CsBr 130
CsC$_n$ 53
CsCl 132
CsCu$_2$Cl$_3$ 25
CsFeF$_4$ 1
CsNH$_2$ 130
Cs(NH$_3$)$_2$C$_n$ 53
Cs$_4$Ni$_3$F$_{10}$ 1
Cs$_3$O 28
Cs$_2$O 28, 274, 277
CsOH 130

CuAgTe$_2$ 134
CuAl$_2$ 9, 33
Cu$_3$AsS$_4$ 38
CuAuTe$_4$ 228
CuBi$_2$O$_4$ 103
CuBiS$_2$ 73, 75
CuBiSe$_2$ 75
CuBiTe$_2$ 75
CuBr$_2$ 278, 279
CuBrSe$_3$ 349
CuCl$_2$ 278, 279
CuCrS$_2$ 39
CuCr$_2$S$_3$Cl 39
CuFeCl$_4$ 276
Cu$_5$FeS$_4$ 38
CuFeS$_2$ 38, 130, 138
CuFe$_2$S$_3$ 38
CuFeS$_2$—Mg(OH)$_2$ 295
Cu$_2$FeSnS$_4$ 38, 132
Cu$_2$GeS$_3$ 38
Cu$_2$HgI$_4$ 38
CuIn$_2$Se$_3$Br 38
CuMgAl$_2$ 4
Cu$_{1.4}$Mg$_{0.6}$(OH)$_3$Cl 281
Cu$_2$MnTe$_2$ 39
Cu$_2$O 138
Cu(OH)$_2$ 267, 268, 281
Cu$_2$(OH)$_3$Br 292, 293
Cu$_2$(OH)$_3$Cl 38, 292, 293
Cu(OH)Cl 285–290
Cu$_2$(OH)$_3$I 292, 293
CuS 29, 135
Cu$_2$Sb 31, 39, 134
Cu$_3$SbS$_4$ 38
CuSbS$_2$ 73, 74
CuSbSe$_2$ 73

FORMULA INDEX

CuSbTe$_2$ 75
CuScS$_2$ 39
CuSe 135
Cu$_2$SiS$_3$ 38
Cu$_4$Te$_3$ 39, 134
Cu$_{3-x}$Te$_2$ 133
Cu$_2$Te 39
CuTe 22, 38, *133*
CuTe$_2$ 223
CuTe$_2$Br 350
CuTe$_2$Cl 349
CuTe—CuS 135
CuTe$_2$I 350
CuTi 1
CuTiCl$_4$ 276
CuVCl$_4$ 276
Cu$_3$VS$_4$ 38
Cu$_4$ZnCl$_6$ 190
Cu$_2$ZnGeS$_4$ 38

DyBr$_3$ 309
DyCl$_3$ 302, 313
DyI$_2$ 274, 276
DyI$_3$ 309
DyOBr 261
DyOCl 261
Dy(OH)$_2$Cl 304
DyOI 262
DySBr 266
DySCl 266
DySF 262
Dy$_2$Te$_5$ 245
DyTe$_3$ 244

ErBr$_3$ 309
ErCl$_3$ 313
ErI$_3$ 309
ErOBr 261
ErOCl 259, 261, 264
Er(OH)$_2$Cl 304
ErOI 262
Er$_3$O(OH)$_5$Cl$_2$ 305
ErSBr 266
ErSCl 266, 267
ErSF 262
ErTe$_3$ 244

EsBr$_3$ 313
EsCl$_3$ 302
EsOCl 261

EuBr$_3$ 302
EuC$_n$ 53
EuFBr 260
EuFCl 260
EuFI 260
EuH$_2$ 120
EuHBr 260
EuHCl 260
EuI$_2$ 29, 271
Eu(NH$_2$)$_2$ 282
Eu(NH$_3$)$_3$C$_n$ 53

EuOBr 261
EuOCl 261
Eu(OH)$_2$Cl 304
EuOI 262
EuSF 262
EuSeF 262

FeAl$_2$S$_4$ 39, 187, 188
Fe$_3$BO$_6$ 39
FeBr$_2$ 275, 276
FeBr$_3$ 310
FeCl$_2$ 275, 276
FeCl$_3$ 310, 313
FeCl$_2$Br 310
Fe$_{0.65}$Cr$_{0.35}$I$_2$ 276
FeF$_3$ 16, 38
FeI$_2$ 275
FeNi$_2$S$_4$ 211
(Fe, Ni)$_3$Te$_2$ 39, 211
FeOBr 266
FeOCl 18, 255, *265*
FeOCl·$\frac{1}{3}$C$_5$H$_5$N 267
Fe(OH)$_2$ 281
Fe$_2$(OH)$_3$Br 292
Fe$_2$(OH)$_3$Cl 281
Fe(OH)Cl 288, 290
Fe(OH)$_2$—CuFeS$_2$ 296
Fe(OH)$_2$·$\frac{1}{4}$FeOCl 291
Fe$_2$(OH)$_3$I 292
Fe$_3$(OH)$_2$(Si$_2$O$_5$)$_2$ 345
Fe$_{3-x}$(OH)$_4$Si$_2$O$_5$ 344
Fe$_3$(OH)$_4$SiFeO$_5$ 344
FeO(NH$_2$) 266, 267
FeO(OCH$_3$) 266
FeO(OH) 265, 268
FePS$_3$ *217*, 312
FePSe$_3$ *217*, 308
Fe$_7$S$_8$ 38
Fe$_3$S$_4$ 38, 210
FeS 38, *132*
FeS—Mg(OH)$_2$ 296
Fe$_7$Se$_8$ 38, 307
Fe$_3$Se$_4$ 210
FeSe 132, 133
FeTe 132, 133
FeTiO$_3$ 38
Fe$_2$WO$_6$ 38

GaAsSe 145
GaCl$_3$ 326
Ga$_3$Ge$_2$Te$_6$ 149
Ga$_2$GeTe$_3$ 148, 149
GaGeTe$_2$ 148, 149
GaInS$_2$ 149
GaInS$_3$ 185
Ga$_2$InSe$_3$ 148
GaInSe$_2$ 149
Ga$_{2/3}$Mo$_2$S$_4$ 39
GaOBr 269
GaOCl 269
GaOF 257
GaOI 257

GaPS$_4$ 23, 38, *140*
Ga$_2$S$_3$ 38, 183
GaS 27, *146*
GaS—GaSe 148
GaSF 257
Ga$_2$Se$_3$ 183, 197, 198
GaSe 146
Ga$_2$Se$_3$·4In$_2$Se$_3$ 187
Ga$_3$Se$_2$Te 149
Ga$_2$SeTe 149
Ga$_2$Te$_3$ 183
GaTe 146
GaTeBr 257
GaTeCl 257
Ga$_{1/2}$V$_2$S$_4$ 39

GdBr$_3$ 309, 313
GdCl$_3$ 302
GdI$_2$ 271
GdI$_3$ 309, 313
GdOBr 261
GdOCl 261
Gd(OH)$_2$Cl 304
GdOI 262
GdSF 262
GdSI 255
GdSeF 263
Gd$_2$Te$_5$ 245
GdTe$_3$ 244

GeAs 146
GeAs$_2$ 144
GeAsSe 351
Ge$_3$Bi$_2$Te$_6$ 38, 204, 205
Ge$_2$Bi$_2$Te$_5$ 194, 204
GeBi$_2$Te$_4$ 38, 198, 203
GeBi$_4$Te$_7$ 205, 206
GeI$_2$ 275, 285
GeNCl 269
GeO 89
GeP 146
GeP$_3$ 86
GeP$_5$ 79
GeS 89, 93
GeS$_2$ 136, *142*
GeSb 146
Ge$_2$Sb$_2$Te$_5$ 205
GeSb$_2$Te$_4$ 203
GeSb$_4$Te$_7$ 38, 205, 206
GeSe 89, 92, 93
GeSe$_2$ 136, *142*
Ge$_2$SeTe 85
GeSnTe$_2$ 86
Ge$_3$Te$_4$ 203
Ge$_2$Te$_3$ 152
GeTe 78, 81, *82*, 89

H$_3$BO$_3$ 43
HBO$_2$ 43
HBS$_2$ 143
HCl$_3$ 48
H$_2$Si$_{14}$O$_{29}$·2(CH$_3$)$_2$SO 346

H$_2$Si$_{14}$O$_{29}$·5H$_2$O 346
H$_2$Te$_2$O$_6$ 116
H$_{0.5}$WO$_3$ 173

HfBr$_2$ 271
HfBr$_4$ 334
HfCl 214
HfCl$_2$ 272
HfCl$_4$ 334
HfI$_2$ 22
Hf$_4$N$_3$ 209
Hf$_3$N$_2$ 197
HfNCl 266
HfO$_2$ 167
Hf$_2$S 235
HfS$_2$ 220, 222
HfS$_3$ 249
Hf$_2$Se 235
HfSe$_2$ 220, 222
HfSe$_3$ 249
HfSe$_3$Te$_2$ 251
HfSiTe 259
HfTe$_2$ 219, 222
HfTe$_3$ 249
HfTe$_5$ 250
HfWSe$_4$ 235

HgAs$_4$S$_8$ 77
HgBi$_4$Se$_8$ 77
HgBr$_2$ 38, *42*, 138, 272
HgCl$_2$ *41*, 272
HgClBr 41
HgGa$_2$Te$_4$ 38
HgI$_2$ 38, 41, *136*, 141
HgPS$_3$ 219
HgPSe$_3$ 219
HgSb$_4$S$_8$ 75
HgSb$_4$Se$_8$ 77

HoBr$_3$ 309
Ho$_2$C 277
HoCl$_3$ 313
HoI$_3$ 309
HoOBr 261
HoOCl 261
Ho(OH)$_2$Cl 304
HoOI 262
Ho$_3$O(OH)$_5$Cl$_2$ 305
HoSBr 266
HoSCl 266
HoSF 262
Ho$_2$Te$_5$ 245
HoTe$_3$ 244

IAlCl$_6$ 119
(IO)$_2$SO$_4$ 119
(IS)$_2$SeO$_4$ 119

InAgI$_2$ 127
In$_{2-x}$As$_x$S$_3$ 184
In$_{2-x}$Sb$_x$S$_3$ 184
InBi 131, 132

InBr 126, 127, 128
InBr$_3$ 311, 314
InCl 125
InCl$_3$ 314
InF 125
InHSO$_4$ 112
InI 126, 127, 128
InI$_3$ 38, 300, 311, 314, *327*
InOBr 266
InOCl 266
InOI 266
InOOH 38, 257
InPS$_4$ 38, 136, 138
In$_{2/3}$PS$_3$ 219
In$_2$S$_3$ 38, 183, 184
InS 146
InSBr 276
InSCl 257, 270, 276
InSO$_3$F 112
InSb 131, 132
In$_2$Se$_3$ 38, *182*, 197, 198
InSe 146
InSeBr 276
InSeCl 270, 276
InSeI 257
In$_3$Te$_4$ 203
In$_2$Te$_3$ 38, 183, 197, 198
In$_2$Te$_5$ 351
InTe 146
InTeBr 257, 270
InTeCl 257, 269, 270
InTeI 270

IrBr$_3$ 314, 315
IrCl$_3$ 38, 300, 314, 315
IrCl$_3$—RuCl$_3$ 313
IrI$_3$ 314, 315
IrTe$_2$ 222

KAl$_2$(OH)$_2$AlSi$_3$O$_{10}$ 2, 346
KAs$_4$O$_6$I 26
KBiO$_2$ 104
KC$_n$ 53, 55, 56
KC$_8$—H 56
KCH$_3$ 130
KCaSi$_4$Cl 7
(K, Cs)C$_8$ 53
KFeS$_2$ 24, 105
KH$_{2/3}$C$_8$ 53
K$_2$Ir(OH)$_6$ 282
K$_3$La(NH$_2$)$_6$ 38
KMg$_3$(OH)$_2$Si$_3$AlO$_{10}$ 346
K$_x$MoS$_2$ 2
K(NH$_3$)$_2$C$_n$ 53
K$_2$NiF$_4$ 1, 347
KOH 125, 128, 130
K$_2$PbO$_2$ 101
K$_2$Pb$_2$O$_3$ 101
K$_2$Pb(SO$_4$)$_2$ 108
K$_2$Pt(OH)$_6$ 282
K$_2$Pt$_4$S$_6$ 1
(K, Rb)C$_n$ 53

KReO$_4$ 169
KSH 130
KSiCl 6
KSiH$_3$ 130
K$_2$Sn(OH)$_6$ 282

LaHTe 260
LaI$_2$ 247, 273, 276
LaI$_{2.4}$ 273
LaI$_3$ 302
La$_2$O$_3$ 39
LaOBr 259, 261
LaOCl 259, 260
LaOF 28, 255, 260
La(OH)CO$_3$ 1
La(OH)$_2$Cl 304
La(OH)$_2$NO$_3$ 302
LaOI 259, 262
La$_2$O$_2$Te 63
LaSF 262
LaSb$_2$Sn 244
LaSeF 262
La$_2$Te$_5$ 245
LaTe$_3$ 244, 246

Li$_3$AlN$_2$ 38
Li$_5$AlO$_8$ 39
(Li, Al)(OH)$_2$·MnO$_2$ 297
Li$_3$Bi 39
LiBiO$_2$ 103
LiC$_6$ 53
LiC$_n$ 53, 54
LiCH$_3$ 130
Li(CH$_3$NH$_2$)$_2$C$_{12}$ 53
Li$_8$CoO$_6$ 38
LiCrS$_2$ 38
Li$_5$Cs(OH)$_6$ 131
LiF 130, 131
LiFeW$_2$O$_8$ 38
Li$_5$GaO$_4$ 38
Li$_2$GeP$_3$ 38
Li$_2$K(OH)$_3$ 131
Li$_{2x}$Mg$_{1-x}$Cl$_2$ 2
Li$_2$MgSn 39
Li$_2$MnO$_3$ 38
LiMnPO$_4$ 39
LiNH$_2$ 38, 130
Li(NH$_3$)$_2$C$_n$ 53
LiNa(OH)$_2$ 131
LiNa$_2$(OH)$_3$ 131
Li$_3$NbO$_4$ 38
LiNb$_3$O$_6$ 38
Li$_2$O 22
LiOH 22, 126, 130, *131*
Li$_2$PbO$_2$ 101
Li$_2$Pb$_2$O$_3$ 101
Li$_2$Pt(OH)$_6$ 38, 280, 282
Li$_3$Rb(OH)$_4$ 131
LiSH 38, 130, 131
LiSbO$_2$ 103
LiSbO$_3$ 38
LiSeH 130

Li_2SnO_3 38
Li_2TeO_3 117
$Li_2Te_2O_5$ 117
Li_2TiO_3 38
Li_xWO_3 182
Li_2ZrF_6 38
Li_2ZrO_3 38
Li_2ZrN_2 39

LnHSe 259
LnHTe 259
Ln Ni 5
LnOBr 259
LnOCl 259
LnOF 258
$Ln(OH)_2Cl$ 302, 306
LnOI 259
$Ln_3O(OH)_5Cl_2$ 306
LnSBr 267
LnSCl 267
Ln_3Se_7 246
Ln_4Te_{11} 246
Ln_3Te_7 246
Ln_2Te_5 245, 246
$LnTe_2$ 242
$LnTe_3$ 242, 246

$LuBO_3$ 43
$LuBr_3$ 309
$LuCl_3$ 313
LuI_3 309
Lu_2O_3 167
LuOBr 259, 261
LuOCl 264
LuOI 262
$Lu_3O(OH)_5Cl_2$ 305
LuSBr 255, 266
LuSCl 266
$LuTe_3$ 244, 246

$MgAl_2O_4$ 39
$Mg_6Al_2(OH)_{16}CO_3\cdot 2H_2O$ 293
$(Mg, Al)(OH)_2$—$CuFeS_2$ 295
$Mg_2Al_2S_5$ 190, 191
$MgAl_2S_4$ 187, 188, 191
Mg_3As_2 38
MgB_2 52
$MgBr_2$ 275
$MgBr_2\cdot 6H_2O$ 283
MgC_2 52
$MgCaP_2Se_6$ 218
$MgCl_2$ 276
$MgCl_2\cdot 6H_2O$ 282, 283
$Mg_{16}Fe_2(OH)_{36}CO_3\cdot 2H_2O$ 294
$Mg_{10}Fe_2(OH)_{24}CO_3\cdot 2H_2O$ 293
$MgGa_2S_4$ 39
MgHBr 263
MgHCl 263
MgHI 263
MgI_2 274, 275
Mg_6MnO_8 38
Mg_3NF_3 38

$Mg(NH_2)_2$ 38, 282
$Mg(NH_3)_2C_n$ 53
MgO 280
$Mg(OH)_2$ 279, 281
$Mg(OH)_2\cdot \frac{1}{4}AlOCl$ 291
$Mg(OH)_2\cdot \frac{1}{4}Al(OH)_3$ 291
$Mg_{10}Fe_2(OH)_{24}CO_3\cdot 2H_2O$ 293
$Mg_2(OH)_3Cl$ 281
$Mg(OH)Cl$ 276
$[Mg_3(OH)_6]Cr(OH)_3\cdot 3H_2O$ 342
$[Mg_4(OH)_8]FeOCl\cdot nH_2O$ 291
$[Mg_4(OH)_8](Ni, Fe)O(OH)$ 291
$Mg_3(OH)_2Si_4O_{10}$ 342
$Mg_3(PO_4)_2$ 39
$MgPS_3$ 217
$MgPSe_3$ 218
Mg_3Sb_2 39
Mg_2SiO_4 39
$Mg_7Si_3O_{12}(OH)_2$ 39
$Mg_5Si_2O_8(OH)_2$ 39
Mg_2Sn 38

$[Mn_2Al_2(OH)_6O_2]AlOCl$ 291
$[Mn_3Al(OH)_7O]AlO(OH)$ 291
$MnAl_2S_4$ 187, 188
Mn_7AsSbO_{12} 39
Mn_2B_2Al 6
MnB_2S_4 47
$MnBr_2$ 275, 276
$MnCl_2$ 276
$(Mn, Cr)I_2$ 279
MnF_4 335
$MnGa_2S_4$ 187
Mn_2Hg_5 1
MnI_2 275
Mn_3O_4 39
Mn_2O_3 38
Mn_2O_7 30, 167
MnOCl 266
$Mn(OH)_2$ 280, 281
$Mn_2(OH)_3Cl$ 281
$Mn(OH)Cl$ 285, 288
$Mn_2(OH)_3I$ 292
$Mn_3(OH)_4Si_2O_5$ 344
$MnO(OH)$ 280, 281
$MnPS_3$ 217, 219
$MnPSe_3$ 218
$MnPSe_3\cdot \frac{1}{3}$ pyridine 219
$MnSCl_2$ 318

MoAlB 7
MoB 5, 29
Mo_2BC 6
$MoBr_3$ 298
$MoCl_2$ 271
$MoCl_3$ 38, 298, 315
$MoCl_4$ 38, 307, 308, 328
$MoCl_5$ 13, 329
$Mo_{16}Ge_2O_{52}$ 177
MoN 4, 5
$MoNCl_3$ 328
$Mo_{18}O_{52}$ 176

$Mo_{16}O_{46}$ 176
MoO_3 16, 17, 28, 30, 167-9, 176
$MoOCl_2$ 317, 349
$Mo(O,F)_3$ 172, 173
$MoOF_4$ 13, 329
$MoO_3 \cdot H_2O$ 168, 175
$MoO_3 \cdot 2H_2O$ 168, 173
$Mo_5O_7(OH)_8$ 173
$Mo_2O_5(OH)$ 172, 173
$MoO_2(OH)$ 172, 173
$MoO(OH)_2$ 173
MoS_2 9-11, *237*
$MoSBr$ 256
Mo_2SN_2 11
$MoSe_2$ 237
$MoSeTe$ 237
$MoSi_2$ 7, 247
$MoTe_2$ 38, 42, *229* 234, 237
$Mo_{16}Ti_2O_{52}$ 177
$(Mo,W)Se_2$ 237
$(Mo,W)Te_2$ 237

$NH_4As_2O_3 \cdot \frac{1}{2}H_2O$ 26
NH_4Br 130, 132
NH_4I 130
NH_4SH 130, 132
$(NH_4)_2SiF_6$ 284

$NaAsO_2$ 104
$(Na, Ba)C_n$ 57
$NaBiO_2$ 104
Na_3BiO_4 38
NaC_n 53
$NaCl$ 18, 38, 78, 130, 194, 282
$NaCrS_2$ 1, 38, 86
$(Na, Cs)C_n$ 53
$NaHF_2$ 296, 297
$NaHg$ 5
$(Na, K)C_n$ 53
$NaNbO_2F_2$ 38
$NaOH$ 125, 128, 130
Na_6PbO_4 38
Na_2PbO_2 100
$Na_2Pb_2O_3$ 101
$Na_2Pt(OH)_6$ 282
$NaSH$ 130
$Na_2Sn(OH)_6$ 280, 282
$NaSrO_2$ 101
$NaTiF_4$ 38
$NaTlPbO_2$ 101
Na_xWO_3 182

Nb_3Br_8 38, 319
$NbBr_3$ 319
$NbBr_5$ 329
Nb_4C_3 209
Nb_3Cl_8 38, 307, 319
$NbCl_4$ 38, 307, 320, 328
$NbCl_5$ 39, 329
NbF_4 335
NbF_5 39, 329
Nb_3I_8 319

NbI_3 298
NbI_4 13, 307
NbI_5 39, 338
$Nb_2Mn_4O_9$ 38
$NbNS$ 267
Nb_2O_5 30, 167
$NbOBr_2$ 317, 349
$NbOCl_2$ 317, 349
$NbOCl_3$ 317
$NbOI_2$ 317, 349
NbS 217
NbS_2 10, 235, *238*
NbS_3 251
NbS_2Br_2 318
$NbSCl$ 272
$NbSCl_2$ 317
NbS_2Cl_2 *316*
$NbSI_2$ 318
$NbSSe$ 235
$NbSe_2$ 10, 11, 222, 223, 235, 238-241
$NbSe_3$ 251, 253
$NbSeBr_2$ 317
$NbSe_2Br_2$ 318
$NbSe_2Cl_2$ 318
$NbSe_2I_2$ 317, 318
$NbSe_2—MoSe_2$ 237
$NbSe_2—NbTe_2$ 227, 235
$NbSe_2—WSe_2$ 235
$NbSiAs$ 258
$NbTe_2$ 222, 223, 226

$NdBr_3$ 302
NdI_2 247, 273
NdI_3 302
$NdOBr$ 261
$NdOCl$ 261
$NdOF$ 260
$Nd(OH)_2Cl$ 304
$NdOI$ 262
$NdSBr$ 255, 258
$NdSF$ 262
$NdSI$ 264
$NdSeF$ 262
Nd_4Te_{11} 245
Nd_3Te_7 245
Nd_2Te_5 30, 243, 245
$NdTe_2$ 29, 243
$NdTe_3$ 30, 243, 244
$NdTeF$ 263

$NiAl_2S_4$ 187, 188
$NiAs$ 18, 38, 186, 194, 210, 214, 219
$NiAs_2$ 159
$NiBr_2$ 276
$NiBr_2 \cdot 6H_2O$ 283
$NiCl_2$ 276
$NiCl_2 \cdot 6H_2O$ 283
$Ni(ClO_4)_2$ 277, 283
NiI_2 274, 276
$NiI_2 \cdot 6H_2O$ 284
Ni_3N 38

Ni(OH)$_2$ 280, 281
Ni$_2$(OH)$_3$Br 281, 292
Ni$_2$(OH)$_3$Cl 281
Ni(OH)Cl 276
Ni$_2$(OH)$_3$I 292
Ni(OH)N$_3$ 281
[Ni$_4$(OH)$_8$]NiO(OH) 291
Ni$_3$(OH)$_2$(Si$_2$O$_5$)$_2$ 345
Ni$_3$(OH)$_4$Si$_2$O$_5$ 344
NiO(OH) 280, 281
NiPS$_3$ 217, 219
NiPSe$_3$ 217
NiPo 222
NiSeTe 222
NiTe 25, 38, 135, 214, 296
NiTe$_2$ 214, 220, 222
Ni$_3$TeO$_6$ 38
NiWO$_4$ 38
(Ni, Zn)(OH)$_2$ 281

NpBr$_3$ 302
NpBr$_4$ 330
NpI$_3$ 302
NpOBr 261
NpO$_2$F$_2$ 338
NpOI 262
NpOS 263
NpOSe 263
NpS$_3$ 249
NpSe$_3$ 249
NpTe$_3$ 244, 246

OsCl$_3$ 312, 313, 315
OsF$_5$ 329
OsO$_4$ 19, 30, 38, 169

P 77, 81, 87, 94
P—As 90
PCl$_5$ 325
PH$_4$Br 130, 132
PH$_4$I 130, 132
PI$_3$ 35, 48
PI$_3$·BBr$_3$ 48
PI$_3$·BI$_3$ 48
P$_2$O$_3$ 62
P$_2$O$_5$ 28, 30, 138, *166*

PaBr$_4$ 328
PaI$_3$ 302
PaI$_4$ 330
PaOS 263
PaOSe 263

Pb$_3$(AsO$_4$)$_2$ 109
PbBeF$_4$ 111
PbBi$_2$O$_4$ 98
PbBiO(PO$_4$) 107
PbBiO(VO$_4$) 107
Pb$_3$Bi$_4$Se$_9$ 204
Pb$_2$Bi$_2$Se$_5$ 38, 203–5
PbBi$_2$Se$_4$ 203, 204
PbBi$_4$Se$_7$ 204, 206

Pb$_2$Bi$_2$Te$_5$ 204
PbBi$_2$Te$_4$ 203, 204
PbBi$_4$Te$_7$ 205, 206
PbBr$_2$ 121
PbCO$_3$ 294
PbCl$_2$ 29, 120, 271
PbCrO$_4$ 111
PbF$_2$ 120, 121
PbF$_4$ 28, 335
PbFBr 260
PbFCl 9, 98, 258, 260
PbFI 260
PbHAsO$_4$ 113, 114
PbHPO$_4$ 113, 114
PbI$_2$ 274–6, 285
PbMnO$_2$(OH) 102
PbMoO$_4$ 111
PbO 28, 78, 89, 95, *97*, 132
Pb$_3$O$_4$ 103
Pb$_2$O$_3$ 102
PbO$_2$ 29, 38, 96
Pb$_2$O$_2$CO$_3$ 98
Pb$_2$O(CrO$_4$) 106
Pb(OH)$_2$ 121
Pb(OH)Br 121
Pb(OH)Cl 120, 121
Pb(OH)I 120, 121
Pb(OH)$_2$·2PbCO$_3$ 294
Pb$_2$O(MoO$_4$) 106
Pb$_2$O(SO$_4$) 105
Pb$_5$O$_4$(SO$_4$) 107
Pb$_8$P$_2$O$_{13}$ 109
Pb$_5$P$_4$O$_{15}$ 109
Pb$_4$P$_2$O$_9$ 109
Pb$_3$(PO$_4$)$_2$ *107*, 109
Pb$_2$P$_2$O$_7$ 109
Pb$_5$(PO$_4$)$_3$F 109
Pb$_{10}$(PO$_4$)$_6$O 109
Pb$_4$(PO$_4$)$_2$SO$_4$ 109
PbS 89, 93, 196
PbS$_2$ 195, 196
Pb$_7$S$_2$Br$_{10}$ 202
Pb$_2$SBr$_2$ 202
Pb$_5$S$_2$I$_6$ 202
Pb$_2$SI$_2$ 202
PbSO$_4$ 110
PbSe 89, 93
Pb$_4$SeBr$_6$ 202
Pb$_2$SeI$_2$ 202
PbSeO$_4$ 111
PbSe—SnSe 93
Pb$_2$SnO$_4$ 103
PbSnS$_2$ 93
PbSnS$_3$ 194
PbTe 89, 93
Pb$_2$TeI$_2$ 194
PbTe—PbBr$_2$ 202
Pb$_3$(VO$_4$)$_2$ 109
PbWO$_4$ 111

PdAsS 159
PdAsSe 159

PdBi 5, 29
PdBi$_2$ 8, 29, 246
PdBr$_2$ 270, 271
PdCl$_2$ 158, 270, 271
PdI$_2$ 271
PdP$_2$ 29, 158
PdPAs 159
PdPS 159
PdPS$_3$ 217
Pd$_3$(PS$_4$)$_2$ 162
PdPSe 159, 161
PdPTe 159
PdS 158
PdS$_2$ *158*, 161
PdSSe 158
PdSbS 159
PdSbSe 159
PdSe$_2$ 29, 158
PdSeTe 222
Pd$_2$Sn$_3$ 213
PdSn$_3$ 9, 36
PdTe 214
PdTe$_2$ 220, 222

PmBr$_3$ 302
PmOBr 261
PmOCl 261
PmOI 262

PrHTe 260
Pr$_2$I$_5$ 273
PrI$_2$ 273, 274, 276
PrI$_3$ 302
PrOBr 261
PrOCl 261
PrOF 260
Pr(OH)$_2$Cl 304
Pr(OH)$_2$NO$_3$ 304
PrOI 262
PrSF 262
PrSI 264
Pr$_3$Se$_7$ 246
PrSeF 262
Pr$_4$Te$_{11}$ 246
Pr$_3$Te$_7$ 246
Pr$_2$Te$_5$ 245
PrTe$_3$ 244

PtAs$_2$ 229
PtBi$_2$ 229
PtBiTe 229
PtBr$_3$ 300
PtBr$_4$ 38
PtCl$_2 \cdot$2NH$_3$ 328
PtCl$_4 \cdot$2NH$_3$ 329
Pt$_2$Ge$_3$ 29, 213
PtGeSe 229
PtI$_4$ 38, 328
PtO$_2$ 28, 222, 223
PtP$_2$ 229
PtPb$_4$ 9, 36
Pt$_4$PbBi$_7$ 229

PtPbTe 229
PtS 158
PtS$_2$ 220
Pt$_3$Sb$_2$ 39
PtSb$_2$ 229
PtSbTe 229
PtSe$_2$ 220
PtSeBr$_2$ 318
PtSeTe 220
Pt$_2$Sn$_3$ 36, 38, 186, *213*
PtSn$_4$ 9, 36
PtSnTe 229
Pt$_3$Te$_4$ *213*, 216
Pt$_2$Te$_3$ *213*, 216
PtTe *213*, 216
PtTe$_2$ 214, 220

PuBr$_3$ 4, 298, 301–3
PuBr$_4$ 330
PuI$_3$ 302
PuOBr 261
PuOCl 261
PuOF 260
PuO$_2$F$_2$ 338
PuOI 262
PuOS 263
PuOSe 263
PuTe$_3$ 244

RbBiO$_2$ 104
RbC$_n$ 53, 55, 56
Rb(NH$_3$)$_2$C$_n$ 53
RbNH$_2$ 130
Rb$_3$O 28
Rb$_2$O$_3$ 29
RbOH 125, 128
Rb$_2$Pb$_2$O$_3$ 101

Re$_3$B 4
ReBr$_3$ 322
ReBr$_5$ 329
ReCl$_3$ 321
ReCl$_4$ 328
ReCl$_5$ 39, 329
ReI$_3$ 322
ReO$_3$ 15, 16
Re$_{1+\delta}$O$_3$ 38
Re$_2$O$_7$ 28, 30, 167, *169*
Re$_2$O$_7 \cdot$2H$_2$O 168
ReOCl$_4$ 329
ReS$_2$ 233
ReSSe 233
ReSe$_2$ 38, 233

RhBr$_3$ 314
RhCl$_3$ 300, 314, 315
RhI$_3$ 314
RhTe$_2$ 222

RuBr$_3$ 298, 300, 312
RuCl$_3$ 312, 313
RuF$_5$ 39, 329

(Ru, Ir)Cl$_3$ 313
RuO$_4$ 167

Sb 77, 90
Sb$_3$As 86
SbAs 85
(Sb, Bi)I$_3$ 311
SbBr$_3$ 324
SbCl$_5$ 329
SbCl$_2$F$_3$ 329
SbI$_3$ 35, 308, 310
Sb$_2$MnO$_4$ 103
SbNbO$_4$ 110, 111
Sb$_2$O$_3$ 35, 62, 63
Sb$_2$O$_4$ 109, 110
SbOCl 257
SbOF 109, 257
SbP$_3$ 87
SbPO$_4$ *109*, 113
Sb$_2$S$_3$ 35, 63, 64, 195
SbSI 257
Sb$_2$S$_2$O 63, *71*
Sb$_2$Se$_3$ 35, 63
SbSiO$_3$F 112
SbTaO$_4$ 110–2
Sb$_4$Te$_3$ 209
Sb$_2$Te$_3$ 35, 63, 197–200
SbTeI 120, 121
Sb$_2$Te$_2$Se 197, 198
SbVO$_4$ 110

ScB$_2$ 52
ScBr$_3$ 309
ScCl 217
ScCl$_3$ 309
ScCrS$_3$ 38
ScI$_3$ 309
Sc$_2$O$_3$ 167
ScOBr 255, 266
ScOCl 266
Sc(OH)$_3$ 16
ScOI 266
ScO(OH) 268
Sc$_2$S$_3$ 38
ScSi 5
Sc$_2$Si$_2$O$_7$ 168

Si 156
SiAs 146, 150
SiAs$_2$ 145
SiBr 156
Si(CH$_3$) 156
SiCl 156
SiF 156
SiH 156, 157
Si(HPO$_4$)$_2 \cdot n$H$_2$O 351
Si$_6$N$_2$ 156
SiN 145, 146
SiNBr 269
SiNCl 269
Si(NHCH$_3$)$_{0.8}$ 156
Si(NHC$_6$H$_5$)$_{0.33}$ 156

SiO 89
SiO$_2$ 22, 141, 342
Si$_6$O$_3$Br$_6$ 157
Si(OCH$_3$)$_{0.8}$Cl$_{0.2}$ 156
Si(OC$_2$H$_5$)$_{0.75}$Cl$_{0.25}$ 156
Si(OC$_3$H$_7$)$_{0.67}$Cl$_{0.33}$ 156
Si$_6$O$_3$H$_6$ 157
SiOSe 269
SiP 146, 151
SiP$_2$ 145
SiPAs 145
SiPCl 139
SiS 89, 156, 157
SiS$_2$ 24, 38, 136, 138, 141
Si(SCl)$_{0.4}$Cl$_{0.6}$ 156
SiSe 89, 156
SiSe$_2$ 141
Si$_2$Te$_3$ 151, 152, 351

SmBr$_3$ 302
SmC$_n$ 53
SmFBr 260
SmFCl 260, 263
SmFI 260
SmI$_3$ 309
SmOBr 261
SmOCl 261
Sm(OH)$_2$Cl 304
SmOI 262
SmSF 262
SmSI 38, 255, 259, 263, 264
SmSeF 262
Sm$_3$Te$_4$ 246
Sm$_3$Te$_7$ 246
Sm$_2$Te$_5$ 245
SmTe 246
SmTe$_3$ 244
SmZrF$_7$ 15

Sn$_4$As$_3$ 209, 210
SnAs 146
SnBi$_2$Te$_4$ 203
SnBr$_4$ 19, 38
Sn(CH$_3$)$_2$F$_2$ 13
SnCl$_2$ 120
SnCrO$_4$ 110
SnF$_4$ 13–16, 38, *334*
SnHPO$_4$ 112–4
Sn(HPO$_4$)$_2 \cdot n$H$_2$O 351
SnI$_4$ 19, 38
Sn$_2$O$_3$ 102
SnO 28, 89, 98, 99
Sn$_4$P$_3$ 209, 210
SnP 146
SnP$_3$ 86
SnPO$_3$F 112
Sn$_4$P$_2$S 210
SnPS$_3$ 217, 218
SnS 78, *87*, 89
Sn$_2$S$_3$ 153, 194
SnS$_2$ 195
Sn$_9$S$_2$Br$_{14}$ 202

FORMULA INDEX

Sn_2SBr_2 202
$Sn_7S_3I_8$ 202
Sn_3SI_4 202
Sn_2SI_2 202
$SnSO_4$ 110, 111
SnS_2—SnO_2 196
SnS_2—$SnSe_2$ 196
SnS_2—ZrS_2 224
Sn_4Sb_5 204
SnSb 146
$Sn_2Sb_2Te_5$ 204
$SnSb_2Te_4$ 203
$SnSb_4Te_7$ 206
SnSe 89, 92
$SnSe_2$ 195
Sn_3SeI_4 202
Sn_2SeI_2 202
SnSe—$SnBr_2$ 202
SnTe 83, 84, 89
SnTe—GeTe 83
SnTe—$SnBr_2$ 202
SnTe—SnI_2 202
$SnWO_4$ 111

$SrBi_2(SiO_4)_2$ 108
$SrBr_2$ 29
$SrCl_2 \cdot 2H_2O$ 332
$Sr_3(CrO_4)_2$ 107
SrFBr 260
SrFCl 260
SrFI 260
SrH_2 120
SrHBr 260
SrHCl 260
SrHI 260
SrI_2 29, 258
Sr_2N 28, 274, 277
Sr_2NH 277
$Sr(NH_2)_2$ 282
$Sr(NH_3)_2C_n$ 53
$Sr_3(PO_4)_2$ 107
$SrPbO_2$ 101
$SrSnO_2$ 101

Ta_3B_4 7
$TaBr_3$ 319
$TaBr_5$ 329
Ta_4C_3 209
Ta_3Cl_8 319
$TaCl_3$ 319
$TaCl_4$ 328
$TaCl_5$ 39, 329
TaF_4 335
$(Ta,Mo)Se_2$ 237
TaNSe 267
Ta_2O_5 30, 167
TaS_2 10, 222, *223*, 226, 235, *238*, 241
TaS_3 251, 252
Ta_2S_2C 197
Ta_2SSe_2 11
$TaSe_2$ 10, 11, 222, *223*, 235, *238*–242
$TaSe_3$ 4, 30, *251*

$Ta(Se,Te)_2$ 227
$TaTe_2$ 222, 223, *226*
$TaVC_2$ 197
$(Ta,W)Se_2$ 238

$TbBr_3$ 309
Tb_2C 277
$TbCl_3$ 302, 303
TbI_3 309
TbOBr 261
TbOCl 261
$Tb(OH)_2Cl$ 304
TbOI 262
TbSF 262
Tb_2Te_5 245
$TbTe_3$ 244

$TcCl_4$ 13
Tc_2O_7 30, 167, *168*
TcS_2 233
$TcSe_2$ 233
$TcTe_2$ 233

Te 79
$TeCl_4$ 38
Te_4O_9 117
Te_2O_5 116
TeO_2 28, 35, *114*
$Te_2O_4 \cdot HNO_3$ 115
$Te_2O_3SO_4$ 116
$TeSiO_4$ 113
$TeTiO_4$ 113
$TeVO_4$ 113

$ThBr_4$ 328
$Th_{0.9}Ge_2$ 5, 29
ThHN 259
ThI_2 270-2, 274
ThI_3 301
ThI_4 330
Th_3N_4 263
ThNBr 259, 263
ThNCl 259, 263
ThNF 28
$Th(NH)_2$ 263, 350
ThNI 259, 263
$ThN(NH_2)$ 263
ThOS 263
ThOSe 263
ThOTe 263
ThPI 264
Th_7S_{12} 202
ThS_2 120
$ThSe_2$ 120
$ThTe_3$ 249

$(Ti,Al)Cl_3$ 313
TiAs 38
TiB_2 52
$TiBr_2$ 274, 275
$TiBr_3$ 299, 310
$TiCl_2$ 275

TiCl$_3$ 299, 307, 308, 310, 312, 313, 316
TiF$_4$ 335
TiI$_2$ 275
TiI$_3$ 36, 38, 298
TiNBr 266
TiNCl 266
TiNI 266
Ti$_4$O$_5$ 38
Ti$_3$O$_5$ 38
Ti$_2$O 1, 28
TiO$_2$ 38, 167
TiOBr 266
TiOCl 266
TiO(OH)$_2$ 178
Ti$_4$S$_5$ 38
TiS 38
TiS$_2$ 219, 220, 221
TiS$_3$ 249
TiSCl$_2$ 318
TiSe$_2$ 220, 221, 226
TiTe$_2$ 220, 221
TiTe$_3$O$_8$ 113
TiZrO$_4$ 113

TlAgI$_2$ 127
TlAlF$_4$ 111
TlAlO$_2$ 105
TlAlS$_2$ 105
TlAsS$_2$ 86
TlAsSe$_2$ 86
TlBF$_4$ 111
Tl$_2$Ba(CrO$_4$)$_2$ 109
TlBi 131
TlBiO$_2$ 97
TlBiO(SO$_4$) 107
TlBiS$_2$ 86
TlBiSe$_2$ 86
TlBiSr(PO$_4$)$_2$ 109
TlBiTe$_2$ 86
TlBr 126, 128
TlCl 126, 128
TlCl$_3$ 314
TlClO$_4$ 111
TlF 96, 122, 126, *128*, 129
TlFeO$_2$ 105
TlFeS$_2$ 105
TlFeSe$_2$ 105
TlGaO$_2$ 105
TlI 5, 122, *126*, 127, 128
TlMnO$_4$ 110
Tl$_4$O$_3$ 123
Tl$_3$O$_4$ 123
Tl$_2$O 28, *122*
Tl$_2$Pb(CrO$_4$) 109
TlPbO(MnO$_4$) 107
Tl$_2$Pb(SO$_4$)$_2$ 109
TlReO$_4$ 111
Tl$_4$S$_3$ 125
Tl$_2$S 123
TlS 146
TlSBr 263
TlSCl 263
TlSI 263
TlSb 131
TlSbO$_2$ 97
TlSbS$_2$ 86
TlSbSe$_2$ 86
TlSbTe$_2$ 86
Tl$_2$Se 125
TlSe 127, 146
Tl$_2$Sr(SO$_4$)$_2$ 108, 109
Tl$_2$Te 125
TlTe 146

TmBr$_3$ 309
TmCl$_3$ 313
TmFCl 260
TmI$_2$ 275
TmI$_3$ 309
TmOBr 261
TmOCl 264
TmOI 262
TmSBr 266
TmSCl 266
TmTe$_3$ 244, 246

UBC 7
UBr$_4$ 329, 330
UCl$_3$ 30, 298, 301, 302
UCl$_4$ 328
UCl$_5$ 39, 329, *339*
UCl$_6$ 39, 341
UF$_5$ 13, 39, 329
UF$_6$ 39
UHN 259
UI$_3$ 301, 302, 303
UI$_4$ 330
UNBr 259, 263
UNCl 263
U(NH)Cl 261
UNI 259, 263
UOBr 261
UOCl 261
UO$_2$F$_2$ 336
UOI 262
UO$_2$(OH)$_2$ 13, *335*
UOS 263
UOSe 263
UOTe 263
UPI 264
UPS 258
US$_2$ 120
US$_3$ 249
USe$_2$ 120
USe$_3$ 249
U$_3$Te$_7$ 246
UTe$_3$ 249
UTe$_5$ 251

VBr$_2$ 275
VBr$_3$ 310
VBr$_2$I 310
V$_4$C$_3$ 209

FORMULA INDEX

VCl_2 275
VCl_3 310
VCl_2Br 310
$VCrO_4$ 39
VF_5 13, 329
VI_2 275
VI_3 310
V_2O_5 28, 167, *181*
VO_2 29, 179
$VOBr$ 266
$VOCl$ 266
$V_2O_5 \cdot 3H_2O$ 168, 182
$V_3O_3(OH)_5$ 179
$V_2O_2(OH)_3$ 179
$V_2O_3(OH)_2$ 180
$VO(OH)_2$ 17, *178*
$VO(OH)_3$ 182
VPS_3 217
$VPSe_3$ 218, 219
V_5S_8 307
VS_2 222
VSe_2 222, 223, 226

WBr_6 341
WC 4, 10, 29
WCl_4 328
WCl_5 329
WCl_6 36, 39, 307, *340*
WN 4, 5, 11
$(W, Nb)Se_2$ 237
WO_3 30, 167, 172, 176
$WOBr_4$ 329
$WOCl_4$ 13, 329
WO_2Cl_2 13, 15, *338*
WOF_4 329
$WO_3 \cdot H_2O$ 176
$WO_3 \cdot 2H_2O$ 174
$W_2O_5(OH)$ 173
WO_2I_2 13, *338*
WS_2 237
$WSBr_4$ 329
$WSCl_4$ 329
WS_2-NbS_2 237, 238
WS_2-TaS_2 237
WS_2-WTe_2 232
WSe_2 237
WSe_2-MoSe_2 237
WSe_2-NbSe_2 237
WSe_2-TaSe_2 237, 238
WSe_2-WTe_2 232, 237
WTe_2 38, 42, *229*, 234

YBr_3 309
Y_2C 28, 277
YCl_3 313
YI_3 309
Y_2O_3 167
$YOBr$ 261
$YOCl$ 260
YOF 255, 258, 260
YOI 262
$Y_3O(OH)_5Cl_2$ 305–7

$Y(OH)_2Cl$ 303, 304
$YPO_4 \cdot 2H_2O$ 120
$YSBr$ 266
$YSCl$ 266
YSF 255, 262
$YSeF$ 255
Y_2Te_5 245
YTe_3 244

$YbBr_3$ 309
YbC_n 53
$YbCl_3$ 313
$YbFBr$ 260
$YbFCl$ 260
$YbFI$ 260
YbH_2 120
YbI_2 275
YbI_3 309
$YbOBr$ 259, 261
$YbOCl$ 264
$YbOI$ 262
$Yb_3O(OH)_5Cl_2$ 305
$YbSBr$ 266
$YbSCl$ 266
Yb_3Se_4 38

$ZnAl_2S_4$ 38, 187
$ZnBr_2$ 136, 139, 276
$ZnCl_2$ 38, 136, 138–140
$ZnFBr$ 269
$ZnFI$ 269
ZnI_2 136, 139, 275, 276
$Zn_5In_2S_8$ 194
$Zn_3In_2S_6$ 39, *190*
$Zn_2In_2S_5$ 39, *190*
$ZnIn_2S_4$ 39, *187*
$Zn_2Mo_3O_8$ 39
ZnO 38
$Zn(OH)_2$ 136, 272, 281
$Zn(OH)Cl$ 286, 287, 290
$Zn(OH)_2-Co(OH)_2$ 281
$Zn_2(OH)_3F$ 281
$Zn(OH)N_3$ 281
$Zn(OH)_2-Ni(OH)_2$ 281
$[Zn_4(OH)_8]ZnCl_2$ 293
Zn_3P_2 38
$ZnPI_3$ 38
$ZnPS_3$ 217
ZnS 38
$ZnS-In_2S_3$ 193
$ZnTe_2$ 223
Zn_4TiS_6 190

ZrB_2 52
$ZrBeSi$ 4, 11
$ZrBr$ 214, 216
$ZrBr_2$ 272
$ZrBr_3$ 310
$ZrBr_4$ 334
$ZrCl$ 214
$ZrCl_2$ 238, 270, 272
$ZrCl_3$ 310

ZrCl$_4$ 38, 333
ZrF$_4$ 30, 328, 335
Zr(HAsO$_4$)$_2 \cdot n$H$_2$O 351
Zr(HPO$_4$)$_2 \cdot n$H$_2$O 351
ZrI$_2$ 272
ZrI$_3$ 13
ZrMoS$_4$ 235
ZrNBr 265, 266
Zr$_2$NCl$_2$ 11
ZrNCl 265, 266
ZrNI 266
ZrN(NH$_2$) 266, 267
ZrNi 5
ZrNiH$_{2.7}$ 5
ZrO$_2$ 29, 167
ZrS$_2$ 220, 221

ZrS$_3$ 249
ZrS$_2$Cl$_2$ 318
ZrS$_2$—SnS$_2$ 224
ZrS$_2$—ZrSe$_2$ 224
ZrS$_3$—ZrSe$_3$ 249
ZrSe$_2$ 220, 221
ZrSe$_3$ 4, 30, *247*
ZrSe$_3$Te$_2$ 251
ZrSe$_2$—ZrTe$_2$ 221
ZrSi$_2$ 5, 29
ZrSiO$_4$ 111
ZrSiS 258
ZrSiTe 259
ZrTe$_2$ 219, 220, 221
ZrTe$_3$ 249
ZrTe$_5$ 251